Four Neotropical Rainforests

La Selva
Barro Colorado Island
M.C.S.E. Reserves
Cocha Cashu
>3000 m.
2000–3000 m.
1000–2000 m.
500–1000 m.
200–500 m.
0 200 400 600 800 1000km
0 100 200 300 400 500 600 miles
Prepared by Hendrik R. Rypkema

Four Neotropical Rainforests

Alwyn H. Gentry, editor

with the editorial assistance of
Myra Guzmán-Teare, and
coordination of avian and herpetological
chapters by James R. Karr and
William E. Duellman

Yale University Press
New Haven and London

Designed by James J. Johnson and set in Times Roman type by Asco Trade Typesetting Limited, Hong Kong. Printed in the United States of America by Vail-Ballou Press, Binghamton, New York.

The paper in this book meets the guidelines of permanence and durability of the Committee on Production Guidelines for Book Longevity of the Council on Library Resources.

10 9 8 7 6 5 4 3 2

Library of Congress of Cataloging-in-Publication Data

Four neotropical rainforests / Alwyn H. Gentry, editor.
p. cm.
"Proceedings of a symposium, organized as part of the Association for Tropical Biology program at the 1987 meeting of the American Institute of Biological Sciences . . . co-sponsored by the Organization for Tropical Studies [and held] at Ohio State University in Columbus on 10–12 August 1987"—Pref.
Includes bibliographical references.
ISBN 0-300-04722-3 (cloth)
0-300-05448-3 (pbk.)
1. Rain forest ecology—Congresses. 2. Rain forest fauna—Congresses. 3. Rain forest plants—Congresses. 4. Rain forests—Congresses. I. Gentry, Alwyn H. II. Association for Tropical Biology. III. American Institute of Biological Sciences. IV. Organization for Tropical Studies.
QH541.5.R27F69 1990
574.5′2642—dc20 90–11909

A catalogue record for this book is available from the British Library.

Contents

PART V Reptiles and Amphibians

PART VI Forest Dynamics

Preface

This book contains the proceedings of a symposium organized as part of the Association for Tropical's Biology program at the 1987 meeting of the American Institute of Biological Sciences. The symposium, cosponsored by the Organization for Tropical Studies, took place at Ohio State University in Columbus on 10–12 August 1987. The objective of the symposium, and of this book, is to compare the faunas, floras, and forest dynamics of what are perhaps the four best-known field stations in the Neotropics: La Selva, Costa Rica; Barro Colorado Island (BCI), Panama; the Cocha Cashu field station in Manu National Park, Peru; and the Minimum Critical Size of Ecosystems (MCSE) study area near Manaus, Brazil.

On behalf of the Association for Tropical Biology, I warmly thank the many people who took part in the symposium and in the subsequent preparation of this volume. Support for the symposium was provided by the Smithsonian Tropical Research Institute, by the World Wildlife Fund-U.S., and by the Organization for Tropical Studies, all of whose assistance is much appreciated. The Missouri Botanical Garden provided secretarial and organizational support. Review comments on contributions were kindly provided by R. Bierregaard, J. Denslow, W. Duellman, L. Emmons, W. R. Heyer, J. Karr, P. Myers, T. Plowman, G. Prance, F. E. Putz, J. V. Remsen, J. Savage, D. Stone, and R. Waide.

Coordination of a project such as this turned out to be somewhat complex logistically. Indeed, gathering so many tropical field biologists in Columbus, Ohio, in the middle of the summer research season in itself represented a major logistical triumph. Similarly, the process of extracting manuscripts and manuscript revisions from authors who frequent far corners of the world as much as do this symposium's contributors was often a real challenge.

Especially appreciated for their role in bringing this project to fruition are the efforts of the four sectional coordinators. Not only did they help to put together the initial program, but they have also functioned as subeditors in preparing this volume. These sectional coordinators, all renowned in their respective fields, are Doctors G. Hartshorn (forest dynamics), J. Karr (birds), J. Eisenberg (mammals), and W. Duellman (reptiles and amphibians). I am also very grateful to Myra Guzmán-Teare, who handled many of the editorial chores associated with preparing this book. Both her efficient help and the support and encouragement of Laura Dooley, Ed Tripp, and Yale University Press were critical elements in its completion.

The Contributors

Dr. Richard O. Bierregaard, Jr., National Museum of Natural History, Smithsonian Institution, NHB 106, Washington, D.C. 20560

Dr. John G. Blake, Natural Resources Research Institute, 3151 Miller Trunk Highway, University of Minnesota, Duluth, MN 55811

Dr. John E. Cadle, Academy of Natural Sciences, 19th and the Parkway, Philadelphia, PA 19103

Dr. David B. Clark, OTS, Aptdo. 676 2050 San Pedro, Costa Rica

Dr. William E. Duellman, Museum of Natural History, University of Kansas, Lawrence, KS 66044

Dr. John F. Eisenberg, Florida State Museum, University of Florida, Gainesville, FL 32611

Dr. Louise H. Emmons, National Museum of Natural History, Division of Mammals, Smithsonian Institution, NHB 390, Washington, D.C. 20560

Dr. Robin B. Foster, Field Museum of Natural History, Roosevelt Road at Lake Shore Drive, Chicago, IL 60605

Dr. Alwyn H. Gentry, Missouri Botanical Garden, P.O. Box 299, St. Louis, MO 63166-0299

Dr. William E. Glanz, Department of Zoology, University of Maine, Orono, ME 04469

Dr. Craig Guyer, Department of Zoology and Wildlife Science, Funchess Hall, Auburn University, Auburn, AL 36849

Dr. Barry Hammel, Missouri Botanical Garden, P.O. Box 299, St. Louis, MO 63166-0299

Dr. Gary S. Hartshorn, World Wildlife Fund, 1250 24th St., NW, Washington, D.C. 20037

Dr. Steven P. Hubbell, Department of Biology, Princeton University, Princeton, NJ 08544

Dr. Roger W. Hutchings H., Department of Ecology, Instituto Nacional de Pesquisas da Amazonia, Caixa Postal 478, Manaus, Amazonas, Brazil

Dr. Charles H. Janson, Department of Biology, State University of New York, Stony Brook, NY 11790

Dr. James R. Karr, Department of Biology, Virginia Polytechnic Institute and State University, Blacksburg, VA 24060

Dr. Egbert Giles Leigh, Jr., Smithsonian Tropical Research Institute, APO Miami 34002-0011

Dr. Diana Lieberman, Department of Biology, University of North Dakota, Grand Forks, ND 58202

Dr. Milton Lieberman, Department of Biology, University of North Dakota, Grand Forks, ND 58202

Dr. Bette A. Loiselle, Department of Zoology, University of Wisconsin, Madison, WI 53706

Dr. Thomas E. Lovejoy, Smithsonian Institution, SI 230, Washington, D.C. 20560

Dr. Jay R. Malcolm, School of Forest Resources and Conservation, University of Florida, Gainesville, FL 32611

Dr. Charles W. Myers, American Museum of Natural History, Central Park West and 79th, New York, NY 10024

Ing. Rodolfo Peralta, Centro Científico Tropical, Apartado 8-3870, San José, Costa Rica

Dr. Ghillean T. Prance, Royal Botanic Gardens, Kew, Richmond, Surrey, TW9 3AB, England

Dr. A. Stanley Rand, Smithsonian Tropical Research Institute, Box 2072, Balboa, Republic of Panama

Dr. Judy M. Rankin-de-Merona, Instituto Nacional de Pesquisas da Amazonia, Caixa Postal 478, Manaus, Amazonas, Brazil

Dr. Scott K. Robinson, Illinois State Natural History Survey, 607 E. Peabody Dr., Champaigne, IL 61820

Dr. Miguel T. Rodrigues, Instituto de Biociencas, Universidade de São Paulo, Caixa Postal 11461-2C9, Cidade Universitaria, São Paulo, Brazil

Dra. Lily B. Rodriguez, APECO, Parque José de Acosta 187, Magdalena, Lima, 17 Peru

Dr. F. Gary Stiles, Universidad de Costa Rica, Escuela de Biología, Ciudad Universitaria Rodrigo Facio, San José, Costa Rica

Dr. John J. Terborgh, Department of Biology and Anthropology, Duke University, Durham, NC 27706

Dr. Don E. Wilson, Department of Mammology, Smithsonian Institution, Washington, D.C. 20560

Dr. S. Joseph Wright, Smithsonian Tropical Research Institute, APO Miami 34002-0011

Dr. Barbara L. Zimmerman, Department of Biological Science, Florida State University, Tallahassee, FL 32306-2043

Four Neotropical Rainforests

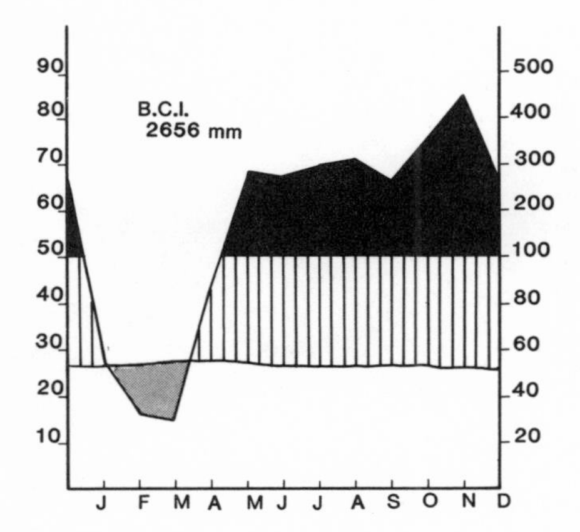
B.C.I.
2656 mm
90
80
70
60
50
40
30
20
10
500
400
300
200
100
80
60
40
20
J F M A M J J A S O N D

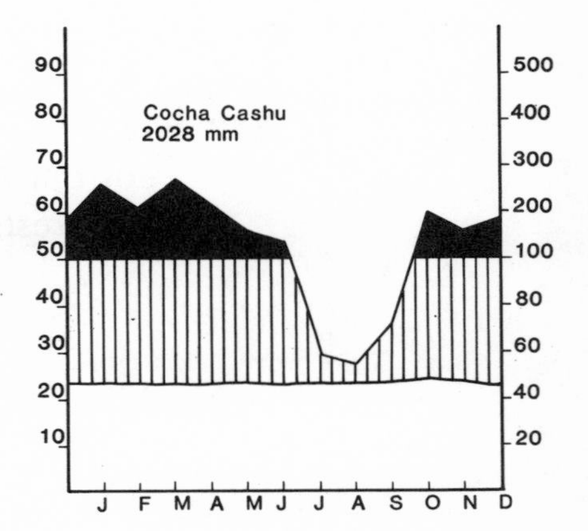
Cocha Cashu
2028 mm
90
80
70
60
50
40
30
20
10
500
400
300
200
100
80
60
40
20
J F M A M J J A S O N D

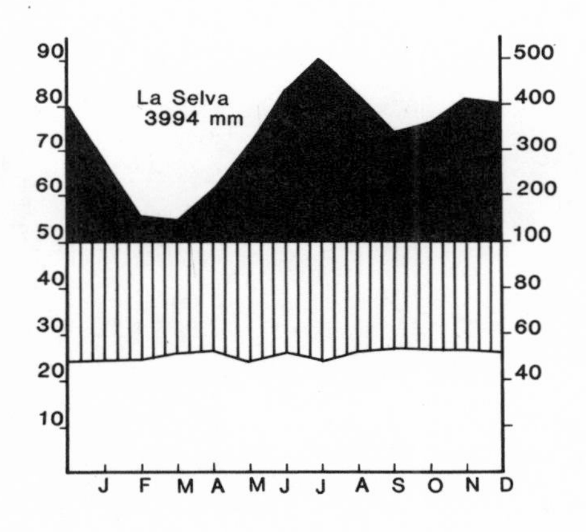
La Selva
3994 mm
90
80
70
60
50
40
30
20
10
500
400
300
200
100
80
60
40
J F M A M J J A S O N D

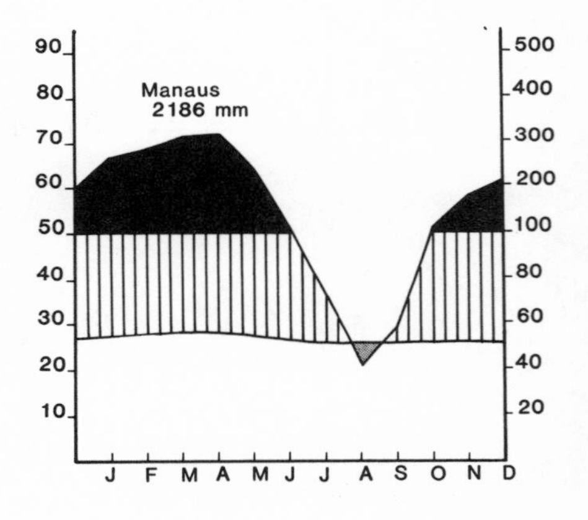
Manaus
2186 mm
90
80
70
60
50
40
30
20
10
500
400
300
200
100
80
60
40
20
J F M A M J J A S O N D

Introduction

ALWYN H. GENTRY

This book introduces four of the most important Neotropical rainforest study sites (Part I) and describes and compares their plants (Part II), birds (Part III), mammals (Part IV), reptiles and amphibians (Part V), and forest dynamics (Part VI). Although detailed introductory notes on each site are presented in Part I, a few remarks on why these sites and organisms were selected for comparison seem appropriate.

The two Central American sites are very well known, and their inclusion in such a comparison needs no justification. La Selva is the premier field station of the Costa Rica–based Organization for Tropical Studies. Barro Colorado Island (BCI), administered by the Smithsonian Tropical Research Institution, has a long history of important contributions to tropical biology. The two Amazonian stations are less well known. Fieldwork in tropical South America has tended to lag behind that in Central America, and only recently are sufficiently focused data from tropical South American sites becoming available to enable intersite comparisons of the kind attempted here. Fieldwork in Manu National Park, Peru, which probably conserves more species than any other such unit in the world, has been until recently, largely the result of a one-man effort by John Terborgh of Princeton University, who, with students and colleagues, has devoted much of his career to studying the biota of the Cocha Cashu field station and its environs. The fourth site, the Minimum Critical Size of Ecosystems Reserve area, is jointly administered by Brazil's Instituto Nacional de Pesquisas da Amazonia and the Smithsonian Institution and has been the flagship project of the World Wildlife Fund-U.S. The baseline data on forest composition and dynamics of the Manaus area, gathered as part of this ambitious attempt to test the effects of reserve size on conservational potential, are now adequate for comparison with the other three Neotropical forests.

Each field station has already generated many publications. Much of the increasingly sophisticated fieldwork carried out on BCI is summarized in a recent work (Leigh et al. 1982), and a volume similarly summarizing research conducted at La Selva is in preparation (McDade et al. 1990). Studies carried out at Cocha Cashu have been summarized in the informally published *Reporte Manu* (Ríos 1985) and by Terborgh (1983). To our knowledge, this volume represents the first time that an effort has been made to compare directly the biotas of a series of tropical forests. Single-site volumes, moreover, have focused on ecological processes and interactions: thus, this volume is also the most complete assessment of the plant and animal species that occur in these forests.

In a sense, the idea for this book dates to the 1982 American Institute of Biological Sciences meeting at Pennsylvania State University. At that meeting, only one contributed tropical biology paper was not based on research conducted either in Costa Rica or on BCI. Since my own field experience suggested strongly that there are important differences between Amazonian and Central American forests, my immediate reaction to this all-too-obvious Central American skew in tropical forest research was to wonder how many emerging generalities about tropical or neotropical forests might turn out to be peculiarities of southern Central America instead. At that time I made a mental note that a broad comparison of similarities and differences between Amazonian and Central American forests would be very much in order. Thus, when I agreed to serve as 1987 ATB program chairman, I decided to organize a symposium focused on this question.

Together these four sites provide a felicitous cross-section of the lowland Neotropics. Of the two Central American sites, one is on the Atlantic slope, one on the drier Pacific slope; one is tropical moist forest with a strong dry season, the other tropical wet forest with virtually no dry season. Of the two Amazonian sites, one is in the poor-soil Guianan shield region, the other on relatively fertile alluvial soil; one is near the base of the Andes, the other in the heart of central Amazonia.

In many ways it would have been preferable to include more sites in this comparison, and during early organizational work on the symposium, inclusion of additional field stations was advocated. Potential tropical dry forest research sites included Chamela in western Mexico (see Lott et al. 1987), Santa Rosa National Park in Guanacaste, Costa Rica (see Janzen 1983), and the Hato Masaguaral (Blohm Ranch) in the Venezuelan Llanos (see Eisenberg 1979). Gaps in the data base for each of these sites, however, precluded complete comparison with the four selected forests. Moreover, the differences between dry and moist forest habitats are so much greater than those among our four sites that the dry forests might more appropriately be addressed in a different forum.

Other moist lowland sites suggested for possible comparison with these four Neotropical forests included El Verde, Puerto Rico (see Odum 1970), rejected because of biotic peculiarities associated with its insular nature; the Rio Palenque Biological Center in western Ecuador (see Dodson and Gentry 1978), rejected because of its tiny size and inadequate data on mammals and forest dynamics: San Carlos de Rio Negro, Venezuela (see Jordan 1982, Klinge and Herrera 1983, Uhl et al., 1987), with good ecosystems data but rudimentary biotic surveys; the Tambopata Nature Reserve, Madre de Dios, Peru (see Erwin 1985), well known for birds and insects but less so for other taxa; and the Los Tuxtlas field station of the Universidad Nacional Autonoma de Mexico (Estrada and Coates-Estrada 1983), marginally tropical at 18°34′N and then inadequately known floristically (but see Ibarra and Sinaca 1987). In addition, such upland field stations as Rancho Grande, Venezuela (see Huber 1986), and San Vito and Monteverde in Costa Rica (see Janzen 1983) were excluded from consideration to avoid the additional complications of altitudinal variation. Most of these sites have also been subjected to greater hunting pressures and other human-induced environmental perturbations than have the selected Neotropical forests.

Even the four forests presented here have significant and sometimes surprising gaps in the available data base. Thus, the lack of primate studies at La Selva precluded a contemplated comparison of primate communities. Similarly, inadequate reptile and amphibian data from Manu (due largely to restrictions on scientific collecting) nearly led herps to be excluded from the symposium; only after considerable arm-twisting did the authors of the Cocha Cashu herp chapter agree to publish their preliminary data. The lack of a floristic inventory in the MCSE study area necessitated the substitution of data, themselves preliminary, from the Reserva Ducke, some eighty kilometers away.

Although an editorial attempt has been made to make the individual chapters as comparable as possible, this has not always proved feasible. In part this is due to differences in the relative levels of taxonomic sophistication of biological subdisciplines. Thus, a complete checklist of the birds of these Neotropical forests could be prepared, as could a somewhat more provisional herp list (see appendixes 15.1, 16.1, and 24.1). However, we are still far from being able to prepare a similar list for vascular plants, many of which have yet to be described and many more of which are only tentatively identified, often only to genus and "morphospecies." Mammals, especially bats, are also incompletely known taxonomically and poorly surveyed. The only one of these four sites with a reasonably complete bat list is La Selva. At La Selva, and presumably elsewhere, bats constitute more than half the mammal species; the other chapters on mammals have been forced to restrict their information to the nonflying minority for their respective fauna. Even far less known are such groups as fungi, nematodes, or insects, where the vast majority of species may be undescribed and whose inclusion in a comparison like this one might be possible only many years in the future.

The main objective of this volume is to focus attention on the similarities and differences among these four Neotropical rainforests. We also hope that calling attention to gaps in our understanding of these forest may stimulate further studies. Our knowledge of the richest and most complex repositories of life on earth is woefully incomplete. In view of the ongoing and accelerating destruction of tropical forests, the reader of this volume can hardly avoid facing the depressing fact that, if such imperfectly known forests are the best studied in the Neotropics, it is unlikely that we will ever understand the many kinds of tropical forests well enough to make rational conservational decisions before most of our world's most priceless natural heritage is forever lost. By emphasizing the distinctiveness of each of these tropical forests, this book will perhaps document the urgent need not only for additional sites of in-depth ecological study but also for more geographically and taxonomically comprehensive inventories to select appropriate sites for study and conservation.

REFERENCES

Dodson, C. H. and A. H. Gentry. 1978. Flora of the Rio Palenque Science Center. Selbyana 4: 1–628.

Eisenberg, J. F. 1979. *Vertebrate Ecology in the Northern Neotropics*. Smithsonian Inst. Press, Washington, D.C.

Erwin, T. L. 1985. Tambopata Reserved Zone, Madre de Dios, Peru: history and description of the reserve. Rev. Peruana Entomol. 27: 1–8.

Estrada, A. and R. Coates-Estrada. 1983. Rain forest in Mexico: research and conservation. Oryx 17: 201–204.

Huber, O. 1986. La Selva Nublada de Rancho Grande, Parque Nacional "Henri Pittier." Editorial Arte, Caracas.

Ibarra M., G. and S. Sinaca C. 1987. Listados floristicos de Mexico, VII. Estacion de Biol. Trop. Los Tuxtlas, Veracruz. Inst. Biol. Univ. Nacional Auton. Mex., Mexico City.

Janzen, D. H. 1983. *Costa Rican Natural History*. Univ. of Chicago Press, Chicago.

Jordan, C. F. 1982. The nutrient balance of an Amazonian rain forest. Ecol. 63: 647–654.

Klinge, H. and R. Herrera. 1983. Phytomass structure of natural plant communities on spodosols in southern Venezuela: the tall Amazon caatinga forest. Vegetatio 53: 65–84.

Leigh, E. G., Jr., A. S. Rand, and D. M. Windsor (eds.). 1982. *The Ecology of a Tropical Forest: Seasonal Rhythms and Long-Term Change*. Smithsonian Inst. Press, Washington, D.C.

Lott, E. J., S. H. Bullock, and J. A. Solis-Magallanes. 1987. Floristic diversity and structure of upland and arroyo forests of coastal Jalisco. Biotropica 19: 228–235.

McDade, L. A., K. S. Bawa, H. A. Hespenheide, and G. S. Hartshorn (eds.). 1990. *La Selva: Ecology and Natural History of a Tropical Rain Forest*. Sinauer, Sunderland, Mass.

Odum, H. 1970. *A Tropical Rain Forest*. U.S. Atomic Energy Commision, Oak Ridge, Tenn.

Ríos, M. A. (ed.). 1985. *Reporte Manu*. Centro de Datos para Conservación, Agraria, La Molina, Peru.

Terborgh, J. 1983. *Five New World Primates: A Study in Comparative Ecology*. Princeton Univ Press, Princeton, N.J.

Uhl, C., K. Clark, N. Dezzeo, and P. Maquirino. 1988. Vegetation dynamics in Amazonian treefall gaps. Ecol. 69: 751–763.

PART I

The Sites

1

La Selva Biological Station: A Blueprint for Stimulating Tropical Research

DAVID B. CLARK

Imagine the volcanic backbone of Central America. Forest-clad volcanoes rise 3,000–4,000 m above the ever wet, ever warm Atlantic coastal plain. Descend the volcanoes' slopes to the point where the foothills meet the flat coastal plains. At this juncture, in the Republic of Costa Rica, sits La Selva Biological Station. La Selva is owned and operated by the Organization for Tropical Studies (OTS), an international consortium of universities and research institutions (appendix 1.1) formed to provide leadership in education, research, and the wise use of natural resources in the tropics (Stone 1988). La Selva is OTS's principal research and teaching facility in lowland tropical rainforests. The climate and vegetation of the area are described in Hartshorn (1983a).

La Selva is a relatively small station (fig. 1.1), but it has recently exerted a major influence on the progress of tropical biology. In the past five years researchers from ninety-five institutions have worked at La Selva. Usage of the station has increased almost fourfold in the past decade (table 1.1), and the number of researchers increased 257% between 1982 and 1988 (table 1.2). The flow of publications and reports on La Selva research has increased similarly. The 1987 AIBS meetings, for example, included twenty-three presentations on La Selva–based research. The varied and numerous investigations carried on at La Selva are an important reason why the tiny republic of Costa Rica is one of the world's most active research sites in tropical field biology (Clark 1985).

How was an isolated country retreat transformed into a world-class

I thank Deborah Clark, Lucinda McDade, Donald Stone, and Gary Hartshorn for numerous ideas and helpful editing.

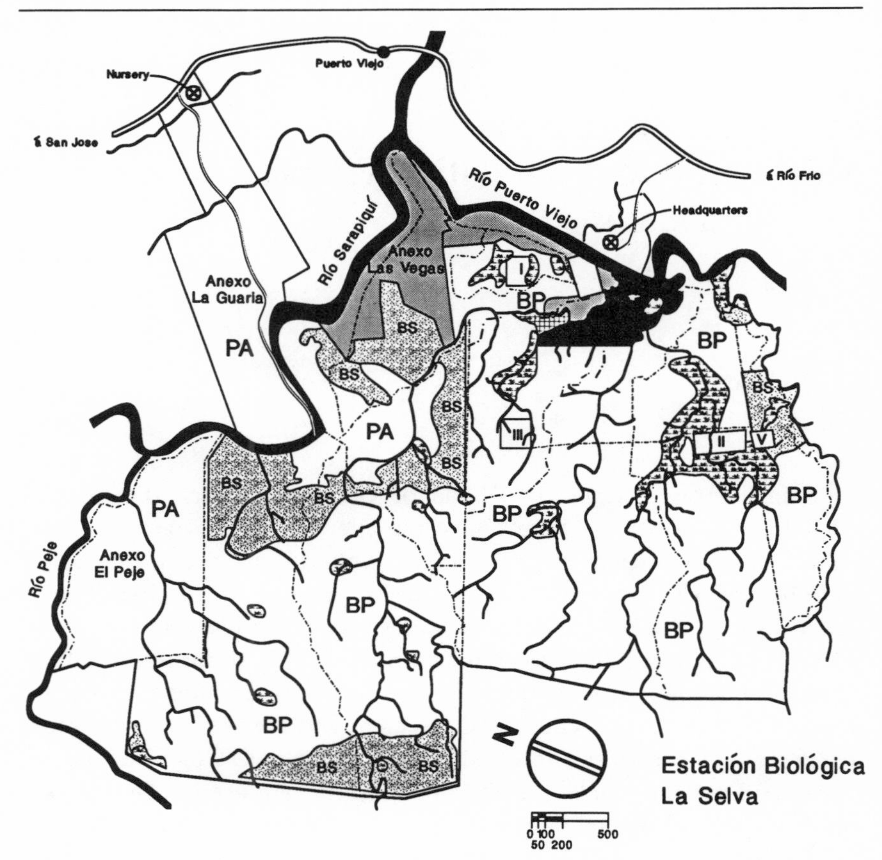

FIGURE 1.1. Component areas of La Selva Biological Station, Costa Rica.

TABLE 1.1 Use of La Selva Biological Station, 1978–1988

Year	*Person-days*
1978	2,856
1979	4,103
1980	5,297
1981	6,068
1982	7,320
1983	7,792
1984	7,784
1985	8,181
1986	9,620
1987	11,004
1988	12,385

TABLE 1.2 Profile of researchers at La Selva, 1982–1988

	1982	*1983*	*1984*	*1985*	*1986*	*1987*	*1988*
United States	48	74	80	95	106	107	112
Costa Rica	9	23	28	27	26	23	19
Other	0	3	3	11	9	11	16
TOTAL	57	100	111	133	141	141	147

research center? I present here ideas and opinions based largely on my experiences during the eight years of serving, with Deborah Clark, as co-director of La Selva. There are surely many ways to develop a successful research facility; biological field stations, like people, all differ. I describe one station, developed by an unusual organization, in a remarkable tropical country. The situation is unique, but I believe that some ideas underlying the development of La Selva have wider applications.

The region of Costa Rica where La Selva is located was inhabited for many centuries by pre-Columbian peoples (Michael Snarskis, p.c.). Artifacts are common on the relatively fertile floodplains of the Sarapiquí and Puerto Viejo rivers. These peoples planted crops on the narrow band of alluvial soils near the rivers and hunted in the more extensive forests away from them. Until recently it was assumed that the forests on these nutrient-poor soils had been little disturbed. The recent discovery of charcoal in soil cores from several areas of the primary forest is forcing a reevaluation of this idea. A project is now underway to evaluate the influence of pre-Columbian peoples on the vegetation of the La Selva region (R. Sanford, p.c.).

The next inhabitants of the area were Costa Rican colonists. Their direct effects on the vegetation were apparently limited primarily to farming (not very successfully) the alluvial soils (Petriceks 1956). In 1953 the property was purchased by Dr. Leslie Holdridge, who named the farm La Selva (the forest). Holdridge was seeking both a weekend retreat and a site to test his ideas of combining perennial crops of different life-forms. Like his Costa Rican predecessors, Holdridge farmed mainly the alluvial soils, leaving most of the primary forest intact.

In 1968, the OTS purchased the 613 ha of Finca La Selva as a lowland tropical rainforest site for teaching and research. The physical plant consisted of a few workers' houses and the Holdridges' farmhouse (fig. 1.2). The only access was via a 4-km boat trip from the village of Puerto Viejo de Sarapiquí, which in turn was accessible only by a bone-crunching

FIGURE 1.2. The station as it appeared during the 1960s; one building housed researchers, students, kitchen, and laboratory facilities. All drinking water was collected off the roof into the tanks on the right. *Photo*: Walter H. Hodge.

half-day drive from Costa Rica's capital, San José. The site lacked electricity, telephone, and laboratory equipment—but a primary forest reserve had now been established.

Through the 1960s and up to mid-1970s the physical plant changed slowly. Generators were added for nighttime electricity (sometimes!). A small prefab schoolhouse was erected as an air-conditioned "laboratory," and additional rudimentary sleeping and eating quarters were constructed. A brief excursion into ecosystem science took place in the early seventies, when researchers from the University of Washington College of Forestry established long-term forest inventory plots. After a few years of sampling, however, the project lapsed. Access to La Selva continued to be problematical, and the station drifted on as a small island of biologists (primarily U.S. graduate students) isolated from Costa Rica in particular and civilization in general.

In 1976, Donald Stone, a professor of botany at Duke University, assumed the position of executive director of OTS. From then until today (1989) Stone has worked tirelessly to develop the research and educational potential of La Selva. He saw clearly the need for improved access, for better laboratory and living facilities, and for protecting a large area of primary forest in the face of regional deforestation.

Rapid development of La Selva began in 1979, when twenty-four-hour electricity from the national electric system reached the site. In that year the first station director was hired (myself and Deborah Clark, splitting the job as co-directors). Between 1980 and 1986 change was rapid. An all-weather road was completed to the east bank property, just across the Puerto Viejo River from the forest. The kidney-threatening, boulder-strewn highway from the highlands to Puerto Viejo was paved. The final step in improving access was the result of several years of Stone's insisting to the OTS Board of Directors and the U.S. National Science Foundation (NSF) that La Selva's future depended on easy, all-weather access. With NSF support, in 1981 a 100-m pedestrian suspension bridge was built connecting the east and west banks of the river.

A microwave telephone now provides easy communication with the outside world. A large, two-storey, air-conditioned laboratory was constructed in 1981. Duplex cabins offer comfortable living space for long-term stays by researchers. The construction of a deep well and a water-tower marked the end of the days of collecting water (and attendent biota and excreta) from the building roofs. In 1986 a spacious new dining room was opened, as well as two twenty-person dormitories and a large new machine and carpentry shop. The original station was remodeled for researcher housing and laboratory space, and the old workshop was converted into an environmental education center. A quantum leap in facilities took place in 1988. OTS received two grants from NSF to build and equip a modern chemical laboratory. The 380-m^2 lab contains a diverse array of analytical equipment, as well as ample sample preparation areas.

Paralleling the development of the physical plant, the land area of La Selva has also expanded. Several small tracts were added in the 1970s to provide natural boundaries. In 1981, OTS purchased a 600-ha farm east of the original tract. This area (the Sarapiquí and La Guaria annexes [fig. 1.1]) contains a variety of human-altered habitats, including secondary forests of different ages, high-graded primary forest, and abandoned pastures, as well as streams, swamps, and a major river (table 1.3). In 1986 the acquisition of the Las Vegas Annex (fig. 1.1), a 68-ha cacao and peach palm farm, provided more habitat for manipulative research and incorporated the last private inholding into the reserve. The final expansion took place in 1987, when OTS acquired the 105-ha Peje Annex, much of which was poor pasture on fairly steep slopes. With this proper-

TABLE 1.3 Distribution of habitats at La Selva

Vegetation	Area (ha)	Area (%)
Primary forest	848	56.2
Moderately high-graded forest	114	7.5
Secondary forest	162	10.7
Early successional pastures	246	16.3
Abandoned plantations	115	7.6
Arboretum and managed habitat areas	8	0.5
Development areas	17	1.1
Total area	1,510	

ty, La Selva now has river or national park boundaries around all forested areas of the reserve.

Even though the area of La Selva has more than doubled in the past six years, at 1,510 ha the reserve is far too small to ensure the long-term preservation of its current biological diversity. This was recognized in the early 1970s, and since then many people have labored to protect a larger conservation unit. The first major step was taken by the government of Costa Rica in 1978 with the establishment of Braulio Carrillo National Park(BCNP) (Pringle 1988). This park protected 32,000 ha of mid- and high-elevation forest 20 km to the south of La Selva, as well as the summit of Barva Volcano and the upper watersheds of several major rivers (fig. 1.3).

The next obvious step was to link La Selva to BCNP. In addition to greatly increasing the ecological security of La Selva, this expanded park would protect the only unbroken transect of primary forest in Central America that spans an elevational gradient from fewer than 100 m to almost 3,000 m. The enormous conservation and research potential of this altitudinal transect was recognized by the U.S. National Research Council's Committee on Research Priorities in Tropical Biology (NRC 1980). The committee recommended La Selva and BCNP as one of four tropical areas where ecosystem research should be concentrated.

Two successive administrations of the government of Costa Rica expressed support for the expanded park. Unfortunately, Costa Rica's sudden and rapidly deepening economic crisis made financial support impossible. The national currency was devalued almost 800 percent against the U.S. dollar; there was simply no money available to compensate landholders within the proposed park. As an emergency measure, in 1982 President Rodrigo Carazo declared that the connecting corridor from

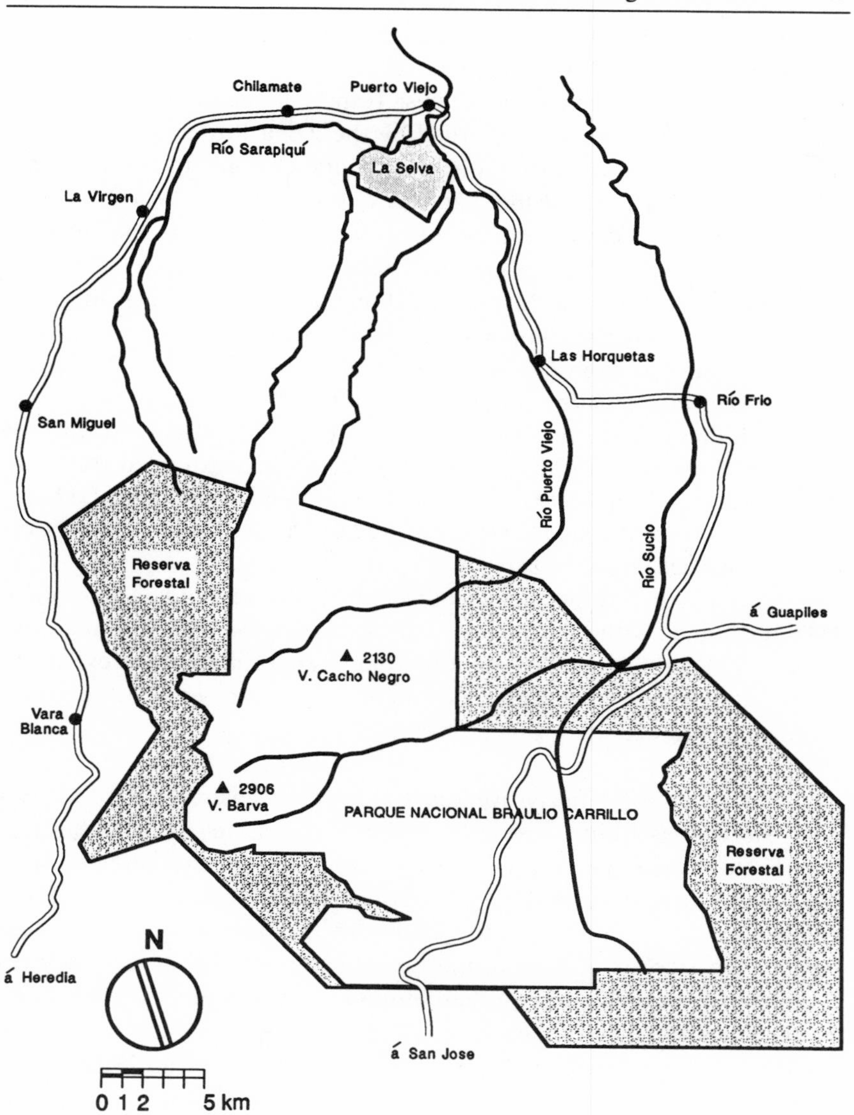

FIGURE 1.3. Regional setting of La Selva Biological Station and Braulio Carrillo National Park.

BCNP to La Selva would be classified a Zona Protectora (protected zone). This zoning allowed landowners to maintain title and use of the land but prohibited changes in land use (especially deforestation).

The interim protection afforded by the Zona Protectora gave time for an international consortium of conservation groups to begin a fund-raising drive. OTS joined the Nature Conservancy, the World Wildlife Fund—U.S., the Costa Rican National Parks Foundation, and the government of Costa Rica in a unique conservation alliance. A $1 million matching grant from the MacArthur Foundation was crucial in giving early momentum to the campaign. Eventually over $2 million was raised, with contributions from over two thousand donors. This was the largest international conservation project in Costa Rica to that date.

With these funds, many properties were quickly purchased. In 1986 President Luis Alberto Monge issued a decree that extended BCNP to the boundaries of La Selva. The area of the expanded park plus La Selva is now 46,000 ha; the area is also bordered by the 91,000-ha Central Cordillera National Forest Reserve.

The campaign to preserve biodiversity in this region has not ended. OTS has joined with Costa Rican and international conservation groups to raise a $1.5-million endowment fund for management of the expanded park. The interest from this endowment will help to ensure the park's permanent protection.

During the past thirty-five years more than seven hundred articles and theses describing research at La Selva have been published. Although an extremely broad range of topics has been addressed, it is possible to identify several major areas of research concentration. From the beginning of research at the station to the present systematic biologists have labored to resolve the taxonomic mysteries of the La Selva biota. Work on such groups as herpetofauna (Savage and Villa 1986) and birds (Stiles 1983) is well advanced. Substantial progress has been made in describing the vascular flora, through the efforts of an NSF-funded Flora of La Selva project (Wilbur 1986). Invertebrates remain nearly a systematic black hole. A few groups have been well studied, including scarab beetles (Beach 1982, 1984; Young 1986a), butterflies (DeVries 1987), ants (E. Wilson, in prep.), and moths (D. Janzen and M. M. Chavarría, in prep.). Most major taxa, however, have not yet been studied. Plans are underway to launch a long-term research effort to describe arthropod diversity at La Selva.

Another prominent focus of research has been plant phenology and reproductive biology. Beginning with community-level phenological analyses (Frankie et al. 1974; Opler et al. 1980), researchers have examined the reproductive biology and phenology of single species or related groups (see Bawa et al. 1985a,b). Patterns of pollination have been intensively studied from the perspective of both plant and pollinator (see Stiles 1975; Kress 1981; Grove 1985; Young 1986b).

It is a fitting tribute to Leslie Holdridge, tropical forester par excellence, that the study of tropical trees and forest dynamics has figured prominently in the development of research at La Selva. In 1969 researchers from the University of Washington College of Forestry established three 4-ha long-term forest inventory plots at La Selva. Now twenty years old, the plots are yielding important data on forest structure and dynamics (Lieberman et al. 1985a,b,c). A related theme has been the importance of natural treefall gaps and their effects on tree regeneration (Hartshorn 1978, 1980; Orians 1982; Barton 1984; Brandani et al. 1988). A multidisciplinary group is seeking to understand the consequences of treefall gaps on processes of nutrient flow, plant demography, herbivory, and microclimate (P. Vitousek, J. Denslow, J. Schultz, and B. Strain). The demographic and ecophysiological effects of gaps on tree regeneration are being addressed in a long-term study of canopy trees with diverse regeneration strategies (Clark and Clark 1987).

The mechanisms by which nutrients enter and cycle through the terrestrial and aquatic systems of La Selva have been studied by numerous researchers. Epiphyllous microorganisms have been shown to contribute fixed nitrogen to their hosts (Bentley and Carpenter 1984). Several canopy trees were found to release substantial quantities of gaseous sulfur (Haines et al., in press). Standing stocks of nitrogen in the soil are among the highest yet reported from the humid tropics (Vitousek and Denslow 1986), and some La Selva creeks have high levels of nitrogen and phosphorus (Pringle et al. 1986). Primary productivity in one stream has been shown to be limited by micronutrients, not by phosphorus or nitrogen (Pringle et al. 1986). Current studies focus on specific mechanisms affecting nutrient flows. One research team is studying the effect of soil biological processes on charge properties and nutrient availability in La Selva's variable-charge clay soils (P. Robertson and P. Sollins). The effects of both wetlands and secondary forest on aquatic nutrient processes are also being evaluated (Paaby 1988; C. Pringle, in prep.).

Since 1980 research on plant ecophysiology has grown rapidly. Plant-water relations and photosynthesis, the relations among light, nutrients, and plant performance in shade houses and in the forest, and the spatial and temporal variability of the light environment have all been addressed. Many of these results are summarized in a volume based on a 1985 conference on Plant Population Biology and Ecophysiology in Mesoamerican Forests, which OTS hosted in Costa Rica (D. A. Clark, Dirzo, and Fetcher 1987).

Much research at La Selva can be categorized under the heading of evolutionary ecology. Such topics as single-species autecology, herbivory, seed dispersal, and predation, community structure of related groups, frugivory, and animal behavior have received much attention. The results of these studies and those outlined above are being synthesized and reviewed in a book summarizing research at La Selva (McDade et al. 1991).

Why do researchers choose to work at a given site? Part of the answer has to do with the availability of different kinds of resources to support research. One of the primary attractions of La Selva is the rich diversity of tropical life in the primary forest. La Selva is an intact, relatively undisturbed forest ecosystem. White-lipped peccaries have disappeared (I. Alvarado, p.c.), but otherwise the main elements of a lowland tropical rainforest fauna are present. Over 400 species of trees and a similar number of bird species have been recorded from La Selva (Hartshorn 1983b; Stiles 1983). The number of moth species is estimated at over 4,000 (D. Janzen and M. M. Chavarria, p.c.), and the vascular plant flora may exceed 2,000 species (Hammel 1986). The vertebrate predator fauna is particularly rich, numbering at least 110 species (Greene 1988).

Another biological resource of La Selva is the availability of a broad variety of habitats for manipulative research. Most of the common primary, secondary, and agricultural ecosystems of the wet Atlantic lowlands are easily accessible. Within the reserve and available for experimental research are secondary forests of different ages, primary forests that have been highgraded at different intensities, abandoned pastures, and permanently open areas around the buildings. Adjoining La Selva are areas in pasture, tree plantations, and different types of agriculture. A significant percentage of the recent research at La Selva has been concentrated in these human-altered systems.

An additional attraction of La Selva is Braulio Carrillo National Park. From 35 m above sea level at the Puerto Viejo River to 2,906 m at the summit of Barva Volcano, one passes successively through tropical wet forest, premontane rain forest, lower montane rain forest, and montane rain forest (Hartshorn and Peralta 1988). The biodiversity of the area has not yet been sampled systematically, but it is estimated that it includes approximately 500 species of birds, 135 species of mammals, and 125 species of reptiles and amphibians, as well as at least 4,000 species of higher plants (Pringle et al. 1984; Michael Grayum, p.c.). The lower reaches of the park are accessible from La Selva by foot or car, and a trail now reaches from the lowlands of La Selva all the way over the divide into the central valley section of the park. Wilderness cabins are available to researchers at several sites.

Braulio Carrillo National Park is a spectacular research area in itself; in addition, it guarantees the long-term survival of La Selva's biodiversity. Secure research sites in an area whose biota is substantially protected over the long-term also favors research.

Many scientific studies require virtually no equipment—only a good question, a pencil, and a piece of paper. Nevertheless, as the physical resources to support research improve, the spectrum of problems that can be addressed expands exponentially. The physical facilities at La Selva are a good reason why research use is increasing rapidly. One of the biggest stimuli to research was the construction in 1983 of the air-conditioned laboratory. The building is outfitted with basic laboratory equipment (microscopes, balances, microcomputers, and so on), supplemented by specialized equipment brought by individual projects. This laboratory has so successfully attracted researchers that it is now completely occupied most of the year. The new soil, water, and tissue analysis laboratory will further increase the range of projects that can be carried out at the station. The juxtaposition of advanced analytical facilities with protected primary and secondary ecosystems can be found at only a handful of sites in the world's wet tropics.

Basic services such as round-the-clock electricity (backed up by emergency generators), a potable water system from a deep well, and comfortable eating and sleeping facilities all expedite research. A well-equipped shop staffed by expert carpenters provides construction and repair services for researchers. Eleven shade houses provide a variety of controlled light environments. Within the reserve an extensive network of trails facilitates access.

The largest databank of La Selva research is the over seven hundred published articles and theses. The bibliography is keyworded and computerized and can be accessed by researchers. Copies of theses and reprints, as well as a selected collection of books on tropical systematics and ecology, are housed in the station library. The Holdridge Arboretum contains over 1,000 identified trees, representing the majority of tree species at La Selva. A small reference herbarium is being expanded. A 1:10,000 soils map was recently completed (Sancho and Mata 1987). Time series of aerial photos of the region are available from the Instituto Geográfico Nacional. Daily rainfall has been recorded since 1957.

Ease of access to La Selva enhances station use greatly. The station is located two hours from Costa Rica's international airport via a paved road. The OTS Costa Rican office offers thrice-weekly shuttle runs. This office also provides support services for researchers, such as assistance with visas and purchasing supplies. A major concern of scientists working overseas is obtaining permits for research and collection, both at the national level and within the chosen field site. OTS arranges Costa Rican permits for visiting scientists and attempts to keep researchers abreast of regulations affecting research. OTS also helps to establish connections between Costa Rican researchers and visiting scientists.

Guidelines for research at La Selva are set out in the Master Plan for Development and Administration, the official operating policy adopted by the OTS Board of Directors. The Master Plan is regularly reviewed by an independent scientific commitee (the La Selva Advisory Committee, LSAC). New regulations are developed and old ones revised as experience accumulates and conditions change. Observational and low-impact projects are approved by the executive director or station director. Projects involving greater impact or long-term allocation of land are referred to LSAC for evaluation. La Selva is zoned into areas of varying research intensities, ranging from observational research to intensive manipulation. Given the range of habitats available, it has been possible to accommodate all requests that comply with the Master Plan environmental impact guidelines.

A field research project depends on successful personal interactions with station support staff, field assistants, technicians, and other scientists. OTS's permanent administrative and technical staff are a valuable resource to visiting scientists. The station directors are active scientists who can offer advice and assistance to researchers. The station staff, including the assistant director, foreman, and accounting personnel, assist researchers with logistic and administrative needs. OTS has facilitated the hiring of Costa Rican biologists as project assistants. All long-term projects at La Selva now have at least one Costa Rican scientist on their staff. In addition, competent field assistants can be hired through OTS from the local town of Puerto Viejo. These assistants are often excellent naturalists, and several have become important contributors to long-term research efforts.

The number of researchers present at La Selva now averages about twenty a day. The investigators are frequently joined by twenty to thirty visiting biology students and faculty. Because La Selva is run by a consortium of universities and research institutions, not one department or ministry, the breadth of disciplines represented is tremendous. Soil scientists, aquatic biologists, systematists, foresters, and evolutionary ecologists all eat in the same dining hall, share the same laboratories and computers, and socialize at the same parties. As the number of researchers increases, so do opportunities for interdisciplinary studies. This intellectual diversity is an important asset of La Selva.

The pace of tropical biological research is likely to continue to increase. Currently OTS trains approximately eighty post-graduate students per year in tropical biology and agricultural ecology. Many alumni return to work in the tropics. The continued interest of the National Science Foundation in supporting tropical research was illustrated in 1987 by the solicitation of proposals for a tropical wet forest long-term ecological research site. International development agencies such as AID (Agency for International Development) and World Bank are also increasing funding in tropical biology, especially conservation and natural resources (Holden 1988). These trends are likely to increase the amount of research carried out at La Selva.

Investigations in some traditional areas of strength at La Selva, such as systematic studies and evolutionary ecology, may continue at more or less the same level. In other fields, however, activity is likely to increase. As infrastructure develops within Braulio Carrillo National Park, particularly improved trails and additional shelters, research use will increase. An effort is already underway to use remote sensing technology to measure such ecophysiological phenomena as evapotransporation at the landscape level (Jeff Luvall, p.c.). The undescribed biota of the park will continue to attract systematists. Patterns of species distribution and community composition will be a major focus of study (see Hartshorn and Peralta 1988; Stiles 1988).

The substantial work already underway in soil and nutrient ecology at La Selva will undoubtedly increase. Much of this research has potential applications to agriculture and forestry, thus inviting funding by development agencies as well as by the NSF. The related field of aquatic biology will also be developed. From the high-gradient oligotrophic streams of BCNP to slack-water high-nutrient swamp waters, the diversity of pristine and human-impacted watersheds in the region offers a tremendous opportunity for comparative ecosystem-level aquatic research. Progress in these fields, as well as in isoenzyme genetics and phytochemistry, will be stimulated by the new analytical laboratory at La Selva.

Research in forestry is poised to increase substantially, supplementing ongoing projects that have concentrated on demography and physiology of tree species in primary forest. A major field trial to screen native species as candidates for land rehabilitation is underway. Illustrative of the way research interactions are fostered at La Selva, this project has already stimulated an NSF proposal to examine long-term changes in soil chemistry during reforestation and an AID proposal to look at soil water chemistry in plantations. A U.S. Department of Agriculture-funded study on photoinhibition will also take advantage of these species trials. The unique role of La Selva in forestry research will likely be in the study of aspects of basic science. Forestry studies will go beyond asking simply which species grows best where, to trying to determine the ecological and physiological bases of species' responses.

A new focus of research is developing around OTS's greater role in relation to regional development. At the request of the government of Costa Rica, La Selva, BCNP, the surrounding national forests, and adjacent agricultural areas were named a World Biosphere Reserve by UNESCO. This ambitious project envisions closer interactions between La Selva, as a teaching and research center, and local and national agencies concerned with regional land use. Whatever mechanisms evolve, a growing interest in applied research at La Selva and in the adjacent community is evident. As with forestry research, La Selva's role is likely to be facilitative and synergistic, linking basic scientists with applied research problems (Clark 1988).

La Selva has grown from an isolated country retreat to a world center for rainforest research in less than twenty years. What has been learned about the development of research that can be applied to other sites?

Throughout its history OTS has depended on the intellectual input of an array of interested scientists. These advisers are volunteers, not staff, so their advice offers OTS an independent evaluation of its policies and projects. Every substantive aspect of research and facilities development at La Selva has benefited from this input.

The marriage of teaching and research at the station has stimulated research tremendously. The OTS courses have introduced hundreds of gung-ho graduate students to La Selva, and many have returned to do their thesis and postdoctoral research. Students have also had a positive effect on resident researchers, exposing them to new ideas and a healthy questioning of ongoing projects.

The Master Plan for management and development has greatly helped the orderly development of La Selva. It has been important in guiding the design and location of new facilities and has served as a vehicle for explaining development of La Selva and for seeking critical evaluation. Finally, published research guidelines give researchers a stable institutional framework within which to develop their research programs.

The importance of facilities in stimulating research has been notable. Each level of improvement in the physical plant and laboratories has been accompanied by an increase in use and in the diversity of research topics. Enlarging the dining and sleeping quarters has permitted more researchers and students at peak periods. Research that relies on sophisticated equipment has clearly benefited by the increase in laboratory facilities. Much of this equipment is dependent on reliable electricity and low-humidity, air-conditioned environments. The recent boom in microcomputers (each lab room averages one micro) is another outgrowth of increased air-conditioned workspace. Research complexity will clearly rise to meet the level of available technical support.

Perhaps the most important of OTS's experiences to be emphasized is the long history of positive relations it has enjoyed with its host country, the Republic of Costa Rica. This special relationship affects research in innumerable ways, from the ease with which research permits can be obtained to the willingness of talented Costa Rican biologists to collaborate in joint projects. Costa Rica is unusually receptive to foreign scientists (Clark 1985), and OTS has benefited greatly from this attitude. For its part, Costa Rica has benefited economically, educationally, and scientifically from the relationship. One-third of current use of La Selva is by Costa Rican biologists and students. Costa Rica's four principal institutions of higher education and research are members of the consortium, which means that in a real sense OTS's operations are part of the national educational and research structure. This interaction is helped by a Costa Rican coordinating committee, composed of OTS representatives from the Costa Rican members, as well as government officials. This committee frequently advises OTS on local issues. The recent signing of a working agreement between the newly created Ministry of Natural Resources, Energy, and Mines and OTS indicates the importance that both organizations place on developing mutual ties.

In its biological diversity, research facilities, and logistic and administrative support, La Selva Biological Station is now an exceptional research resource for the scientific community. Countries in the world's lowland wet tropics are changing rapidly in the face of social, political, economic, and ecological pressures. Only a handful of well-equipped research sites next to large intact reserves are likely to be sustained. Because of OTS's pivotal role in tropical ecology and the dedication of Costa Rica to its national park system, La Selva has a good chance to continue as one of these sites.

REFERENCES

Barton, A. M. 1984. Neotropical pioneer and shade-tolerant tree species: do they partition tree-fall gaps? Trop. Ecol. 25: 196–202.

Bawa, K. S., D. R. Perry, and J. H. Beach. 1985a. Reproductive biology of tropical lowland rain forest trees. I. Sexual systems and incompatibility mechanisms. Amer. J. Bot. 72: 331–345.

Bawa, K. S., S. H. Bullock, D. R. Perry, R. E. Coville, and M. H. Grayum. 1985b. Reproductive biology of tropical lowland rain forest trees. II. Pollination systems. Amer. J. Bot. 72: 346–356.

Brandani, A., G. S. Hartshorn, and G. H. Orians. 1988. Internal heterogeneity of gaps and tropical tree species richness. J. Trop. Ecol. 4: 99–119.

Beach, J. H. 1982. Beetle pollination of *Cyclanthus bipartitus* (Cyclanthaceae). Amer. J. Bot. 69: 1074–1081.

———. 1984. The reproductive biology of the peach or "pejibaye" palm (*Bactris gasipes*) and a wild congener (*B. porschiana*) in the Atlantic lowlands of Costa Rica. Principes 28: 107–119.

Bentley, B. L. and E. J. Carpenter. 1984. Direct transfer of newly-fixed nitrogen from free-living epiphyllous microorganisms to their host plant. Oecologia 63: 52–56.

Clark, D. B. 1985. Ecological field studies in the tropics: geographical origin of reports. Bull. Ecol. Soc. Amer. 66: 6–9.

———. 1988. The search for solutions: research and education at the La Selva Biological Station and their relation to ecodevelopment. In F. Almeda and C. Pringle (eds.). *Tropical Rainforests: Diversity and Conservation*. Calif. Acad. Sci., San Francisco, pp. 209–224.

Clark, D. A. and D. B. Clark. 1987. Análisis de la regeneración de árboles del dosel en bosque muy húmedo tropical: aspectos teóricos y prácticos. Rev. Biol. 34 (supp.): 41–54.

Clark, D. A., R. Dirzo, and N. Fetcher (eds.) 1987. Ecología y ecofisiología de plantas en los bosques mesoamericanos. Rev. Biol. Trop. 35 (suppl. 1): 1–234.

DeVries, P. J. 1987. *The Butterflies of Costa Rica and Their Natural History*. Princeton Univ. Press, Princeton, N.J.

Frankie, B. W., H. G. Baker, and P. A. Opler. 1974. Comparative phenological studies of trees in tropical wet and dry forests in the lowlands of Costa Rica. J. Ecol. 62: 881–919.

Greene, H. W. 1988. Species richness in tropical predators. In F. Almeda and C.

Pringle (eds.), *Tropical Rainforests: Diversity and Conservation*, Calif. Acad. Sci., San Francisco, pp. 259–280.

Grove, K. F. 1985. Reproductive biology of Neotropical wet forest understory plants. Ph.D. diss., Univ. of Iowa.

Haines, B. L., M. Black, J. Fail, Jr., L. McHargue, and G. Howell. In press. Potential sulfur gas emissions from a tropical rainforest and a southern Appalachian deciduous forest. In T. C. Hutchinson (ed.), *Effects of Acidic Deposition on Forests, Wetlands, and Agricultural Ecosystems*. Springer-Verlag, New York.

Hammel, B. E. 1986. Characteristics and phytogeographical analysis of a subset of the flora of La Selva (Costa Rica). Selbyana 9: 149–155.

Hartshorn, G. S. 1978. Tree falls and tropical forest dynamics. In P. B. Tomlinson and M. H. Zimmerman (eds.), *Tropical Trees as Living Systems*. Cambridge Univ. Press, London, pp. 617–638.

———. 1980. Neotropical forest dynamics. In J. Ewel (ed.), *Tropical Succession*. Biotropica 12 (suppl.): 23–30.

———. 1983a. Plants. In D. H. Janzen (ed.), *Costa Rican Natural History*. Univ. of Chicago Press, Chicago, pp. 118–157.

———. 1983b. Checklist of tress. In D. H. Janzen (ed.), *Costa Rican Natural History*. Univ. of Chicago Press, Chicago, pp. 158–183.

Hartshorn, G. and R. Peralta. 1988. Preliminary description of primary forests along the La Selva-Volcán Barva altitudinal transect, Costa Rica. In F. Almeda and C. Pringle (eds.), *Tropical Rainforests: Diversity and Conservation*. Calif. Acad. Sci., San Francisco, pp. 281–295.

Holden, C. 1988. The greening of the World Bank. Science 240: 1610.

Kress, W. J. 1981. Reproductive biology and systematics of Central American *Heliconia* (Heliconiaceae). Ph.D. diss., Duke Univ.

Lieberman, D., M. Lieberman, R. Peralta, and G. S. Hartshorn. 1985a. Mortality patterns and stand turnover rates in a wet tropical forest in Costa Rica. J. Ecol. 73: 915–924.

Lieberman, D., M. Lieberman, G. S. Hartshorn, and R. Peralta. 1985b. Growth rates and age-size relationships of tropical wet forest trees in Costa Rica. J. Trop. Ecol. 1: 97–109.

———. 1985c. Small-scale altitudinal variation in lowland wet tropical forest vegetation. J. Ecol. 73: 505–516.

McDade, L. A., K. S. Bawa, H. A. Hespenheide, and G. S. Hartshorn. 1991. *La Selva: Ecology and Natural History of a Neotropical Rainforest*. Univ. of Chicago Press, Chicago.

National Research Council. 1980. Committee on Research Priorities in Tropical Biology. Research priorities in tropical biology. Nat. Acad. Sci., Washington, D.C.

Opler, P. A., G. W. Frankie, and H. G. Baker. 1980. Comparative phenological studies of treelet and shrub species in tropical wet and dry forests in the lowlands of Costa Rica. J. Ecol. 68: 167–188.

Orians, G. H. 1982. The influence of tree-falls in tropical forests in tree species richness. Trop. Ecol. 23: 255–279.

Paaby, P. 1988. Light and nutrient limitation of primary producers in a Costa Rican lowland stream. Ph.D. diss., Univ. of California, Davis.

Petriceks, J. 1956. Plan de ordenación del bosque de la Finca "La Selva." M.A. thesis, Instituto Interamericano de Ciencias Agrícolas, Turrialba, Costa Rica.

Pringle, C. M. 1988. History of conservation efforts and initial exploration of the lower extension of Parque Nacional Braulio Carrillo, Costa Rica. In F. Almeda and C. M. Pringle (eds.), *Tropical Rainforests: Diversity and Conservation.* Calif. Acad. Sci., San Francisco, pp. 225–241.

Pringle, C. I. Chacón, M. Grayum, H. Greene, B. Hartshorn, G. Schatz, G. Stiles, C. Gómez, and M. Rodrígues. 1984. Natural history observations and ecological evaluation of the La Selva Protection Zone, Costa Rica. Brenesia 22: 189–206.

Pringle, C. M., P. Paaby-Hanson, P. D. Vaux, and C. R. Goldman. 1986. *In situ* nutrient assays of periphyton growth in a lowland Costa Rican stream. Hyrobiol. 134: 207–213.

Sancho, F. and R. Mata. 1987. Estudio detallado de suelos. Organization for Tropical Studies, San José, Costa Rica.

Savage, J. and J. Villa R. 1986. Introduction to the herpetofauna of Costa Rica. Soc. Study. Amphib. Rept. Contrib. Herpetol. 3: 1–207.

Stiles, F. G. 1975. Ecology, flowering phenology, and hummingbird pollination of some Costa Rican *Heliconia* species. Ecol. 56: 285–301.

———. 1983. Checklist of birds. In D. Janzen (ed.), *Costa Rican Natural History.* Univ. of Chicago Press, Chicago, pp. 530–544.

———. 1988. Altitudinal movements of birds on the Caribbean slope of Costa Rica: implications for conservation. In F. A. Almeda and C. M. Pringle (eds.), *Tropical Rainforests: Diversity and Conservation.* Calif. Acad. Sci., San Francisco, pp. 243–258.

Stone, D. E. 1988. The Organization for Tropical Studies (OTS): a success story in graduate training and research. In F. A. Almeda and C. M. Pringle (eds.), *Tropical Rainforests: Diversity and Conservation.* Calif. Acad. Sci., San Francisco, pp. 143–187.

Vitousek, P. M. and J. S. Denslow. 1986. Nitrogen and phosphorus availability in treefall gaps of a lowland tropical rainforest. J. Ecol. 74: 1167–1178.

Wilbur, R. L. 1986. The vascular flora of La Selva Biological Station, Costa Rica—introduction. Selbyana 9: 191.

Young, H. J. 1986a. Pollination of *Dieffenbachia longispatha*: effects of beetles on reproductive success, gene flow, and gender. Ph.D. diss., SUNY, Stony Brook.

———. 1986b. Beetle pollination of *Dieffenbachia longispatha* (Araceae). Amer. J. Bot. 73: 931–944.

APPENDIX 1.1. Member institutions of the Organization for Tropical Studies in 1989

Auburn University
City University of New York
Cornell University
Duke University
Florida International University
Harvard University
Indiana University
Instituto Technológico de Costa Rica
Louisiana State University
Michigan State University
Museo Nacional de Costa Rica
North Carolina State University
Ohio State University
Pennsylvania State University
Rutgers University
Smithsonian Institution
Stanford University
State University of New York at Stony Brook
Texas A&M University
Tulane University
Universidad de Costa Rica
Universidad Nacional
University of Arizona
University of California at Berkeley
University of California at Davis
University of California at Irvine
University of California at Los Angeles
University of Chicago
University of Colorado
University of Connecticut
University of Florida
University of Georgia
University of Hawaii
University of Iowa
University of Kansas
University of Maryland
University of Miami
University of Michigan
University of Minnesota
University of Missouri at Columbia
University of Missouri at St. Louis
University of North Carolina
University of Utah
University of Puerto Rico
University of Washington
University of Wisconsin
Washington University
Yale University

2

Barro Colorado Island and Tropical Biology

EGBERT GILES LEIGH, JR., AND S. JOSEPH WRIGHT

Barro Colorado Island (BCI), Panama, the subject of more than fifteen hundred scholarly publications, may be the most studied tract of tropical forest in the world. It is perhaps best known for studies in natural history, ecology, and ethology, centering on vertebrates, insects, plants, and their relationships. Large-scale integrated studies at the community level, now in progress, complement the diversity of projects carried out by graduate students, postdoctoral fellows, and visiting scientists.

The 1,500 ha of BCI (9°9′N, 79°51′W) were isolated from the surrounding mainland after the Chagres River was dammed in 1910 (McCullough 1977, 589) to create Gatun Lake and complete the Panama Canal. BCI is now administered by the Smithsonian Tropical Research Institute (STRI) as part of the Barro Colorado Nature Monument (BCNM). The Panama Canal treaties of 1977 gave STRI custodianship of the BCNM, which was established under the Western Hemisphere Convention on Nature Protection and Wildlife Preservation of 1940. The BCNM also includes five nearby mainland peninsulas (fig. 2.1), three of which adjoin Panama's Parque Nacional Soberania. BCI has accommodations for thirty-five visitors, air-conditioned office and laboratory space, an insectary, a screened growing house with a plant physiology laboratory, a herbarium, a small reference library, and equipment for general use, such as freeze driers, autoclaves, drying ovens, a muffle furnace, analytical balances, microcomputers, and small boats for access to mainland sites. These facilities will be augmented when a new, fully equipped laboratory is completed in 1992. Launch service to BCI is provided twice a day from the roadhead at Gamboa (where STRI also maintains a small dormitory), allowing mainlanders to visit BCI for the day and researchers on BCI to

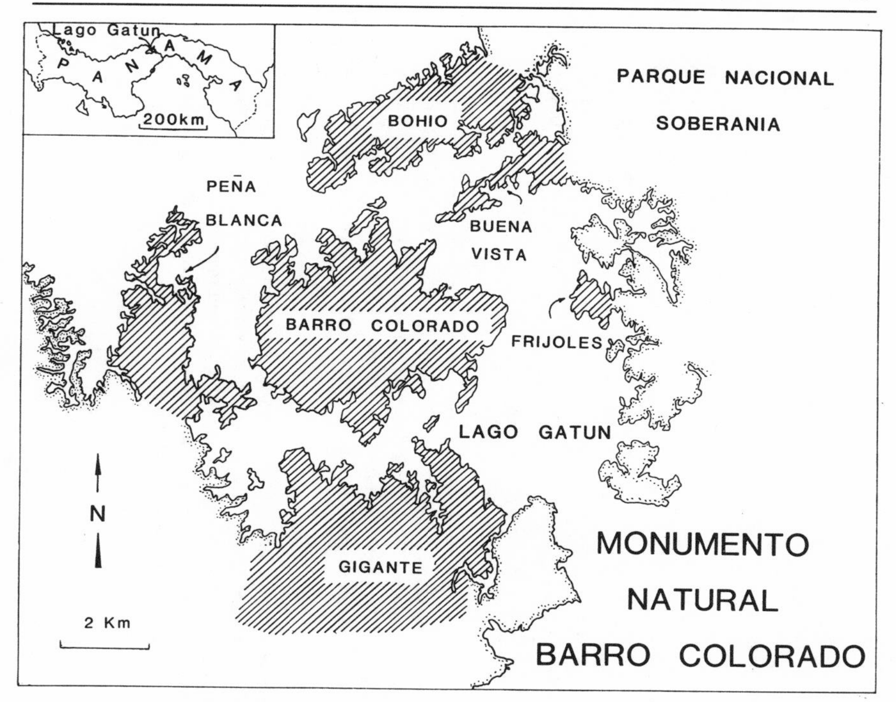

FIGURE 2.1. Map of Barro Colorado Nature Monument. This reserve includes Barro Colorado Island and five mainland peninsulas. Inset: position of the nature monument within the Republic of Panama.

consult STRI's extensive library in Panama City. Further information on facilities, availability of space, and charges for their use may be obtained from the Visitors Desk, Smithsonian Tropical Research Institute, Apartado 2072, Balboa, Republic of Panama. Information on predoctoral and postdoctoral fellowships may be obtained from the education coordinator at the same address.

To set the work on BCI in perspective, we provide some background on climate, vegetation, and soil. We then show how the diversity of projects on BCI has provided a community-level understanding of the forest. We consider two main themes: the seasonal rhythms of the forest and their roles in limiting animal populations, and why there are so many kinds of plants and animals in the tropics.

Barro Colorado Island and its environs are covered by tropical moist forest whose canopy can attain a height of 35–40 m. Judging by the amount of litter fall and the amount of light reaching the forest floor, the

TABLE 2.1. Litter fall and light interception in different tropical forests

	Litter Fall: g (dry weight)/m^2 per year							
	Leaves	(% liana)	Fruit	Total	LAI[1]	Light[2] penetration (%)	Height of sensor (cm)	Understory light in sunflecks (%)
BCI, Panama	789	11.8	127	1,366	7–8	0.3	61	0.43
La Selva, Costa Rica				915		0.9–1.5	70	0.44
Central Amazonia, Brazil	441			742				
Makokou, Gabon	650	36.0		1,390		3.0[3]	130	
Pasoh, Malaysia	703	5.0	38	1,082	7–8	0.3	0	

Sources: BCI: litter fall is from Wright and Cornejo (1990b), leaf area index from Leigh and Windsor (1982), and light data from A. P. Smith and P. F. Becker (p.c.). La Selva: litter fall from D. W. Cole, quoted by Leigh and Windsor (1982); light data from Chazdon and Fetcher (1984: table 1) and Chazdon (1986: table 3). Brazil: litter fall from Luizao and Schubart (1987). Gabon: litter fall from Hladik (1978), light data from Hladik and Blanc (1987). Malaysia: all data from Kira (1978). Percent liana leaves in Pasoh litter is assumed equal to percent of liana leaves among the leaves weighed from a harvested plot.

1. LAI is Leaf Area Index, the number of m^2 of leaf per m^2 of ground.
2. Percent light penetration is measured as photosynthetically active radiation, except for Gabon, where radiation between 400 and 1,100 nm wavelength was measured instead. At La Selva, Makokou, and Pasoh, % light penetration is an average for shady areas; on BCI, it is the median of the rainy season averages for 400 sites, spaced 2.5 m apart along a line in mature forest, passing through light gaps and shady places.
3. Judging by Johnson and Atwood (1970: table 4), penetration of photosynthetically active radiation to 130 cm above ground level in Gabon would probably lie between 1.5% and 2%. Some far-red, but little photosynthetic, radiation passes through leaves.

forest on BCI is as productive, and suffers as intense competition for light, as tropical forests elsewhere (table 2.1; see also Leigh and Windsor 1982; Alexandre 1982).

On the southwest half of BCI, the forest has been little disturbed since the Spanish conquest. Some stands have not been cleared for two thousand years (Piperno 1989). The remainder of BCI was much disturbed during the French attempt to build a canal: its forest is clearly secondary (Kenoyer 1929; Foster and Brokaw 1982).

The island receives an average of 2,600 mm of rain a year. A sharp dry season usually begins in December and ends in April: median rainfall during the first thirteen weeks of the year is only 88 mm (compiled from Windsor et al. 1989:app.). In most years, nearly all BCI's streams cease flowing by the end of the dry season, when exposed soil is dry and deeply cracked and soil water potentials are as low as −1.6 MPa, or −16 bars (Wright and Cornejo 1990a; Becker et al. 1988). In the open, the average annual temperature is 27 °C, and the average diurnal temperature range is 8 °C (Dietrich et al. 1982).

The soils of BCI vary with the underlying geologic formations (Woodring 1958; M. Keller, p.c.). An andesite (R. F. Stallard, p.c.) flow underlies a central plateau of some 3 km^2, whose highest point is 165 m above sea level. The soils of the plateau are well-weathered oxisols, as poor in nitrogen and phosphorus as the oxisols near Manaus, Brazil, and probably much poorer in potassium (M. Keller and R. F. Stallard, p.c.). Occasional "weathering spheroids" of andesite occur in the soils of this plateau, always a meter or more below the surface: these spheroids are surrounded by roots, as if competition for nutrients is fairly intense (R. F. Stallard, p.c.). Sedimentary formations of terrestrial, marine, and volcanic origins underlie the remainder of BCI (Woodring 1958). Relatively rich alfisols have developed on the terrestrial and marine sediments surrounding the central plateau. A second well-weathered oxisol, derived from sedimentary rock of volcanic origin, covers the eastern lobe of the island (M. Keller, p.c.). Conrad Stream, draining the deeply weathered soils of BCI's plateau, is hardly richer in cations than streams of western Amazonia, and hardly richer in nitrogen and phosphorus than streams of the nutrient-poor regions of central and eastern Amazonia. Lutz stream, on the other hand, which drains relatively fertile slopes near the laboratory clearing on BCI, is richer in cations than is the Amazon at Iquitos, which drains the rapidly eroding eastern slopes of the Andes (R. F. Stallard, p.c.: see table 2.2).

Curiously, the litter that falls on the "infertile" plateau is unusually

TABLE 2.2 Comparison of nutrient levels in streams and rivers of Amazonia and BCI

	Ca^{2+} (mg/L)	Mg^{2+} (mg/L)	K^{+} (mg/L)	inorganic N (mg/L)	inorganic P (mg/L)
Amazonia					
Lower Amazon River (Obidos)	5.0	0.9	0.9	0.10	0.009
Eastern Amazon Tributaries[1] (Xingu, Tapajos, Trombetas)	1.6	0.5	0.8	0.04	0.0012
Central Amazonia (Rio Negro at Manaus)	0.35	0.14	0.4	0.06	0.0022
Western Amazonia Tributaries[1] (Purus, Jurua, Javari)	4.0	0.8	0.9	0.08	0.007
Andean Amazon (Iquitos)	20.0	2.2	1.2	0.16	0.022
Amazon rain: showers over river[2]	0.03	0.012	0.03	0.04	—
Amazon rain: rainfall over Manaus[3]	0.14	0.12	—	0.28	0.003
Barro Colorado Island					
Conrad stream	2.6	1.0	0.5	0.07	0.002
Lutz stream (at weir)	50.0	4.0	2.2	0.08	0.010
BCI rain, early wet season[2]	0.10	0.08	0.09	0.20	0.0026

Source: Data from R. F. Stallard

Notes: Most Amazon water samples were taken during the period of high discharge. All BCI stream samples were taken during times of free flow during the rainy season. Inorganic nitrogen includes nitrate, nitrite, and ammonia.

1. Lakes at mouths of the eastern and western Amazon tributaries may explain the low N and P values
2. These figures are based on a limited number of samples
3. These data are from an anonymous author in Amazoniana (1972), 3: 186–98

rich in nitrogen, phosphorus, and calcium (Leigh and Windsor 1982), and the flora of BCI is considered representative of tropical forest on fertile soil (Foster and Brokaw 1982). We have no explanation for this remarkable contradiction.

Tree diversity is lower on BCI than in Amazonia (Gentry, this volume). On BCI, an average hectare of old forest contains 55 species among 158 trees twenty centimeters diameter at breast height and over, whereas in Manaus 214 such trees include 126 species (Foster and Hubbell 1989:table 5). Most canopy trees and lianas on BCI have ranges extending to amazonia, whereas most shrubs and epiphytes on BCI belong to more rapidly speciating groups with centers of diversity in the northern Andes (Gentry 1982a).

The Rhythms of the Forest

Seasons of scarcity and abundance alternate for frugivores and herbivores on BCI (Smythe 1970; Foster 1982a; Leigh and Windsor 1982). This seasonal rhythm is governed by the times when plants produce fruit and new leaves. For the forest as a whole, fruit and new leaves are both most abundant at or near the beginning of the rains. A second peak in fruit and leaf production occurs in the middle of the rainy season; a third peak in leaf production, restricted primarily to evergreen trees and understory shrubs, occurs at the beginning of the dry season (Foster 1982a; Leigh and Windsor 1982; T. M. Aide 1988).

Smythe (1970) suggested that seasonal shortages of fruit and new leaves play a fundamental role in limiting animal populations on BCI. In the season of fruit shortage, for example, the fruit supply is demonstrably too low to maintain populations of manakins and terrestrial frugivores. Manakins stop breeding in the season of shortage, spend more time foraging, and eat more insects (Worthington 1989). In the season of shortage, agoutis abandon their young at an earlier age, spend more time foraging, consume more stored food, grow more slowly, and are more likely to die (Smythe et al. 1982). The abundance of squirrels on BCI fluctuates from year to year as if governed by the effect of food supply on reproductive success (Giacalone et al. 1989). Evidence of one or more of these types suggests that on BCI most mammal populations are limited by seasonal shortage of food (Leigh et al. 1982).

Food limitation does not necessarily imply death by starvation. It can, instead, render animals more vulnerable to parasites, as in howler monkeys (Milton 1982), or to predators, as in agoutis (Smythe et al. 1982).

To be sure, food limitation does not explain everything. The densities of territorial adults of some species of antwrens are nearly constant, fixed by the number of suitable territories (Greenberg and Gradwohl 1986); here, regulation is social. Dry weather depresses insect numbers (Levings and Windsor 1984), and birds appear to play a crucial role in protecting plants from insect herbivores (Leigh and Windsor 1982). The rhythms of some frog populations are governed by the seasonal availability of predator-free bodies of water for their tadpoles (Rand and Myers, this volume). Nevertheless, the seasonal shortage of fruit and new leaves is decisive for many mammals, as well as for many insects (Wolda 1978a).

To assess the reliability of food supplies for animals, we must decipher the cues for flowering, fruiting, and leaf flush in plants. For example, no mid-rainy-season peak of fruit production occurred in 1970, and many frugivorous mammals died. The dry season of 1970 was unusually wet, and Foster (1982b) believed that either flowering or pollination was inhibited as a result. He also found that previous famine years tended to follow wet dry seasons. Other evidence also suggested that moisture availability is an important cue for plant phenology on BCI. Leaf fall increases in apparent response to the onset of the dry season (Haines and Foster 1977; Leigh and Smythe 1978; Wright and Cornejo 1990a), and many species flower and flush leaves in apparent response to the onset of the rainy season (Croat 1978; Leigh and Windsor 1982; Coley 1982).

"Rain flowers" provide another example for which moisture availability seems to govern phenology. These species typically flower synchronously after a dry season rain. On BCI, the shrub *Hybanthus prunifolius* flowers synchronously several days after a heavy rain in February, March, or April. Augspurger (1982) induced flowering in February and March by watering the roots of the shrubs concerned just once. Individuals watered every third day through most of the dry season did not flower, however, as if bud break in *Hybanthus* requires a dry period followed by release from water stress. Unfortunately, Augspurger's experiment stopped before a rain that could have triggered flowering occurred.

One of us, SJW, has undertaken a large-scale manipulation of soil moisture availability to determine how soil moisture affects the seasonal rhythms of the forest on BCI. Two 2.25-ha plots of mature forest are being irrigated throughout the dry season, starting in 1985–86 and ending in 1989–90. Irrigation keeps the soil in the root zone saturated (over

−0.04 MPa) throughout the year, while in nearby control plots soil moisture potential falls to −1.6 MPa late in the dry season. The effect on plant water status is equally dramatic. Late in the dry season, midday water potentials of plants in the irrigated plots are similar to rainy season values recorded for some of the same species by Rundel and Becker (1987) and are significantly greater than values recorded during the dry season for these species in the control plots (Wright and Cornejo 1990b).

Preliminary results from this experiment indicate that increasing soil moisture content has little effect on the phenology of most trees, treelets, and lianas. Fruit production has not failed in the irrigated plots. For most species of trees and lianas, the timing of flowering is indistinguishable in irrigated and control plots, although irrigation did delay leaf fall in *Dipteryx panamensis*, greatly delayed flowering in *Tabebuia guayacan*, and disrupted the synchrony of flowering within and between *Tabebuia* individuals (Wright and Cornejo 1990a). Irrigation had no effect on the seasonal rhythm of leaf fall in the forest as a whole (Wright and Cornejo 1990b). Moreover, patterns of leaf fall in many species of trees and lianas were scarcely affected (Wright and Cornejo 1990a).

Yet some shrubs have responded to irrigation. Irrigation has advanced leaf and flower production by two to four weeks in several species of *Psychotria*. After three years of irrigation, synchrony of their leaf production is steadily declining, suggesting the breakdown of an endogenous circannual rhythm normally reset and synchronized by seasonal changes in humidity or soil moisture (SJW, unpubl. data). Other shrubs, however, have not responded. Irrigation inhibited gregarious flowering in *Hybanthus* less than 2 m tall, but taller *Hybanthus* flowered simultaneously in control and irrigated plots (SJW, unpubl. data), in remarkable contrast with the predictions of Augspurger (1982).

We do not yet understand what aspects of the seasonal rhythm in climate control plant phenology. In singling out rainfall, early students of BCI may have repeated the error of the early students of dipterocarps in Malaysia. There, the drop in nocturnal air temperature following the end of prolonged, unusually high insolation appears to trigger gregarious flowering (Ashton et al. 1988): the end of a prolonged drought had previously been thought to be the cue.

The timing or intensity of flowering may depend not only on the climate of the moment but also on the intensity of flowering in the previous year. On BCI, *Faramea occidentalis* treelets that put out unusually many flowers one year tend to put out unusually few the next, and vice versa (E. W. Schupp, p.c.). The same appears true for fruit production in the canopy tree *Dipteryx panamensis* (Smithsonian Environmental Sciences Program, unpubl. data). Once again, the clearest example is provided by Malaysian dipterocarps: the stimulus required for gregarious flowering is weaker, the longer the time elapsed since the previous episode of gregarious fruiting (Van Schaik 1986). Obviously, we need to learn far more about the physiological mechanisms governing the timing of flowering, fruiting, and leaf flush.

What role do these rhythms play in the life of the forest? Some believe that natural selection on seasonal rhythms of fruit and leaf production "assembles" a community-wide rhythm that protects the forest from its herbivores and helps seed dispersal (Sabatier 1985; Leigh and Windsor 1982; Charles-Dominique et al. 1981; Foster 1982a). Chan (1980) and Janzen (1974) have argued that the community-wide synchrony of fruiting in Malaysian dipterocarp forests reflects adaptive design. There, trees of many species fruit synchronously in apparent response to a common signal (Ashton et al. 1988), but flowering, though necessarily a response to the same signal, is much less synchronous, perhaps allowing more efficient use of pollinators (Chan and Appanah 1980). Fruits produced at times other than a synchronized mass fruiting are usually damaged (Chan 1980). By contrast, Gautier-Hion et al. (1985) see no sign of adaptive design in the phenological rhythms of rain forest in Gabon.

On BCI, natural selection imposed by herbivores reinforces the seasonal rhythms of many plants. Understory shrubs on BCI concentrate their leaf production near the beginning of the dry season and again during the peak of canopy leaf flush at the beginning of the rains (Aide 1988). By synchronizing leaf production in this way, plants create a temporary surfeit of new leaves, many of which survive because there are too few herbivores to eat them (Coley 1982). Shrubs that produce new leaves out of turn are more likely to lose them to herbivores (Aide 1988). Natural selection within populations for herbivore avoidance thus strengthens the community-wide rhythm.

On BCI, the best evidence for adaptive design in the flowering and fruiting times of species comes from Augspurger's (1981, 1982) studies of *Hybanthus prunifolius*. Individuals that flower out of synchrony attract fewer pollinators, more seed predators, and mature one-tenth as many seeds as individuals flowering in synchrony with many conspecifics. Seasonal rhythms may originally have evolved in response to seasonal drought, but they persist in its absence and play a fundamental role in limiting animal populations, even in ever wet forests (Fogden 1972; Terborgh 1986).

The Diversity of Tropical Communities

Our discussion of the rhythms of the forest drew on results from fields as diverse as ethology, plant physiology, and community ecology. Understanding why there are so many kinds of plants and animals in the tropics depends on a similar variety of approaches.

Trade-Offs and Specialization

The understanding of biotic diversity begins with an understanding of the trade-offs that lead to specialization. On BCI, "pioneer" trees are specialized for treefall gaps of differing sizes (Brokaw 1987). *Trema micrantha* survives only in large gaps, but they are the tallest plants in these gaps, growing up to 7 m per year. *Cecropia insignis* survive only in large and mid-sized gaps but are the tallest plants in mid-sized gaps, growing up to 5 m per year. *Miconia argentea* survives in small gaps but grows no more than 2.5 m per year. This specialization appears to be enforced by a trade-off between growing quickly in large gaps and surviving in smaller ones (see also Oberbauer and Strain 1984). This trade-off implies that the "jack of all trades is master of none": no one species can dominate, or even persist, in gaps of all sizes. Each species does best in a particular type of gap, governed in part by its adjustment of the trade-off between rapid growth in well-lit large gaps and survival in poorly lit small gaps.

Hogan (1986) has analyzed the mechanisms which, in two midstory palms, enforce the trade-off between surviving economically in the shade and using sunlight fully when in the open. *Scheelea* leaves photosynthesize more effectively (fix more carbon per unit leaf area) than *Socratea*'s, *Scheelea*'s leaves overlap more, and they are less horizontal. Thus *Scheelea* needs more light to grow at all but uses abundant light more effectively. Yet *Socratea* leaves photosynthesize more effectively in the shade than *Scheelea*'s, and *Socratea*'s leaf arrangement is more suited to using dim light effectively. Mulkey (1986) has analyzed the trade-off between adaptability to different light levels and economical survival in the shade among three species of understory bamboo. The one species flexible enough to grow leaves with increased photosynthetic capacity when in the sun has leaves needing more light to "break even" when grown in the shade.

Predatory robberflies (Asilidae) face an analogous trade-off between foraging in light gaps and foraging in the shade (Shelly 1984). Basking in light gaps increases body temperature, allowing quicker take-off and speedier, more maneuverable flight, but the physiology that makes an insect more effective in light gaps makes it more torpid, and less effective, in the shade.

The antbirds that follow swarms of army ants over the forest floor, consuming the insects flushed by these ants, also face a trade-off. The antbirds most capable of dominating their competitors at antswarms feed at the head of large swarms, where the most insects are flushed, but they cannot feed away from antswarms and are absent from the smallest swarms, which flush the fewest insects. Less dominant species, which feed at smaller swarms and at the less productive margins of large swarms, can feed more effectively away from antswarms (Willis and Oniki 1978).

Specialization, Population Fluctuations, and Extinction

If, as in the examples above, the jack of all trades is master of none, selection leads to specialization. The more specialized a species, however, the more sensitive its members will be to environmental variation, and the more its population will fluctuate in response to this variation. This suggests that species specialize as much as the variability, or predictability, of the environment allows: the more stable the environment, the more species can coexist therein (Levins 1968).

Although rainfall is unpredictable at any latitude, tropical environments are more stable than their temperate-zone counterparts because seasonal variation in temperature is much lower in the tropics (Allee 1926). Does greater stability of tropical environments reduce population fluctuations or enhance species diversity? To answer this question, the Smithsonian's Environmental Sciences Program monitors fluctuations in aspects of the physical environment, and appropriate biotic responses, on BCI, at a reef flat on the Caribbean coast of Panama, and at aquatic and terrestrial sites near Edgewater, Maryland (Windsor 1975; Correll 1975).

The amplitudes of population fluctuation vary enormously on BCI. The normally abundant understory lizard *Anolis limifrons* fluctuates greatly in numbers, more than any lizard yet reported from the temperate zone (Andrews and Rand 1982, 1989), whereas the densities of territorial adults in some antwren populations are nearly constant (Greenberg and Gradwohl 1986). In general, insect populations fluctuate as much on BCI as they do in the temperate zone (Wolda 1978). The evidence from BCI does not suggest that tropical populations vary less than their temperate-zone counterparts.

There does appear, however, to be a connection between the amplitude and rapidity of a population's fluctuations and its prospects of extinction. Since Barro Colorado is an island, the relation between population fluctuations and susceptibility to extinction is of great interest (Leigh 1982). The understory bird species that have disappeared from Barro Colorado since it became an island are those whose mainland populations vary most or whose individuals are shortest-lived. Curiously, rarity has little to do with it (Karr 1982, this volume). More generally, homeotherms have disappeared more rapidly from BCI than poikilotherms (Wright 1981).

Although specialization is a volatile concept, difficult to define operationally, Willis (1974) feels that on BCI those ant-following antbirds that are more specialized are the ones whose populations fluctuate more, and the ones that were most specialized to (most dependent on) feeding at antswarms are the ones that have now disappeared from BCI.

Environmental Stability and Plant Diversity

Trade-offs between quick growth in large gaps and survival in small ones, between efficient use of bright sunlight and economical existence in the shade, or between effective use of water on wet soils and minimizing water loss on dry ones, allow trees to specialize to different sizes of light gaps, different strata of the forest, or different soil types. Nevertheless, many of the most common tree species on BCI are distributed without regard to availability of soil moisture or the time elapsed since the last major treefall (Hubbell and Foster 1986).

Trade-offs can enhance tree diversity by a different means, however. Trade-offs between detoxifying different compounds cause specialization on a grand scale among those insects that eat seeds or seedlings (Janzen 1975; Gilbert 1979). Plants can escape their enemies by inventing novel defenses.

Such pests as seed and seedling predators may also enhance species diversity by creating space for trees of other species to grow between a tree and its offspring (Gillett 1962). A tree can reduce the damage to its seeds and seedlings by dispersing its seeds effectively. The advantages of seed dispersal have been shown for several species on BCI and elsewhere (Howe et al. 1985; Augspurger 1984; Wright 1983; Clark and Clark 1985). Dispersal is not always advantageous. Schupp (1987) found that where there are 100–200 adult *Faramea occidentalis* treelets per ha, seeds dispersed 5 m from parent crowns were two to six times more likely to survive and germinate than seeds under the parent, but that in tracts with

300 adults per ha, dispersal reduced a seed's survival prospects by 30%. Because 11 of the 50 ha in the plot of old forest on BCI mapped by Hubbell and Foster (1986) each contain over 275 adult *Faramea*, dispersal must often be disadvantageous for this species. Seed dispersal nevertheless appears advantageous for most kinds of plants on BCI. In their 50-ha plot, where every stem 1 cm or more in diameter has been mapped, Hubbell and Foster (this volume) find that saplings whose nearest neighboring canopy tree is a conspecific grow more slowly, and die more quickly, than those whose nearest canopy neighbor is of another species.

If pest pressure is to allow so many more kinds of trees in the tropics, then pest pressure must be higher there. Is this true? One might expect so, since tropical pest populations suffer less from "harsh" seasons than their temperate-zone counterparts. In the Neotropics, moreover, tree diversity is highest in those lowland rainforests where rainfall is least seasonal (Gentry 1982b) and where pest populations presumably suffer least from harsh seasons. Nevertheless, Coley (1982) noted that insect damage to leaves is no higher on BCI than in various temperate-zone forests. This observation is not necessarily inconsistent with the idea that pest pressure contributes to latitudinal gradients in plant diversity. Pest pressure may be more influential in the tropics, but damage may be reduced by the diversity of tropical plants or by increased investment in defense. Neither proposition will be easy to demonstrate.

Interdependence and Diversity

Another feature enhancing tropical tree diversity is the complex web of mutualisms that plants have evolved with pollinators, dispersers, and the like. These mutualisms can be quite subtle. The understory shrub *Hybanthus* must flower in synchrony to create a blaze of bloom sufficient to attract its pollinators from the canopy: in turn, synchronized flowering enhances survival of the seeds (Augspurger 1982). Two vegetatively different species of the understory herb *Costus* have evolved similar flowers, cooperating, as it were, to provide sufficient flowers to command the attention of their shared pollinator (Schemske 1981). By the way its fruit are designed, packaged, and displayed, *Virola* attracts toucans to disperse its fruits while repelling other frugivores (Howe 1982). Toucans carry these seeds far enough away that the seeds' chances of survival, germination, and establishment are forty-four times higher than those of seeds under the crown of the parent (Howe et al. 1985).

Mutualisms among plants and the dispersers of their seeds or the defenders of their leaves can be strict. Seeds of the palm *Astrocaryum* cannot survive to germinate unless agoutis peel the flesh from them and bury them (Smythe 1989). These mutualisms can also be perverted in strange ways. The pioneer tree *Croton billbergianus* uses extrafloral nectaries to attract ants to patrol its leaves and defend them from herbivores. Caterpillars of the riodinid butterfly *Thisbe irenea* consume the fluid from these nectaries and transform it into a fluid that the ants prefer. The ants accordingly defend the caterpillars rather than the leaves that the caterpillars eat (P. DeVries and I. Baker, 1989).

Populations of dispersers, seed-eaters, and seedling-browsers vary asynchronously. The incredible variety of seed and seedling defenses, modes of pollination, and dispersal strategies, many possible only in the tropics, may cause different trees to achieve maximum reproductive success in different years. Theory suggests that such "temporal sorting" of reproduction enhances tree diversity (Chesson and Warner 1981). This idea warrants further study.

How general are the conclusions one might draw from BCI? Barro Colorado has been an island since 1914. Large predators (pumas, jaguars, and harpy eagles) are essentially absent from the island. The island does not include the diversity of habitats available to mobile animals in some mainland areas. Moreover, BCI represents part of a narrow isthmus, which never had the diversity of plants and animals characteristic of core areas in Amazonia. What difference do these limitations make?

The obvious point of comparison is Manu National Park, Peru (Terborgh, this volume). The floristics of the two sites are similar (Foster, this volume), as are their annual rates of fruit fall and their total biomass of frugivores (Terborgh 1986). Both places are noteworthy for their abundance of mammals, and in both, fruit shortage plays a dominant role in limiting frugivore populations (Terborgh 1986). Manu, however, has jaguars, pumas, and harpy eagles, although the abundance of ocelots is the same ($0.8/km^2$) in both places (Emmons 1987; Glanz, this volume), and both sites harbor several large eagles and hawk-eagles. In Manu, pumas eat many agoutis and pacas, the two mammal species that are unusually abundant on BCI, but the abundances of most other mammals on BCI accord reasonably well with those of other Neotropical sites where jaguars do live (Glanz, this volume). Absence of big cats seems an insufficient basis for denying the conclusions BCI suggests concerning the significance of seasonal rhythms of fruit and leaf production or the causes of tropical species diversity.

The restricted habitat diversity of BCI seems, at first sight, a more valid reason to limit the generality of conclusions drawn from the island. Barro Colorado has nothing comparable to the amazing diversity of successional habitats, with a diversity of inhabitants to match, created by the wanderings of the Rio Manu. So stunning a diversity of habitats, however, is not to be found in upland sites such as Lovejoy and Bierregaard (this volume) describe. Nor was there such diversity of habitats near BCI before 1800. The valley of the Chagres, which was relatively narrow near Barro Colorado, gave its river far less room to wander than Manu now has.

Studies on BCI have played an important role in the development of tropical biology. Today, the legacy of sixty-five years of research is available to build on, providing a perspective essential to interpreting the significance of current research and an unparalleled foundation for attacking new questions. More than thirteen hundred publications on the biota of BCI are available for consultation on that island: these are indexed to species.

Until recently, the successes of BCI were based largely on skillful fieldwork. Recent additions to the BCNM and projected improvements in laboratory facilities will allow new problems to be tackled and will enable a more detailed analysis of processes and mechanisms. The mainland peninsulas include areas of early successional forest and offer opportunities for large-scale field experiments. The new facilities will place modern laboratories at the edge of the Neotropical forest.

REFERENCES

Aide, T. M. 1988. Herbivory as a selective agent on the timing of leaf production in a tropical understory community. Nature 336: 574–575.

Alexandre, D. Y. 1982. Etude de l'éclairement du sous-bois d'une forêt dense humide sempervirente (Tai, Côte d'Ivoire). Acta Oecologia, Oecologia Generalis 3: 407–447.

Allee, W. C. 1926. Measurement of environmental factors in the tropical rain forest of Panama. Ecol. 7: 273–302.

Andrews, R. M. and A. S. Rand. 1982. Seasonal breeding and long-term population fluctuations in the lizard *Anolis limifrons*. In E. G. Leigh, Jr., A. S. Rand, and D. M. Windsor (eds.), *The Ecology of a Tropical Forest: Seasonal Rhythms and Long-Term Changes*. Smithsonian Inst. Press, Washington, D.C., pp. 405–412.

———. 1989. Adición: nuevas percepciones derivadas de la continuación de un estudio a largo plazo de la lagartija *Anolis limifrons*. In E. G. Leigh, Jr., A. S.

Rand, and D. M. Windsor (eds.), *Ecología de un bosque tropical.* Smithsonian Tropical Research Inst., Balboa, Panama, pp. 477–479.

Ashton, P. S., T. J. Givnish, and S. Appanah. 1988. Staggered flowering in the Dipterocarpaceae: new insights into floral induction and the evolution of mast fruiting in the aseasonal tropics. Amer. Nat. 132: 44–66.

Augspurger, C. K. 1981. Reproductive synchrony of a tropical shrub: experimental studies on effects of pollinators and seed predators on *Hybanthus prunifolius* (Violaceae). Ecol. 62: 775–788.

———. 1982. A cue for synchronous flowering. In E. G. Leigh, Jr., A. S. Rand, and D. M. Windsor (eds.), *The Ecology of a Tropical Forest: Seasonal Rhythms and Long-Term Changes.* Smithsonian Inst. Press, Washington, D.C., pp. 133–150.

———. 1984. Seedling survival of tropical tree species: interactions of dispersal distances, light-gaps, and pathogens. Ecol. 65: 1707–1712.

Becker, P., P. E. Rabenold, J. R. Idol, and A. P. Smith. 1988. Water potential gradients for gaps and slopes in a Panamanian tropical moist forest's dry season. J. Trop. Ecol. 4: 173–184.

Brokaw, N. V. L. 1987. Gap-phase regeneration of three pioneer tree species in a tropical forest. J. Ecol. 75: 9–19.

Chan, H. T. 1980. Reproductive biology of some Malaysian dipterocarps. II: Fruiting biology and seedling studies. Malaysian Forester 43: 438–451.

Chan, H. T. and S. Appanah. 1980. Reproductive biology of some Malaysian dipterocarps. I: Flowering biology. Malaysian Forester 43: 132–143.

Charles-Dominique, P., M. Atramentowicz, M. Charles-Dominique, H. Gérard, A. Hladik, C. M. Hladik, and M. F. Prévost. 1981. Les mammifères frugivores arboricoles nocturnes d'une forêt guyanaise: inter-relations plantes-animaux. Terre et Vie 35: 341–435.

Chazdon, R. L. 1986. Light variation and carbon gain in rain forest understorey palms. J. Ecol. 74: 995–1012.

Chazdon, R. L. and N. Fetcher. 1984. Photosynthetic light environment in a lowland tropical rain forest in Costa Rica. J. Ecol. 72: 553–564.

Chesson, P. L. and R. R. Warner. 1981. Environmental variability promotes coexistence in lottery competitive systems. Amer. Nat. 117: 923–943.

Clark, D. A. and D. B. Clark. 1984. Spacing dynamics of a tropical rain forest tree: evaluation of the Janzen-Connell model. Amer. Nat. 124: 769–788.

Coley, P. D. 1982. Rates of herbivory on different tropical trees. In E. G. Leigh, Jr., A. S. Rand, and D. M. Windsor (eds.), *The Ecology of a Tropical Forest: Seasonal Rhythms and Long-Term Changes.* Smithsonian Inst. Press, Washington, D.C., pp. 123–132.

Correll, D. L. 1975. *1974 Environmental Monitoring and Baseline Data: Temperate Studies, Rhode River, Maryland.* Mimeograph. Smithsonian Inst., Washington, D.C.

Croat, T. B. 1978. *Flora of Barro Colorado Island.* Stanford Univ. Press, Stanford.

DeVries, P. J. and I. Baker. 1989. Butterfly exploitation of an ant-plant mutualism: adding insult to herbivory. J. N.Y. Entomol. Soc. 97: 332–340.

Dietrich, W. E., D. M. Windsor, and T. Dunne. 1982. Geology, climate and hydrology of Barro Colorado Island. In E. G. Leigh, Jr., A. S. Rand, and D. M. Windsor (eds.), *The Ecology of a Tropical Forest: Seasonal Rhythms and Long-Term Changes*. Smithsonian Inst. Press, Washington, D.C., pp. 21–46.

Emmons, L. H. 1987. Comparative feeding ecology of felids in a Neotropical rainforest. Behav. Ecol. Sociobiol. 20: 271–283.

Fogden, M. P. L. 1972. The seasonality and populations of equatorial forest birds in Sarawak. Ibis 114: 307–342.

Foster, R. B. 1982a. The seasonal rhythm of fruitfall on Barro Colorado Island. In E. G. Leigh, Jr., A. S. Rand, and D. M. Windsor (eds), *The Ecology of a Tropical Forest: Seasonal Rhythms and Long-Term Changes*. Smithsonian Inst. Press, Washington, D.C., pp. 151–172.

———. 1982b. Famine on Barro Colorado Island. In E. G. Leigh, Jr., A. S. Rand, and D. M. Windsor (eds.), *The Ecology of a Tropical Forest: Seasonal Rhythms and Long-Term Changes*. Smithsonian Inst. Press, Washington, D.C. pp. 201–212.

———. (This volume). The floristic composition of the Rio Manu floodplain forest.

Foster, R. B. and N. V. L. Brokaw. 1982. Structure and history of the vegetation of Barro Colorado Island. In E. G. Leigh, Jr., A. S. Rand, and D. M. Windsor (eds.), *The Ecology of a Tropical Forest: Seasonal Rhythms and Long-Term Changes*. Smithsonian Inst. Press, Washington, D.C., pp. 67–81.

Foster, R. B. and S. P. Hubbell. 1989. Estructura de la vegetación y composición de especies en un lote de cincuenta hectáreas de la isla de Barro Colorado. In E. G. Leigh, Jr., A. S. Rand, and D. M. Windsor (eds.), *Ecología de un bosque tropical*, Smithsonian Tropical Research Inst., Balboa, Panama, pp. 141–151.

Gautier-Hion, A., J. M. Duplantier, L. Emmons, F. Feer, P. Heckest-Weiler, A. Moungazi, R. Quris, and C. Sourd. 1985. Coadaptation entre rythmes de fructification et frugivorie en forêt tropicale humide du Gabon: mythe ou realité? Rev. Ecol. (Terre et Vie) 40: 405–434.

Gentry, A. H. 1982a. Neotropical floristic diversity: phytogeographical connections between Central and South America, Pleistocene climatic fluctuations, or an accident of the Andean orogeny? Ann. Mo. Bot. Gard. 69: 557–593.

———. 1982b. Patterns of Neotropical plant species diversity. Evol. Biol. 15: 1–84.

———. (This volume). Floristic similarities and differences between southern Central America and upper and central Amazonia.

Giacalone, J., W. E. Glanz, and E. G. Leigh, Jr. 1989. Fluctuaciones poblacionales a largo plazo de *Sciurus granatensis* en relación con la disponibilidad de frutos. In E. G. Leigh, Jr., A. S. Rand, and D. M. Windsor (eds.), *Ecología de un bosque tropical*. Smithsonian Tropical Research Inst., Balboa, Panama, pp. 331–335.

Gilbert, L. E. 1979. Development of theory in the evolution of insect-plant interactions. In D. Horn, R. Mitchell, and D. Stairs (eds.), *Analysis of Ecological Systems*. Ohio State Univ. Press, Columbus, pp. 117–154.

Gillett, J. B. 1962. Pest pressure, an underestimated factor in evolution. Syst. Assoc. Publ. 4: 37–46.

Glanz, W. E. (This volume). Mammal densities: how unusual is the community on Barro Colorado Island, Panama?

Greenberg, R. and J. Gradwohl. 1986. Constant density and stable territoriality in some tropical insectivorous birds. Oecologia 69: 618–625.

Haines, B. and R. B. Foster. 1977. Energy flow through litter in a Panamanian forest. J. Ecol. 65: 147–155.

Hladik, A. 1978. Phenology of leaf production in rain forest of Gabon: distribution and composition of food for folivores. In G. G. Montgomery (ed.), *Ecology of Arboreal Folivores*. Smithsonian Inst. Press, Washington, D.C., pp. 57–71.

Hladik, A. and P. Blanc. 1987. Croissance des plantes en sous-bois de forêt dense humide (Makokou, Gabon). Rev. Ecol (Terre et Vie) 42: 209–234.

Hogan, K. P. 1986. Plant architecture and population ecology in the palms *Socratea durissima* and *Scheelea zonensis* on Barro Colorado Island, Panama. Principes 30: 105–107.

Howe, H. F. 1982. Fruit production and animal activity in two tropical trees. In E. G. Leigh, Jr., A. S. Rand, and D. M. Windsor (eds.), *The Ecology of a Tropical Forest: Seasonal Rhythms and Long-Term Changes*. Smithsonian Inst. Press, Washington, D.C., pp. 189–199.

Howe, H. F., E. W. Schupp, and L. C. Westley. 1985. Early consequences of seed dispersal for a Neotropical tree (*Virola surinamensis*). Ecol. 66: 781–791.

Hubbell, S. P. and R. B. Foster. 1986. Commonness and rarity in a Neotropical forest: implications for tropical tree conservation. In M. E. Soulé (ed.), *Conservation Biology: The Science of Scarcity and Diversity*. Sinauer, Sunderland, Mass.

———. (This volume). The floristic composition of the Barro Colorado Island forest.

Janzen, D. H. 1974. Tropical blackwater rivers, animals, and mast fruiting by the Dipterocarpaceae. Biotropica 6: 69–103.

———. 1975. Interactions of seeds and their insect predators/parasitoids in a tropical deciduous forest. In P. W. Price (ed.), *Evolutionary Strategies of Parasites, Insects and Mites*. Plenum Press, New York, pp. 154–186.

Johnson, P. L. and D. M. Atwood. 1970. Aerial sensing and photographic study of the El Verde rain forest. In H. T. Odum and R. Pigeon (eds.), *A Tropical Rain Forest*. U.S. Atomic Energy Commission, Washington, D.C., pp. B-63–78.

Karr, J. R. 1982. Population variability and extinction in the fauna of a tropical land-bridge island. Ecol. 63: 1975–1978.

———. (This volume). The avifauna of Barro Colorado Island and the Pipeline Road, Panama.

Kenoyer, L. A. 1929. General and successional ecology of the lower tropical rain forest at Barro Colorado Island, Panama. Ecol. 10: 201–222.

Kira, T. 1978. Community architecture and organic matter dynamics in tropical lowland rain forests of Southeast Asia, with special reference to Pasoh Forest, West Malaysia. In P. B. Tomlinson and M. H. Zimmermann (eds.), *Tropical Trees as Living Systems*. Cambridge Univ. Press, Cambridge, pp. 561–590.

Leigh, E. G., Jr. 1982. Introduction: the significance of population fluctuations. In E. G. Leigh, Jr., A. S. Rand, and D. M. Windsor (eds.), *The Ecology of a Tropical Forest: Seasonal Rhythms and Long-Term Changes*. Smithsonian Inst. Press, Washington, D.C., pp. 435–440.

Leigh, E. G., Jr., A. S. Rand, and D. M. Windsor (eds.). 1982. *The Ecology of a Tropical Forest: Seasonal Rhythms and Long-Term Changes*. Smithsonian Inst. Press, Washington, D.C.

———. 1989. *Ecología de un bosque tropical*. Smithsonian Tropical Research Inst., Balboa, Panama.

Leigh, E. G., Jr., and N. Smythe. 1978. Leaf production, leaf consumption, and the regulation of folivory on Barro Colorado Island. In G. G. Montgomery (ed.), *The Ecology of Arboreal Folivores*. Smithsonian Inst. Press, Washington, D.C., pp. 33–50.

Leigh, E. G., Jr., and D. M. Windsor. 1982. Forest production and regulation of primary consumers on Barro Colorado Island. In E. G. Leigh, Jr., A. S. Rand, and D. M. Windsor (eds.), *The Ecology of a Tropical Forest: Seasonal Rhythms and Long-Term Changes*. Smithsonian Inst. Press, Washington, D.C., pp. 111–122.

Levings, S. C. and D. M. Windsor. 1984. Litter moisture content as a determinant of litter arthropod distribution and abundance during the dry season on Barro Colorado Island, Panama. Biotropica 16: 125–131.

Levins, R. 1968. *Evolution in Changing Environments*. Princeton Univ. Press, Princeton, N.J.

Lovejoy, T. E., and R. O. Bierregaard, Jr. (This volume). Central Amazonian forests and the Minimum Critical Size of Ecosystems Project.

Luizao, F. J. and H. O. R. Schubart. 1987. Litter production and decomposition in a terra-firme forest of central Amazonia. Experientia 43: 259–264.

McCullough, D. 1977. *The Path between the Seas*. Simon and Schuster, New York.

Milton, K. 1982. Dietary quality and demographic regulation in a howler monkey population. In E. G. Leigh, Jr., A. S. Rand, and D. M. Windsor (eds.), *The Ecology of a Tropical Forest: Seasonal Rhythms and Long-Term Changes*. Smithsonian Inst. Press, Washington, D.C., pp. 273–289.

Mulkey, S. S. 1986. Photosynthetic acclimation and water-use efficiency of three species of understory herbaceous bamboo (Gramineae) in Panama. Oecologia 70: 514–519.

Oberbauer, S. F. and B. R. Strain. 1984. Photosynthesis and successional status of Costa Rican rain forest trees. Photosyn. Res. 5: 227–232.

Piperno, D. R. 1989. Fitolitos, arqueología y cambios prehistóricos de la vegetación en un lote de cincuenta hectáreas de la Isla de Barro Colorado. In E. G. Leigh, Jr., A. S. Rand, and D. M. Windsor (eds.), *Ecología de un bosque tropical,* Smithsonian Tropical Research Inst., Balboa, Panama, pp. 153–156.

Rand, A. S., and C. W. Myers. (This volume). The herpetofauna of Barro Colorado Island, Panama: an ecological summary.

Rundel, P.W. and P. F. Becker. 1987. Cambios estacionales en las relaciones hídricas y en la fenología vegetativa de plantas del estrato bajo del bosque tropical de la Isla de Barro Colorado, Panamá. Rev. Biol. Trop. 35(suppl. 1): 71–84.

Sabatier, D. 1985. Saisonnalité et determinisme du pic de fructification en forêt guyanaise. Rev. Ecol. (Terre et Vie) 40: 289–320.

Schemske, D. W. 1981. Floral convergence and pollinator sharing in two bee-pollinated tropical herbs. Ecol. 62: 946–954.

Schupp, E. W. 1987. Studies on seed predation of *Faramea occidentalis*, an abundant tropical tree. Ph.D. diss., Univ. of Iowa.

Shelly, T. E. 1984. Comparative foraging behavior of Neotropical robber flies (Diptera: Asilidae). Oecologia 62: 188–195.

Smythe, N. 1970. Relationships between fruiting seasons and seed dispersal in a Neotropical forest. Amer. Nat. 104: 25–35.

———. 1989. Seed survival in the palm *Astrocaryum standleyanum*: evidence for dependence upon its seed dispersers. Biotropica 21: 50–56.

Smythe, N., W. E. Glanz, and E. G. Leigh, Jr. 1982. Population regulation in some terrestrial frugivores. In E. G. Leigh, Jr., A. S. Rand, and D. M. Windsor (eds.), *The Ecology of a Tropical Forest: Seasonal Rhythms and Long-Term Changes*. Smithsonian Inst. Press, Washington, D.C., pp. 227–238.

Terborgh, J. 1986. Community aspects of frugivory in tropical forests. In A. Estrada and T. H. Fleming (eds.), *Frugivores and Seed Dispersal*. W. Junk, Dordrecht, pp. 371–384.

———. (This volume). An overview of research at Cocha Cashu Biological Station.

Van Schaik, C. P. 1986. Phenological changes in a Sumatran rain forest. J. Trop. Ecol. 2: 327–347.

Willis, E. O. 1974. Populations and local extinctions of birds on Barro Colorado Island, Panama. Ecol. Monogr. 44: 153–169.

Willis, E. O. and Y. Oniki. 1978. Birds and army ants. Ann. Rev. Ecol. Syst. 9: 243–263.

Windsor, D. M. (ed.). 1975. *1974 Environmental Monitoring and Baseline Data, Tropical Studies*. Unpublished report, Smithsonian Inst., Washington, D.C.

Windsor, D. M., A. S. Rand, and W. M. Rand. 1989. Caracteristicas de la precipitación en la Isla de Barro Colorado. In E. G. Leigh, Jr., A. S. Rand, and D. M. Windsor (eds.), *Ecología de un bosque tropical*. Smithsonian Tropical Research Inst., Balboa, Panama, pp. 53–71.

Wolda, H. 1978a. Seasonal fluctuations in rainfall, food, and abundance of tropical insects. J. Anim. Ecol. 47: 367–381.

———. 1978b. Fluctuations in abundance of tropical insects. Amer. Nat. 112: 1017–1045.

Woodring, W. P. 1958. Geology of Barro Colorado Island, Canal Zone. Smithsonian Misc. Coll. 135(3): 1–39.

Worthington, A. H. 1989. Comportamiento de forrajeo de dos especies de saltarines en respuesta a la escasez estacional de frutos. In E. G. Leigh, Jr., A. S. Rand, and D. M. Windsor (eds.), *Ecología de un bosque tropical*. Smithsonian Tropical Research Inst., Balboa, Panama, pp. 285–304.

Wright, S. J. 1981. Intra-archipelago vertebrate distributions: the slope of the species-area relation. Amer. Nat. 118: 726–748.

———. 1983. The dispersion of eggs by a bruchid beetle among *Scheelea* palm seeds and the effect of distance to the parent palm. Ecol. 64: 1016–1021.

Wright, S. J. and F. H. Cornejo. 1990a. Seasonal drought and the timing of flowering and leaf fall in a Neotropical forest. In K. Bawa and M. Hadley (eds.), *The Reproductive Ecology of Tropical Forest Plants*. UNESCO, Paris.

———. 1990b. Seasonal drought and leaf fall in a tropical forest. Ecol., *in press*.

3

An Overview of Research at Cocha Cashu Biological Station

JOHN TERBORGH

Cocha Cashu Biological Station is located in the lowland sector of Manu National Park in southeastern Peru at 11°54′S and 71°22′W. The station was founded in 1969–70 through the joint efforts of biologists representing La Molina University in Lima and the Frankfurt Zoological Society in West Germany. Its creation significantly predated the formal establishment of Manu Park in 1973.

The initial purpose of the station was to facilitate an investigation of the black caiman (*Melanosuchus niger*), South America's largest and commercially most valuable crocodilian. At the time, the black caiman had been exploited to exhaustion in nearly every river system on the continent. Even the Manu River had been harvested in the mid-1960s, but only the larger adults had been removed. Although in a battered condition, the Manu population nevertheless remained one of the last in existence. Following completion of a small station building, a German biologist, Kai Christian Otte, spent two years at Cocha Cashu (1970–1972), marking and observing caimans with a Peruvian assistant. At the end of this period, Otte moved to the vicuña project at Pampa Galeras in the highlands, and the station at Cocha Cashu was abandoned.

It is a pleasure to acknowledge the encouragement of generations of officials in Peru's General Directorate of Forestry and Fauna, and Manu National Park manifested in the unfailing issuance of research permits to Cocha Cashu investigators over the last 15 years. The program at Cocha Cashu would simply not have been possible without mutual confidence in this continuing relationship. Financial support for the research reported herein was gratefully received from many organizations, including the National Science Foundation, National Geographic Society, New York Zoological Society and World Wildlife Fund—U.S. My thanks extend also to the many individuals, too numerous to name, who contributed to the results reported above.

By entirely fortuitous good fortune, a party that included Howard and Nicholas Brokaw, Robin Foster, and me stumbled upon the empty building in an exploratory excursion the following year. The park had just received official status but was effectively unknown outside a small circle of conservationists. I had learned about it through friends in Peru but had little idea of what to expect.

Arriving at Cocha Cashu, I was struck not only by the beauty of the place, a sparkling lake set amid a majestic forest, but by an abundance of wildlife that exceeded anything I had seen in 10 years of travels in the Peruvian Amazon. The combination was irresistible, and I resolved to transfer my research program to the site, provided this met with the approval of Peruvian authorities. I am happy to say that it did.

The environment of Manu Park owes its pristine quality to its location in the remote southeastern corner of Peru. The nearest population center, Cuzco, offers a small market for fruit, vegetables, and wood that is easily satisfied from closer sources. The larger markets of Lima and Arequipa are reached by a grueling four-to-five-day trip that extracts a heavy toll from both trucks and drivers. Only the most valuable products, principally such prime hardwoods as *Cedrela*, *Swietenia*, and *Cedrelinga*, can justify the heavy costs of transport to such distant destinations. These severe economic limitations have inhibited development in the region, notwithstanding the pressures of poverty and overpopulation in adjacent highlands. In the eyes of this observer, the factors most likely to influence the short-term future of this corner of Peru are the possible discovery of mineral deposits and the price of petroleum.

The always slow and expensive, and during the rains unpredictable, transportation in and out of the Manu region poses challenges to the investigator who wishes conduct research at Cocha Cashu. We buy our food, fuel, and hardware in Cuzco and, with the best of luck, arrive at the field station five days later, after two arduous days and a night in the back of a chartered truck and three decidedly more pleasant days plying the river in an open 14-m dugout freight canoe. The accommodations that greet the visitor are elementary and rustic. There is no permanent staff and no cook; such luxuries are beyond our means. There is no running water and there are no beds. Everyone sleeps in tents in the forest. The buildings provide space only for work, storage, and eating. Cooking duty rotates through the group, and senior scientists are not exempt. Our contact with the outside world is via two-way radio. One or two seasonal employees help maintain the trails and buildings, but that is all. It is a place where you learn to do everything yourself and to invent in the face of necessity. Nevertheless, we normally accomplish our goals, and people seldom complain about lack of convenience.

Since the mid-1970s, when Cocha Cashu received only a handful of visitors a year, the research station has gradually expanded to accommodate an ever increasing flow of investigators. This flow has lately reached a level of over forty per year, for a total of about 3,200 person-days. In 1981, with $5,000 from World Wildlife Fund—U.S., we added a second building—a kitchen/dining hall—and in 1986–87, with help from an anonymous donor, we put up a third structure to provide space for a classroom/library, labs, and offices. These two construction grants constitute the only external financial inputs the station has received since the original building was erected in 1969–70. Otherwise, it has been strictly a do-it-yourself proposition. Operating costs of less than $5,000 a year are met entirely through a modest per diem charge levied on foreign (non-Peruvian) investigators. A minimal stock of equipment—boats, outboard motors, a solar electrical system, a two-way radio, telemetry equipment, a microscope, mammal traps, a small generator—is the accumulated legacy of generations of research grants.

The participation of Peruvian students in the scientific work at Cocha Cashu has increased steadily from the arrival of the first pioneer in 1976 to the point that we are currently accommodating fifteen to twenty a year. These are engaged at various levels: as assistants to more senior investigators, as independent researchers carrying out projects for theses leading to *licenciado* status, and as graduate students enrolled in foreign doctorate programs. (Peru offers no graduate-level education in biology.)

The publication list for the station runs to more than two hundred titles in both Spanish and English. Recently, the Conservation Data Center at La Molina University published an extensive volume in Spanish summarizing the results of projects carried out at Cocha Cashu (Rios 1986). Though the research covers a wide range of topics from anthropology to zoology, the emphasis has been in a few areas: (1) community ecology of birds and mammals, (2) interactions between plants and vertebrate consumers, (3) studies of the flora and successional processes in the vegetation, and (4) endangered species. Research on invertebrates has been scanty, while much-needed studies on soils, nutrient cycling, and aquatic biology have not been undertaken to date. A prohibition against collecting animal specimens in the park has discouraged work on small mammals, reptiles, amphibians, and fish.

An obvious tilt toward large vetebrates is justified by the lamentable fact that Cocha Cashu is one of the few research sites in the Neotropics to offer a completely undisturbed ecosystem with a full complement of both top predators and their prey. Ongoing autecological studies of such species as the giant otter, black caiman, jaguar, side-necked terrapin, spider monkey, razor-billed curassow, and various macaws receive high priority at Cashu because there are few alternative sites at which they would be possible.

The Physical Setting

The 10-km^2 study area at Cocha Cashu is but a tiny speck in the immensity of the 1,532,000-ha Manu Park. The park's southern boundary runs along the crest of the eastern Andes through alpine grasslands that extend to over 4,000 m elevation. To the north, the mountains fall away precipitously into a jumbled, and still unexplored, topography of jagged foothills and deep gorges gouged by raging clearwater torrents. There are no weather stations in this portion of the park, but a nearby locality at the base of the mountains, Quince Mil, registers over 10 m of annual rainfall, the highest in Peru. From the foothill region, the park extends out onto the Amazonian plain, where the climate becomes drier and more seasonal the farther one moves from the mountains. The combination of major topographic and climatic gradients, along with a varied geology, create a mosaic of environments.

At Cocha Cashu we are able to sample only a fraction of this complexity, representing the lowland riparian zone. The station itself is located in the 6-km-wide meander belt of the Manu River. Four complete years of data indicate an average rainfall of just over 2,000 mm. This is seasonally distributed, most of it falling between November and May. Dry season months (June–October) normally receive less than 10 cm, though year-to-year variation in the intensity and duration of the dry season is considerable.

The soils of the meander belt are composed of young alluvial silt and sand carried down from the Andes. Ten years of measurements on a bend in the river near Cocha Cashu indicate an average loss of about 25 m every rainy season, a rate which implies that meander loops may extend as much as 2.5 km per century. If this extrapolation is anywhere near the mark, the entire meander belt is swept out and rejuvenated on the order of every 500–1,000 years. Everywhere one finds indications of the river's past activities—raised levees, backwater depressions, ancient channels, lakes, swamps, and forests of varying successional age. The environment is anything but homogeneous (Salo et al. 1986).

The meander belt cuts through a previous alluvial horizon, elevated about 30 m above the present river level. Once a nearly horizontal plain, the topography of these sandy "uplands" is now extremely irregular, having been dissected by innumerable small streams. Although the vegetation of these upland terraces must be far older than that of the meander belt, superficial appearances convey the opposite impression. Frequent treefalls result in a canopy of generally low stature and an understory tangled with vines. Nearly a third of the terrain is occupied by patchy stands of bamboo within which trees are scattered or lacking, suggesting a dynamical cycle of alternation between forest and bamboo. This and other aspects of the vegetation have been described elsewhere and will not be reviewed here (Terborgh 1983, 1985).

Plant Resources and Animal Populations

Investigators at Cocha Cashu have devoted great effort to documenting the availability of plant resources to the animal community (Terborgh 1983; Terborgh et al. 1986; Terborgh and Stern 1987). Fruits and seeds are the most important resources in this environment, sustaining about 85% of a total mammal biomass of circa 1,800 kg/km^2, and about 64% of a total avian biomass of circa 200 kg/km^2. Herbivory is a minor way of life in this forest, at least among vertebrates (Terborgh 1986b).

In a pattern that is remarkably parallel to the one documented at Barro Colorado Island (BCI), fruiting activity follows a pronounced seasonal rhythm (Smythe 1970; Foster 1982; Terborgh 1983). Flowering begins in earnest during the mid-dry season (July) and continues at a high level into the early rainy season (November). Fruiting follows after variable delays, reaching an early peak in October and a second peak in mid-rainy season sometime between January and March (fig. 3.1). Fruit is generally abundant from September through April, and I have calculated elsewhere that the supply during this time greatly exceeds reasonable estimates of animal consumption (Terborgh 1986b).

A drastically different situation prevails from May through August, during the rainy-dry transition and the early dry season. Estimates of consumption then greatly exceed the measured availability of fruit, and many frugivorous birds and mammals confront an inadequate food supply. Some birds disappear during this time; those that remain, along with many frugivorous mammals, must alter their diets (Terborgh 1986b). Several birds and a number of small mammals subsist on nectar, though

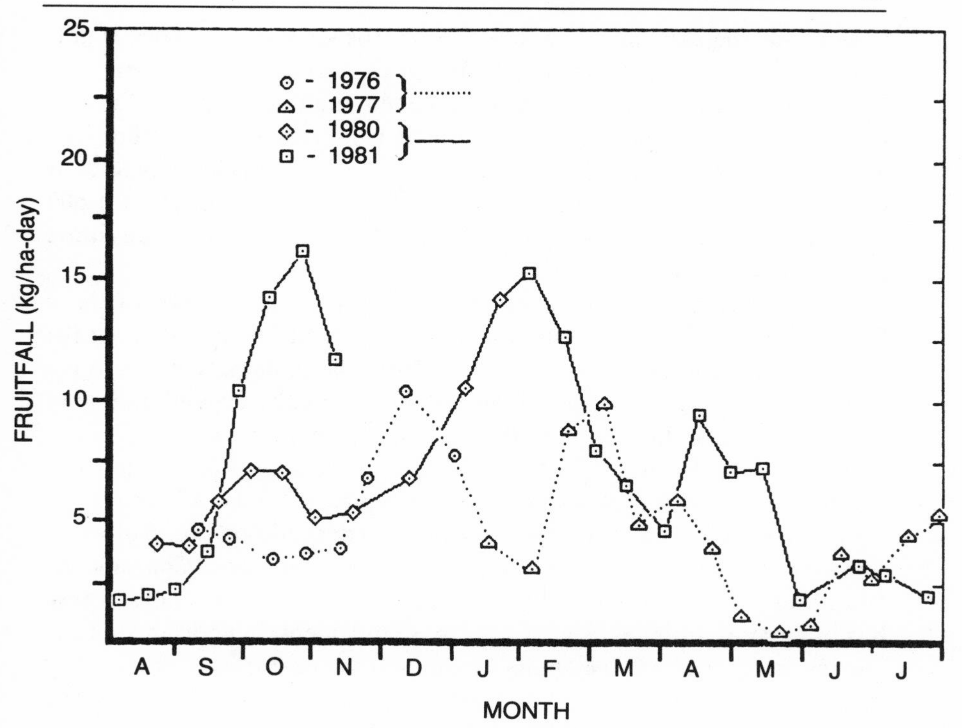

FIGURE 3.1. Seasonal rhythm of fruitfall at Cocha Cashu (from Terborgh 1986c).

they show no evident adaptations for doing so (Janson et al. 1981; Terborgh and Stern 1987). Other birds and small mammals rely on arthropods, the supply of which undergoes far less seasonal variation than that of plant reproductive parts (Terborgh et al. 1986). Large mammals resort to feeding on palm nuts, pith, foliage, and whatever fruit can be found, principally figs (Terborgh 1986a).

Scarcity thus leads to a consistent response, one that was first predicted theoretically by MacArthur and Pianka (1966). With their normal diets in short supply, animals begin to exploit low-quality foods that are otherwise eschewed. Which of a limited selection of low-quality resources a particular animal seeks seems to depend on its body size, physical strength, or digestive capacity, because these attributes determine whether it can exploit the resource more efficiently than other species (Terborgh 1985b). In a normal year, the switch to less preferred, and presumably less nutritious, resources assures a high level of survival throughout the dry season, although not an easy survival, since significant weight losses may be incurred (Wilson-Goldizen et al., in press).

Even though many frugivores broaden their diets during annual periods of scarcity, the number of available resources is small. Palm nuts, nectar, and figs are the principal ones, estimated to sustain nearly 85% of the animal biomass (Terborgh 1986b). Out of a flora of more than 1,500 species, a mere dozen, or less than 1%, produce these so-called keystone plant resources (Terborgh 1986a,b).

Several types of observations suggest that the abundance of these keystone resources may set the carrying capacity of the environment for frugivorous birds and mammals. Few birds or mammals raise young between May and August. Some primates greatly expand the area searched for food during this time, whereas others drastically curtail their daily activity periods, presumably to conserve energy (Terborgh 1983; Wright 1985). The consumption of poor-quality foods, such as pith, succulent petioles, and rotten wood suggests a desperation borne of nutritional stress. At the corresponding time of year on Barro Colorado Island, Milton (1982) has noted excess mortality in howler monkeys. These bits of circumstantial evidence point to a fundamental importance of the few resources that tide the animals over until the next time of plenty. But is this strong enough evidence on which to base a claim that keystone resources establish the carrying capacity of the environment for frugivorous vertebrates? The question is open to debate, and as well as we shall see, the answer does not have to be the same for all species. Although food may typically be scarce during the dry season, the scarcity may only rarely reach the critical point of causing starvation, and thus the role of population limitation may fall to the predators.

The Role of Predators in the Cocha Cashu Ecosystem

At Cocha Cashu, one cannot take a walk after a rain without discovering fresh cat tracks in the trail or spend a week with a troop of squirrel monkeys without witnessing raptor attacks. When the jaguar roars at night, spider monkeys and capybaras bark in terror. What impact do these predators have?

A recent study of jaguar, puma, and ocelot at Cocha Cashu suggests that the impact of these felids on terrestrial mammals is substantial (Emmons 1987). Mammalian prey species are taken by these cats in almost precisely the same proportions as they occur in censuses. Forest felids behave as searchers in the optimal foraging sense. Prey are havested as encountered, and no species seems immune. (No quantitative information on raptor predation is available, but the presence of some thirty-five species in the environment suggests a major collective impact [Terborgh et al. 1984]).

Emmons (1987) calculates that large felids (jaguar and puma) annually harvest 8% of the standing crop of terrestrial mammals weighing more than 1 kg. This is similar to the fraction of prey biomass harvested by all predators combined on the Serengeti Plain (8–9%; Schaller 1972), and also accords with an estimate for tiger in Nepal (8–10%; Sundquist 1981).

That terrestrial mammals are taken at Cocha Cashu in proportion to their relative abundances in the environment suggests the potential of strong indirect interactions among prey species, mediated through predators. Species with high reproductive rates, such as peccary and capybara, will suffer the highest annual per capita losses to predators. The presence of fecund prey will maintain the predators at relatively high densities, which in turn will depress the populations of less fecund species. In other words, species with the highest birth rates are best able to maintain their numbers in the presence of predators and in so doing impose an incidental burden on other members of the community. The upshot of this is that under a constant level of predation, some prey species will maintain population densities much closer than others to their intrinsic carrying capacities (mean densities in the absence of predation).

To look at this in another way, recall that maximum sustained yield is attained at a density equal to one-half the carrying capacity of a population that obeys the logistic growth curve. If predators were perfectly efficient, they would reduce the population densities of their prey to the maximum sustainable (MSY) level, but given the multiplicity of prey species present in the Amazonian ecosystem, their varied fecundities, and the probable operation of the indirect effects described above, few prey populations are likely to be maintained at the MSY level. Some may be held to much lower densities, whereas others may persist at densities closer to their intrinsic carrying capacities. If this interpretation of Emmons's results is valid, it could be anticipated that prey species would respond in widely varying degrees to a general release from predation.

This prediction can be tested by comparing the population densities of terrestrial mammals at Cocha Cashu to those of BCI. The environment of BCI is very similar to that at Cocha Cashu. BCI is at nearly the same

TABLE 3.1 Densities of terrestrial mammals of adult weight > 1 kg at Cocha Cashu, Peru, and Barro Colorado Island, Panama

	Individuals/km²		BCI/ CC
	Cocha Cashu	*Barro Colorado Island*	
Didelphis marsupialis (opossum)	20	47	2.4
Dasypus sp. (armadillo)	4	53	13
Silvilagus brasiliensis (rabbit)	+	7	ca. 7
Dasyprocta sp. (agouti)	5	100	20
Agouti paca (paca)	4	40	10
Nasua narica (coati)	+	24	ca. 20
Tapirus sp. (tapir)	<0.5	0.5	>1.0
Tayassu tajacu (peccary)	6	9	1.5
Mazama + *Odocoileus* (deer)	2.6	2.7	1.0

Source: Compiled from Emmons (1984) and Glanz (1982)

latitude (9° versus 11°) and receives only slightly more rainfall (2,500 versus more than 2,000 mm) in a similar regime of seven wet months and five dry months per year. Perhaps most important, the annual fruitfall at both localities is approximately 2 tons (wet weight) per ha-year (Smythe 1970; Terborgh 1983). We might thus expect the intrinsic carrying capacities of the two sites to be similar.

One major difference offsets these similarities—BCI lacks felid predators larger than the ocelot. From Emmons's (1987) work at Cocha Cashu we know that ocelots rarely take prey weighing more than 1 kg. To ascertain the impact of large cats on their prey, we can compare the densities of terrestrial mammals weighing more than 1 kg as adults in the two localities (table 3.1). The density of nearly every species is greater on BCI, in some cases in excess of an order of magnitude. Agoutis, pacas, and coatis are especially numerous on BCI, or especially scarce at Cashu, depending on how one views it. The implication is that at Cashu these animals are held by predation to a fraction of their intrinsic carrying capacities and that nearly all other terrestrial prey species in the community are less abundant than they would be in the absence of big cats. Predators thus seem to be reducing the populations of some of their prey to far less than MSY densities, affirming the operation of indirect effects.

Two major inferences can be drawn from this conclusion. One that Emmons emphasized strongly is that felids and human hunters compete head-on for the same prey species and that in the presence of felids, the game supply available to hunters may be drastically reduced, and vice versa. This brings us to the unwelcome conclusion that if we wish to manage Neotropical forests at MSY levels for game, it will have to be at the expense of maintaining big cats at abnormally low densities. A combination of hunting for meat and trapping for pelts is the obvious prescription.

A second, more general inference relates to what used to be called the balance of nature, though the phrase is no longer in fashion. Neotropical fauna evolved in the presence of a full complement of predators, and many of the component species may be adapted in unsuspected ways to a life with predators. Disease, famine, and even heavy parasite loads are manifestations of overcrowding that may seldom affect populations held to low densities by predation. Where major predators have been eliminated, and nowadays this is nearly everywhere near human habitation, prey species may be exposed to a host of selection pressures never before encountered in their evolutionary histories. Absent normal predation, some species become excessively abundant—agoutis, pacas, and coatis on BCI, rats in cities around the world, racoons and opossums in U.S. suburbs—and these in turn can wreak havoc on other components of the ecosystem. When nature's system of checks and balances is held in abeyance, we discover that songbirds become extinct on BCI and in suburban parks and woodlands in our eastern states (Willis 1974; Terborgh and Winter 1980; Wilcove 1985). What other repercussions may follow we do not yet know. All this should provide ample warning that predators are an essential component of healthy ecosystems and that if our species is allowed to continue its global campaign against them, we can expect unforeseen and unwanted consequences.

REFERENCES

Emmons, L. H. 1984. Geographic variation in densities and diversities of nonflying mammals in Amazonia. Biotropica 16: 210–222.

———. 1987. Comparative feeding ecology of felids in a Neotropical rainforest. Behav. Ecol. Sociobiol. 20: 271–283.

Glanz, W. E. 1982. The terrestrial mammal fauna of Barro Colorado Island: censuses and long-term changes. In E. G. Leigh, Jr., A. S. Rand, and D. M.

Windsor (eds.), *The Ecology of a Tropical Forest: Seasonal Rhythms and Long-Term Changes*. Smithsonian Inst. Press, Washington, D.C., pp. 455–468.
Goldizen, A. W., F. Cornejo, D. T. Porras, R. Evans, and J. Terborgh. 1988. Seasonal food shortage and weight loss in relation to timing of births in saddle-backed tamarins (*Saguinus fuscicollis*). J. Anim. Ecol. 57: 893–901.
Holt, R. D. 1984. Spatial heterogeneity, indirect interactions, and the coexistence of prey species. Amer. Nat. 124: 377–406.
Janson, C. H., J. Terborgh, and L. H. Emmons. 1981. Non-flying mammals as pollinating agents in the Amazonian forest. Biotropica 13 (suppl.): 1–6.
MacArthur, R. H. and E. Pianka. 1966. On optimal use of a patchy environment. Amer. Nat. 100: 603–609.
Milton, K. 1982. Dietary quality and population regulation in a howler monkey population. In E. G. Leigh, Jr., A. S. Rand, and D. M. Windsor (eds.), *The Ecology of a Tropical Forest: Seasonal Rhythms and Long-Term Changes*. Smithsonian Inst. Press, Washington, D.C., pp. 273–289.
Ríos, M. A. (ed.). 1986. *Reporte Manu*. Centro de Datos para la Conservacíon, Univ. Agraria, La Molina, Peru.
Salo, J., R. Kalliola, I. Häkkinen, Y. Mäkinen, P. Niemelä, M. Puhakka, and P. D. Coley. 1986. River dynamics and the diversity of Amazon lowland forest. Nature 322: 254–258.
Schaller, G. B. 1972. *The Serengeti Lion*. Univ. of Chicago Press, Chicago.
Smythe, N. 1970. Relationships between fruiting seasons and seed dispersal methods in a Neotropical forest. Amer. Nat. 104: 25–35.
Sundquist, M. E. 1981. The social organization of tigers (*Panthera tigris*) in Royal Chitawan National Park, Nepal. Smithsonian Contrib. Zool. 336: 1–98.
Terborgh, J. 1983. *Five New World Primates: A Study in Comparative Ecology*. Princeton Univ. Press, Princeton, N.J.
———. 1985a. Habitat selection in Amazonian birds. In M. L. Cody (ed.), *Habitat Selection in Birds*. Academic Press, New York, pp. 311–338.
———. 1985b. The ecology of Amazonian primates. In G. T. Prance and T. E. Lovejoy (eds.), *Amazonia*. Pergamon Press, Oxford, pp. 284–304.
———. 1986a. Keystone plant resources in the tropical forest. In M. E. Soulé (ed.), *Conservation Biology: The Science of Scarcity and Diversity*. Sinauer, Sunderland, Mass., pp. 330–344.
———. 1986b. Community aspects of frugivory in tropical forests. In A. Estrada and T. H. Fleming (eds.), *Frugivores and Seed Dispersal*. W. Junk, Dordrecht, pp. 371–384.
Terborgh, J., C. H. Janson, and M. Brecht. 1986. Cocha Cashu: su vegetación, clima y recursos. In M. A. Ríos (ed.), *Reporte Manu*. Centro de Datos para la Conservación, Univ. Agraria, La Molina, Peru, pp. 1–18.
Terborgh, J. and M. Stern. 1987. The surreptitious life of the saddle-backed tamarin. Amer. Sci. 75: 260–269.
Terborgh, J. and B. Winter. 1980. Some causes of extinction. In M. E. Soulé and B. A. Wilcox (eds.), *Conservation Biology: An Evolutionary-Ecological Perspective*. Sinauer, Sunderland, Mass., pp. 119–133.

Wilcove, D. S. 1985. Nest predation in forest tracts and the decline of migratory songbirds. Ecol. 66: 1211–1214.

Willis, E. O. Populations and local extinctions of birds on Barro Colorado Island, Panama. Ecol. Monogr. 44: 153–169.

Wright, P. C. 1985. The costs and benefits of nocturnality for *Aotus trivirgatus* (the night monkey). Ph.D. diss., CUNY.

4

Central Amazonian Forests and the Minimum Critical Size of Ecosystems Project

THOMAS E. LOVEJOY AND
RICHARD O. BIERREGAARD, JR.

The study sites of the Minimum Critical Size of Ecosystems (MCSE) Project are located 70–90 km north of Manaus (2°30′S, 60°W) (figs. 4.1, 4.2) in the Distrito Agropecuario, an area of approximately 500,000 ha of relatively undisturbed, upland Amazonian rain forest that is being developed by the Manaus Free Trade Zone Authority (SUFRAMA). Government-subsidized activities in the district include cattle ranching, timbering, rubber plantations, and farming. By Amazonian standards, the area has been strictly controlled, although some weekend hunting by both farm workers and residents of Manaus occurs. The study areas are protected by presidential decree no. 91.884 (5 November 1985).

Manaus is the location of Brazil's National Institute for Amazon Research (Instituto Nacional de Pesquisas da Amazonia, INPA). INPA has a 20,000-ha reserve about 50 km north of Manaus in which are situated three small watersheds. These are known as the Bacia Modelo and are set aside for hydrological studies.

Closer to Manaus lies the 10,000-ha Reserva Ducke with forest that is similar to the MSCE sites. The forests at Ducke have been subjected to

Many researchers, field volunteers and our field crew have contributed to the success of the project to date. To all of them we are grateful.

The MCSE project has been supported by the World Wildlife Fund, the Instituto Nacional de Pesquisas da Amazônia (INPA), the Instituto Brasileiro de Desenvolvimento Florestal (IBDF), and grants from the National Park Service (Cooperative Agreement CX-0001-9-0041), the A. W. Mellon Foundation, the Sequoia Foundation, the Tinker Foundation, and the Weyerhauser Family Foundation. This is publication number 43 in the Minimum Critical Size of Ecosystems Project (Projeto Dinámica Biológica de Fragmentos Florestais) Technical Series.

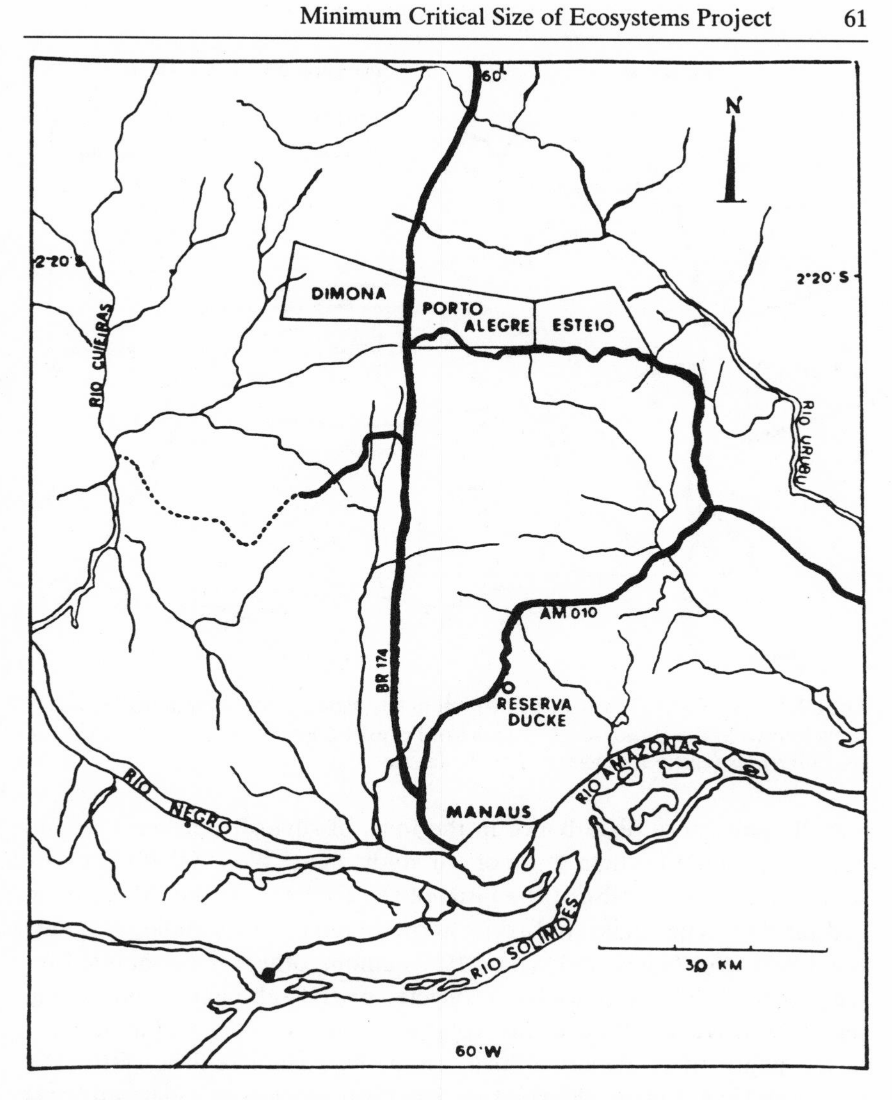

FIGURE 4.1. Manaus and three cattle ranches (Dimona, Porto Alegre, and Esteio), where the Minimum Critical Size of Ecosystems Project reserves are located.

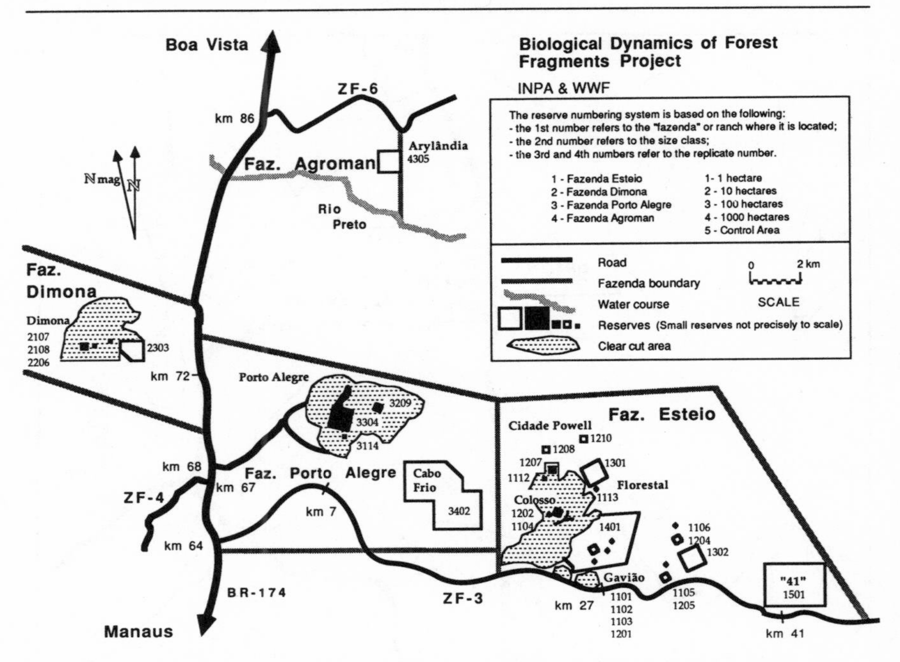

FIGURE 4.2. The location of the Minimum Critical Size of Ecosystems Project (now known as the Biological Dynamics of Forest Fragments Project) reserves on four *fazendas* (cattle ranches) north of Manaus.

considerably more disturbance in the form of silviculture, hunting, and even oil exploration than those of our study areas. Willis (1976) has presented a preliminary list of the birds of the Reserva Ducke, Magnusson and his colleagues have studied its herpetofauna (for example, Magnusson 1987), and Penny and Arias (1982) among others have sampled insects there. Plants have also been collected by INPA botanists and visiting scientists at Ducke. Prance (this volume) has compiled a preliminary list of the plants of the reserve. Aside from these studies, little systematic work has been done in the reserve. The most extensive faunal and floral inventories in the Manaus region have been carried out in the MCSE reserves. Some results of the inventories of mammals, herps, and birds are presented in later chapters, along with floristic data from the Reserva Ducke.

Ubiquitous charcoal in the soil suggests that the MCSE area was inhabited prehistorically (Fearnside, p.c.; Prance and Schubart 1977). More recently the region has been largely uninhabited. Although the

TABLE 4.1 Taxa and questions under study at the MCSE Project

Subproject	*Principal investigator(s)*
Woody plants*	Rankin**
Lecythidaceae*	Becker, Mori, van Roosmalen
Reproductive biology of *Clusia**	Mesquita
Microclimate/plant physiology*	Kapos**
Understory birds*	Bierregaard**
Mixed-species flocks*	Powell**
Ant-following birds	Harper**
Birds of gaps and edges	Quintela**
Herpetofauna*	Zimmerman**
Primates*	Rylands,** van Roosmalen
Small, nonflying mammals*	Emmons, Malcolm**
Bats*	Gribel
Butterflies*	Brown,** Hutchings
Ants*	Benson**
Scarab beetles	Klein**
Euglossine bees*	H. and G. Powell**
Solitary bees and wasps*	L. Oliveira
Drosophila	Martins**
Leafcutter ants	Vasconcelos**
Termites*	Souza

Notes: * Ongoing projects
** Cited in References

forest was probably exploited for rosewood, the fauna still includes such large predators as jaguar, puma, tapir, and harpy eagles.

In about 1970, construction of a highway (BR-174) north from Manaus to Boa Vista in Roraima was begun and provided the first real access to the area. The area did not suffer from rapid deforestation, the normal fate of a tropical forest when a road is opened from a nearby city, because of effective control imposed by SUFRAMA.

The Minimum Critical Size of Ecosystems Project was initiated in 1979 as a joint Brazilian (INPA) and U.S. (World Wildlife Fund) project to study the effects of habitat fragmentation on rainforest ecosystems. In particular, the project focuses on how scale (size of forest fragments) and rates of species loss relate to subsequent species composition (Lewin 1984; Lovejoy 1980; Lovejoy et al. 1983, 1984, 1986). A broad range of taxa are being studied under the auspices of the project (table 4.1).

Researchers working on the project and spending time in the city generally stay in project housing in Manaus and have access to laboratory facilities at INPA. Transportation to the sites, a two-to-four-hour drive in four-wheel drive vehicles, is available three or four times a week. Several rustic but relatively comfortable camps are maintained for sleeping (in hammocks), dining (cooks prepare meals at the larger camps), and field manipulation of data and specimens. The camps do not have electricity.

Permission to work on the project is granted by the project's management committee. Prospective researchers must arrange for Brazilian collaborators and should have at least partial funding support. Proposals that require funding from the project must be submitted by 1 January for inclusion in the project's fiscal year (1 July–30 June).

Studies at the MCSE forests were initiated more recently than at the three other research areas described in this book. The experimental aspects of the research project themselves (see Lovejoy et al. 1983, 1984, 1986) are not relevant to the concerns of this volume, but the pre-isolation information necessary to identify and interpret the change induced by fragmentation provides some data for comparison with the other Neotropical sites described in this volume.

Climate, Geology, and Soils

The mean annual temperature in Manaus is 26.7 °C. Monthly means fluctuate only about 2 °C (Anon. 1978). On rare occasions a cool air mass from more southerly latitudes, locally termed a *friagem*, can drop temperatures to 17 °C. In an east-west transect from the Amazon's estuary to the Andes annual rainfall decreases from the coast to the area of Santarem, about 600 km east of Manaus, and then climbs steadily to the western edge of the basin. A thirty-year average for annual rainfall at Manaus is 2,186 mm. Other cited annual means for the Manaus area range from 1,900 to 2,400 mm. The rainiest months tend to be March and April, with 300 mm each. Months during the pronounced dry season—July, August, and September—normally receive less than 100 mm each (Anon. 1978).

Before the major orogeny of the Andes in the Miocene, the Amazon basin is thought to have drained to the west, entering the Pacific near Guayaquil, Ecuador. Subsequent uplift of the Andes is thought by some to have formed a giant lake, bounded on the west by the newly rising Andes and on the east by the Brazilian and Guiana shields (Jenks 1956; Beurlen 1970; Goulding 1980). During the Tertiary, extensive sediments accumulated up to 2.5 km deep in the central basin. By the Pleistocene the Amazon had cut through the eastern barrier in the vicinity of Obidos and has subsequently drained to the Atlantic.

The MCSE sites are located in the Amazon's meander plain through these ancient sediments. The study areas are near no major river, so there are no recent alluvial deposits. Major reworking of forest and soils by river meanders, common in western Amazonia (for example, at Manu) (Salo et al. 1986), is likewise not a factor in the local geology. The soils lie on top of deep sediments that have been subjected to long periods of leaching and are generally poor in nutrients. In situ leaching, rather than sheet erosion, appears to be a major factor in determining the current soil characteristics.

The soils in the vicinity of the study areas are sandy or clayey yellow latosols in Brazilian terminology (Camargo 1979), also known as xanthic ferralsols (FAO/UNESCO 1974). The formation that accompanies the major rivers through the central basin is known as the Barreira, or more recently as the Alter do Chao (Santos 1984). The soils are deep, acid, and yellowish brown and drain well (Sombroek 1984).

Studies in the nearby model basin (Bacia Modelo) watershed are summarized in Acta Amazônica (12:4 [1982]) and indicate the tightness of the nutrient cycling in these forests. Input of nutrients from precipitation actually exceeds nutrient loss via the streams leaving the watersheds (Franken and Leopoldo 1984).

The Forest

In general the forests are 30–37 m in height with emergent trees to 45–50 m; some true rainforest giants occasionally reach 55 m. The average height of the first branch is 11.6 m, with a range of 3–32 m (Rankin-de-Merona et al., in press). Based on a 15 ha sample there is a mean of 647 ± 51 stems/ha at 10 cm DBH (diameter at breast height) or greater, with a basal area of 30.55 ± 3.5 m^2 per hectare (Rankin-de-Merona et al., this volume). The forests are low in epiphytic growth. The understory has an abundance of stemless palms (*Astrocaryum* spp., *Attalea* spp.). Fittkau and Klinge's (1973) estimate of fresh biomass of a forest in the region was 730.7 metric tons/ha and 255 for below-ground root systems, for a total of 985.7 metric tons/ha.

Prance discusses the floristics of the Manaus region in chapter 8. Rankin-de-Merona et al. (in press) report on preliminary floristic analyses of individuals with a DBH ≥ 10 cm from noncontiguous plots (most 1 ha or larger) totaling 31 ha and extending over about 40 km of the area sampled by the MCSE Project reserves. The samples contain 56 families, with 738 identified species and 923 species when obvious morphospecies are included. The 4 most abundant families are the Burseraceae, Sapotaceae, Leguminosae, and Lecythidaceae. The most species-rich families analyzed to date are the Sapotaceae (72 species, with work still in progress), Chrysobalanaceae (69 +/− 11 species), Annonaceae, and Lecythidaceae. Manaus is the most diverse location known for the Sapotaceae (A. Teixeira, p.c.). Several other families, including the Lauraceae, Leguminosae, Elaeocarpaceae, Euphorbiaceae, and Moraceae, have not yet been studied by specialists and may prove to be among the most species-rich in the study area.

We do not yet know what constitutes a sufficient sample size to compare the tree communities of different Amazonian forests. Rankin and Ackerly (1987) suggest that 10–20 ha may be needed to characterize the species composition and population structures for the more abundant families that do not show clumped distributions. Well in excess of 30 ha may be needed for some of the more species-rich families, such as Sapotaceae, which contains mostly rarer species. The species-area curve for Chrysobalanaceae reaches an asymptote at about 60 ha of noncontiguous sampling (ibid.).

Seasonality

Five years' continuous collection of fertile material and phenological data indicate that flowering peaks at the end of the dry season and in the beginning of the rainy season, from August to November. There is also a minor peak at the height of the rainy season, in March and April. Fruiting peaks in the rainy season from November to January, although some flowers and fruits are present year-round.

Although Haffer (1988) suggests that Amazonian forest birds show a peak of reproductive activity in the rainy season (usually the austral spring), our observations (unpubl. data) show that most species, particularly those of mixed-species flocks (Powell, p.c.), breed during the austral winter, which is the driest and hottest part of the year at Manaus. Some species breed all year long, and a few breed in the rainy season, but in general, the peak of reproductive activity occurs from June through September.

For insects, Penny and Arias's (1982) pioneering data from thirteen months of trapping at the Reserva Ducke suggest generally higher populations in December and April. Whereas the Ducke populations showed considerably less fluctuation than has been shown for Panama (Wolda 1978), some families (Miridae, Carabidae, Dascillidae, and

Elateridae) from the Ducke study appear to have strongly seasonal abundances in samples.

Species Richness

There are 352 bird species on the three ranches where most of the project reserves are situated (Stotz and Bierregaard, in press; Bierregaard, this volume); these data are based on mist-netting in the understory, supplemented by observations in surrounding forest, second growth, and pastures. The total includes 32 species not on the Reserva Ducke list (Willis 1976), which in turn lacks 85 species recorded from the MCSE study areas (Stotz and Bierregaard, in press). Of the MCSE total, 16 species characteristic of open water (rivers, large streams) and about 60 species typical of pasture and young second growth should be deducted, leaving about 275 species regularly found in primary forest. This is well below the 368 forest bird species reported for Manu (Terborgh et al. 1984).

Mammals (Emmons 1984), birds (Bierregaard et al. 1987; Stotz and Bierregaard, in press), and woody plants (van Roosmalen, p.c.) are representative of the Guianan subset of the Amazonian flora and fauna. Mammalian densities at the MCSE sites fluctuate widely from year to year (Malcolm 1988) and are markedly lower than at Manu, on occasion showing the lowest known population levels for Neotropical forest mammals (Emmons 1984). Mammalian species richness is also lower than in areas of western Amazonia studied by Emmons (1984).

Seven of nine species of primates known to be in the region are present, including several species that also occur in western Amazonia. The home range of *Cebus apella*, which occurs both at Manaus and at Manu, is roughly ten times larger at Manaus than at Manu, where it is 50–70 ha (Spironelo, unpubl. data; Terborgh 1983). Similarly, at Manaus mixed-species bird flocks have home ranges of about 12–14 ha (Powell 1985), or about three times the area used by comparable flocks at Manu (Munn and Terborgh 1979).

The pattern of fewer species and lower density of animals in the Manaus area forests than in western Amazonia would appear to be tied to the generally poorer soils and presumably lower productivity at Manaus (Gentry and Emmons 1987). The idea of a link between productivity and species richness (for example, Rosenzweig 1968, Walter 1985), though not new, is clearly supported by these results.

Detailed ecological studies at Manaus and Manu could eventually take us beyond this suggestive correlation to an understanding of the relation between productivity and diversity, as well as other important ecosystem-level interactions. Comparisons of long-term studies of Neotropical forests of roughly similar climatic regimes and biogeographic origin, such as are now possible between Manaus and Manu, should lead to insights into population dynamics and interdependencies among species. Such an increase in our understanding of these complex and threatened forests is essential if we are to counsel wisely those charged with their preservation.

REFERENCES

Anon. 1978. Projeto Radambrasil; Folha SA20 Manaus, p. 261. Ministério de Minas e Energia: Depto. Nacional de Produção Mineral, Rio de Janeiro.

Benson, W. W. and A. Harada. 1988. Local diversity of tropical and temperate ant faunas (Hymenoptera Formicariidae). Acta Amazônica 18: 275–289.

Beurlen, K. 1970. *Geologie von Brasilien*. Gebruder Borntraeger, Berlin.

Bierregaard, R. O., Jr. (This volume). Species composition and trophic organization of the understory bird community in a Central Amazonian terra firme forest.

Bierregaard, R. O., Jr. and T. E. Lovejoy. Effects of isolation of understory bird communities in Amazonian forest fragments. Acta Amazônica, *in press*.

Bierregaard, R. O., Jr., D. F. Stotz, L. H. Harper, and G. V. N. Powell. 1987. Observations on the occurrence and behavior of the crimson fruit crow (*Haematolerus militaris*) in central Amazonia. Bull. Brit. Ornithol. Club 107: 134–137.

Brown, K. S., Jr. Comparisons between Lepidoptera faunas in reserves of different size and degree of isolation. Acta Amazônica, *in press*.

Camargo, M. N. 1979. Sistema de classificação dos solos Brasileiros. EMBRAPA–SNLCS, Rio de Janeiro.

Emmons, L. H. 1984. Geographic variation in densities and diversities of non-flying Amazonian mammals. Biotropica 16: 210–222.

FAO/UNESCO. 1971. *Soil Map of the World at 1:5,000,000*. Vol. 1: *Legend*. UNESCO, Paris.

Fittkau, E. J. and H. Klinge. 1973. On biomass and trophic structure of the central Amazonian rainforest ecosystem. Biotropica 5: 2–14.

Franken, W. and P. R. Leopoldo. 1984. Hydrology of catchment areas of central Amazonian forest streams. In H. Sioli (ed.), *The Amazon: Limnology and Landscape Ecology of a Mighty Tropical River and Its Basin*. W. Junk, Dordrecht, pp. 501–519.

Gentry, A. H. and L. H. Emmons. 1987. Geographic variation in fertility, phenology, and composition of the understory of Neotropical forests. Biotropica 19: 216–227.

Goulding, M. 1980. *The Fishes and the Forest: Explorations in Amazonian Natural History*. Univ. of California Press, Berkeley and Los Angeles.
Haffer, J. 1988. Vogel Amazoniens: okologie, brutbiologie und artenreichtum. J. Ornithol. 129: 1–53.
Harper, L. H. 1987. The conservation of ant-following birds in small Amazonian forest fragments. Ph.D. diss., SUNY, Albany.
———. Birds and army ants (*Eciton burchelli*): observations on their ecology in undisturbed forest and isolated reserves. Acta Amazônica, *in press*.
Jenks, W. F. (eds.). 1956. *Handbook of South American Geology: An Explanation of the Geologic Map of South America*. Geol. Soc. Amer. Memoir 65.
Kapos, V. Effects of isolation on the water status of forest patches in the Brazilian Amazon. J. Trop. Ecol., *in press*.
Klein, B. C. 1987. The effects of forest fragmentation on dung and carrion beetle (Scarabaeinae) communities in Central Amazonia. M.S. thesis, Univ. of Florida, Gainesville.
Lewin, R. 1984. Parks: how big is big enough? Science 225: 611–612.
Lovejoy, T. E. 1980. Discontinuous wilderness: minimum areas for conservation. Parks 5(2): 13–15.
Lovejoy, T. E., R. O. Bierregaard, Jr., J. M. Rankin, and H. O. R. Schubart. 1983. Ecological dynamics of forest fragments. In S. L. Sutton, T. C. Whitmore, and A. C. Chadwick (eds.), *Tropical Rain Forest: Ecology and Management*. Blackwell, Oxford, pp. 377–384.
Lovejoy, T. E., J. M. Rankin, R. O. Bierregaard, Jr., K. S. Brown, Jr., L. H. Emmons, and M. van der Voort. 1984. Ecosystem decay of Amazon forest remnants. In M. H. Nitecki (ed.), *Extinctions*. Univ. of Chicago Press, Chicago, pp. 295–325.
Lovejoy, T. E., R. O. Bierregaard, Jr., A. B. Rylands, J. R. Malcolm, C. E. Quintela, L. H. Harper, K. S. Brown, Jr., A. H. Powell, G. V. N. Powell, H. O. R. Schubart, and M. B. Hays. 1986. Edge and other effects of isolation on Amazon forest fragments. In M. E. Soulé (ed.), *Conservation Biology: The Science of Scarcity and Diversity*. Sinauer, Sunderland, Mass., pp. 257–285.
Malcolm, J. R. 1989. Small mammal abundances in isolated and non-isolated primary forest reserves near Manaus, Brazil. Acta Amazônica 18: 67–83.
Martins, M. B. Invasão de fragmentos florestais por espécies oportunistas de *Drosophila* (Diptera, Drosophilidae). Acta Amazônica, *in press*.
Munn, C. A. and J. W. Terborgh. 1979. Multi-species territoriality in Neotropical foraging flocks. Condor 81: 338–347.
Penny, N. D. and J. R. Arias. 1982. *Insects of an Amazon Forest*. Columbia Univ. Press, New York.
Powell, A. H. and G. V. N. Powell. 1987. Population dynamics of euglossine bees in Amazonian forest fragments. Biotropica 19: 176–179.
Powell, G. V. N. 1985. Sociobiology and adaptive significance of interspecific foraging flocks in the Neotropics. Ornithol. Monogr. 36: 713–732.
Prance, G. T. (This volume). The floristic composition of the forests of central Amazonian Brazil.
Prance, G. T. and H. O. R. Schubart. 1977. Nota preliminar sobre a origem das

campinas abertas de areia branca do baixa Rio Negro. Acta Amazônica 8(4): 567–570.
Quintela, C. E. 1985. Forest fragmentation and differential use of natural and man-made edges by understory birds in Central Amazonia. M.S. thesis, Univ. of Illinois, Chicago.
Rankin-de-Merona, J. M. and D. D. Ackerly. 1987. Estudos populacionais de árvores em florestas fragmentadas e as implicações para conservação in situ das mesmas na floresta tropical da Amazônia central. Pap. Avul. Inst. de Pesquisas e Estudos Florestais da Escola Superior de Agricultura "Luiz de Queiroz." Univ. de São Paulo, Piracicaba, S.P. 35: 47–59.
Rankin-de-Morona, J. M., R. H. Hutchings, and T. E. Lovejoy. (This volume). Tree mortality and recruitment over a five-year period in undisturbed upland rainforest of the Central Amazon.
Rankin-de-Morona, J. M., G. T. Prance, M. F. Silva, W. A. Rodrigues, and M. Uehling. Resultados preliminares de um levantamento florestal de 31 ha de terra firme na Amazônia central: Descrição geral da vegetação e dados taxonômicos. Acta Amazônica, *in press*.
Rosenzweig, M. L. 1968. Net primary productivity of terrestrial communities. Amer. Nat. 102: 67–74.
Rylands, A. B. and A. Keuroghlian. 1988. Primate populations in continuous forest and forest fragments in Central Amazonia. Acta Amazônica 18: 291–307.
Salo, J., R. Kalliola, I. Häkkinen, Y. Mäkinen, P. Niemelä, M. Puhakka, and P. D. Coley. 1986. River dynamics and the diversity of Amazon lowland forest. Nature 322: 254–258.
Santos, J. O. S. 1984. A parte setentrional do craton Amazônico (escudo das Guianas) e a bacia Amazônica. In C. Schobbenhaus, D. de A. Campos, G. R. Derze, and H. E. Asmus (eds.), *Geologia do Brasil: texto explicativo do mapa geológico do Brasil e da área oceânia adjacente incluindo depósitos minerais, escala 1:2,500,000*. Depto. Nacional da Produção Mineral, Brasilia, pp. 57–91.
Sombroek, W. G. 1984. Soils of the Amazon region. In H. Sioli (ed.), *The Amazon: Limnology and Landscape Ecology of a Mighty Tropical River and Its Basin*. W. Junk, Dordrecht, pp. 521–535.
Stotz, D. F. and R. O. Bierregaard, Jr. The birds of the fazendas Porto Alegre, Dimona and Esteio north of Manaus, Amazonas, Brazil. Rev. Bras. Biol, *in press*.
Terborgh, J. 1983. *Five New World Primates: A Study in Comparative Ecology*. Princeton Univ. Press, Princeton, N.J.
Terborgh, J., J. W. Fitzpatrick, and L. Emmons. 1984. Annotated checklist of bird and mammal species of Cocha Cashu Biological Station, Manu National Park, Peru. Fieldiana (Zool.) 21: 1–29.
Vasconcelos, H. 1987. Atividade forrageira, distribuição de colônias de sauvas (*Atta* spp.) em uma flôresta de Amazonia Central. M.S. thesis, Inst. Nacional de Pesquisas da Amazonia and Fundação da Univ. de Amazonas.
Walter, H. 1985. *Vegetation of the Earth*, 3d ed. (trans.). Springer-Verlag, New York.

Willis, E. O. 1977. Lista preliminar das aves da parte noroeste e áreas vizinhas da Reserva Ducke. Rev. Bras. Biol. 37: 585–601.
Wolda, H. 1978. Seasonal fluctuations in rainfall, food and abundance of tropical insects. J. Anim. Ecol. 47: 369–381.
Zimmerman, B. L. and R. O. Bierregaard, Jr. 1986. Relevance of the equilibrium theory of island biogeography with an example from Amazonia. J. Biogeogr. 13: 133–143.

PART II

Floristics

5

The Distribution of Diversity among Families, Genera, and Habit Types in the La Selva Flora

BARRY HAMMEL

For most biologists working at La Selva, research has been hindered by the lack of an identification guide to its diverse flora. Early synecological work, especially, was throttled by the poor state of baseline floristic knowledge of the forest. Although three permanent plots (12.4 ha total) representing the common forest types at La Selva (plateau, swamp, and ridge) were established in 1969, published work has dealt little with forest demography (see Lieberman et al. 1985). Only now are results appearing (Lieberman and Lieberman 1987) from studies measuring size, abundance, and distribution of all species (10 cm DBH or more) in these plots. Plant ecologists working at La Selva have concentrated mostly on the physiological, demographic, or reproductive autecology of a few common species. Other field botanists have concentrated on improving the baseline floristic knowledge. This chapter summarizes information about the flora of La Selva in terms of the numbers and kinds of species present for comparison with other forests (see Gentry, this volume; Hartshorn and Hammel 1991).

Collecting for the La Selva Flora project began in summer 1979 (Hammel and Grayum 1982) and continued with resident collectors (for about forty months on-site) through 1984. From previous work, mostly by G. Hartshorn and P. Opler, a checklist accounting for 710 species was available at the beginning of the project. By 1982, 850 of an estimated 1,500 species collected had been identified. The current checklist (Wilbur et al., 1987) enumerates approximately 1,740 species of vascular plants, including 66 cultivated species.

The time spent on the La Selva floristic inventory and numbers of collections made (circa 20,000) should be weighed against the fact that nearly two-thirds of La Selva's 1,500 ha are primary forest, and roughly one-third of that (the Western Annex) was not acquired until 1982. Future workers will undoubtedly find species unaccounted for in the published flora, especially if intensive plot studies are undertaken in primary forest on the Western Annex and near the southern boundary of the original property. Since publication of the flora of Barro Colorado Island (BCI) (Croat 1978) more than 25 additional species (Foster and Hubbell, this volume) have been found on the island, and BCI comprises less primary forest of a less diverse type (moist rather than wet) with a much longer history of collecting. Nevertheless, by the end of the 1979–84 collecting campaign at La Selva, additions to the flora were infrequently found and were very rare species or ephemeral pasture weeds.

Treatments for the families will be published as soon as they are ready in Selbyana, with all treatments to be bound in a volume when completed (Wilbur et al. 1986). Seven families (Atwood 1988; Hammel 1986a–f) and treatments for ferns and fern allies have now been published (Grayum & Churchill 1989a,b). In addition, according to the newest checklist (Wilbur et al. 1987), keys for most remaining families have been prepared.

The La Selva Biological Station lies across a transition between tropical premontane and tropical wet forest life zones (Hartshorn 1983) in the Caribbean lowlands of northeastern Costa Rica with an elevational range of 30–200 m. The average annual rainfall is nearly 4 m. The site is usually described as having a continuous wet season or no effective dry season; however, a distinctly drier season occurs January–April. Physical characteristics of the site are presented in more detail by Clark (this volume), Hammel and Grayum (1982), Hartshorn (1983), Lieberman et al. (1985), Hammel (1986g), and Wilbur et al. (1986). (This chapter is based on the current checklist [Wilbur et al. 1987], previous work [Hammel 1986g], personal observation [Hammel, unpubl. data], and consultation with several contributors to the flora. It deals only with the occurrence of species, not their abundance or biomass.)

Categories of flora generally follow those of Foster and Hubbell (this volume) for BCI but need qualification. The disparity of taxonomic opinion concerning fern families is avoided by considering ferns and fern allies as one group: pteridophytes. For this reason, "ferns" represent the most diverse "family." Fabaceae here includes all three subfamilies of legumes. Moraceae and Cecropiaceae are treated separately. All "woody" terrestrial species (including palms and tree ferns) with mature individuals greater than 4 m tall are considered trees. Those 4 m or less are considered shrubs. In addition to such plants as most orchids, bromeliads, and species of *Anthurium*—that is, true epiphytes—strangler figs and such root-climbing plants as *Monstera* and species of *Philodendron*, *Asplundia*, and *Drymonia* are also counted as epiphytes. Small herbaceous tendril-climbing vines or scandent plants are treated as herbs. Except for one fern (*Salpichlaena*), one gymnosperm (*Gnetum*), one aroid (*Heteropsis*), one bamboo (*Elytrostachys*); and one palm (*Desmoncus*) only large, woody, dicot vines are treated as lianas.

As pointed out by Foster and Hubbell (this volume), only native species should be considered for comparison with other areas. Recently introduced paleotropical species (mostly grasses), as well as a number of pantropical or widespread Neotropical species—*Portulaca oleracea* L., *Momordica charantia* L., *Emelia sonchifolia* (L.) DC, and *Hamelia patens* Jacq.—that are known at La Selva only from pastures, laboratory clearings, or other sites of much human traffic and disturbance, are treated separately as weeds. A few "accidental" native herbs, such as *Pothomorphe umbellata* (L.) Miq., found only once along the river and otherwise restricted to higher elevation are also treated as weeds. Approximately one-fourth of the 220 weeds at La Selva are paleo- or pantropical. The remaining 1,450 native "forest" species grow in such natural habitats as swamps, forest edges, treefall gaps, and river and stream sides, as well as in the various types of forest at La Selva. The ferns, Orchidaceae, Araceae, Rubiaceae, Melastomataceae, Fabaceae, Piperaceae, Euphorbiaceae, Moraceae, and Arecaceae are the most speciose families of natural habitats at La Selva (fig. 5.1). If only the paleo- and pantropical weeds are ignored, Poaceae, Asteraceae, and Cyperaceae would also rank among the fifteen msot diverse families. *Piper*, *Psychotria*, *Philodendron*, *Miconia*, *Anthurium*, *Thelypteris*, *Inga*, *Peperomia*, *Ficus*, and *Pleurothallis* are the ten genera with most species at La Selva (fig. 5.2).

As within most floras, tropical and temperate, herbs are the single most speciose life form. Herbaceous stem anatomy predominates even if trees, shrubs and lianas are combined as "woody" life forms and contrasted with herbaceous life-forms; most epiphytes at La Selva are also herbaceous. Considering only the plants of natural habitats, epiphytes and herbs rank equally in numbers of species, and each is more diverse than any of the other three categories. La Selva is unusual among well-

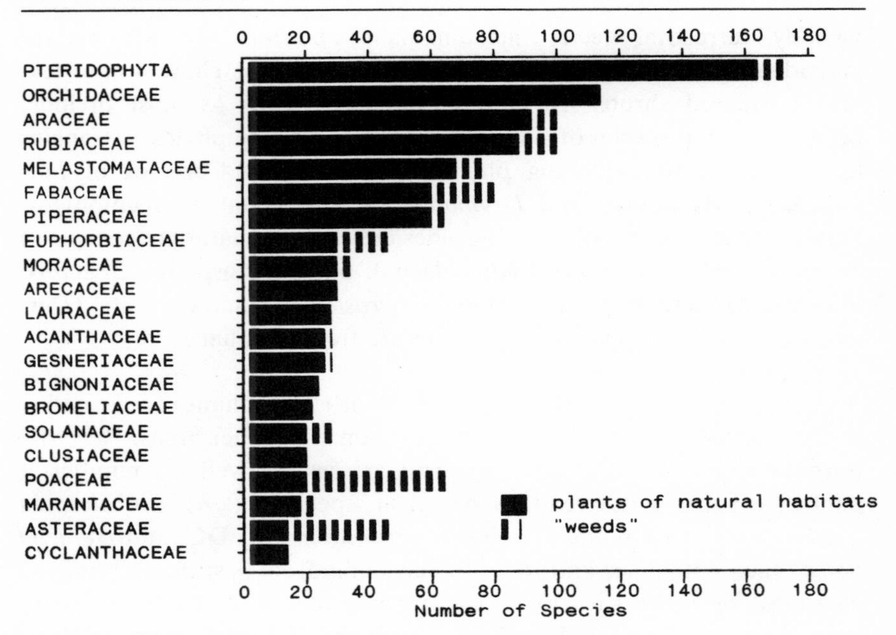

FIGURE 5.1. Families with the most species at La Selva.

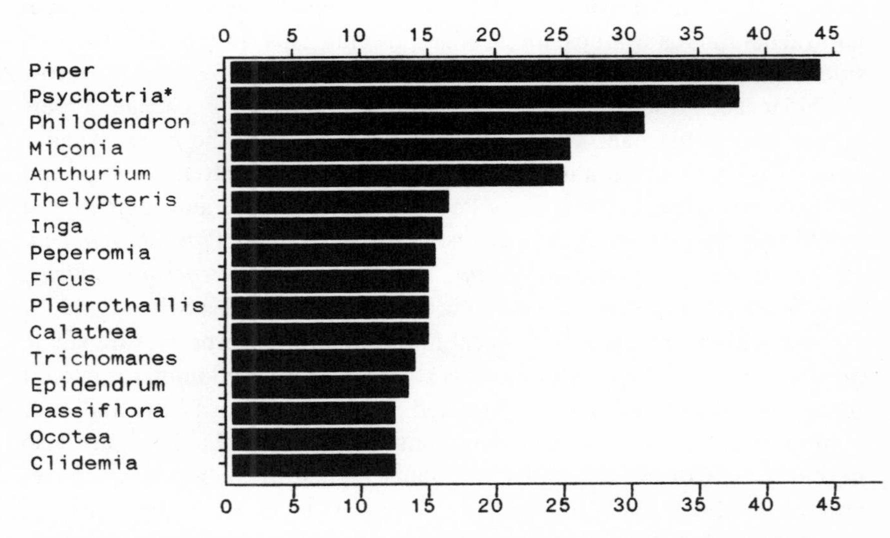

FIGURE 5.2. Genera with the most species at La Selva. **Psychotria* includes *Cephaelis* (5 spp.).

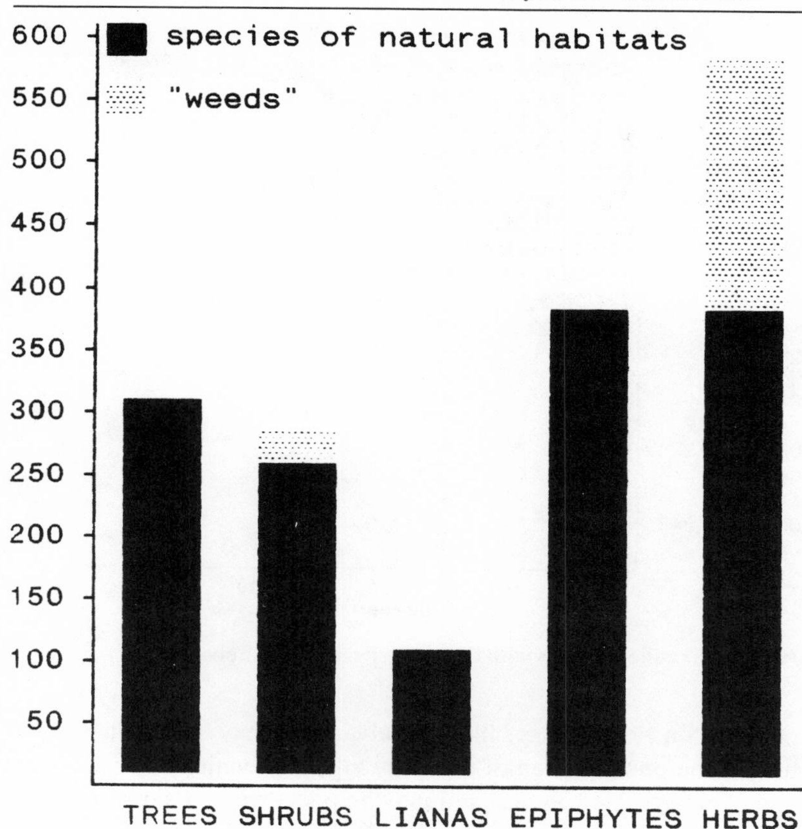

FIGURE 5.3. Numbers of species in different habit types at La Selva.

studied Neotropical areas in the great diversity of its epiphyte flora (Gentry and Dodson 1987; Gentry, this volume); 25% of all species of natural habitats are epiphytes (fig. 5.3).

The tree flora, dominated by one legume species (*Pentaclethra macroloba*), also derives much of its diversity from the Fabaceae (fig. 5.4). The high diversity of legumes is typical of lowland Neotropical forests (Gentry 1988), but the ranking of Lauraceae, Rubiaceae, and Annonaceae well above the Moraceae (even including Cecropiaceae) is unusual. Even though La Selva has more species of trees overall than BCI or the Rio Palenque Science Center (RPSC) in Ecuador, it has fewer species of *Cecropia* (Hammel 1986g). Although the Lauraceae commonly rank among the ten most speciose families in lowland Neotropical forests (Gentry 1986; this volume), their rank as second only to legumes is un-

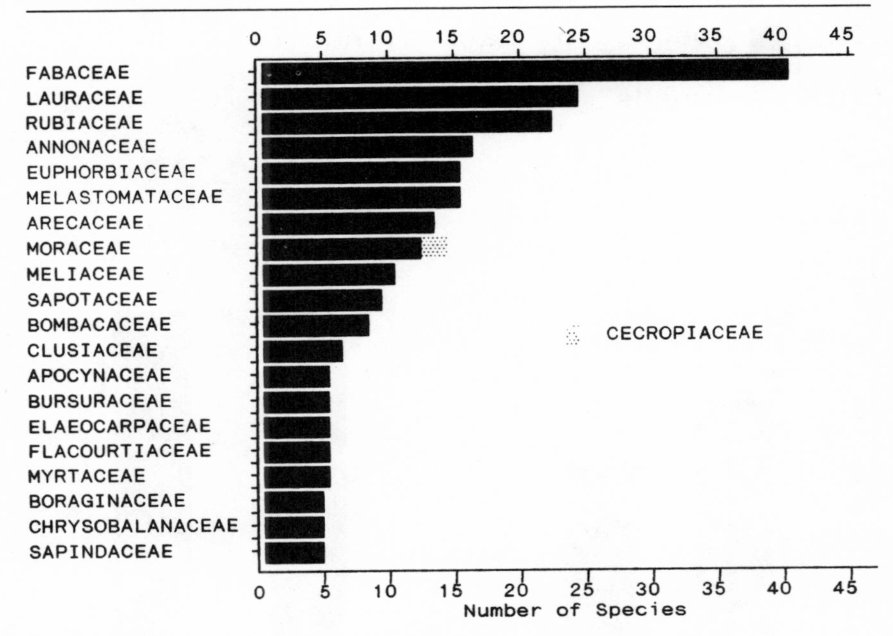

FIGURE 5.4. Families of trees with the most species at La Selva.

usual, at least on rich-soil sites like La Selva. This may be another manifestation of the phenomenon of normally higher elevation species at La Selva (see Gentry, this volume) and may help to explain its overall high floristic diversity; four species of Lauraceae are rare at La Selva but much more common at higher elevations. Melastomataceae, Piperaceae, and Rubiaceae, with their three very large genera—*Miconia*, *Piper*, and *Psychotria*—account for nearly 50 percent of the diversity of the shrub flora (fig. 5.5). The top-ranked liana families—Bignoniaceae, Fabaceae, Sapindaceae, and Malpighiaceae—and lianas in general are not particularly diverse at La Selva (fig. 5.6).

The ferns and two families of monocots—Orchidaceae and Araceae—account for 60% of the epiphyte species (fig. 5.7). Most of the epiphytes are true epiphytes, germinating above the ground and never establishing terrestrial contact. Among these, orchids and ferns are most diverse. About 25% of the species of epiphytes are climbers that apparently germinate on the ground, climb by their roots, and may eventually become detached from the ground. The Araceae (primarily *Philodendron*) are by far the most speciose family of climbers. About twenty-five species (less than 10% of the total) are stranglers (*Ficus* and *Clusia*)

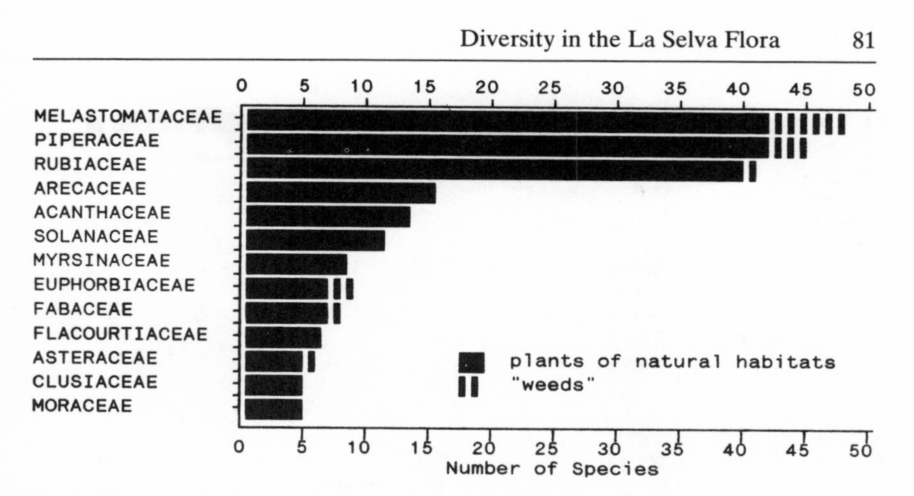

FIGURE 5.5. Families of shrubs with the most species at La Selva.

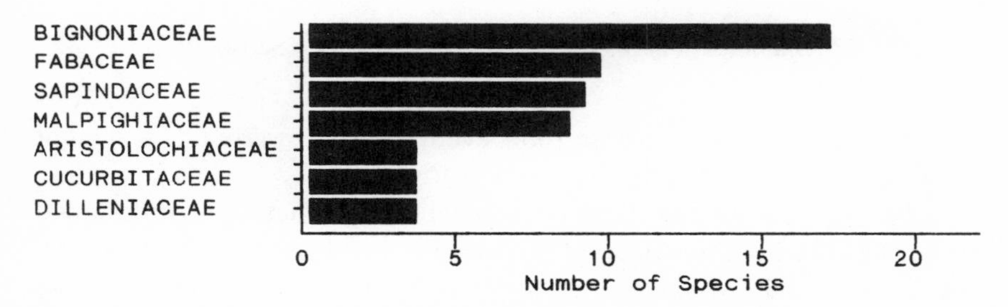

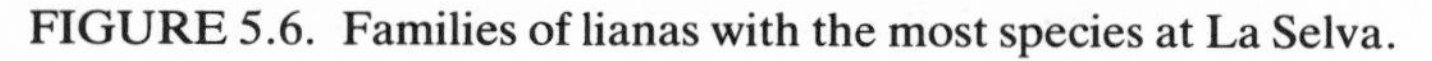
FIGURE 5.6. Families of lianas with the most species at La Selva.

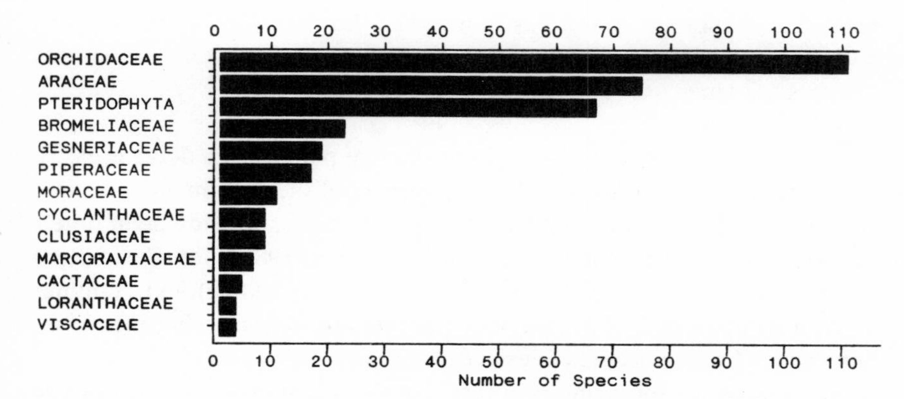

FIGURE 5.7. Families of epiphytes with the most species at La Selva.

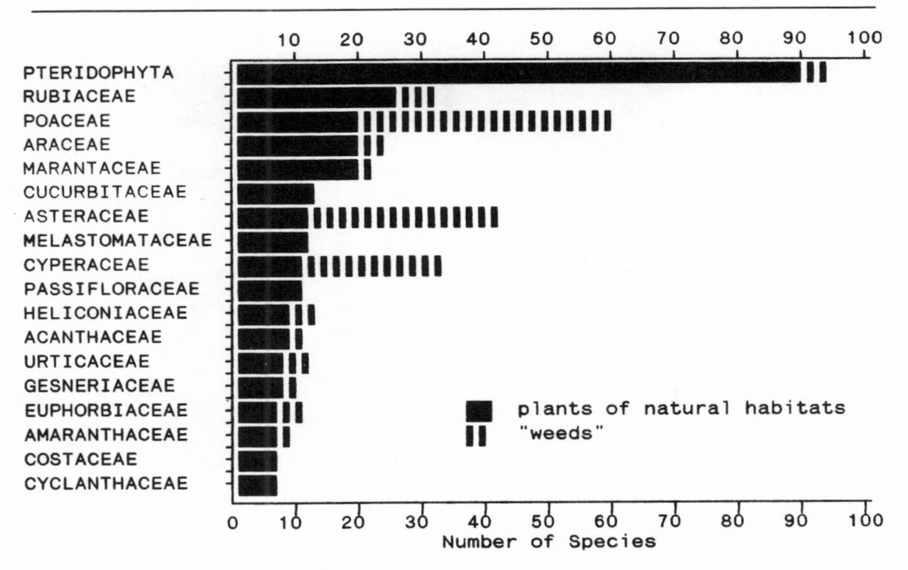

FIGURE 5.8. Families of herbs with the most species at La Selva.

or other hemiepiphytes, whose roots eventually reach ground, and eight species are parasites (Loranthaceae and Viscaceae).

The ranking of the most diverse families of herbs is confounded somewhat by the abundance of herbaceous weeds among the Poaceae, Asteraceae, and Cyperaceae. But the ferns are clearly the most diverse group of herbs and have over twice as many native species in natural habitats as any single family of flowering plants (fig. 5.8). The Rubiaceae, Poaceae, Araceae, Marantaceae, Cucurbitaceae, Asteraceae, Melastomataceae, Cyperaceae, and Passifloraceae follow the ferns in numbers of herb species. About 17% (circa sixty-five species) of the herbs of natural habitats are vines.

Information on geographic distribution of plants at La Selva is easily available only for those families already published or in manuscript. However, a floristic affinity with South America has already been noted for southern Central America by Gentry (1978, 1982), for Costa Rica by Standley (1937), and for La Selva by Hammel (1986g) and Hammel and Grayum (1982). An analysis of six families (Cyclanthaceae, Marantaceae, Cecropiaceae, Clusiaceae, Lauraceae, and Moraceae)—representative of the primary forest—showed that roughly 85% of the species were known also from Panama or South America, whereas only about about 45% (essentially the widespread species, known from both Central and South America) were known also from Nicaragua or further north. About 40% of the species known outside Costa Rica were known only from areas to the south, but less than 2% of them were known only to the north (Hammel 1986g). Almost 45% of the species in this group of six families appear to be endemic to Central America, and more than 10% were endemic to Costa Rica. Recent work on the Nicaraguan flora has increased the number of "southern" species in these families known from that country, but the number of "northern" species known from La Selva has not increased. These general trends are also true for several other large families (such as Rubiaceae, Araceae) for which manuscripts for the flora are available.

The grasses probably represent an extreme; most (85%) of the native New World species at La Selva are widespread, and only about 15% appear to be endemic to Central America (Judziewicz and Pohl 1984). For this family, the La Selva flora represents neither a northern nor a southern limit for any species.

Perhaps the most striking aspect of the La Selva flora is its great diversity of epiphytes, which account for 25% of the flora. This diversity is primarily invested in the orchids, aroids, and ferns. In fact, the category "plants growing or climbing on others"—including vines and lianas—accounts for one-third of plant diversity at La Selva. The ferns, orchids, and aroids account for nearly 25% of taxonomic diversity. The number of species of Lauraceae among the trees at La Selva also seems unusually high for a lowland forest and may be part of a phenomenon of normally higher elevation taxa descending to very wet lower elevation forests (see Gentry, this volume). Discounting the approximately 300 cultivated or "weedy" species, nearly 1,450 species at La Selva—the total flora of natural habitats—could be considered endangered in the sense that this type of forest is rapidly disappearing. With increased exploration and intensive inventories of small areas we can hope to have a much clearer picture of the present forest demography and the phytogeographic history of this once extensive habitat.

REFERENCES

Atwood, John T. 1988. Orchidaceae. In R. L. Wilbur (ed.), *The Vascular Flora of La Selva Biological Station, Costa Rica*. Selbyana 10: 76–145.

Clark, D. (This volume). La Selva Biological Station: a blueprint for stimulating tropical research.

Croat, T. 1978. *Flora of Barro Colorado Island*. Stanford Univ. Press, Stanford.

Foster, R. B. and S. P. Hubbell. (This volume). Floristic composition of the Barro Colorado forest.

Gentry, A. H. 1978. Floristic knowledge and needs in Pacific tropical America. Brittonia 30: 134–153.

———. 1982. Phytogeographic patterns as evidence for a Choco refuge. In G. Prance (ed.), *Biological Diversification in the Tropics*. Columbia Univ. Press, New York, pp. 112–136.

———. 1986. Sumario de patrones fitogeográficos neotropicales y sus implicaciones para el desarrollo de la Amazonía. Rev. Acad. Col. Cienc. Ex. Fis. Nat. 16: 101–116.

———. 1988. Changes in plant community diversity and floristic composition on environmental and geographical gradients. Ann. Mo. Bot. Gard. 75: 1–34.

———. (This volume). Floristic similarities and differences between southern Central America and Upper and Central Amazonia.

Gentry, A. H. and C. Dodson. 1987. Diversity and biogeography of Neotropical vascular epiphytes. Ann. Mo. Bot. Gard. 74: 205–233.

Grayum, M. H. and H. W. Churchill. 1989a. The vascular flora of La Selva Biological Station, Costa Rica. Lycopodiophyta. Selbyana 11: 61–65.

———. 1989b. The vascular flora of La Selva Biological Station, Costa Rica. Polypodiophyta. Selbyana 11: 66–118.

Hammel, B. E. 1986a–f. Cecropiaceae (et seq.). In R. L. Wilbur (ed.), *The Vascular Flora of La Selva Biological Station, Costa Rica*. Selbyana 9: 192–259.

———. 1986g. Characteristics and phytogeographical analysis of a subset of the flora of La Selva (Costa Rica). Selbyana 9: 149–155.

Hammel, B. E. and M. Grayum. 1982. Preliminary report on the flora project of La Selva Field Station, Costa Rica. Ann. Mo. Bot. Gard. 69: 420–425.

Hartshorn, G. 1983. Plants: introduction. In D. H. Janzen (ed.), *Costa Rican Natural History*. Univ. of Chicago Press, Chicago, pp. 118–183.

Hartshorn, G. and B. Hammel. 1991. An introduction to the flora and vegetation of La Selva. In L. A. McDade, K. S. Bawa, H. A. Hespenheide, & G. S. Hartshorn (eds.), *La Selva: Ecology and Natural History of a Neotropical Rainforest*. Univ. of Chicago Press, Chicago.

Judziewicz, E. J., and R. W. Pohl. 1984. Grasses of La Selva, Costa Rica. Contrib. Univ. Wisc. Herb. 1: 1–86.

Lieberman, D. and M. Lieberman. 1987. Forest tree growth and dynamics at La Selva, Costa Rica (1969–1982). J. Trop. Ecol. 3: 347–358.

Lieberman, M., D. Lieberman, G. Hartshorn, and R. Peralta. 1985. Small-scale altitudinal variation in lowland wet forest tropical forest vegetation. J. Ecol. 73: 505–516.

Standley, P. C. 1937. Flora of Costa Rica. Field Mus. Nat. Hist., Bot. Ser. 18: 5–63.

Wilbur, R. 1987. A preliminary checklist to the vascular plants of La Selva Biological Field Station, Costa Rica. 12th version. Typescript, 1–56.

Wilbur, R. et al. 1986. The vascular flora of La Selva Biological Station, Costa Rica: introduction. Selbyana 9: 191.

6

The Floristic Composition of the Barro Colorado Island Forest

ROBIN B. FOSTER AND STEPHEN P. HUBBELL

To compare the flora of different tropical forests it is imperative that appropriate segments of the flora are chosen. Without selective partitioning, differences in floras often have little meaning and cannot provide answers to our questions about variation between tropical forests. Barro Colorado Island (BCI), in the lowland forest area of Panama, has the most thoroughly studied and documented flora of any research area in the Neotropics, allowing relatively clear separation of subsets or segments of the flora, whether by habitat, area, life-form, or other means of partitioning. In this chapter we analyze the floristic composition of that subset of the flora that occurs inside the forest proper. We examine the forest flora both in reference to the island's flora as a whole and in such a way that it can be compared with other Neotropical forests (Gentry, this volume).

Barro Colorado is an island formed by the creation of Gatun Lake in 1914 for the Panama Canal. It has been a biological research station since 1923 and is administered by the Smithsonian Tropical Research Institute. A more complete description of the semideciduous forest can be found in Croat (1978) and Leigh et al. (1982). An inventory of the trees and shrubs (more than 1 cm DBH) in a 50-ha plot on the relatively flat central plateau of the island was begun in 1980 and is described in more detail in Hubbell and Foster (this volume).

We wish to thank 150+ people from eight countries for helping census the plants in the 50-ha plot. Financial and logistical support were provided by the National Science Foundation, World Wildlife Fund, Smithsonian Tropical Research Institute, Field Museum of Natural History, and University of Iowa.

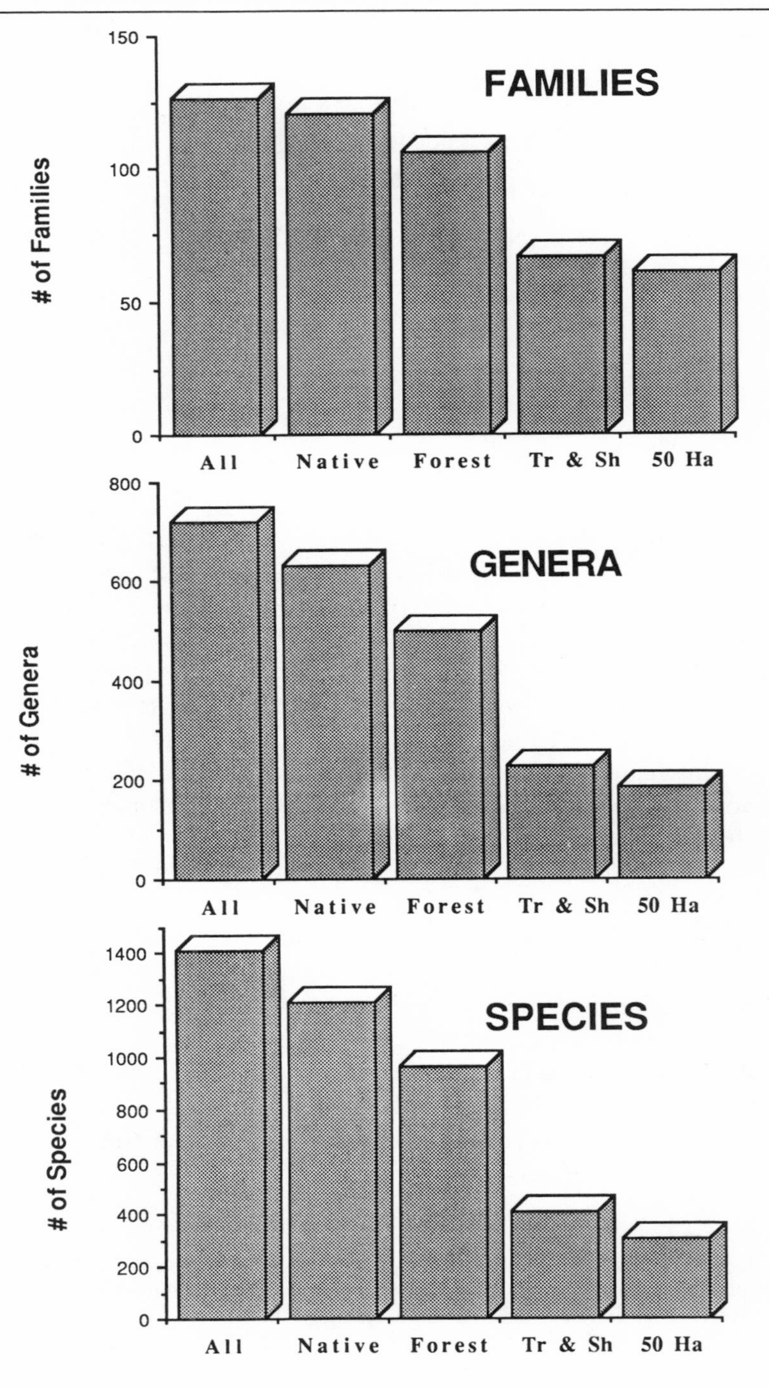
FAMILIES
# of Families
150
100
50
0
All
Native
Forest
Tr & Sh
50 Ha
GENERA
# of Genera
800
600
400
200
0
All
Native
Forest
Tr & Sh
50 Ha
SPECIES
# of Species
1400
1200
1000
800
600
400
200
0
All
Native
Forest
Tr & Sh
50 Ha

The Island Flora

The flora listed in Croat (1978), plus species discovered since then, consists of 126 families, 719 genera, and 1,407 species (fig. 6.1). Sixty-five of these species are cultivated plants, and 140 (mostly weedy species of the clearings and lake edge) had either disappeared from the island by the time of Croat's survey or were originally included in the flora (Standley 1933) by mistake. The remaining flora of native species and naturalized "weedy" species comprises 120 families, 633 genera, and 1,207 species.

This "native" or "natural" flora should generally be used when comparing total BCI flora with any other tropical area that includes cleared or successional habitats. The mere presence of these cleared areas and a long shoreline contributes to the preponderance of herb species in the flora (fig. 6.2). Frequent human travel to the island from a mainland that is one of the most important national and international migration routes in the world is probably another important contributor to the high diversity of weed species. Nevertheless, it is surprising that so much of the noncultivated current flora (21%) is attributable to such a small area of the island (less than 1%).

The geographic affinities of the flora have been discussed in Croat (1978), ecological affinities in Foster and Brokaw (1982). It has a mixture of elements from the wet Atlantic coast, the dry Pacific coast, and those seemingly restricted to an intermediate moisture regime. Within areas of the same or similar moisture regime, the flora has strong similarity to other areas with volcanic or sedimentary substrate of Oligocene to Miocene age, though not the recent limestone of the central Bayano and Darien basin. Areas just 5–10 km to the north of the island (Parque Soberania) on much older Cretaceous rock have a strikingly different flora.

Of the 633 BCI genera, 72% are also found in the rich soils of the Manu area of Peru nearly 2,500 km to the south and across the equator

←

FIGURE 6.1. Subsets of the Barro Colorado flora. *All:* taxa listed in Croat 1978 plus species discovered up to 1986. *Native:* taxa of native and naturalized "weedy" plants now known from BCI—that is, excluding cultivated plants and others no longer or only guessed to be on the island. *Forest:* taxa of plants known to occur in forested parts of the island—that is, excluding plants of shoreline and human clearings. *Tr & Sh:* taxa of trees and shrubs (more than 1 cm DBH) included among the forest taxa. *50 Ha:* the taxa of trees and shrubs found in the 50-hectare mapped plot in the forested center of the island.

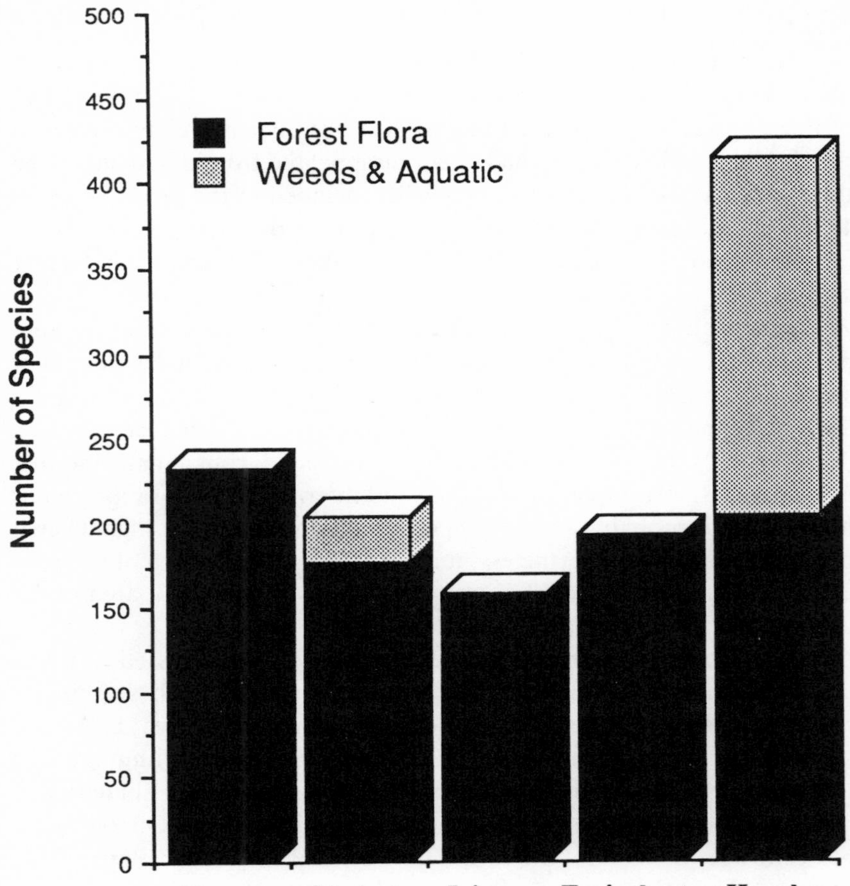

FIGURE 6.2. Life-forms of the Barro Colorado flora, native and naturalized. Weeds of human clearings and aquatics constitute a large fraction of the herb species. *Trees:* woody plants, freestanding as adults (includes stranglers) and greater than 10 cm DBH. *Shrubs:* woody, freestanding plants greater than 1 cm DBH but less than 10 cm DBH. *Lianas:* woody climbers reaching at least 10 m in canopy as adults. *Epiphytes:* herbs and woody plants that germinate on and require support from other woody plants. *Herbs:* terrestrial herbaceous plants, terrestrial woody plants less than 1 cm DBH, and vines reaching less than 10 m in canopy as adults.

(see Foster, this volume). Considering only woody plants (327 genera), the figure rises to 82% shared with Manu. Of the species on BCI, 271, or 21%, are also found in Manu.

The Forest Flora

Plants that can at least sometimes be found within the forested area of the island on a regular basis, be it in high, closed canopy forest, in treefall gaps, along streams, or colonizing streamside landslides, are all included here as the forest flora, consisting of 106 families, 499 genera, and 966 species (fig. 6.1).

This is the appropriate flora to compare with other tropical forests with minimal human clearing or river and lake edge, and for comparing tropical forests per se, as in this book. It excludes most human-associated weed species, as well as strictly aquatic species. It is in some ways surprising that so many of these weedy species cannot colonize the large gaps within the forest. The explanation may lie in problems of dispersal. The weedy species that appear are nearly always those with an effective wind-dispersal mechanism.

At the familial level, the forest flora (fig. 6.3) is dominated by three taxa of terrestrial herbs and epiphytes: the ferns (as one family), orchids, and aroids. A large proportion of these taxa is restricted to the more humid conditions in the deep ravines on the island. Among the trees (fig. 6.4), which dominate the biomass of the vegetation, the Leguminosae and Moraceae are by far the most speciose, as they are throughout much of the Neotropics. Among the shrubs (fig. 6.5), the Rubiaceae and Melastomataceae predominate in number of taxa, although the Violaceae is also important ecologically mainly due to one superabundant species, *Hybanthus prunifolius*, which makes up about a sixth of all woody plant individuals ≥ 1 cm DBH. The most Speciose families of lianas (fig. 6.6) are the Bignoniaceae, Leguminosae, Sapindaceae, and Malpighiaceae; of epiphytes (fig. 6.7), the Orchidaceae, ferns, and Araceae; and of terrestrial herbs (fig. 6.8), primarily ferns.

The size of genera represented in the forest (fig. 6.9) is not especially large in comparison to other tropical forests, and the four that stand out—*Psychotria* (Rubiaceae), *Inga* (Leguminosae), *Piper* (Piperaceae), and *Ficus* (Moraceae)—are speciose through most of the Neotropics, the last two primarily on rich soils, the first two on both rich and moderately poor soils.

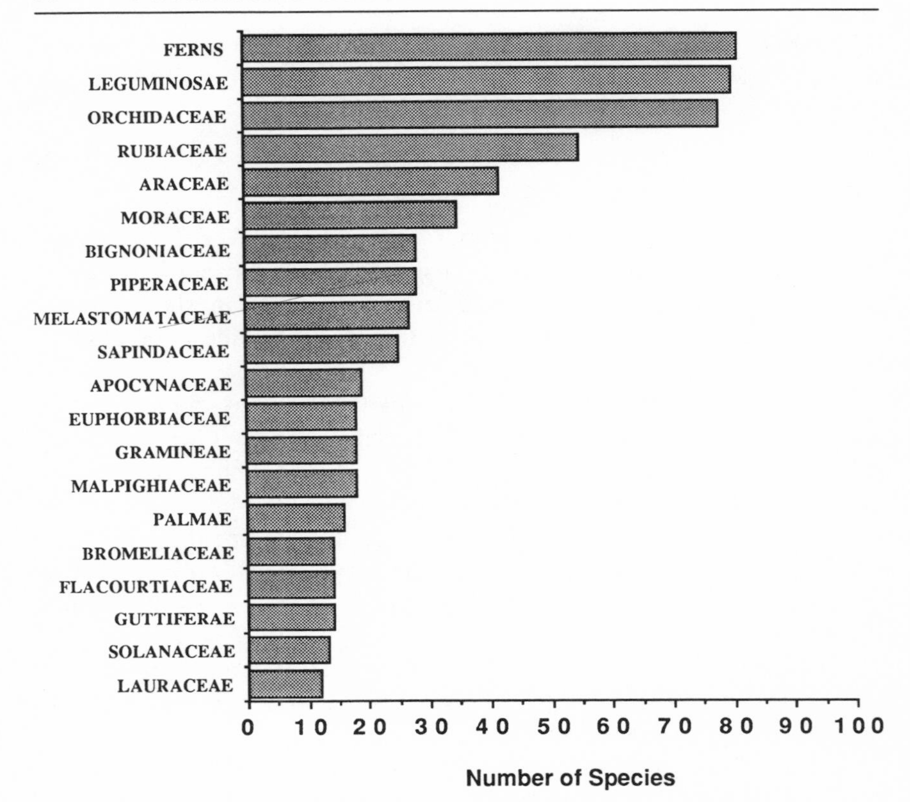

FIGURE 6.3. Largest families of the Barro Colorado forest flora, including all vascular plants. "Ferns" includes all Pteridophyta as one family.

The Trees and Shrubs on a Fifty-Hectare Plot

We may also examine the ecological importance of the BCI forest species by drawing on data from a 50-ha plot (Hubbell and Foster 1986). Of the 409 species of trees and shrubs known in the forest flora, three-fourths (306) occur in the 50-ha plot. In other words, our sample of only 3% of the forest area contains 75% of the island's forest species. The 306 species, though a large sample, might not be representative of floristic composition of trees and shrubs for the rest of BCI. Do special characteristics keep 104 species from occurring in the plot?

The only substantial difference in family representation between the plot and the rest of the forest is the relatively smaller proportion of Melastomataceae in the plot (fig. 6.10). In life-form, the species not found in the plot differ in having a greater preponderance of shrubs to

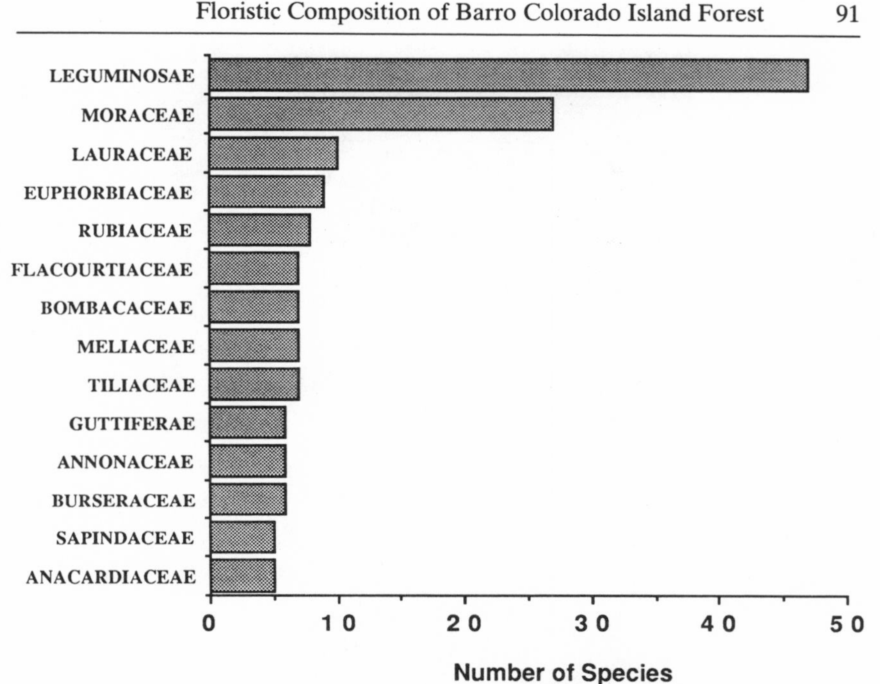

FIGURE 6.4. Tree families of the Barro Colorado forest flora. Shows only those families with five or more species represented.

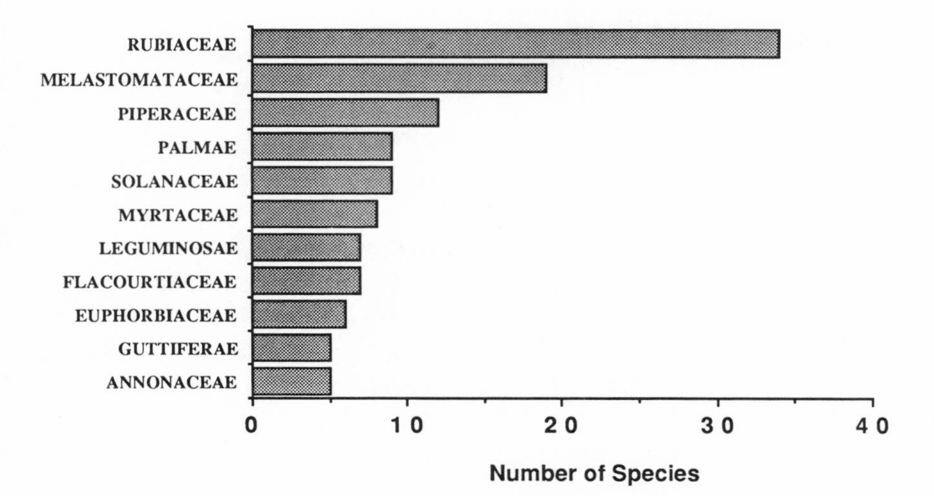

FIGURE 6.5. Shrub families of the Barro Colorado forest flora (> 5 species).

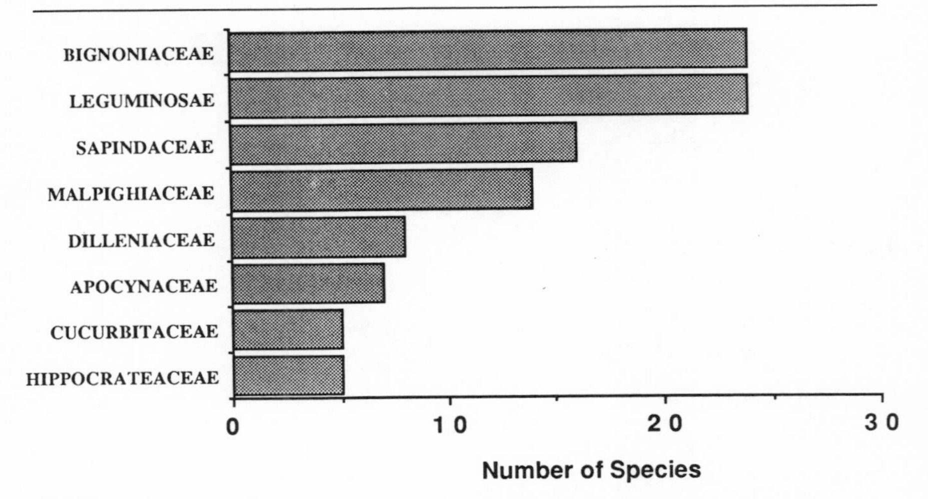

FIGURE 6.6. Liana families of the Barro Colorado forest flora (> 5 species).

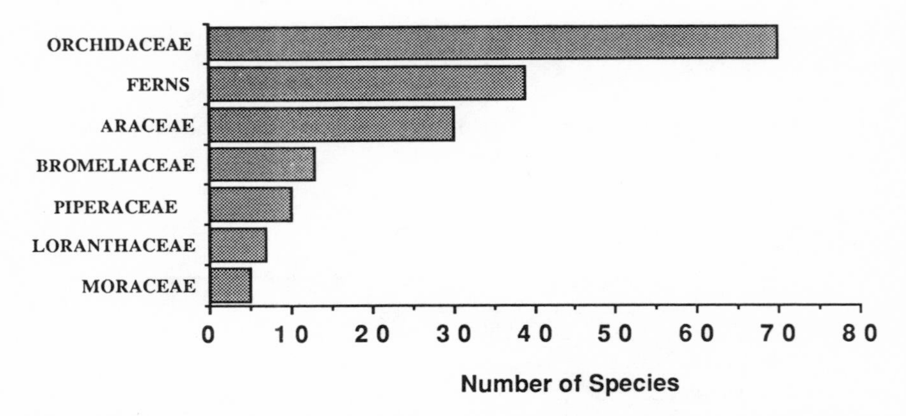

FIGURE 6.7. Epiphyte families of the Barro Colorado forest flora (> 5 species).

trees (60 versus 44 species). Sixty-three (61%) of the species not in the plot are listed in the forest flora as very rare. Of the remaining species, 30 are apparently restricted to the young second growth of the island, 5 are known only from deep ravines, and 2 are known only from swamps. Thus, only 4 species that are not rare and not obviously restricted to some special habitat are missing from the plot. (Differences in soil from different geological substrate are suspected in these cases.) Many of the rare species are also found in wet ravines, but it is uncertain whether they are restricted to this habitat. Many melastome shrub species are restricted to

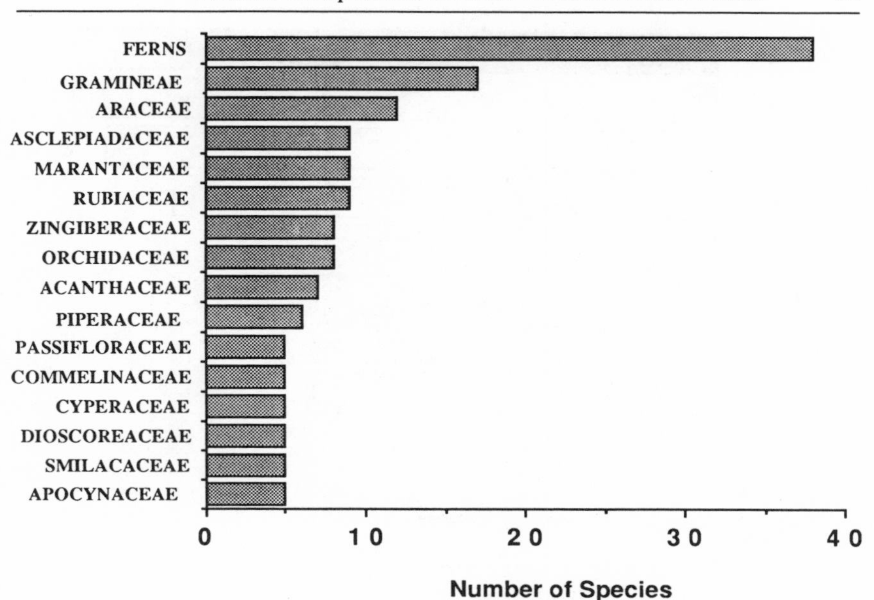

FIGURE 6.8. Herb families of the Barro Colorado forest flora (> 5 species).

second growth and wet ravines. We can conclude that species are missing from the plot for the combined reasons of extreme rarity—that our plot by chance did not happen to include them—and restriction to habitats not well represented on the plot, which is on the high plateau of BCI and covered largely by old forest (see also Hubbell and Foster 1986).

The Trees and Shrubs on One Hectare

Single hectares within the 50-ha plot contain an average of 176 species of trees and shrubs greater than 1 cm DBH (fig. 6.11)—a surprisingly high proportion (43%) of all the tree and shrub species in the BCI forest. Because liana stems are less dense and herbs are more dense than trees and shrubs in a hectare, one would predict that 1 ha would be less representative of the lianas on the island but would have a better representation of herbs. If herbs are far more restricted to wet ravines, however, the location of the plot on the plateau would preclude a large representation of this life-form.

The hectares with the maximum and minimum numbers of species greater than 1 cm DBH are only about 7% different from the mean. For successively larger stem size classes, diversity drops and the maximum

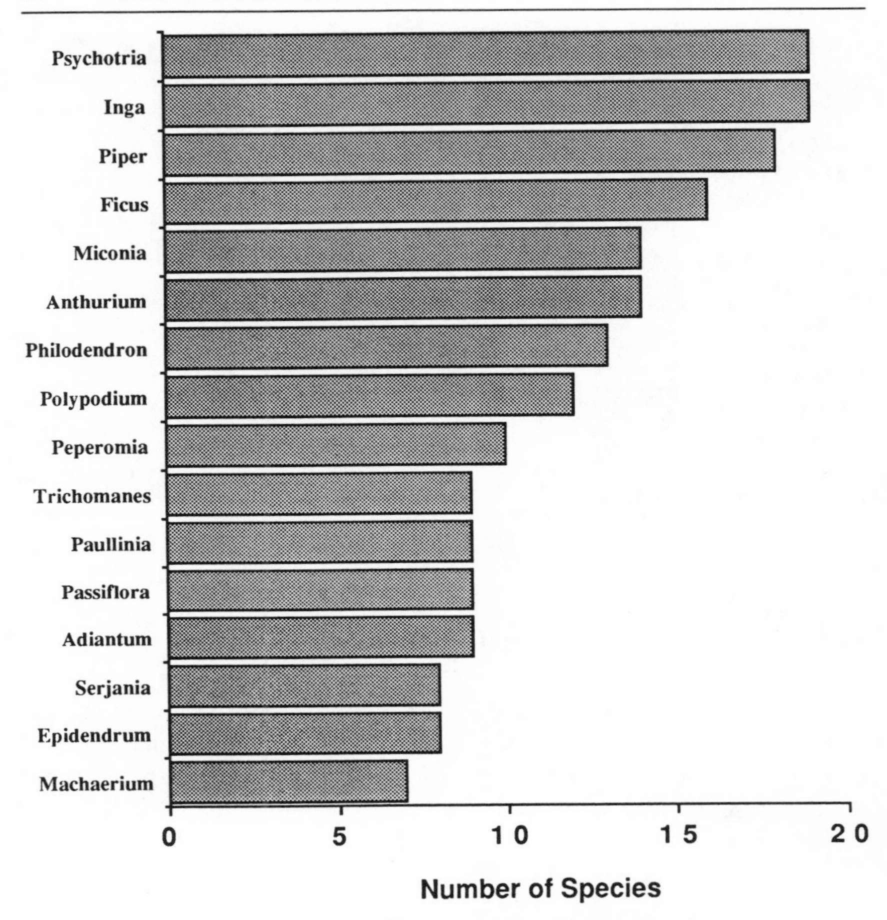

FIGURE 6.9. Largest genera of the Barro Colorado forest flora.

and minimum hectares in diversity differ proportionately more from the mean. For trees greater than 30 cm DBH (canopy trees) the number of species in 1 ha may differ from the mean hectare by 40%. This means that estimates of the species diversity of larger size classes based on one or a few ha must be treated with considerable caution when trying to compare forests.

Of the mean number of 4,710 stems greater than 1 cm DBH in 1 ha, 1,861 stems (40%) are of adult size. This was calculated from each species based on the stem diameter at which that species is estimated to reach reproductive adulthood (at least 10% of full-grown reproductive capacity). Though not a rigorous measure at present, it does indicate that

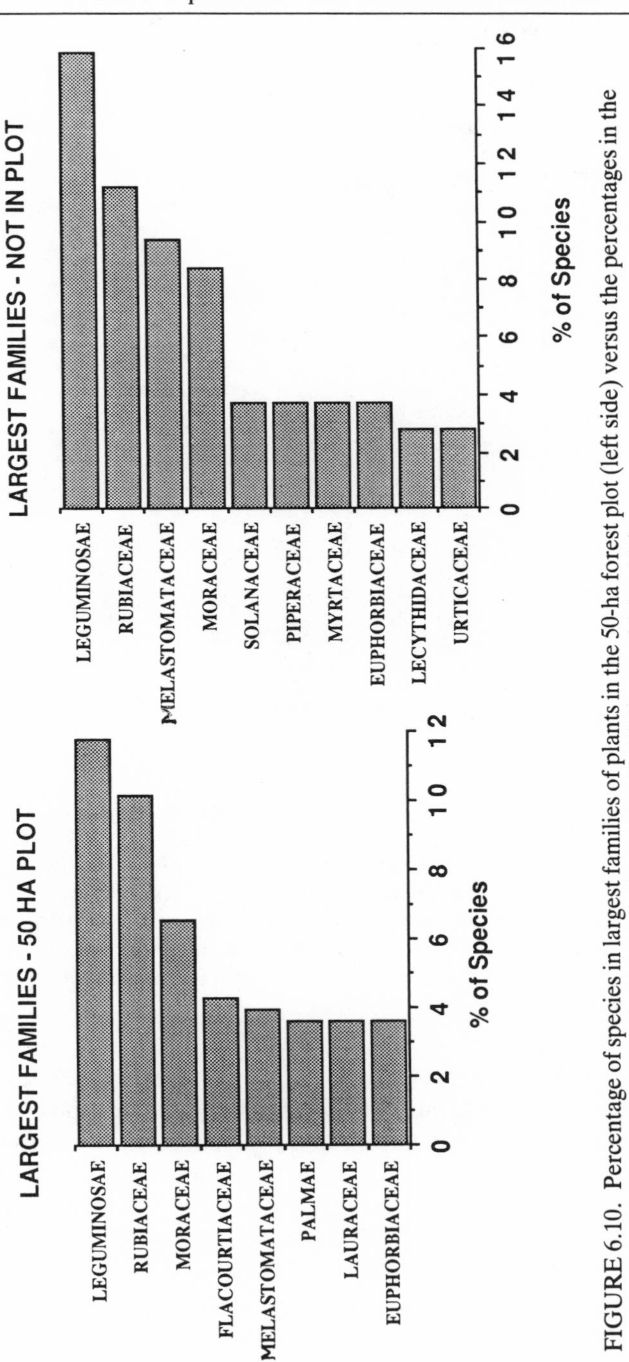

FIGURE 6.10. Percentage of species in largest families of plants in the 50-ha forest plot (left side) versus the percentages in the largest families for Barro Colorado forest species *not* in the 50-ha plot (right side).

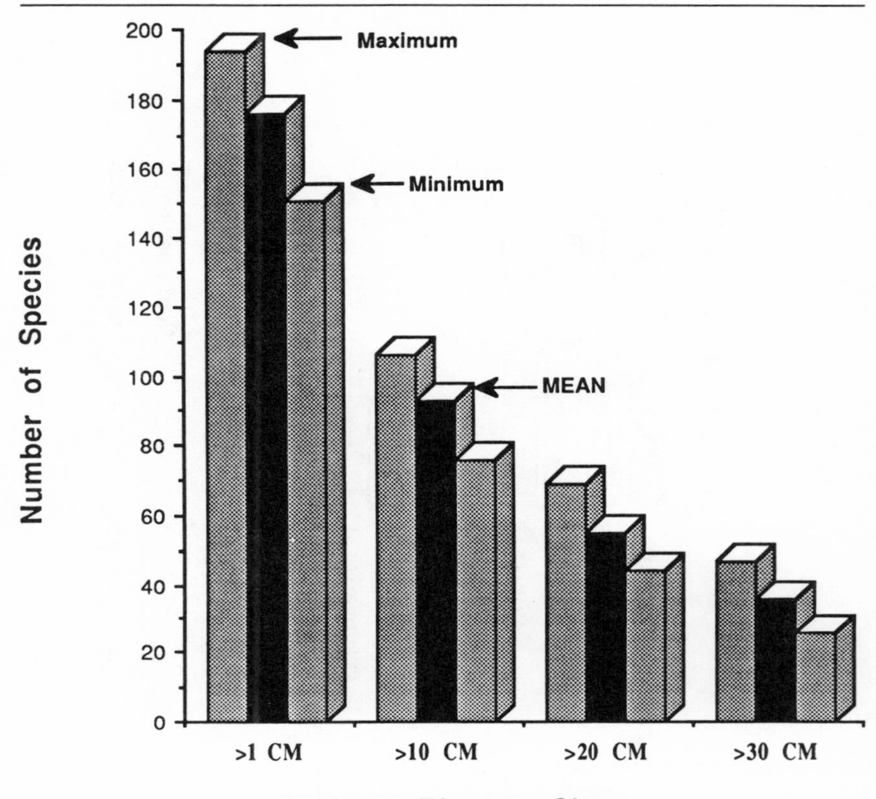

FIGURE 6.11. Species per hectare (means and extremes) in the Barro Colorado 50-ha plot. The black column for each size class of woody plants is the mean number of species/hectare for 50 contiguous hectares in a 1,000 m × 500 m plot. The gray column to the left of each mean is the maximum species richness found for a single hectare; the gray column to the right is the minimum species richness found for a single hectare.

from half to two-thirds of the stems in our plot are juveniles. Any use of plot species lists for local conservation purposes must obviously take this into consideration, since species well represented only as juveniles or only as adults may be as vulnerable to disappearance as very rare species.

Changes in the Barro Colorado Flora

It is difficult to be sure how many kinds of plants are on BCI. Since the publication of Croat's Flora in 1978, at least 25 species have been added to the flora of the island. Fifteen of these were added during the survey of the 50-ha plot, resulting in a preponderance of new trees and shrubs. But any equally intensive survey of lianas, herbs, and epiphytes is also certain to reveal a proportionate increase in species of those life-forms. The new additions were certainly growing on the island at the time of Croat's survey and presumably for hundreds of years before that.

Through both chance encounter and thorough surveys of individual stems, the known flora of BCI will continue to increase. There will also be real immigrants to the island from the adjacent mainland, and some plants will certainly become extinct. By contrast, some rare or inconspicuous plants may never be recognized again on the island and may be presumed extinct even if still present. For all practical purposes, we will never know the real gains and losses taking place in the flora of the island. This contrasts with serious attempts at estimating and understanding the loss of BCI's bird and mammal species (Karr, this volume; Glanz, this volume). With a substantial area of permanent plots in different habitats and forest ages on the island, however, we could determine under what conditions mean numbers of plant species per unit area are dropping or rising.

Another kind of change has taken place in the BCI flora. As much as 20% of the forest species in Croat's flora may have had their names changed within the two decades since the Flora was written. The scientific names of the plants on Barro Colorado are probably more stable than are those for much of the Neotropics, due to the relatively intense activity by the Missouri Botanical Garden on the flora of Panama. Better understanding of genera through monographic work on the ever growing collections from Central and South America nevertheless continues to result in new circumscription of species. Substantial changes have already taken place in the names of the Meliaceae, Lauraceae, and Palmae, as well as throughout many smaller families. Though this might seem a tremendous nuisance to the ecologist, it significantly improves our understanding of plant biogeography (usually by expanding the range of a species) and ultimately enhances our knowledge of ecological-evolutionary relationships and definition of conservation priorities.

The published flora of Barro Colorado must be broken into subsets to be comparable to other tropical floras and forests. This is mostly because of the large fraction of species associated with human-made clearings on the island. Even after all the truly weedy aquatic plants are removed from consideration, herbs and epiphytes make up a substantial fraction of the forest flora. A sample of 50 ha, or even only 1 ha, contains a significant fraction of the forest tree and shrub species. The plants not included in a sample of 50 ha are almost entirely either very rare or restricted to a habitat poorly represented in the sample. The family composition of the forest flora is typical of much of the Neotropics, especially that on more nutrient-rich soils. Knowledge of the flora continues to grow, but an accurate understanding of the real gains and losses to the flora of BCI as a whole is exceedingly difficult to obtain.

REFERENCES

Croat, T. 1978. *Flora of Barro Colorado Island*. Stanford Univ. Press, Stanford.

Foster, R. B. (This volume). The floristic composition of Rio Manu floodplain forest.

Foster, R. B. and N. V. L. Brokaw. 1982. Structure and history of the vegetation on Barro Colorado Island. In E. G. Leigh, Jr., A. S. Rand, and D. M. Windsor (eds.), *The Ecology of a Tropical Forest: Seasonal Rhythms and Long-Term Changes*. Smithsonian Inst. Press, Washington, D.C., pp. 67–81.

Gentry, A. H. (This volume). Floristic similarities and differences between southern Central America and upper and central Amazonia.

Glanz, W. (This volume). Neotropical mammal densities: how unusual is the community on Barro Colorado Island, Panama?

Hubbell, S. P. and R. B. Foster. 1987. Commonness and rarity in a Neotropical forest: implications for tropical tree conservation. In M. E. Soulé (ed.), *Conservation Biology: The Science of Scarcity and Diversity*. Sinauer, Sunderland, Mass., pp. 205–231.

———. (This volume). Structure, dynamics, and equilibrium status of old-growth forest on Barro Colorado Island.

Karr, J. R. (This volume). The avifauna of Barro Colorado Island and the Pipeline Road, Panama.

Leigh, E. G., A. S. Rand, and D. M. Windsor (eds.). 1982. *The Ecology of a Tropical Forest: Seasonal Rhythms and Long-Term Changes*. Smithsonian Inst. Press, Washington, D.C.

Standley, P. 1933. The flora of Barro Colorado Island, Panama. Contrib. Arnold Arboretum 5.

7

The Floristic Composition of the Rio Manu Floodplain Forest

ROBIN B. FOSTER

Intact Amazonian floodplain forest is rare, and studies of the flora on such floodplains are few (Balslev et al. 1986). Rio Manu, a meandering, white-water tributary of the Rio Madeira in southeastern Peru, provides an undisturbed example of such a flora. In this chapter I consider the lowland part of the river between 300 and 400 m elevation from the river mouth (Boca Manu) to the tributary Rio Sotileja (for a general discussion of this area, see Terborgh 1983). Judging from the density of animals and the growth rate of trees, the soil must be among the most fertile in the tropics. A description and checklist of the vegetation and flora of the lowland area of Manu, including the Alto Madre de Dios drainage, is being prepared.

For purposes of appropriate comparison with the other Neotropical

For permission to collect in Manu National Park I thank the Direcion General Forestal y de Fauna in Lima and the Jefatura del Parque Manu in Salvacion. For field assistance in collecting on the Rio Manu, I thank especially Barbara d'Achille, Alejandro de la Cruz, Carol Augspurger, Howard and Nick Brokaw, Patricia Wright, and Catherine Toft; and for indirect assistance through plant collections incidental to other research, I thank especially John Terborgh and Charles Janson. For assistance in collection data entry, I thank Tyana Wachter and Elsa Meza. For field financial and logistical assistance, I thank especially John Terborgh, the Field Museum of Natural History, the Missouri Botanical Garden, the Jefatura del Parque Nacional Manu, Luis Yallico, Alejandro de la Cruz, the Museo de Historia Natural Lima, Magda Chanco, Ramon Ferreyra, Emma Cerrate, Joaquina Alban, Blanca Leon, and Howard Clark. For assistance in processing plant collections, I thank the Botany Department of the Field Museum, Tyana Wachter, Grace Russell, Bill Grime, Nora Murphy, Kevin Swagle, Freddy Robinson, Birthel Atkinson, Missouri Botanical Garden, Kathleeen Cook, and Myra Guzman-Teare. For assistance in plant identification, I thank overworked specialists too numerous to mention here, and especially my colleagues at the Field Museum: Mike Dillon, Bob Stolze, Mike Huft, and Tim Plowman.

forests covered in this book, I consider only those plants occurring in forest on the recent lowland floodplain of the river, not the older terraces and low hills, though they abut the river. The recent floodplain is on the richest of alluvial soils, subject in parts to yearly inundation and in large areas to the highest floods, which may only occur once a century. Also included is area up to approximately 5 m above the highest level of flooding. Most collections have been within a 5-km radius of the Cocha Cashu station (see Terborgh, 1983, and this volume), and collections elsewhere on the river floodplain are mostly of species also found near Cocha Cashu.

I first began collecting on the Rio Manu in 1973 and made subsequent collecting trips in 1974, 1976, 1980, 1984, and 1986. These collections have been augmented by those of Terborgh, Janson, Russell, Fitzpatrick, Moskovits, Torres, Wright, Emmons, Wilson, and McFarland while pursuing the food plants of various vertebrates near Cocha Cashu. Gentry made significant collections at Cocha Cashu in 1979 and 1983, adding especially to the list of canopy liana species.

Most collections are from June–December, which might lead one to believe that we are missing a large segment of the flora. A disproportionately large number of species either flower or fruit during this period, however, so we often have a species represented by only flowering or fruiting condition, not both. In addition, neither I nor Gentry have had qualms about collecting plants in sterile condition, many of which are identifiable to species. The surveys of plots in the vicinity of Cocha Cashu (Foster 1986, Foster, this volume; Foster unpubl. data; Gentry 1985, 1988; Gentry and Terborgh, this volume) leave no doubt that we have collections of virtually all common species of trees and shrubs, and certainly many of the rarer species.

The known flora of the Manu vicinity, extending up to humid puna vegetation at 3,500 m, consists of 2,874 species in 1,006 genera and 153 families (Foster 1987). To these should be added species represented in collections from the higher elevations of the Cosñipata Valley that I have not seen. Collections from the Rio Manu, including the older terraces and low hills adjacent to the river from Boca Manu to the vicinity of the Rio Sotileja tributary, have produced 1,856 species in 751 genera and 130 families. The recent, low-lying floodplain contains 1,372 species in 637 genera and 119 families.

The flora within the floodplain can be separated into species representing three distinct habitats. The "beach" category (123 species) includes all herbs, herbaceous vines, and shrubs that germinate and reach maturity on the beach or exposed river edge. In addition to flat beaches, it includes *Tessaria* and *Gynerium* thickets, as well as steep eroding riverbanks. The "cocha" category (34 species) includes all aquatic herbs and shrubs in or on the exposed edges of oxbow lakes or other permanently flooded depressions. The "forest" category (1,215 species, 515 genera, 107 families) includes all the vascular plants in areas dominated by trees and is the primary focus here. Successional areas, including fig swamps, are classed as forest when the trees reach 20 cm DBH. Such habitats as forest dying back from flooding have unusual combinations of beach, cocha, and forest species and are not easily categorized.

Using conspicuous physiognomic characteristics and the presence or absence of certain dominant species, one can distinguish successional communities within the forest, such as *Cecropia-Ochroma*, *Ficus-Cedrela*, *Ceiba-Dipteryx-Poulsenia*, and *Calycophyllum-Calophyllum* communities. From a floristic point of view, however, it is difficult to characterize these communities unless one considers only canopy-level or emergent tree species. Beneath the canopy dominants can be found the juveniles of an array of floodplain species. This results in a much greater floristic similarity and slower gradation between these communities than is apparent to the casual observer.

The periodic formation of levees or dikes (separated by swales or depressions) as the river meanders over its floodplain creates a smaller-scale habitat distinction that is not used here (see Foster, this volume). Species restricted to the depressions are few. Some, notably *Cedrela odorata* (Meliaceae), seem restricted primarily to the highest levees (or the swales between closely packed, high levees), where their root systems presumably are not subject to prolonged flooding.

The life-form spectrum (fig. 7.1) of the floodplain flora reveals the considerable contribution of the beach and cocha communities to the herb category. The forest flora by itself is heavily weighted toward trees and shrubs relative to epiphytes and terrestrial herbs. This is in strong contrast to Barro Colorado Island (BCI) (Foster and Hubbell, this volume) and La Selva (Hammel, this volume), where species of epiphytes and herbs are as well represented as trees and shrubs. There seem to be at least two explanations for this. First, there has been less emphasis on collection of herbs and epiphytes at Manu. Second, the low floodplain of the Manu lacks the year-round rainfall of La Selva or the moist ravines to which many of the BCI herb and epiphyte species are restricted; occa-

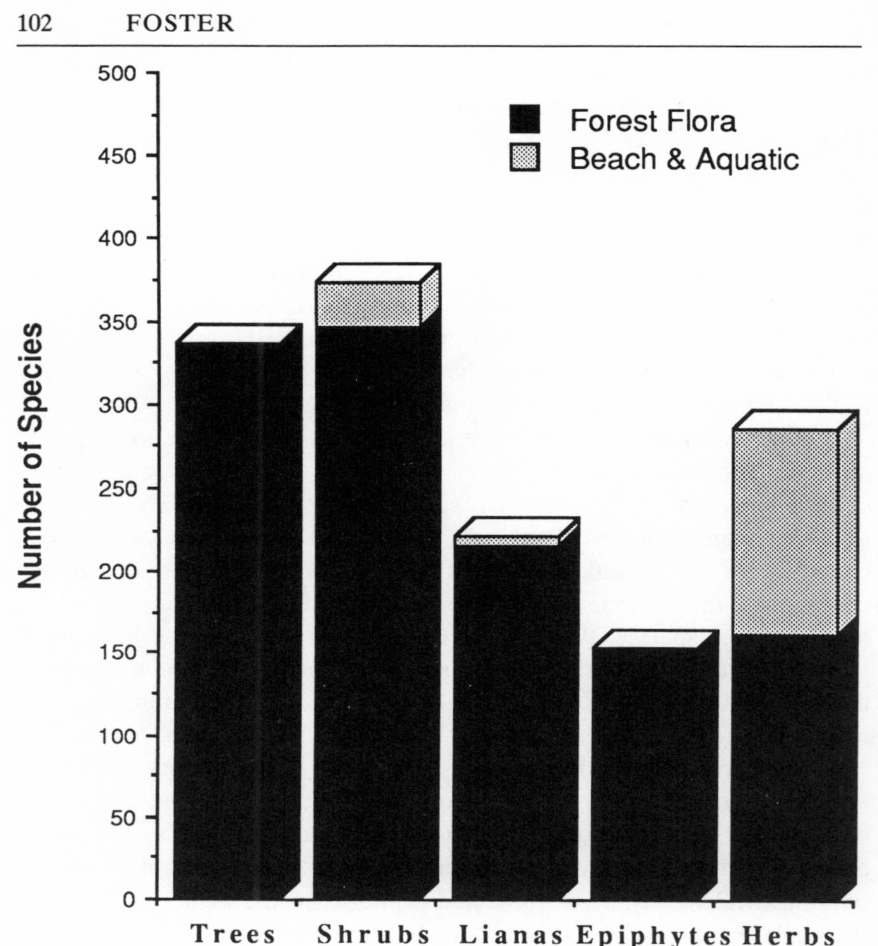

FIGURE 7.1. Life-forms of the Manu floodplain flora. Excluded are species from the higher, unflooded, river terraces and low hills and the few cultivated plants. The subset of beach and aquatic species is shown in gray. *Trees:* woody plants freestanding as adults (includes stranglers) and greater than 10 cm DBH. *Shrubs:* woody, freestanding plants greater than 1 cm DBH but less than 10 cm DBH. *Lianas:* woody climbers reaching at least 10 m in canopy as adults. *Epiphytes:* herbs and woody plants that germinate on and require support from other woody plants. *Herbs:* terrestrial herbaceous plants, terrestrial woody plants less than 1 cm DBH, and vines reaching less than 10 m in canopy as adults.

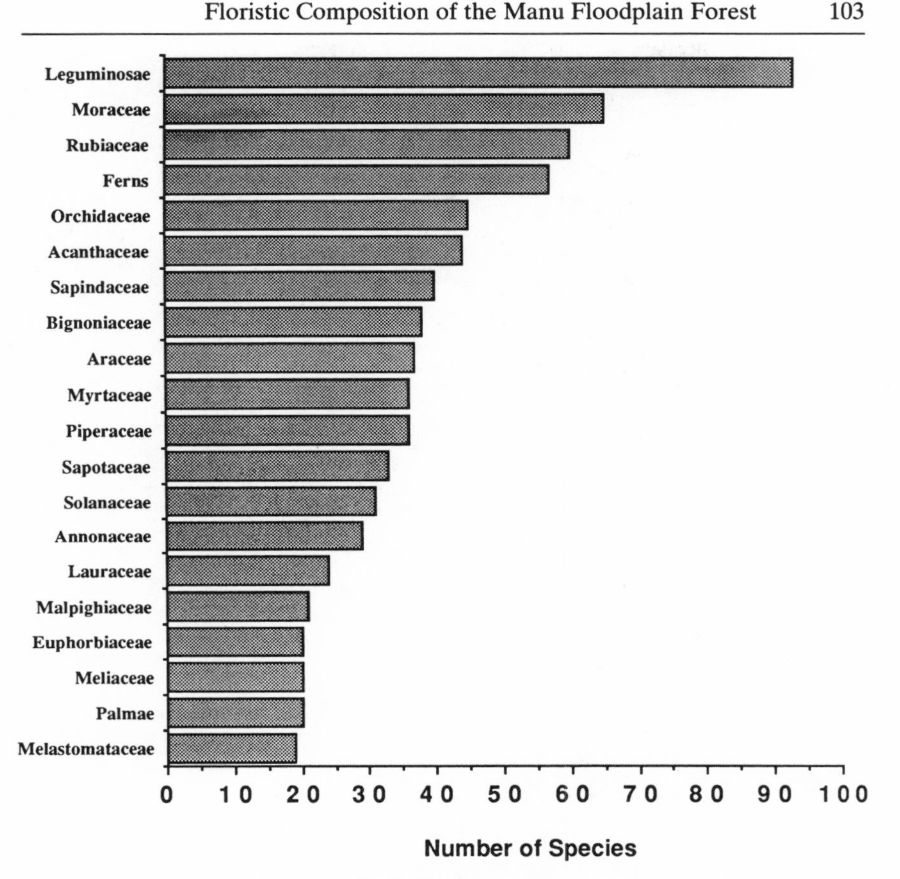

FIGURE 7.2. Largest families of the Manu floodplain forest flora, including all vascular plants but excluding the beach and aquatic species. "Ferns" includes all Pteriodophyta as one family.

sional drought stress on the relatively flat Manu floodplain may thus play a limiting role. Annual flooding might also limit terrestrial herbs, but only in the lower parts of the floodplain forest. After the old terraces and low hills along the Manu are better collected for herbs and epiphytes, the flora of the Rio Manu forest, including such habitats, may well have a life-form spectrum much more similar to that of BCI, but with many more species in each category.

In the familial composition of the floodplain forest flora (fig. 7.2) the Leguminosae are clearly the most important taxon, with over 90 species (and more than 140 species in the entire Rio Manu flora). However, even the legumes make up only 8 percent of the species. A dozen other fami-

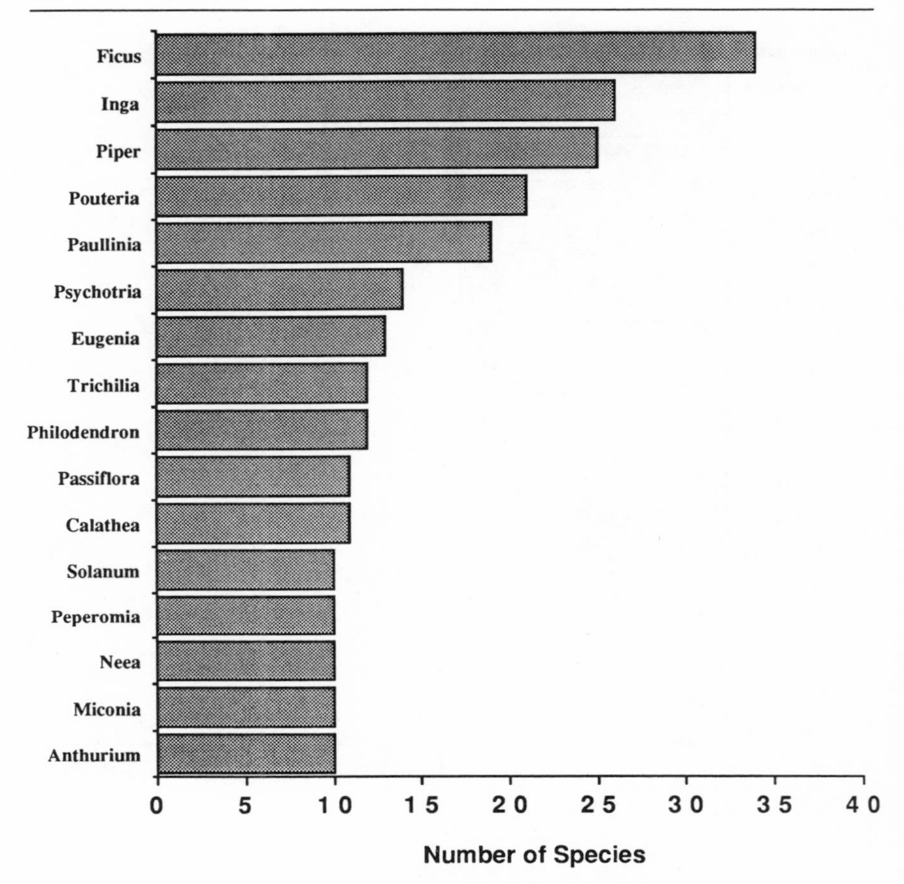

FIGURE 7.3. Largest genera of the Manu floodplain forest flora.

lies have more than 30 species each, the most prominent being Moraceae, Rubiaceae, and Pteridophyta (treated as one family).

Each genus averages only 2.4 species in the floodplain forest flora. Some genera, however, are strikingly speciose. *Ficus* has 35 species (probably a world record for one locality), and *Inga*, *Piper*, *Pouteria*, and *Paullinia* all have more than 15 species (fig. 7.3).

Among the tree species (fig. 7.4), Leguminosae, Moraceae, Sapotaceae, and Lauraceae predominate. The most speciose shrub families (fig. 7.5) are Rubiaceae, Myrtaceae, Piperaceae, and Solanaceae. The four largest families of lianas (fig. 7.6) are Bignoniaceae, Sapindaceae, Leguminosae, and Malpighiaceae, the same four as on BCI (Foster and Hubbell, this volume). Similarly, among the epiphytes (fig. 7.7), Orchi-

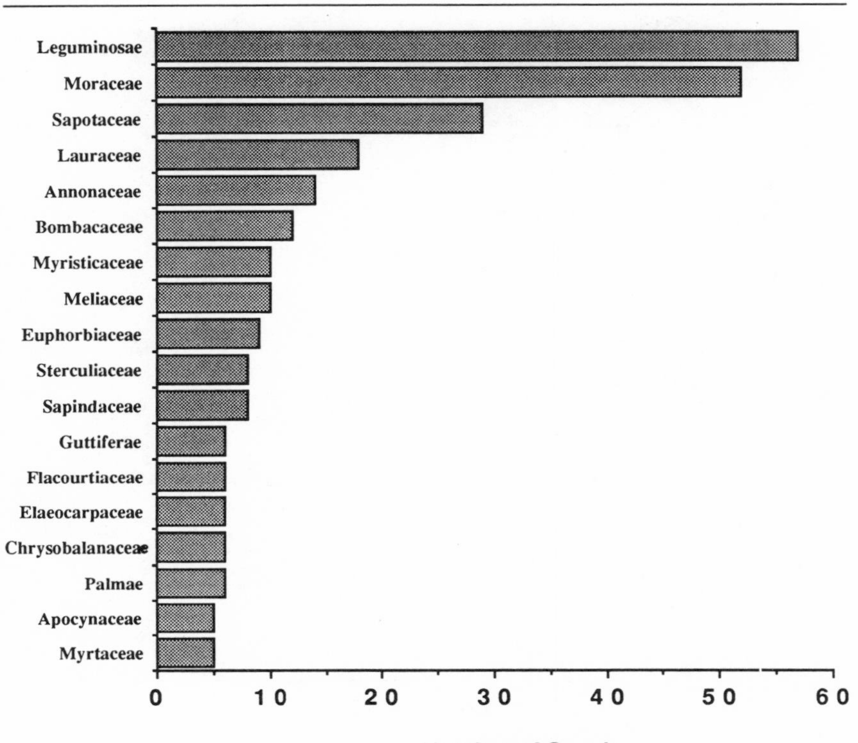

FIGURE 7.4. Tree families of the Manu floodplain forest flora. Shows only those families with five or more species represented. See fig. 7.2 for life-form definitions.

daceae, Pteridophyta, and Araceae predominate at both sites. The most speciose terrestrial herb families (fig. 7.8) are Pteridophyta, Acanthaceae, and Marantaceae.

All of southern Amazonian Peru, and especially the Madre de Dios drainage, which includes parts of adjacent Brazil and Bolivia, is poorly collected, and the taxonomic status of the plants is often in doubt. It seems premature to analyze the biogeographic relations of the Manu floodplain flora, if indeed such analyses are useful. There are clear affinities and overlap with all parts of the lowland Neotropics from Mexico to Argentina. A few generalizations and salient points, however, are worth mentioning.

Most striking is that this is a rich-soil flora. As noted in Foster and

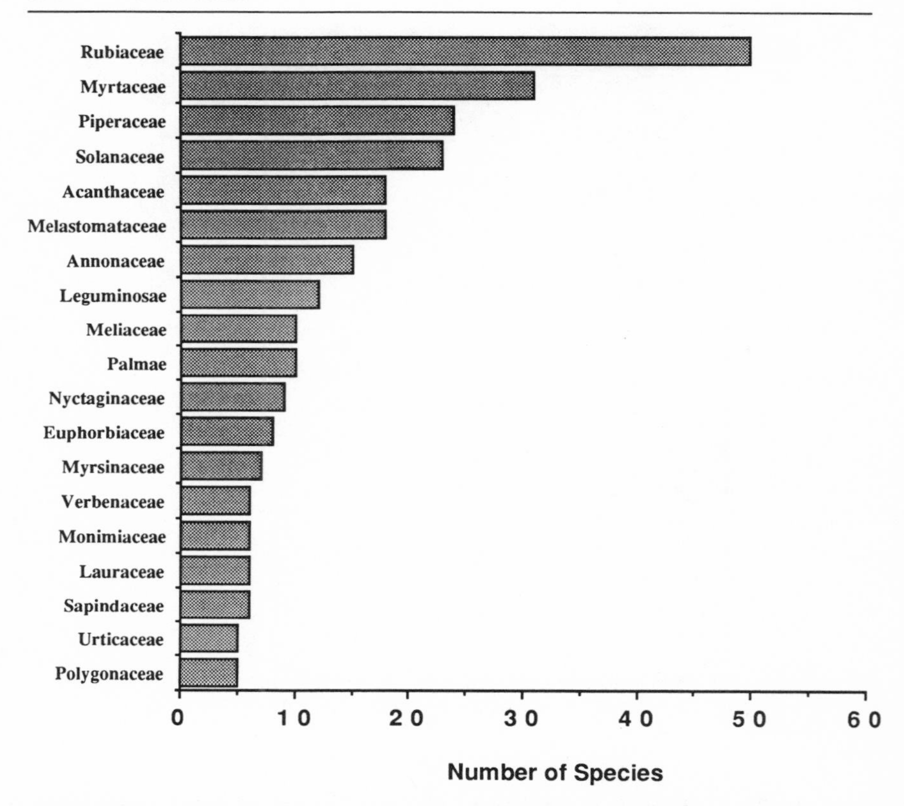

FIGURE 7.5. Shrub families of the Manu floodplain forest flora (more than 5 species).

Brokaw (1982) and Foster and Hubbell (this volume), the flora of BCI has 72 percent of its genera (82 percent of woody genera) and at least 22 percent (271) of its species in common with Manu. The Manu floodplain flora probably will have a similarly high overlap in floristic composition with the forests on rich sedimentary hills and remnants of floodplain (varzea) forest throughout much of Amazonia and even Central America. Its closest affinities will presumably be with other areas like BCI that have distinct dry seasons. The seasonality presumably selects for species that can tolerate or evade occasional severe drought.

With few exceptions, the most speciose families of the Manu floodplain (fig. 7.2) are typical not only of rich-soil forest but of virtually all moist lowland Neotropical areas. What makes the rich-soil flora distinct are the relative absence of such families as Lecythidaceae, Chrysobala-

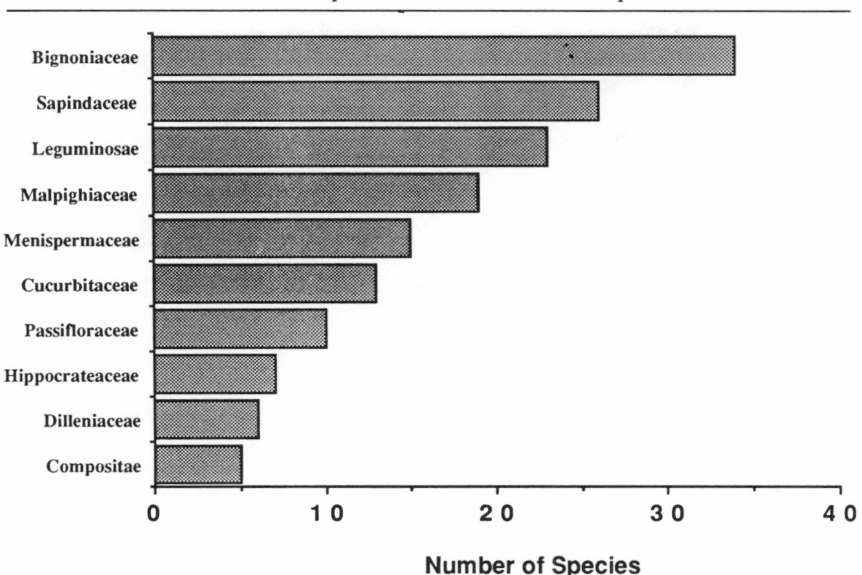

FIGURE 7.6. Liana families of the Manu floodplain forest flora (more than 5 species).

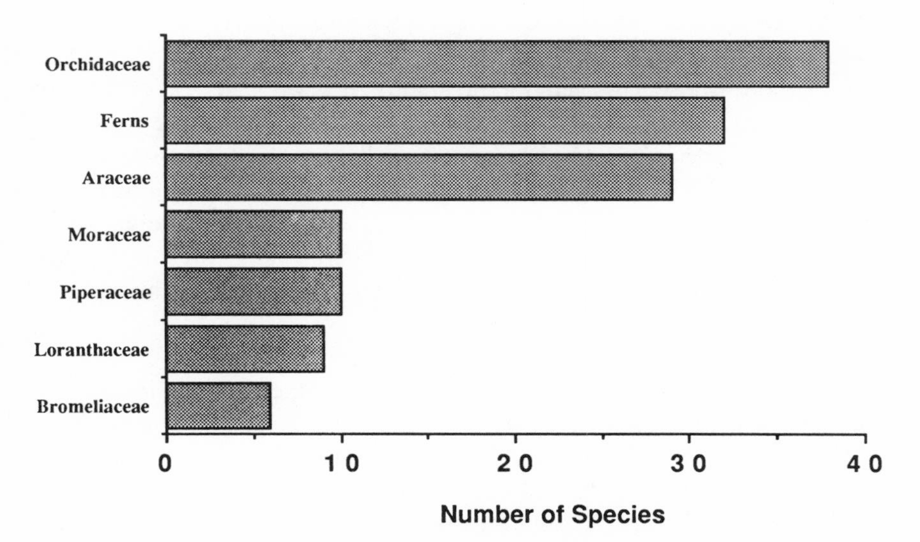

FIGURE 7.7. Epiphyte families of the Manu floodplain forest flora (more than 5 species).

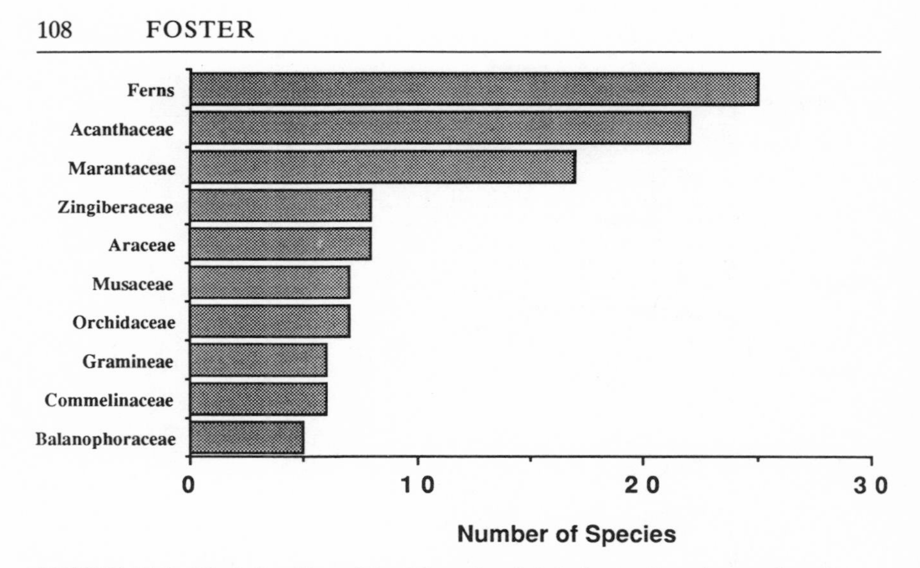

FIGURE 7.8. Herb families of the Manu floodplain forest flora (more than 5 species).

naceae, Vochysiaceae, and Burseraceae; the diversity of such genera as *Ficus* (Moraceae), *Inga* (Leguminosae), *Piper* (Piperaceae), *Paullinia* (Sapindaceae), *Heliconia* (Musaceae); and the presence of such widespread species as *Ceiba pentandra* (Bombacaceae), *Ficus insipida*, *Poulsenia armata* (Moraceae), *Hura crepitans*, *Acalypha diversifolia* (Euphorbiaceae), and *Tectaria incisa* (fern).

If there is a specific geographic affinity, it is with the rest of the Rio Madeira drainage, including the Beni in Bolivia and the Acre in Brazil. Many Manu specimens seem intermediate between different species described from the Ucayali drainage to the north and the Beni to the south, calling into question the validity of species at two ends of a gradual cline. It also appears that there are several similar, but perhaps more rapid, cases of gradation from a lowland Amazonia species through the foothills to an Andean lower-montane species.

The large stands of 40 m tall, 1–2 m diameter *Cedrela odorata* (Meliaceae), and large individuals of the rare *Swietenia macrophylla* (Meliaceae) at the edge of the river, are perhaps the most unusual aspect of the Manu floodplain flora. The presence of these valuable timber trees set Manu apart from virtually all other Neotropical floodplains, where these species have already been removed for local and international commerce. *Cedrela* and *Swietenia* were once probably conspicuous elements of all rich-soil Amazonian floodplains.

The recent floodplain of Manu has a conspicuous abundance of species with fruits that appear to be adapted primarily to dispersal by mammals (Foster 1986), as compared with other Neotropical forests, including BCI and even the old terraces and low hills of Manu. Most striking in the older forest are the large, yellow to orange-red fruits, somewhat acid with a thick but removable pericarp and preferred by the larger monkey species (a syndrome described by Janson 1983). Such fruits are produced in a wide spectrum of families and are especially significant in the families Sapotaceae and Menispermaceae. The high frequency of fruits dispersed by mammals and great densities of primates and other mammals on the Manu floodplain (Terborgh 1983; Janson and Emmons, this volume) are presumably related, the animals helping to maintain high densities of these plant species through dispersal, and the availability of these fruits contributing to the high densities of primates and other mammals. The proportionally much greater species richness of Sapotaceae, Menispermaceae, and Annonaceae at Manu than on BCI (see Hubbell and Foster, this volume) may reflect in part the long history and intensity of clearing on BCI (Piperno 1989), which reduced the populations and diversity of species dispersed by mammals.

Some other groups are also unusually well represented on the Manu floodplain. There are 6 species of Balanophoraceae, apparently a record number for one site. In the lowland Neotropics these colorful and rarely collected root parasites are found mostly in the younger stages of forest development. Other groups that seem disproportionately speciose on the Manu floodplain in comparison with other rich soil areas are Acanthaceae, *Calyptranthes* (Myrtaceae), *Neea* (Nyctaginaceae), *Alibertia*, and *Randia* (Rubiaceae). The much greater proportionate richness of Sapotaceae, Menispermaceae, and Annonaceae than is found on BCI (see Hubbell and Foster, this volume) may reflect more the centuries of hunting and clearing in that part of Panama, which reduced the populations and diversity of species dispersed by mammals.

Although some unusual plant species were first described from Manu, most will probably turn up elsewhere in this region of the Amazon if sufficient collecting is done before the forests disappear. The first record of a lowland Amazonian Buxaceae, a new species of *Styloceras* (*S. brokawii*), was from Manu (Gentry and Foster 1981), where it is an abundant dioecious shrub in younger forest; however, this species is now also known from the Beni of Bolivia. A large, opposite-leaved Lauraceous tree, locally frequent in the older Manu floodplain forest, was described as a new species of the genus *Caryodaphnopsis* (*C. fosteri*), a genus previously known only from subtropical China and North Vietnan (Van der Werff 1985); this species, however, is now known from several other parts of Peru, and other species in the genus have been discovered elsewhere in tropical America.

At least two dozen additional species at Manu appear to be undescribed; they await more complete material or additional study in order to be described with certainty as new. It is premature to conclude that any species is endemic to the Manu floodplain, but the rate of forest removal in the region may make certain that a large percentage of the Manu flora will eventually become "endemic" to Manu by default.

REFERENCES

Balslev, H., J. Luteyn, B. Ollgaard, and L. B. Holm-Nielsen. 1987. Composition and structure of adjacent unflooded and floodplain forest in Amazonian Ecuador. Opera Bot. 92: 37–57.

Foster, R. B. 1986. Dispersal and the sequential plant communities in Amazonian Peru floodplain. In A. Estrada and T. H. Fleming (eds.), *Frugivores and Seed Dispersal*. W. Junk, Dordrecht, pp. 357–370.

———. 1987. Plantas del Parque Manu: 1987 checklist. Unpubl. report.

———. (This volume). Long-term change in the successional forest community of the Rio Manu floodplain.

Foster, R. B. and N. V. L. Brokaw. 1982. Structure and history of the vegetation on Barro Colorado Island. In E. G. Leigh, Jr., A. S. Rand, and D. M. Windsor (eds.), *The Ecology of a Tropical Forest: Seasonal Rhythms and Long-Term Changes*. Smithsonian Inst. Press, Washington, D.C., pp. 67–81.

Foster, R. B. and S. P. Hubbell. (This volume). The floristic composition of the Barro Colorado Island forest.

Gentry, A. 1985. *Reporte Manu*. M. A. Ríos (ed.), Centro de Datos para la Conservación, Univ. Agraria, La Molina, Peru, pp. 2/1–2/24.

———. 1988. Changes in plant community diversity and floristic composition on environmental and geographical gradients. Ann. Mo. Bot. Gard. 75: 1–34.

Gentry, A. and R. B. Foster. 1981. A new *Styloceras* (Buxaceae) from Peru: a phytogeographic missing link. Ann. Mo. Bot. Gard. 68: 122–124.

Gentry, A. H. and Terborgh, J. (This volume). Composition and dynamics of the Cocha Cashu mature floodplain forest.

Hammel, B. (This volume). The distribution of diversity among families, genera, and habit types in the La Selva flora.

Janson, C. H. 1983. Adaptation of fruit morphology to dispersal agents in a Neotropical forest. Science 219: 187–188.

Janson, C. H. and L. H. Emmons. (This volume). Ecological structure of the nonflying mammal community at Cocha Cashu Biological Station, Manu National Park, Peru.

Piperno, D. R. 1989. Fitolitos, arqueología y cambios prehistoricos de la vegetacíón en un lote de cincuenta hectáreas de la Isla de Barro Colorado. In E. G. Leigh, Jr., A. S. Rand, and D. M. Windsor (eds.), *Ecología de un bosque tropical*. Smithsonian Tropical Research Inst., Balboa, Panama.

Terborgh, J. 1983. *Five New World Primates: A Study in Comparative Ecology*. Princeton Univ. Press, Princeton, N.J.

———. (This volume). An overview of research at Cocha Cashu Biological Station.

Van der Werff, H. and H. G. Richter. 1985. *Caryodaphnopsis* Airy Shaw, a genus new to the Neotropics. Syst. Bot. 10: 166–173.

8

The Floristic Composition of the Forests of Central Amazonian Brazil

GHILLEAN T. PRANCE

The Reserva Florestal Adolpho Ducke is approximately 80 km from the Minimum Critical Size of Ecosystems (MCSE) study sites in Amazonian Brazil. The MCSE sites have been surveyed floristically only for trees, and inventory of other habit groups has barely begun. Since it is on similar soil and has many of the same species, the Reserva Ducke is discussed here as a floristic representation of Central Amazonian forests.

In 1954, toward the end of a distinguished career in Amazonian botany, Brazilian botanist Adolpho Ducke expressed his concern to preserve the Amazon forest and suggested that a reserve be created by INPA (the National Amazon Research Institute). The government of Amazonas donated a 100-km^2 tract of land at km 26 of the recently built Manaus-Itacoatiara highway (AM-10) (fig. 8.1). The donation became official in February 1963, although INPA had already begun silvicultural work in the area by 1960. At first, Ducke's advice to conserve an area was not followed, and a portion of the reserve was used as the experimental site of the new silvicultural section of INPA. In 1962, INPA began work there on forest enrichment and forest nurseries and established plantations of useful native tree species. This continued until 1972, when INPA

The fieldwork on which this chapter is based was carried out through the support of several National Science Foundation grants, most recently BSR 84-09536, and in cooperation with the Instituto Nacional de Pesquisas da Amazônia (INPA), Manaus, and the Conselho Nacional de Desenvolvimento Científico e Technológico (CNPq). I am grateful to William A. Rodrigues and Marlene F. da Silva for stimulating work on the Reserva Ducke, to Bruce Nelson for assistance in the field and herbarium, David G. Campbell for carrying out computer analysis of data on his forest inventory programs, Anne Reilly for input of data, and David Johnson for preparing figures and reviewing the manuscript.

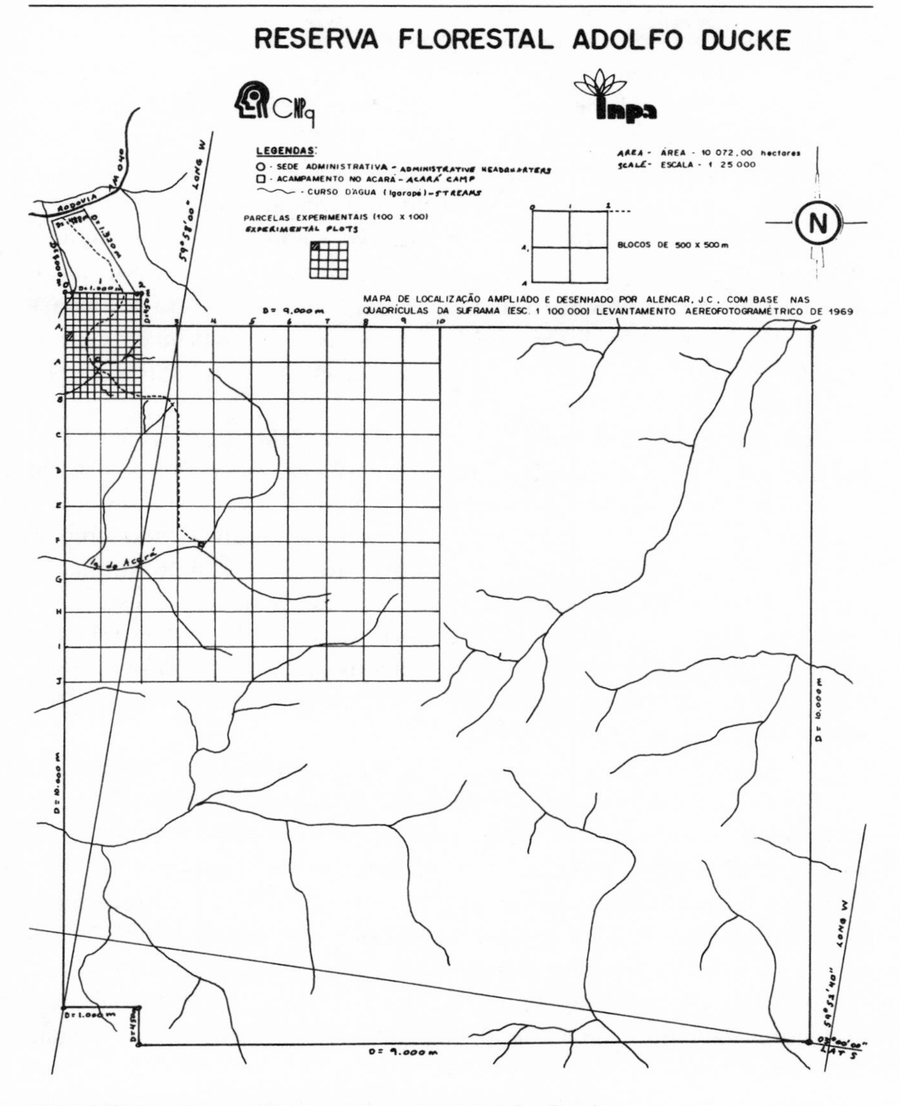

FIGURE 8.1. Map of Reserva Florestal Adolpho Ducke.

obtained another reserve in the SUFRAMA (Free Port) agricultural zone north of Manaus. Then all new silvicultural research was transferred to the new reserve, and the Reserva Ducke was declared a biological reserve. Only experiments already underway in 1972 have since been continued and encompass less than 2% of the total area. In addition, an area of 1,479 ha was invaded and partially cut by a Manaus industrialist; more recently, a small part at the rear of the reserve was invaded by inhabitants of Cidade Nova, an extension of the city of Manaus. The cut portion of the invaded area is now patrolled and is returning to mature secondary forest, which makes for interesting comparisons with the primary vegetation. Over 90% (90 km^2) of the reserve remains intact and untouched by either experiments or invasion.

With its convenient proximity to Manaus and the scientists at INPA, the Reserva Ducke is one of the most intensively researched patches of Amazonian forest in many disciplines. A meteorological station established in the reserve in 1974 has produced continuous rainfall and temperature data since then. In 1980, a 50-m-tall tower was built within the forest to gather microclimatic data from all levels of the forest. Since 1965, five individuals each of 100 species of trees have been the subject of a phenological study, with almost continuous gathering of data on their flower, fruit, and leaf production (Araújo 1970; Alencar et al. 1979; Prance and Mori 1979, 88–98).

The Reserva Ducke is covered mainly by tropical rainforest on terra firme latosol. It also includes several small streams that flood their banks and create swamps in the rainy season. These swamps contain many species characteristic of várzea forest, such as *Allantoma lineata* (Mart. ex Berg) Miers (Lecythidaceae) and *Mauritia flexuosa* L. f. (Arecaceae). Near the Igarapé Acará are several areas of white sand where the vegetation is a dense caatinga forest with such characteristic species as *Rhabdodendron macrophyllum* (Spruce ex Benth.) Huber (Rhabdodendraceae), *Humiria balsamifera* (Aubl.) St. Hil. (Humiriaceae), and *Pagamea macrophylla* Spruce (Rubiaceae). Finally, there are the areas of capoeira (secondary forest), where invasions and experiments have disturbed the primary forest. In 1987, INPA, in collaboration with the New York Botanical Garden and the Royal Botanic Gardens, Kew, launched a plan to prepare a florula of the Reserva Ducke over the next five years. This will intensify botanical study of the area.

The data presented here all derive from within or near the Reserva Ducke. The primary source of data is a computer printout of label data from the INPA herbarium from the 825 species that have been collected so far in the reserve. These data are by no means complete, since detailed, systematically organized plant collection in the reserve is just beginning. The specimen sample is, however, adequate to draw general conclusions about regional floristics. The second source of data is the distribution of species of Chrysobalanaceae, Lecythidaceae, and a few other small plant families I have studied. The distributions of all species known in the Reserva Ducke were plotted to analyze their phytogeography. The final data set is the inventory of 1 ha of forest on terra firme of the same type as in the reserve at a site 2 km outside the northeast boundary of the reserve. All trees of 15 cm DBH or more were collected and identified, and standard inventory analyses of the data were made, such as calculation of family importance values, relative frequencies, densities, and dominances, using methods outlined in Mori et al. (1983).

The herbarium collections from the Reserva Ducke entered into the *Programa Flora* computerized data base of the Amazonian herbaria include 825 species. This sample is certainly not complete, and not all the names have been confirmed at this early stage, but the general floristic trends discussed here are unlikely to change as further collections are made. A spot check of the data base made in the INPA herbarium for Chrysobalanaceae, Dichapetalaceae, Humiriaceae, and Lecythidaceae, revealed in each case that about 25 percent of the species known to occur in the reserve are represented only by more recent collections or identifications, and thus are not in the data bank. If this figure is correct, the number of species of vascular plants in the reserve is about 1,030. To date, however, collecting has concentrated on woody groups, especially trees, because of the forest studies carried out in the reserve. The number of species of herbaceous groups and epiphytes will probably increase the species list considerably beyond 1,030, yet the central Amazonia epiphyte and shrub flora is definitely low compared with western Amazonian and Central American sites.

Data from the Reserva Ducke species list are summarized by family in table 8.1; 96 families and 825 species are known to occur in the reserve. The most common family, with 13.2% of the species (109), is the Leguminosae sens. lat., and all three legume segregate families (Fabaceae, Mimosaceae, and Caesalpiniaceae) are among the ten most diverse families. The species of the ten most common families—Sapotaceae (48), Fabaceae (43) Rubiaceae (42), Chrysobalanaceae (40), Lauraceae (38), Annonaceae (36), Moraceae (35), Mimosaceae (33), Lecythidaceae (30),

TABLE 8.1 Vascular plants from Reserva Ducke

Family	Species (*N*)	Habits
Sapotaceae	48	Trees
Fabaceae	43	Trees, vines
Rubiaceae	42	Shrubs, trees
Chrysobalanaceae	40	Trees
Lauraceae	38	Trees
Annonaceae	36	Trees
Moraceae	35	Trees, stranglers
Mimosaceae	33	Trees
Lecythidaceae	30	Trees
Caesalpiniaceae	28	Trees
Melastomataceae	26	Shrubs, trees
Arecaceae	25	Trees, acauls
Orchidaceae	24	Epiphytes, herbs
Apocynaceae	22	Trees, vines
Euphorbiaceae	21	Trees, shrubs
Myristicaceae	20	Trees
Bignoniaceae	18	Lianas, trees
Burseraceae	18	Trees
Aspleniaceae	16	Herbs, epiphytes
Flacourtiaceae	15	Shrubs, trees
Clusiaceae	12	Stranglers, lianas, shrubs, trees
Bombacaceae	10	Trees
Humiriaceae	10	Trees
Araceae	9	Epiphytes, herbs
Sapindaceae	9	Trees, vines
Meliaceae	8	Trees
Myrtaceae	8	Trees, shrubs
Violaceae	8	Trees
Vochysiaceae	8	Trees
Hymenophyllaceae	7	Herbs, epiphytes
Icacinaceae	7	Shrubs, trees
Loganiaceae	7	Lianas, shrubs
Olacaceae	7	Trees
Bromeliaceae	6	Epiphytes, herbs
Combretaceae	6	Trees
Poaceae	6	Herbs
Sterculiaceae	6	Trees
Adiantaceae	5	Herbs, epiphytes
Anacardiaceae	5	Trees
Ebenaceae	5	Trees
Ochnaceae	5	Shrubs, herbs
Piperaceae	5	Shrubs
Cucurbitaceae	5	Vines
Malpighiaceae	4	Trees, vines
Menispermaceae	4	Lianas, shrubs
Monimiaceae	4	Lianas, shrubs

TABLE 8.1 (*Continued*)

Family	Species (*N*)	Habits
Quiinaceae	4	Trees
Rutaceae	4	Trees
Simaroubaceae	4	Trees
Solanaceae	4	Shrubs, epiphyte
Boraginaceae	3	Trees, shrubs
Caryocaraceae	3	Trees
Connaraceae	3	Vines
Convolvulaceae	3	Lianas
Gentianaceae	3	Saprophytes, herb
Polypodiaceae	3	Herbs, epiphytes
Schizaeaceae	3	Herbs
Anisophyllaeaceae	2	Trees
Asteraceae	2	Herbs
Burmanniaceae	2	Saprophytes
Celastraceae	2	Trees
Cyatheaceae	2	Trees, herbs
Cyperaceae	2	Herbs
Dichapetalaceae	2	Trees
Erythroxylaceae	2	Shrubs
Gesneriaceae	2	Herb, epiphyte
Hugoniaceae	2	Trees
Lacistemataceae	2	Shrubs
Loranthaceae	2	Hemiparasites
Marcgraviaceae	2	Lianas
Passifloraceae	2	Vines
Polygalaceae	2	Herbs
Rhabdodendraceae	2	Tree, shrub
Styracaceae	2	Trees
Tiliaceae	2	Trees
Triuridaceae	2	Saprophytes
Zingiberaceae	2	Herbs
Acanthaceae	1	Herb
Araliaceae	1	Tree
Capparaceae	1	Tree
Dilleniaceae	1	Liana
Dioscoreaceae	1	Vine
Duckeodendraceae	1	Tree
Eriocaulaceae	1	Herb
Gnetaceae	1	Liana
Lentibulariaceae	1	Herb
Liliaceae	1	Herb
Lycopodiaceae	1	Herb
Marattiaceae	1	Herb
Onagraceae	1	Herb
Opiliaceae	1	Tree
Oxalidaceae	1	Herb
Peridiscaceae	1	Tree

TABLE 8.1 (*Continued*)

Family	Species (*N*)	Habits
Rapateaceae	1	Herb
Sabiaceae	1	Tree
Selaginellaceae	1	Herb
Theaceae	1	Shrub
Xyridaceae	1	Herb
Totals 96 families, 825 species		

Source: Programa Flora computerized data base of Amazonian herbaria

and Caesalpiniaceae (28)—are almost all trees, with the exception of 3 species of vines in the Fabaceae and many shrubs in the Rubiaceae. Many of the families below the top ten have a more even distribution between growth forms. The three most common epiphyte families are Orchidaceae (24), Araceae (9), and Bromeliaceae (6), although each total includes terrestrial species. Epiphytes are not a strong component of the central Amazonian flora except in areas of caatinga forest on white sand soils. Three families, the Burmanniaceae, Gentianaceae, and Triuridaceae, have saprophytes, with two saprophytic species each. The most species-rich genera are given in table 8.2. Interestingly, nearly all are tree genera, not shrubs or epiphytes, as in the other areas.

The important families in the Reserva Ducke are the same as the those in the MCSE project, where the families with the greatest numbers of tree species are Sapotoceae (83 species), Chrysobalanaceae (51), Burseraceae (49), Annonaceae (39), and Rubiaceae (30) (Rankin de Merona et al., in press). The MCSE figure for Rubiaceae is lower than that of the Reserva Ducke because the MCSE project includes only trees of 10 cm DBH or more, whereas many Rubiaceae are shrubs.

Many species of Caryocaraceae, Chrysobalanaceae, Dichapetalaceae, and Lecythidaceae are known to occur in terra firme forest of the Reserva Ducke (see figs. 8.2–11). Most are widespread within Amazonia, and their overall distribution almost duplicates the Amazonian phytogeographic region. Five main distribution patterns, however, are shown by species that occur in the Reserva Ducke: (1) widespread Amazonian species, some of which extend beyond the region to Central America or Atlantic coastal Brazil; (2) western Amazonian species that are at the extreme eastern part of their range around Manaus; (3) eastern Amazo-

TABLE 8.2 Species-rich genera of Reserva Ducke

Genus	Species (*N*)	Habit	Family
Licania	21	Trees	Chrysobalanaceae
Inga	17	Trees	Mimosaceae
Protium	14	Trees	Burseraceae
Eschweilera	13	Trees	Lecythidaceae
Swartzia	13	Trees	Fabaceae
Aniba	12	Trees	Lauraceae
Miconia	12	Small trees	Melastomataceae
Ocotea	11	Trees	Lauraceae
Casearia	10	Shrubs, trees	Flacourtiaceae
Couepia	10	Trees	Chrysobalanaceae
Palicourea	10	Shrubs	Rubiaceae
Pouteria	10	Trees	Sapotaceae
Sloanea	10	Trees	Elaeocarpaceae
Virola	10	Trees	Myristicaceae
Bactris	9	Shrubs	Arecaceae
Iryanthera	8	Trees	Myristicaceae
Brosimum	7	Trees	Moraceae
Hirtella	7	Shrubs, trees	Chrysobalanaceae
Guatteria	7	Trees	Annonaceae
Micropholis	7	Trees	Sapotaceae
Duguetia	6	Trees	Annonaceae
Geonoma	6	Shrubs	Arecaceae
Lecythis	6	Trees	Lecythidaceae
Licaria	6	Trees	Lauraceae
Mouriri	6	Trees	Melastomataceae
Trichomanes	6	Herb, epiphytes	Hymenophyllaceae

nian species that are at the western extreme of their range; (4) species that extend from the Guianas to Manaus, but not south of the Amazon River; (5) Central Amazonian endemics; and (6) Manaus endemics (see table 8.3). There are also a few disjunctions involving Manaus species.

The distribution of all 37 terra firme species of Chrysobalanceae and 27 of Lecythidaceae that occur in the Reserva Ducke are analyzed, also showing their occurrence in other 5° × 5° squares. The greater number of Manaus species of Chrysobalanaceae occur to the west and in the upper Rio Negro region, indicating a much stronger relationship of the Central Amazon flora to that of the upper Rio Negro and western Amazonia for that family (fig. 8.2). A strong Guianas relationship in the Manaus species of Lecythidaceae is evident, however (fig. 8.3).

As would be expected, the várzea species of the Reserva Ducke tend to be widespread with no local endemism (fig. 8.4). The white sand

TABLE 8.3 Distribution patterns of terra firme forest species from selected plant families in Reserva Ducke

1. Widespread species
Caryocar glabrum (Aubl.) Persoon
Connarus ruber (P. & E.) Planchon
Couepia Bracteosa Benth.
C. canomensis (Mart.) Benth. ex Hook.f.
C. guianensis Aubl.
Couratari multiflora (Smith) Eyma
C. stellata A. C. Smith
Eschweilera coriacea (DC.) Mart. ex Berg
E. decolorans Sandw.
E. micrantha (Berg) Miers
E. parviflora (Aubl.) Miers
E. pedicellata (Richard) Mori
Hirtella hispidula Miq.
H. physophora Mart. & Zucc.
Lecythis pisonis Cambess.
L. zabucaja Aubl.
Licania apetala (E. Mey.) Fritsch
L. canescens R. Ben.
L. caudata Prance
L. egleri Prance
L. heteromorpha Benth.
L. hypoleuca Benth.
L. longistyla (Hook.f.) Fritsch
L. micrantha Miq.
L. octandra (Hoffmgg. ex R. & S.) Kuntze

2. Western species
Caryocar pallidum A. C. Smith
Couepia elata Ducke
C. longipendula Pilg.
Eschweilera tessmannii R. Knuth
Hirtella duckei Hub.
H. rodriguesii Prance
Licania bracteata Prance
L. unguiculata Prance

3. Eastern Species
Couepia robusta Hub.
Eschweilera apiculata (Miers) A. C. Smith
Hirtella piresii Prance
Licania impressa Prance
L. rodriguesii Prance

4. Manaus-Guiana species
Connarus grandifolius Planchon
Corythophora rimosa Rodrigues
Eschweilera collina Eyma
E. subglandulosa (Steud. ex Berg) Miers
Lecythis poiteaui Berg
L. coriacea Benth.
L. laxiflora Fritsch

L. reticulata Prance
Parinari excelsa Sabine
P. montana Aubl.
Pseudoconnarus macrophyllus (P. & E.) Radlk.
Rhabdodendron amazonicum (Spruce ex Benth.) Huber
Tapura amazonica P. & E.
T. guianensis Aubl.

5. Manaus endemics
Couepia magnoliifolia Benth. ex Hook.f.
Eschweilera atropetiolata Mori
E. amazonicaformis Mori
E. rodriguesiana Mori
Hirtella fasciculata Prance
Lecythis prancei Mori

6. Central Amazonia
Gustavia elliptica Mori
Hirtella myrmecophila Pilg.
Lecythis retusa Spruce ex Berg

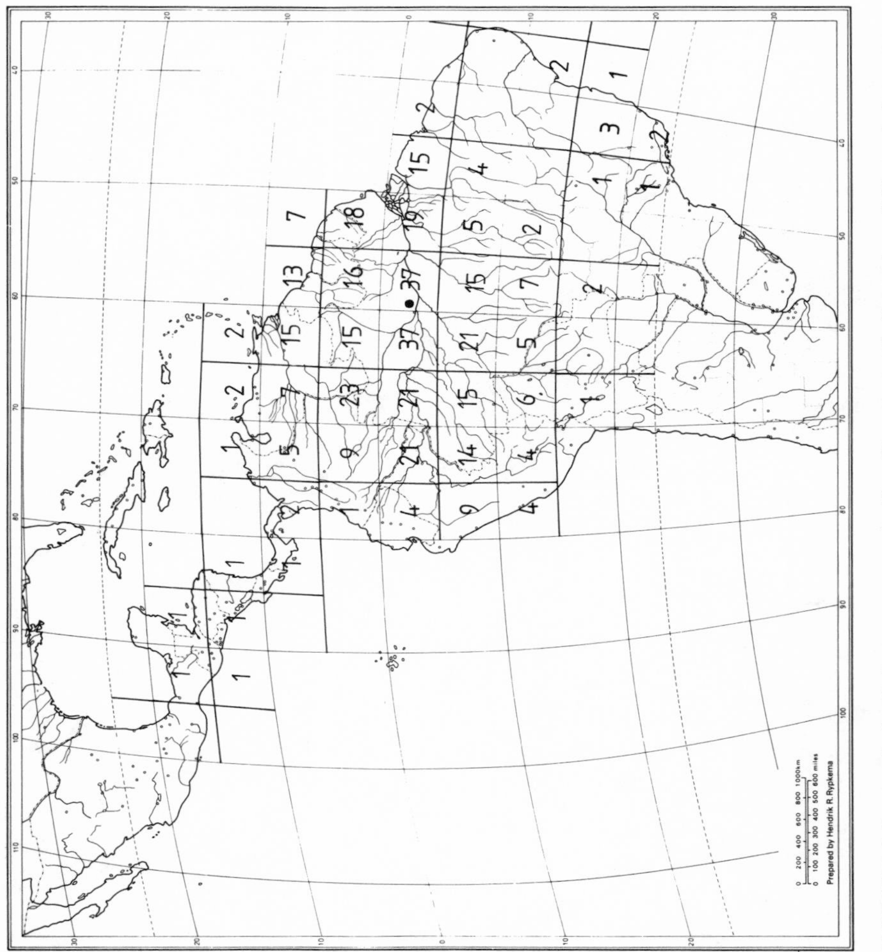

FIGURE 8.2. Analysis of the occurrence in 5° × 5° squares of all species of Chrysobalanaceae known to be present in Reserva Ducke.

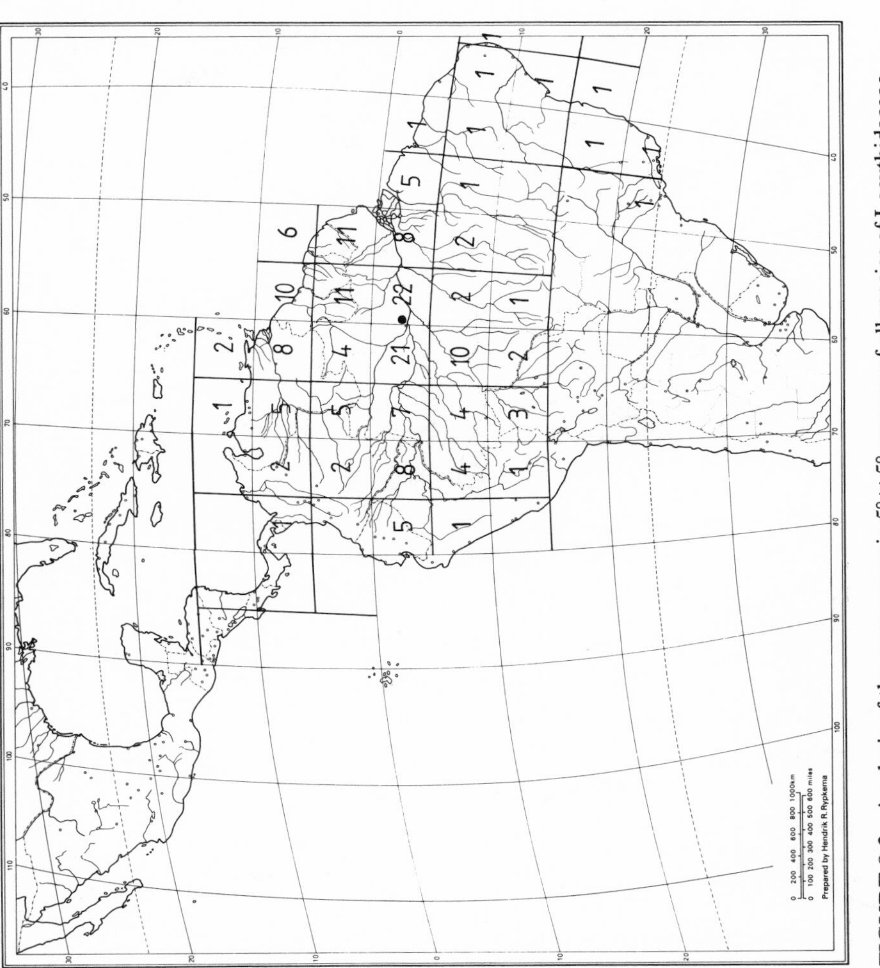

FIGURE 8.3. Analysis of the occurrence in 5° × 5° squares of all species of Lecythidaceae known to be present in Reserva Ducke.

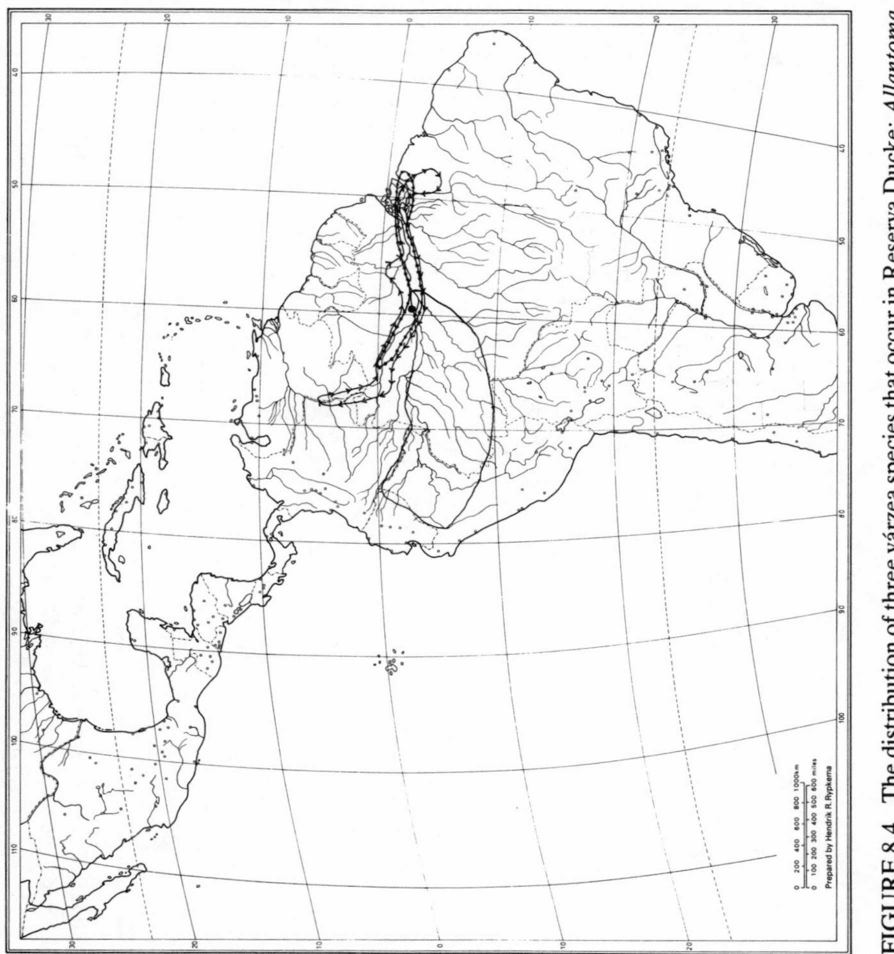

FIGURE 8.4. The distribution of three várzea species that occur in Reserva Ducke: *Allantoma lineata* (Mart. ex Berg) Miers (Lecythidaceae), (line with arrows), *Couepia ulei* Pilg. (Chrysobalanaceae) (solid line), and *Eschweilera amazonica* Kunth (Lecythidaceae) (beaded line).

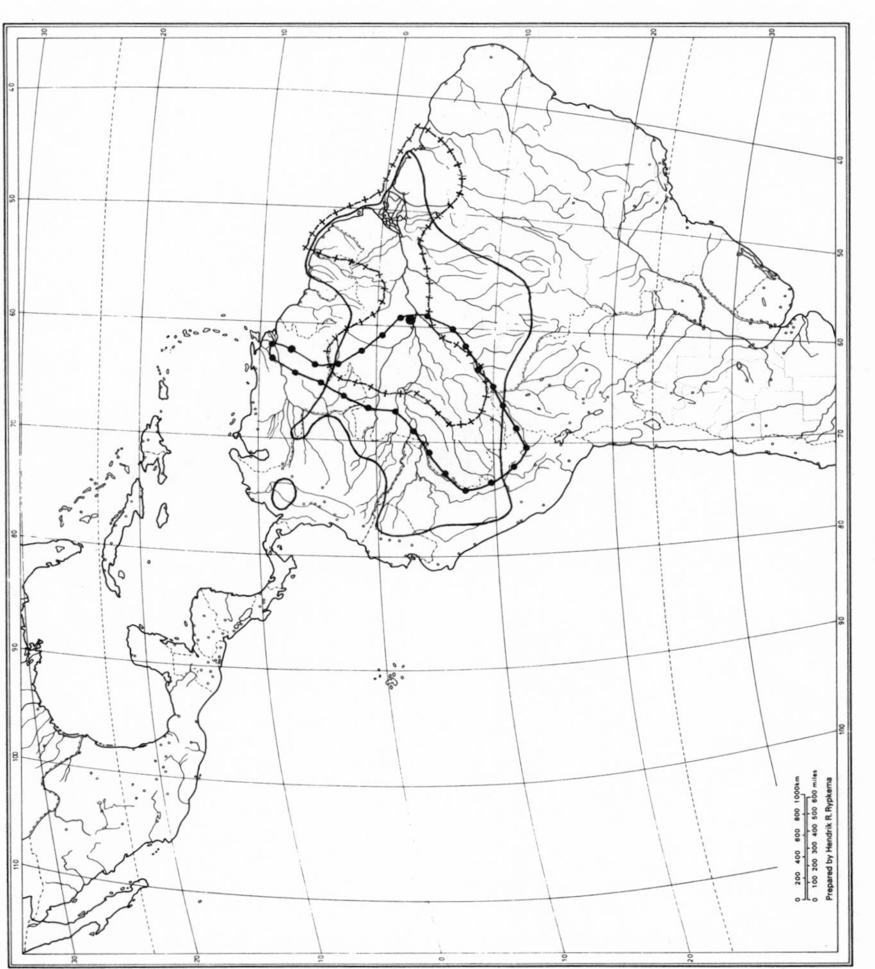

FIGURE 8.5. Distribution of the three species of *Caryocar* that occur in Reserva Ducke: *Caryocar glabrum* (Aubl.) Pers. (solid line), *C. pallidum* A. C. Smith (beaded line) and *C. villosum* (Aubl.) Pers. (line with crosses).

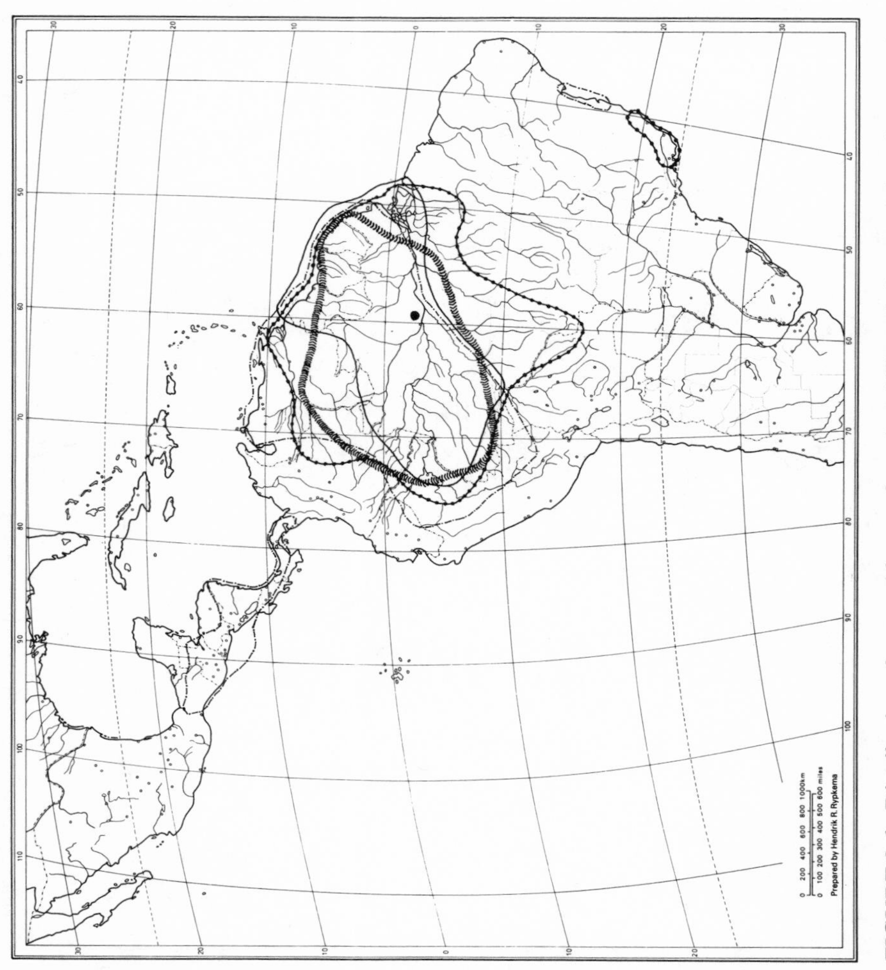

FIGURE 8.6. Distribution of some widespread species of Chrysobalanaceae that also occur in Reserva Ducke: *Couepia bracteosa* Benth. (solid line) *Hirtella physophora* Mart. (hatched line), *Licania hypoleuca* Benth. (dot & dash line), and *Parinari excelsa* Sabine (beaded line).

caatinga species are mostly widespread, although some local endemics, such as *Rhabdodendron macrophyllum*, also occur.

Some examples of widespread species include all three species of *Caryocar* (fig. 8.5), *Eschweilera coriacea* (DC.) Mart. ex Berg, which is found in Panama, Colombia, and throughout Amazonia; *Licania hypoleuca* Benth. (fig. 8.6), which ranges from Mexico to Atlantic coastal Brazil; *Licania longistyla* Hook. f., in Panama and throughout Amazonia; and *Tapura amazonica* P. & E., a common species of Dichapetalaceae in Amazonia and the Planalto of central Brazil. Most of the larger plant families represented in the reserve have a few of these extremely widespread species.

Manaus species share the strongest relationship with western and northwestern Amazonia (fig. 8.7). Typical species include *Licania unguiculata* Prance and *Hirtella rodriguesii* Prance, which are both distributed from west of Iquitos, Peru, to Manaus. This pattern is found in all four families studied and occurs in many others.

Several species, such as *Couepia robusta* Hub., *Eschweilera apiculata* (Miers) A. C. Smith, and *Hirtella piresii* Prance, are eastern Amazonian species that reach the westernmost part of their range around Manaus (fig. 8.8). Considerably fewer species, however, have this type of range than from the west. One reason may be the drier climate belt that divides eastern Amazonia and passes through the Santarém and Trombetas river region.

A strong element of the Manaus flora is species that extend to the Guianas, especially French Guiana (fig. 8.9). Examples include *Corythophora rimosa* Rodrigues, *Lecythis poiteaui* Berg, and *Licania laxiflora* Fritsch. These species do not extend south of the Amazon River.

The forests north of Manaus have been pinpointed as a regional center of endemism by Prance (1974, 1982), and in spite of recent collecting efforts, many rainforest species remain known only from the region of Manaus (figs. 8.10, 8.11). Examples include *Eschweilera atropetiolata* Mori, *E. amazonicaformis* Mori, and *Hirtella fasciulata* Prance. *Anaxagorea manausensis* Timmerman is known only from the Reserve Ducke (Maas and Westra 1985).

For the Chrysobalanaceae, 40 species have been collected in the Reserva Ducke, 32 of which are also present in the MCSE reserves. Fifty-eight species of Chrysobalanaceae have been collected in the MCSE reserves, including 26 that have not been collected in the Reserva Ducke. Eighteen of these 26 have distributions strongly centered in the Guianas. The species found in the MCSE reserves show the same patterns as those

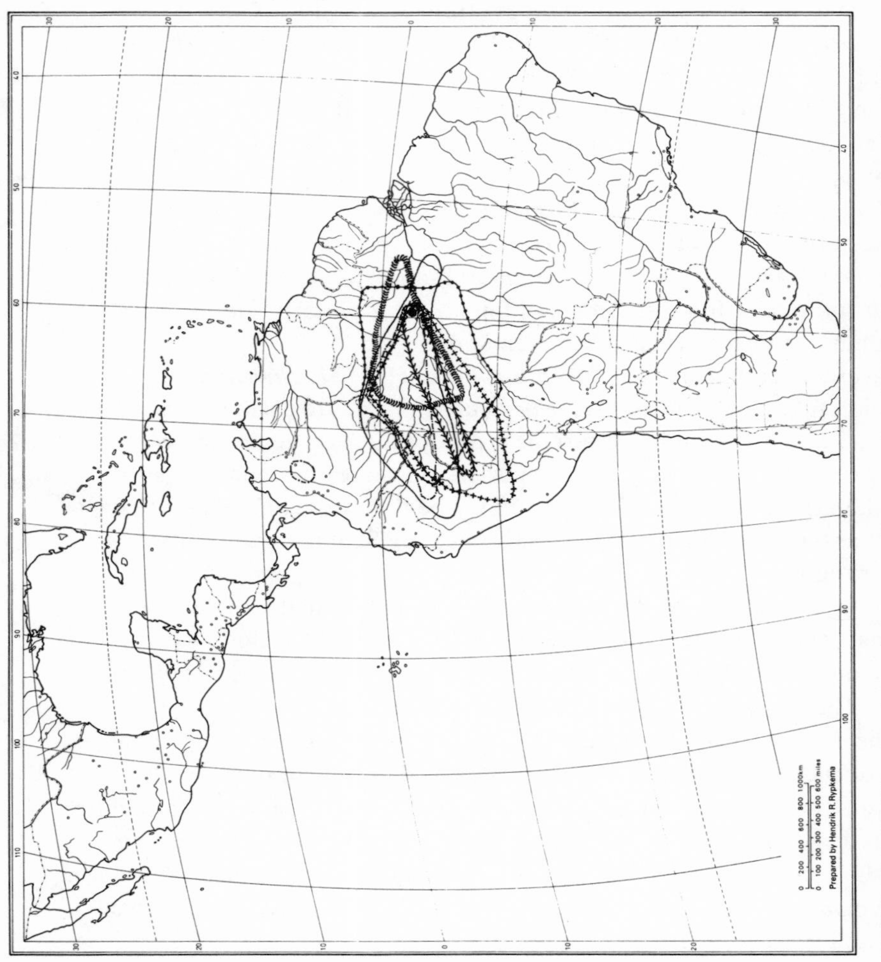

FIGURE 8.7. Western Amazonian terra firme species of Chrysobalanaceae that also reach as far east as Reserva Ducke: *Couepia elata* Ducke (line with arrows), *Couepia longipendula* Pilg. (hatched line), *Hirtella duckei* Pilg. (beaded line), *H. rodriguesii* Prance (crosses), *Licania bracteata* Prance (dot & dash line), and *L. unguiculata* Prance (solid line).

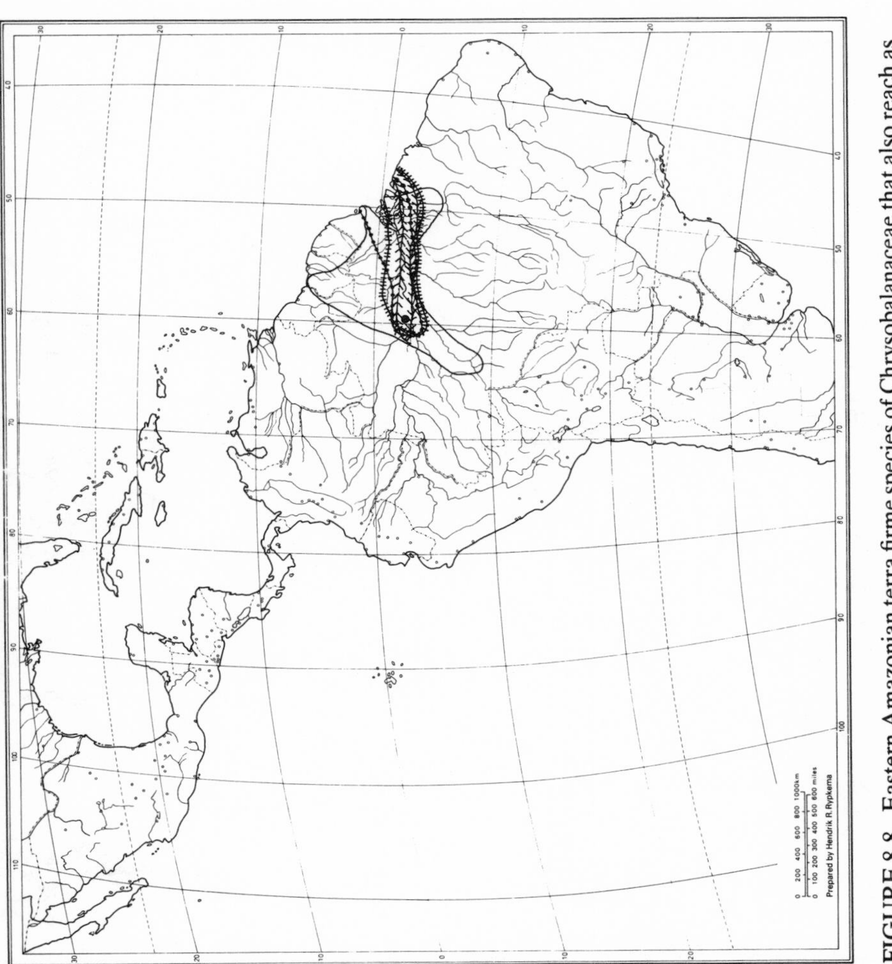

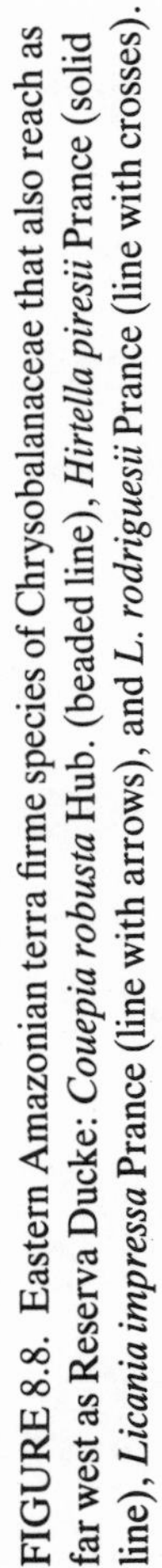

FIGURE 8.8. Eastern Amazonian terra firme species of Chrysobalanaceae that also reach as far west as Reserva Ducke: *Couepia robusta* Hub. (beaded line), *Hirtella piresii* Prance (solid line), *Licania impressa* Prance (line with arrows), and *L. rodriguesii* Prance (line with crosses).

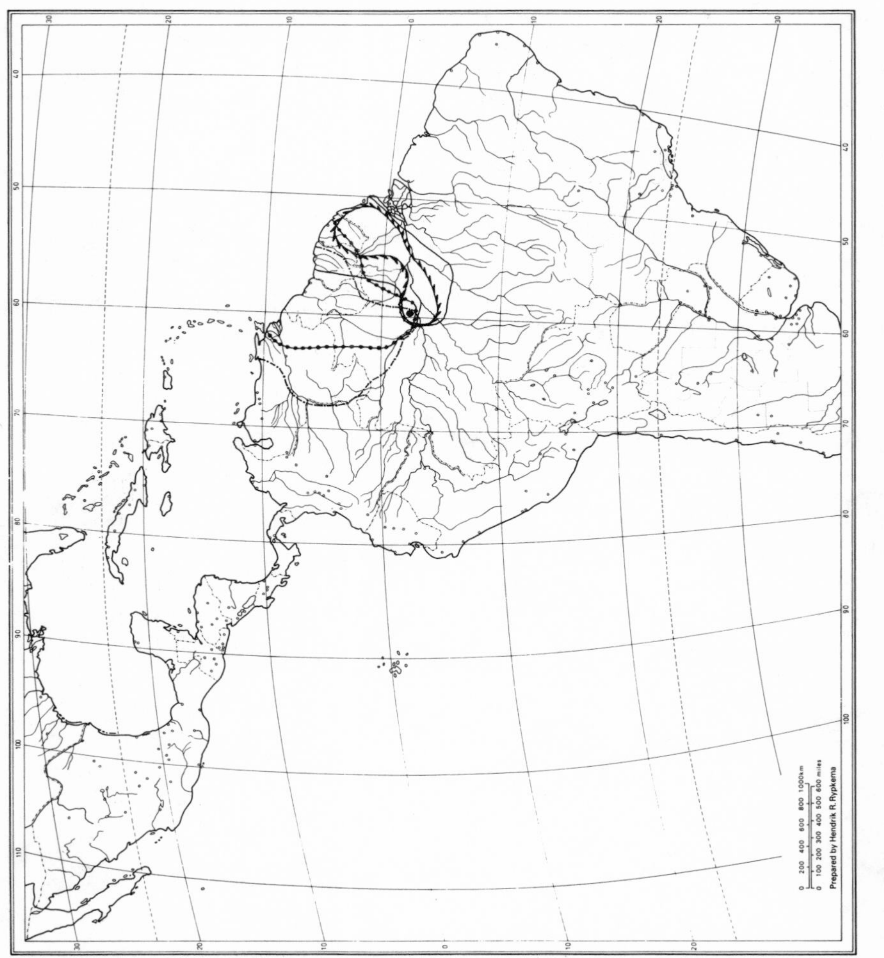

FIGURE 8.9. Guianan species of Lecythidaceae that occur in Reserva Ducke: *Corythophora rimosa* Rodr. (line with fins), *Eschweilera collina* Eyma (beaded line), *E. subglandulosa* (Steud. ex Berg) Miers (dot & dash line), and *Lecythis poiteaui* Berg (solid line).

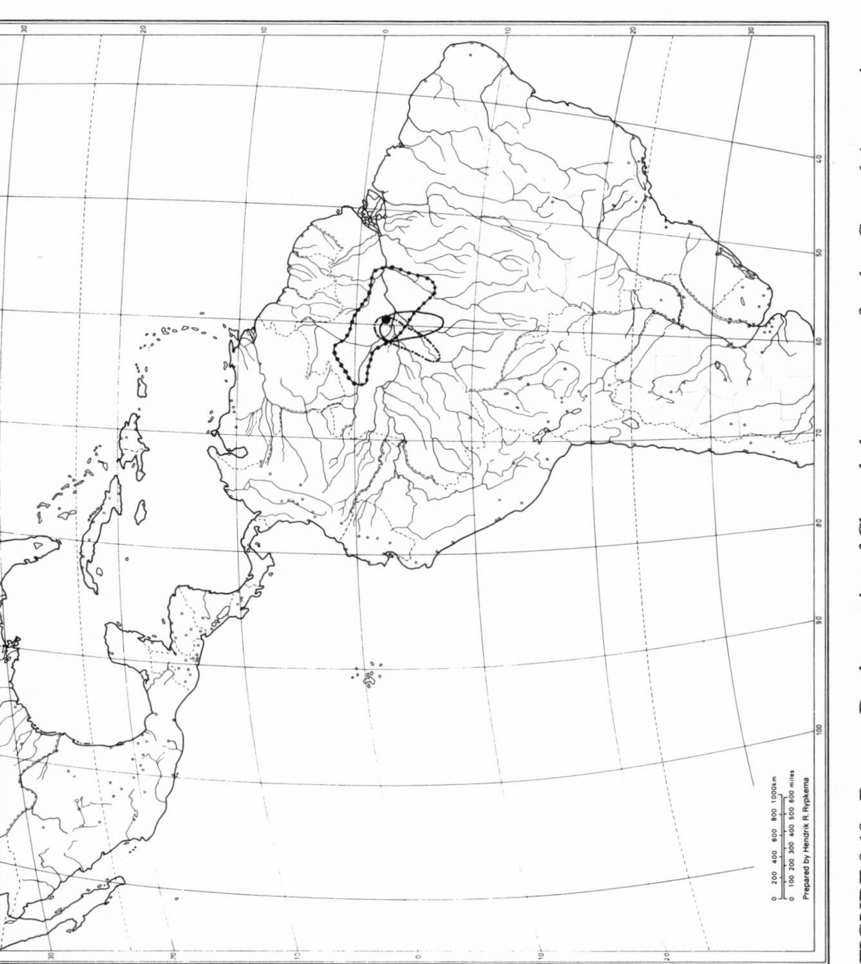

FIGURE 8.10. Reserva Ducke species of Chrysobalanaceae confined to Central Amazonia: *Hirtella fasciculata* Prance (solid line), *H. myrmecophila* Pilg. (beaded line), and *Couepia magnoliifolia* Benth. ex Hook.f. (dot & dash line).

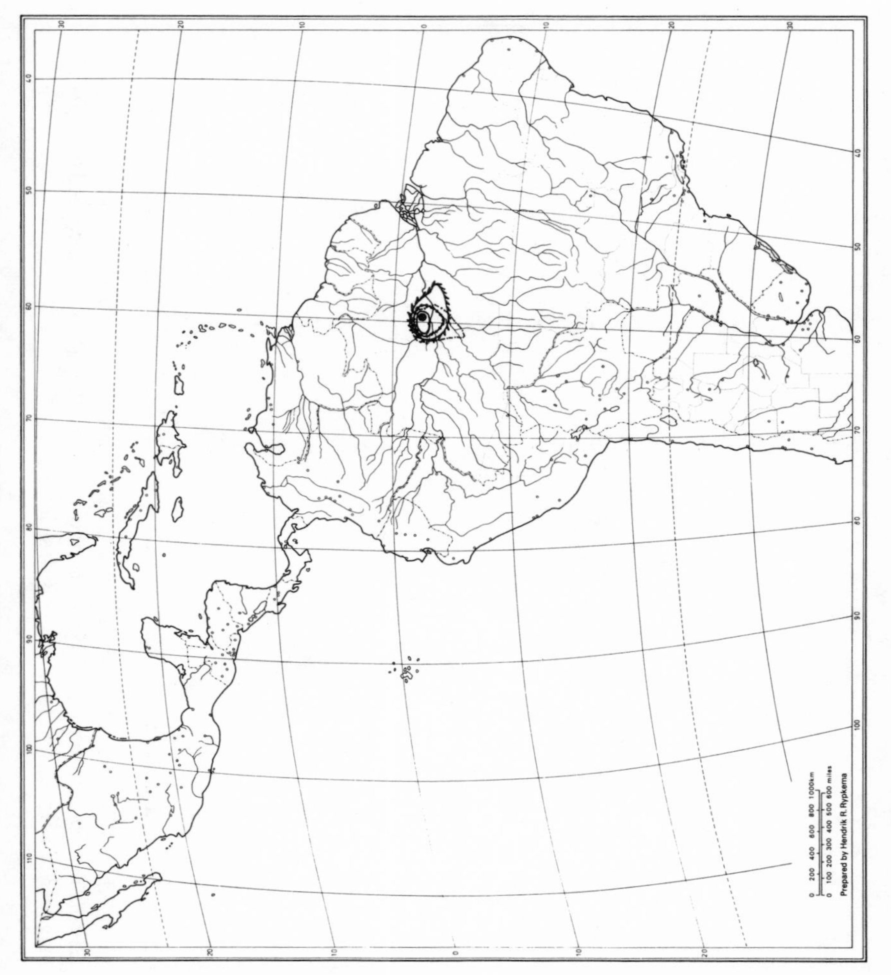

FIGURE 8.11. Reserva Ducke species of Lecythidaceae confined to Central Amazonia: *Eschweilera amazonicaformis* Mori (solid line), *E. atropetiolata* Mori (beaded line), *E. rodriguesiana* Mori (solid line), *Gustavia elliptica* Mori (dot & dash line), *Lecythis prancei* Mori (solid line), and *L. retusa* Spruce ex Berg. (line with fins).

in the Reserva Ducke except that there is a much stronger Guiana element. Some typically Guianan species of Chrysobalanaceae, for example, do not occur in the Reserva Ducke but are found in the MCSE reserves: *Couepia caryophylloides* R. Ben., *Hirtella obidensis* Ducke, *Licania majuscula* Sagot, *L. rufescens* Kl. ex Fritsch, *L. sandwithii* Prance, and *Parinari campestris* Aubl. These are all Guianan species, and they show a trend that is present in the species from other families (see also table 8.4). The link between the MCSE reserves and the Guianan flora is as strong as that with northwestern Amazonia.

Data also come from the forest inventory of 1 ha of forest at the Centro de Pesquisas de Cacau (CEPEC) reserve about 2 km outside the Reserva Ducke; the CEPEC forest is physiognomically identical to the reserve forest. The results of that inventory were reported in Prance et al. (1976), but computer analyses of the relative dominance, family importance values, and other measures were lacking. For summary of these values, see tables 8.5 and 8.6. The families that dominate the forest in terms of importance differ slightly from those dominating in absolute number of species. By far the most important family is the Lecythidaceae, with an importance value (FIV) of 51.72, or 17.24% of the total FIV of 300. Half of the FIV is contributed by just 7 families: Lecythidaceae (51.72), Moraceae (24.46), Sapotaceae (20.41), Burseraceae (19.22), Caesalpiniaceae (16.37), Chrysobalanaceae (13.93), and Vochysiaceae (12.72). Most of the important families in the sample are also those with high species numbers in the Reserva Ducke, with the exception of the Vochysiaceae, which is important because of the large size of the trees rather than the number of species or relative frequency of individuals (table 8.5).

Eleven woody plant families emerge as most important in the central Amazonian forest on the basis of combined consideration of inventory and collection data: Sapotaceae (48 species, FIV 20.41), Lecythidaceae (30, 51.72), Moraceae (35, 24.46), Caesalpiniaceae (28, 16.37), Burseraceae (18, 19.22), Chrysobalanaceae (40, 13.93), Lauraceae (38, 11.01), Melastomataceae (26, 9.97), Mimosaceae (33, 8.53), Fabaceae (43, 7.47), and Annonaceae (36, 7.65). At the species level, *Eschweilera odora* is by far the most abundant in the forest on the 1-ha inventory site (table 8.7).

The abundance of different families on the Reserva Ducke is very similar to that of the MCSE reserves. Rankin-de Merona et al. (in press) found that, based on the number of individuals present in a 10-ha plot, Burseraceae, Sapotaceae, Leguminosae, Lecythidaceae, Lauraceae, and

TABLE 8.4 Comparison of Chrysobalanaceae species of Reserva Ducke and MCSE Reserves and their distribution

	Ducke	MCSE	Distribution
Couepia bracteosa Benth.	+	+	Widespread
C. canomensis (Mart.) Benth. ex Hook.f.	+	+	Widespread
C. caryophylloides R. Ben	−	+	Guianas
C. cideana Prance	−	+	E. Amazonia
C. elata Ducke	+	+	W. Amazonia
C. excelsa Ducke	−	+	Guianas
C. glabra Prance	−	+	Manaus endemic
C. guianensis Aubl.	+	+	Guianas, w. Amazonia
C. habrantha Standley	−	+	
C. longipendula Pilg.	+	+	W. Amazonia
C. magnoliifolia Benth. ex Hook.f.	+	+	Manaus endemic
C. obovata Ducke	+	+	
C. parillo DC.	−	+	Guiana, w. Amazonia
C. racemosa Benth. ex Hook.f.	+	−	
C. robusta Hub.	+	+	E. Amazonia
C. sandwithii Prance	−	+	Guianas
C. ulei Pilg.	+	+	E. Amazonia
Hirtella bicornis Mart. & Zucc.	−	+	Widespread
H. duckei Hub.	+	−	W. Amazonia-shrub
H. fasciculata Prance	+	−	Manaus endemic
H. hispidula Miq.	+	+	Widespread
H. myrmecophila Pilg.	+	−	Widespread
H. obidensis Ducke	−	+	
H. physophora Mart. & Zucc.	+	−	Widespread-shrub
H. piresii Prance	+	−	E. Amazonia
H. rodriguesii Prance	+	+	W. Amazonia
Licania affinis Fritsch	−	+	Guianas, Central America
L. apetala (E. Mey.) Fritsch	+	+	Widespread
L. blackii Prance	−	+	
L. bracteata Prance	+	+	W. Amazonia

L. canescens R. Ben.	+	+	Widespread
L. caudata Prance	+	+	Widespread
L. coriacea Benth.	+	–	Manaus-Guiana
L. egleri Prance	+	+	Widespread
L. elliptica Standl.	–	+	
L. gracilipes Taub.	+	–	
L. guianensis Aubl.	–	+	Guianas, W. Amazonia
L. heteromorpha Benth.	+	+	Widespread
L. hirsuta Prance	+	+	
L. hypoleuca Benth.	+	+	Widespread
L. impressa Prance	+	+	E. Amazonia
L. kunthiana Hook.f.	–	+	Widespread
L. latifolia Benth.	–	+	Widespread
L. laxiflora Fritsch	+	+	Manaus-Guiana
L. longistyla (Hook.f.) Fritsch	+	+	Widespread
L. majuscula Sagot	–	+	Guianas
L. micrantha Miq.	+	+	Widespread
L. minutiflora (Sagot) Fritsch	–	+	Guianas
L. niloi Prance	+	+	Widespread
L. oblongifolia Standl.	+	+	Widespread
L. octandra (Hoffmgg. ex. R. & S.) Kuntze	+	+	Widespread
L. pallida Spr. ex Sagot	+	+	Widespread
L. polita Benth.	–	+	Guianas
L. reticulata Prance	+	+	Widespread
L. rodriguesii Prance	+	+	E. Amazonia
L. refescens Klotzsch	–	–	Guianas
L. sandwithii Prance	–	–	Guianas
L. sprucei Hook.f.	–	+	Guianas
L. unguiculata Prance	+	+	W. Amazonia
Parinari campestris Aubl.	–	+	Guianas
P. excelsa Sabine	+	+	Widespread
P. montana Aubl.	+	+	Widespread
P. parvifolia Sandw.	–	+	Guianas
P. rodolphii Hub.	–	+	Guianas

TABLE 8.5 Family data for trees ≥ 15 cm DBH on 1 ha of terra firme forest at Manaus-Itacoatriara Highway km 30

Family	Species (N)	Trees (N)	Basal area	Relative frequency	Relative density	Relative dominance	FIV
Annonaceae	7	8	2,911	4.22	2.25	1.18	7.65
Apocynaceae	3	5	3,046	1.80	1.40	1.24	4.14
Arecaceae	1	8	1,824	0.60	0.28	0.74	1.62
Bombacaceae	2	10	9,457	1.20	3.09	3.84	8.13
Burseraceae	10	27	13,835	6.02	7.58	5.62	19.22
Caesalpinaceae	9	26	13,310	4.74	8.74	5.41	18.89
Caryocaraceae	1	1	6,079	0.60	0.28	2.47	3.35
Cecropiaceae	1	1	227	0.60	0.28	0.09	0.97
Chrysobalanaceae	10	16	8,414	6.02	4.49	3.42	13.93
Clusiaceae	2	2	428	1.20	0.56	0.17	1.93
Combretaceae	1	1	572	0.60	0.28	0.23	1.11
Dichapetalaceae	1	2	2,005	0.60	0.56	0.81	1.97
Duckeodrendaceae	1	6	10,397	0.60	1.69	4.23	6.52
Elaeocarpaceae	2	2	3,066	1.80	2.25	1.25	5.30
Euphorbiaceae	4	5	2,750	2.41	1.40	1.12	4.93
Fabaceae	3	6	8,302	1.80	1.69	3.37	6.86
Flacourtiaceae	3	3	4,918	1.80	2.25	2.22	6.27
Humiriaceae	4	5	6,571	2.41	3.81	2.67	8.89
Lauraceae	7	10	9,804	4.22	2.81	3.98	11.01
Lecythidaceae	18	68	53,590	10.84	19.10	21.78	51.72
Melastomataceae	5	8	5,655	5.42	2.25	2.30	9.97
Meliaceae	4	8	4,202	2.41	2.25	1.71	6.37
Mimosaceae	8	8	3,602	4.82	2.25	1.46	8.53
Monimiaceae	2	2	630	1.20	0.56	0.26	2.02
Moraceae	13	27	18,488	7.83	8.15	7.51	23.49
Myristicaceae	3	4	1,737	1.80	1.12	0.71	3.63
Myrtaceae	4	7	2,981	2.41	1.97	1.21	5.59

Nyctaginaceae	2	11	6,560	1.20	2.81	2.67	6.68
Ochnaceae	1	2	1,690	0.60	0.56	0.69	1.85
Olacaceae	2	4	5,819	1.20	1.12	2.36	4.68
Quiinaceae	1	1	254	0.60	0.28	0.10	0.98
Rubiaceae	3	3	1,813	1.80	0.84	0.74	3.38
Sapindaceae	2	2	2,189	1.20	0.56	0.89	2.65
Sapotaceae	12	28	14,533	6.63	7.87	5.91	20.41
Simaroubaceae	1	2	254	0.60	0.28	0.10	0.98
Sterculiaceae	1	1	227	0.60	0.28	0.09	0.97
Violaceae	3	6	2,839	1.80	1.97	0.97	4.74
Vochysiaceae	7	10	11,825	4.22	3.09	4.81	12.12
TOTAL	165	346	246,801	100.00	100.00	100.00	300.00

TABLE 8.6 FIV, in descending order, for 1 ha of forest, Manaus-Itacoatiara Highway km 30, and total number of species in Reserva Ducke

Family	Species (*N*)	FIV *I*	Cumulative *Total FIV*	Reserva *Ducke Species* (*N*)
Lecythidaceae	18	51.72	51.72	30
Moraceae	13	24.46*	76.18	35
Sapotaceae	11	20.41	96.59	48
Burseraceae	10	19.22	115.81	18
Caesalpiniaceae	7	16.37	132.18	28
Chrysobalanaceae	10	13.93	146.11	40
Vochysiaceae	8	12.72	158.83	8
Lauraceae	7	11.01	169.84	38
Melastomataceae	5	9.97	179.81	26
Mimosaceae	8	8.53	188.34	33
TOTAL		188.34		

Note: * Includes 0.97 for Cecropiaceae

TABLE 8.7 Most important species of 1-ha inventory at Manaus-Itacoatiara Highway km 30

	Individuals (*N*)	Importance
Eschweilera odora (Poepp. ex Berg) Miers	26	16.94
Scleronema micranthum (Ducke) Ducke	10	9.10
Duguetia caudata R. E. Fries	1	7.69
Duckeodendron cestroides Kuhlm.	6	7.51
Corythophora rimosa Rodrigues	6	6.90
Eperua bijuga Mart. ex Benth.	7	6.47
Sapotaceae 1	6	6.02
Oenocarpus bacaba Mart.	9	5.19
Neea cf. altissima P. & E.	5	4.30
Brosimum parinarioides Ducke	4	4.27
Minquartia guianensis Aubl.	3	4.00

Moraceae were the six most common families. All of these families rank high in FIV and abundance in the Reserva Ducke.

Various biological studies have been carried out in the study site, though a detailed analysis has not yet been made. The dispersal of all trees on the 1-ha study site has been studied, but the dissertation with that information is as yet unpublished (F. Ehrendorfer, p.c.). In conclusion,

TABLE 8.8 Bark exudates in 1-ha sample of forest containing 571 trees of ≥ 5 cm diameter

Type	*Frequency*	*Percent*
none	379	66.73
resin	68	11.97
latex	60	10.56
phenolic	46	8.09
oil	7	1.23
scented wood resin	2	0.35
orange wood latex	2	0.35
essential oil	1	0.18
yellow phenol	1	0.18
red phenol	1	0.18
viscous white latex	1	0.18
TOTAL	568	100.00
Total with bark exudates		33.66

however, by far the greatest number of tree species are adapted to short-range dispersal mainly by dropping their diaspores to the ground, with secondary dispersal then being effected by agoutis and other rodents on the forest floor. Interestingly, the striking exceptions to this rule are in species that reach above the canopy, mainly lianas in such families as Bignoniaceae and Malpighiaceae, and in emergent trees, where wind dispersal predominates. There is an especially notable difference in dispersal mechanisms of emergent trees and those of the canopy and subcanopy. The most common emergents in the Lecythidaceae, for example, are wind-dispersed *Cariniana micrantha* Ducke and *Couratari tauari* Berg, whereas canopy and subcanopy species of that family are dispersed by mammals. In the Apocynaceae, the emergent is an *Aspidosperma*, which also has winged seeds. Species of Vochysiaceae in the area are also emergents with winged seeds.

For the obvious bark exudates of the tree species of 5 cm diameter or more in the 1-ha sample, see table 8.8. In the 33.66% of the trees with exudates, resins (11.91%) are slightly more common than latex (10.51%). These data show the importance of these bark chemicals as predator defense in tropical rainforest.

The Reserva Ducke, in the center of Amazonia, is also at the crossroads of many plant distribution patterns. It is thus particularly interesting for floristic study. The planned florula of the reserve will be useful not only for research around Manaus but throughout Amazonia.

REFERENCES

Alencar, J. C., R. A. Almeida, and N. P. Fernandes. 1979. Fenologia de espécies florestais tropical úmida de terra firme na Amazônia Central. Acta Amazônica 9(1): 163–198.

Araújo, V. C. 1979. Fenologia de essências florestais amazônicas 1. Boletin do INPA 4: 1–25.

Maas, P. J. M. and L. Y. Th. Westra. 1985. Studies in Annonaceae. II. A monograph of the genus *Anaxagorea* A. St. Hil., pt. 2. Bot. Jahrb. Syst. 105: 145–204.

Mori, S. A., B. M. Boom, A. M. de Carvalho, and T. S. dos Santos. 1983. Southern Bahian moist forests. Bot. Rev. 49: 155–232.

Prance, G. T. 1973. Phytogeographic support for the theory of Pleistocene forest refuges in the Amazon Basin, based on evidence from distribution patterns in Caryocaraceae, Chrysobalanaceae, Dichapetalaceae and Lecythidaceae. Acta Amazônica 3(3): 5–28.

———. 1982. Forest refuges: evidence from woody angiosperms. In G. T. Prance (ed.), *Biological Diversification in the Tropics*. Columbia Univ. Press, New York, pp. 137–158.

Prance, G. T. and S. A. Mori. 1979. The actinomorphic-flowered Lecythidaceae, Lecythidaceae 1. Flora Neotropica 21: 1–268.

Prance, G. T., W. A. Rodrigues, and M. F. da Silva. 1976. Inventário florestal de um hectare de mata de terra firme, km 30 da Estrada Manaus-Itacoatiara. Acta Amazônica 6(1): 9–35.

Rankin-de Merona, J. et al. In press. Preliminary tree inventory results from upland rainforest of the Central Amazon. Acta Amazônica.

9

Floristic Similarities and Differences between Southern Central America and Upper and Central Amazonia

ALWYN H. GENTRY

Although I originally intended to focus on the differences between Central American and Amazonian floras, after analyzing the available data I am more impressed with their similarities. There are indeed many more tree species in Amazonia, but their absence is compensated for in Central America by more kinds of epiphytes, herbs, and shrubs. Nevertheless, each of the four Neotropical rainforests considered in this volume has distinctive and interesting floristic features. Generalities about "tropical forests" or "Neotropical forests" must thus be made with extreme caution.

Here I will attempt to compare the floras of the four study areas with several data sets compiled from Hammel (this volume, p.c.), Foster (this volume, p.c.), Croat (1978), and Prance (this volume, p.c.). These comparisons are: (1) the largest families at each site; (2) the families present (or absent) at each site; (3) asymmetrical distributions in widespread families; (4) the largest genera at each site; (5) habit composition of the four floras; and (6) tree plot data. After noting some of the between-site similarities and differences in each parameter, I will try to explain why some of these differences exist.

It is increasingly clear that Neotropical forests are put together very similarly floristically, generally with the same families speciose at all sites (table 9.1). The dozen largest families are virtually the same at all four

I thank the National Science Foundation (BSR-8607113), National Geographic Society (3783-88), and Mellon Foundation for current support of the ongoing field and herbarium studies on which this overview is based, and T. Plowman and G. Prance for review comments. Plant lists and other data for the various sites were kindly made available by R. Foster, B. Hammel, and G. Prance.

TABLE 9.1 Largest families, including weeds, in local florulas

La Selva[1]		BCI[2]		Cocha Cashu[3]		Ducke[4]	
Family	Spp.	Family	Spp.	Family	Spp.	Family	Spp.
(Ferns)	169	Leguminosae	112	Leguminosae	102[5]	Leguminosae	104
Orchidaceae	114	(Mimosaceae)	37	(Mimosaceae)	48	(Mimosaceae)	33
Araceae	99	(Caesalpiniaceae)	16	(Caesalpiniaceae)	16^{+}	(Caesalpiniaceae)	28
Rubiaceae	99	(Fabaceae)	59	(Fabaceae)	28^{+}	(Fabaceae)	43
Leguminosae	79	(Ferns)	102	Moraceae	66	Sapotaceae	48
(Mimosaceae)	31	Orchidaceae	90	Rubiaceae	65	Rubiaceae	42
(Caesalpiniaceae)	10	Gramineae	79	(Ferns)	65	Chrysobalanaceae	40
(Fabaceae)	38	Rubiaceae	66	Orchidaceae	45	Lauraceae	38
Piperaceae	79	Araceae	46	Acanthaceae	44	(Ferns)	37
Melastomataceae	71	Compositae	42	Sapindaceae	40	Annonaceae	36
Gramineae	63	Moraceae	36	Bignoniaceae	38	Moraceae	35
Compositae	48	Melastomataceae	35	Araceae	38	Lecythidaceae	30
Euphorbiaceae	42	Piperaceae	32	Solanaceae	38	Melastomataceae	26
Cyperaceae	32	Bignoniaceae	29	Myrtaceae	37	Palmae	25
Moraceae	31	Cyperaceae	28	Piperaceae	36	Orchidaceae	24
Palmae	30	Euphorbiaceae	28	Sapotaceae	34	Apocynaceae	22
Gesneriaceae	27	Sapindaceae	26	Annonaceae	30	Euphorbiaceae	21
Acanthaceae	27	Solanaceae	23	Euphorbiaceae	29	Myristicaceae	20
Lauraceae	27	Apocynaceae	20	Compositae	27	Burseraceae	18
Solanaceae	26	Malpighiaceae	20	Lauraceae	24	Bignoniaceae	18

Bignoniaceae	24	Bromeliaceae	18	Gramineae	24	Flacourtiaceae	15
Bromeliaceae	23	Flacourtiaceae	16	Malpighiaceae	21	Guttiferae	12
Guttiferae	22	Palmae	16	Palmae	20	Bombacaceae	10
Marantaceae	21			Melastomataceae	20		
				Meliaceae	20		
121 families		118 families		119 families[6]		88 families	
1,668 species		1,320 species		1,370 species[6]		825 species	

Notes: 1. Hammel, p.c. (includes introduced weedy species, unlike Hammel, this volume)
2. Croat 1978
3. Foster, p.c., this volume (includes beach species and aquatics, unlike Foster, fig. 7.2)
4. Prance, this volume
5. 8 generic indets. not assigned to subfamily
6. The total for lowland Rio Manu, including the forest on latosol, is 130 families and 1,856 species

sites, except that weedy taxa (composites, sedges, grasses) are underrepresented in Amazonia and poor-soil specialist trees are overrepresented at Manaus. Half of the top 20 families are shared by all sites—ferns, Orchidaceae, Rubiaceae, Leguminosae, Melastomataceae, Moraceae, Euphorbiaceae, Palmae, and Bignoniaceae. These may be viewed as the predominant Neotropical plant families. Five more families are in the top twenty everywhere except Manaus—Araceae, Gramineae, Piperaceae, Compositae, and Solanaceae.

Moreover, few families are speciose at only one site. Just nine enter the list of twenty most speciose families at one site: Marantaceae and Gesneriaceae at La Selva; Myrtaceae and Meliaceae at Cocha Cashu; and Chrysobalanaceae, Lecythidaceae, Myristicaceae, Burseraceae, and Bombacaceae at Manaus.

Excluding ferns, where familial limits are not agreed upon, 153 vascular plant families are represented in at least one of the four field stations. Nearly half of these (66 families) are present at all four sites, and this number will increase as collecting artifacts are eliminated with more thorough fieldwork, especially in Amazonia. Of the taxa not represented at all sites, the predominant distributional pattern, shown by 29 families, is presence in La Selva, Barro Colorado Island (BCI), and Cocha Cashu but absence at Ducke. Most of these families are herbs, and many are probably included here due to inadequate collecting in Central Amazonia. Some typical families with this distribution are Amaranthaceae, Asclepiadaceae, Cactaceae, Caricaceae, Commelinaceae, Cyclanthaceae, Hippocrateaceae, Labiatae, Malvaceae, Marantaceae, Musacae, Nyctaginacae, Nymphaeaceae, Polygonaceae, Scrophulariaceae, Ulmacae, Verbenaceae, and Vitaceae. The next most predominant pattern is to occur only at La Selva. The 12 families that occur only at La Selva make this the most distinctive site at the family level. These are mostly middle elevation cloud forest families that come down to sea level in the very wet conditions of that field station as they do in the Colombian Choco (Gentry 1986b). Some typical examples are Aquifoliaceae, Caprifoliaceae, Chloranthaceae, Clethraceae, Cruciferae, Ericaceae, Loasaceae, Magnoliaceae, Podocarpaceae, and Symplocaceae. Since the same distributional phenomenon occurs in appropriate ecological conditions in South America (Gentry 1986b), we may assume that this distinctive floristic element at the northernmost station is due to ecology rather than biogeography even though several of these are Laurasian families.

Nine families have been collected only at Ducke. These are Duckeodendraceae, Rapateaceae, Rhabdodendraceae, Peridiscaceae, Triuridaceae, Xyridaceae, Linaceae, Eriocaulaceae, and Styracaceae. Most of these families are poor-soil Guayana shield specialists or are at least most diverse in that region (or in similar regions of the Brazilian shield, in the case of Xyridaceae and Eriocaulaceae).

Seven families are represented only at BCI. Three of these are aquatics—Ceratophyllaceae, Hydrocharitaceae, Typhaceae—not surprising in view of the extensive aquatic habitat provided by Gatun Lake. The others are a miscellaneous assemblage—Actinidaceae, Iridaceae, Trigoniaceae, and Turneraceae. At least the last is clearly a collection artifact, since Turnera is a common weed near Manaus.

In contrast to the other three field stations, Manu has relatively few uniquely represented families. The only five families recorded solely at Cocha Cashu are Balanophoraceae, Buxaceae, Lemnaceae, Rosaceae, and Salicaceae.

All other combinations of sites have only one to three species that are uniquely shared between them, except BCI and Cocha Cashu, which are united by fairly good soil and a pronounced dry season and uniquely share six families (Alismataceae, Rafflesiaceae, Staphyleaceae, Proteaceae, Cochlospermaceae, and Bixaceae). In spite of their geographic proximity, BCI and La Selva uniquely share only three families, aquatic Pontederiaceae and herbaceous predominantly Laurasian Caryophyllaceae and Portulacaceae.

The rest of the families represented at at least one field station include: La Selva plus Cocha Cashu—two families (Cycadaceae and Hernandiaceae); BCI plus Ducke—three families (Burmanniaceae, Lentibulariaceae, Theaceae); La Selva plus Ducke—one family (Sabiaceae); Cocha Cashu plus Ducke—two families (Opiliaceae and Caryocaraceae), a surprisingly low number since more uniquely South American families might have been expected; all but Cocha Cashu—three families (Gentianaceae, Humiriaceae, Rhizophoraceae); all but BCI—four families (Dichapetalaceae, Icacinaceae, Oxalidaceae, Quiinaceae)—these are mostly quintessentially South American families that manage to get to La Selva but not more seasonal BCI (from this distributional pattern, also shown at the generic level, we might generalize that the flora of La Selva is more South American in nature than is that of geographically closer BCI); and all but La Selva—only a single family (Ebenaceae).

Distributional Differences within Families

It seems appropriate to generalize that ecology may be generally more important than geography in determining the nature of a site's flora, somewhat contrary to my original expectation. If the four sites are roughly categorized as wet and aseasonal on good soil (La Selva), poor soil and a strong dry season (Manaus), and good to medium-poor, soil with a strong dry season (Manu and BCI), most of the floristic differences seem predictable from known patterns (Gentry 1982, 1986, 1988b).

Taxa partial to cloud forest are much more prevalent at La Selva, which is floristically the most distinctive of the four centers. Ferns, the largest taxon (169 species) make up 11% of the La Selva flora versus only 4–5% of the Cocha Cashu and Ducke floras (where ferns and allies are in third and sixth place, respectively, in familial diversity) and 8% of the BCI flora, where ferns and allies are second only to legumes in diversity. Orchidaceae, the largest seed plant family at La Selva (and in the world, Gentry and Dodson 1987), shows exactly the same pattern with 116 species (7% of the flora) as compared to 45 at Cocha Cashu (fifth place) and 24 at Manaus (twelfth place); again, BCI is intermediate with 90 species (third place). Araceae ranks third at La Selva with 100 species but decreases progressively to 42 (sixth place) at BCI, 38 (eighth place) at Cocha Cashu, and a mere 9 at Manaus. Other families unusually well represented at La Selva and poorly represented at Manaus are Melastomataceae, Piperaceae, Bromeliaceae, Gesneriaceae, Acanthaceae, Cyclanthaceae, Solanaceae, Guttiferae, and Marantaceae. All of these taxa are predominantly epiphytes, herbs, or shrubs. Among families apparently underrepresented at La Selva are Leguminosae, especially caesalpinoids, and Moraceae.

At the other extreme, the Manaus site has less similarity of its dominant families with the other sites than they do with one another. Five of the ten families represented on only one site's "top twenty" list are at Manaus—Chrysobalanaceae, Lecythidaceae, Myristicaceae, Burseraceae, and Bombacaceae. All of these are trees, as discussed below; Manaus is conspicuously lacking in taxa of herbs and epiphytes.

Cocha Cashu is characterized by the richest flora of all four sites even though it is not very distinctive floristically. More families (130) and about 200 more species are known from the Cocha Cashu area (including latosol areas) than from runner-up La Selva (table 9.1). Although the area included is larger and less precisely defined, it has also been less intensively studied, so the biases should tend to cancel each other out. Moreover, the flora of the region is extremely rich, with almost 3,000 species now known for Manu Park as a whole, in spite of very incomplete collecting in most areas and somewhat spotty intercalation of the identified species into the current Manu Park list (Foster, this volume, p.c.).

Fewer families are represented only at Cocha Cashu than at other sites, and the only two families uniquely represented in the Cocha Cashu "top twenty" list are Myrtaceae and Meliaceae. Interestingly, the Cocha Cashu flora seems more like that of the two Central American sites, on similar relatively rich soil (Foster and Brokaw 1982; Foster, this volume), than like the Central Amazonian site, with its very different, much poorer soils. It is probably also more like the Central American sites floristically than it is like the Iquitos region, which, though nearby in Amazonian Peru, is characterized by a habitat mosaic that includes poor soils and seasonally inundated forests like those of Central Amazonia (Gentry 1985, 1986a).

Bignoniaceae make the case for a Central American floristic relationship clearly. Ten of the 34 Cocha Cashu species are found more or less throughout the Neotropics, and 8 species are found only in Central America and rich-soil parts of upper Amazonia, for a total of 18 species that range into Central America. This contrasts with just 16 species that are found only in South America. Of these 16, 6 are widespread in Amazonia and the Orinoco region (2 also ranging into coastal Brazil), 6 are restricted to more-or-less rich soil areas of upper Amazonia, 2 are mostly in the Brazilian cerrado and adjacent dry areas, and 2 are striking disjunctions from southeastern Brazil (Gentry 1985).

Barro Colorado Island appears as something of the odd man out in this analysis. Its woody flora is characterized by underrepresentation of many families like Lauraceae, Annonaceae, and Sapotaceae. There is a significant complement of epiphytes, but unlike La Selva, they are inadequate to compensate for lack of tree taxa (as compared to Amazonia) and the site is floristically rather impoverished. Although the known flora is sizable, this is largely accounted for by a more complete census of weeds; the families that are overrepresented at BCI are weedy ones (Gramineae, Cyperaceae, Compositae).

Generic Analysis

At the generic level, there are striking similarities among three of the field stations, but Manaus is distinctly different (table 9.2). *Piper* is the most speciose genus at the Central American stations, as well as at Rio Palenque, Ecuador (Dodson and Gentry 1978), and is in the top three at Cocha Cashu. *Psychotria* is the second largest genus at La Selva and BCI and sixth at Cocha Cashu. *Inga* and *Ficus* are the largest genera at Cocha Cashu; *Inga* is third and *Ficus* fourth in number of species at BCI, and

TABLE 9.2 Largest genera of vascular plants, including weeds, in local florulas

Rio Palenque[1]	N	La Selva[2]	N	BCI[3]	N	Cocha Cashu[4]	N	Manaus (Ducke)[5]	N
Piper	22	Piper	44	Piper	21	Ficus	34	Licania	21
Ficus	18	Psychotria	38	Psychotria	20	Inga	26	Inga	17
Solanum	18	Philodendron	31	Inga	18	Piper	25	Protium	14
Peperomia	15	Anthurium	25	Ficus	16	Pouteria	21	Eschweilera	13
Philodendron	15	Miconia	25	Miconia	14	Paullinia	19	Swartzia	13
Pleurothallis	15	Thelypteris	17	Polypodium	13	Psychotria	14	Aniba	12
Anthurium	14	Inga	16	Philodendron	13	Miconia	13	Miconia	12
Epidendrum	14	Peperomia	16	Epidendrum	13	Eugenia	13	Ocotea	11
Maxillaria	13	Calathea	15	Anthurium	12	Philodendron	12	Casearia	10
Thelypteris	11	Ficus	15	Cyperus	11	Trichilia	12	Couepia	10
Heliconia	10	Pleurothallis	15	Passiflora	11	Calathea	11	Palicourea	10
Columnea	9	Trichomanes	14	Solanum	11	Passiflora	11	Pouteria	10
Drymonia	9	Passiflora	13	Panicum	10	Thelypteris	11	Sloanea	10
Passiflora	8	Ocotea	13	Peperomia	10	Peperomia	10	Virola	10
Adiantum	7	Clidemia	13	Trichomanes	9	Neea	10	Iryanthera	8
Cyperus	7	Cyperus	12	Adiantum	9	Anthurium	10	Brosimum	7
Calathea	7	Panicum	12	Paspalum	9	Solanum	10	Hirtella	7
Cordia	7	Heliconia	12	Desmodium	9	Epidendrum	9	Micropholis	7
Hyptis	7	Epidendrum	12	Paullinia	9	Adiantum	8	Guatteria	7
Ocotea	7	Solanum	12	Serjania	9	Justicia	8	Duguetia	6

Psychotria	7	Asplenium	11	Thelypteris	8	Alibertia	8	Licaria	6
Pilea	7	Diplazium	11	Heliconia	8	Calyptranthes	8	Mouriri	6
Inga	6	Tillandsia	11	Calathea	8			Lecythis	6
Tillandsia	6	Maxillaria	10	Maxillaria	8			Trichomanes	6
		Drymonia	10	Dioscorea	7				
		Selaginella	9	Machaerium	7				
		Polypodium	9	Casearia	7				
		Monstera	9						
		Costus	9						
		Clusia	9						
		Conostegia	8						
		Pilea	8						

Notes: 1. Dodson and Gentry 1978
2. Hammel, this volume, p.c.
3. Croat 1978
4. Foster, this volume, p.c.
5. Prance, this volume, p.c.

TABLE 9.3 Major families in 1-ha tree plots in Amazonia (sites 1–11), on BCI, and at La Selva

	Cocha Cashu	Manaus (≥ 15 cm DBH)[1]	Tambopata Alluvial	Tambopata Swamp	Tambopata Sandy 1	Tambopata Sandy 2	Tambopata Clayey 2	Tambopata Clayey 1	Yanamono	Mishana	Cabeza de Mono	La Selva[2]	La Selva[2]	BCI (5 ha, ≥ 20 cm DBH)[3]	BCI Ha. 4,4[4]	BCI Ha. 1,2[4]
Leguminosae	27	19	22–23	ca. 23	25–26	19	30	ca. 28	42–44	37–38	30–31	28	30	14	14	12
Moraceae	18	13	16	ca. 23	ca. 20	18	ca. 20	18	24	15	5–6	9	13	12	8	7
Annonaceae	18	7	13	ca. 10	11	10	10–11	14–15	12	15	7	7	8	4	2	2
Lauraceae	15–16	7	14	ca. 20	ca. 18	14	ca. 20	8–10	15	30	22	13	7	5	5	5
Sapotaceae	14	11	10	7	8	7	10	ca. 12	16	16	16	6	7	3	3	2
Meliaceae	11	4	7	1	0	1	4	5	11	7	4	8	8	2	4	4
Myristicaceae	7	3	7	6	6	7–8	6	6	17–19	13–14	7–8	2	2	2	2	2
Bombacaceae	7	2	3	1	3	1	1	4	6	2	1	2	3	7	3	3
Sapindaceae	7	2	1	1	3	2	1	1	4	4	4	3	2	3	2	1
Euphorbiaceae	6	4	6	4	6	7–8	3	4	18	15	11	5	10	3	3	3
Palmae	4	0	6	7	5–6	6	5	5	6	6	1	4	5	2	4	2

Bignoniaceae	4	0	4	1	3	4	6	7	1	1	4	0	2	3	2	1
Nyctaginaceae	3–4	2	6	3–4	3	2	3–4	ca. 5	5	3	3	2	1	1	1	1
Chrysobalanaceae	3	10	2	4–5	5–6	5	2	1	3	9	7	2	1	3	2	2
Rubiaceae	2	3	3–4	2	6	5	1	4	13–14	9	2	6	7	5	5	4
Melastomataceae	1	5	0	5	7–8	7	0	2	0	2	0	4	3	1	1	0
Burseraceae	1	10	3	4–5	3	3	4	4–5	9	13	12–13	3	3	5	4	5
Lecythidaceae	1	18	3	2	2–3	3	3	4	7–9	5	8–9	3	3	1	1	1
Vochysiaceae	0	8	0	0	0	0	1	0	0	1	0	0	0	0	0	0
Flacourtiaceae	3	3	3	3	3	2	3	5	8	1	1	5	7	6	6	6

Source: Data from Genty 1988a, b, except as indicated; mostly plants ≥ 10 cm DBH

Notes: 1. Prance et al. 1976

2. D. and M. Lieberman, p.c. (work supported by NSF grant BSR 811 7507)

3. Thorington et al. 1982

4. Hubbell, p.c.; plot 4, 4 is most species rich in 50-ha plot; plot 1, 2 is average (92 spp.; excludes lianas)

≡ = Numbers for most speciose families

═ = second most speciose families

— = third to tenth most speciose families

these two genera tie as the seventh most species rich genus at La Selva. Both *Inga* and *Ficus* are also among the largest genera at Rio Palenque and *Inga* is second at Ducke. With the exception of *Ocotea* at La Selva and *Pouteria* and *Trichilia* at Cocha Cashu, *Inga* and *Ficus* are the only tree genera included on the list of largest genera at any site other than Manaus. *Philodendron* and *Anthurium* are the third and fourth most species rich genera at La Selva, and both are also among the top dozen at the other sites (including Rio Palenque); interestingly, at every site there are a few more species of *Philodendron* than of *Anthurium*. *Miconia* is the fourth or fifth most species rich genus at each Central American site and among the dozen most speciose genera in Amazonia. *Thelypteris*, *Adiantum*, *Trichomanes*, *Polypodium*, *Asplenium*, and *Diplazium* are the most speciose fern genera, again repeating at the different field stations. The most speciose orchid genera are *Epidendrum*, *Maxillaria*, and *Pleurothallis*; except at Cocha Cashu and Ducke, at least two of these are among the largest genera. *Peperomia* is among the dozen most speciose genera at every station except Manaus (and fourth at Rio Palenque). In addition to genera mentioned above, *Calathea*. *Passiflora*, *Heliconia*, and *Solanum* are among the twenty most speciose genera everywhere but Manaus. Excluding Manaus, the only additional genera to enter the top twenty at any site are *Paspalum*, *Desmodium*, *Paullinia*, and *Serjania* (nine species each) at BCI, *Clidemia* and *Ocotea* at La Selva, and *Paullinia*, *Pouteria*, *Eugenia*, *Calyptranthes*, *Neea*, *Trichilia*, and *Justicia* at Cocha Cashu. Manaus has a completely different complement of large genera than the other sites, with twenty-one genera of trees, two genera of shrubs, and the fern *Trichomanes* included in the largest Ducke genera listed in table 9.2.

The most striking aspect of these patterns is that even though the same genera are speciose at each site (except Manaus), in most cases the species overlap from site to site is rather little. That the same few genera have speciated profusely in different regions can hardly be coincidental. The distinct nature of the Central Amazonian flora with its predominance of large genera of trees is also noteworthy.

Habit Differences

A striking difference among these field stations is in the representation of different habit groups in their floras (table 9.3 and fig. 9.1). The most obvious anomaly is the high percentage of trees in the Manaus data set. Unfortunately, this is the least complete floristic list (and may have

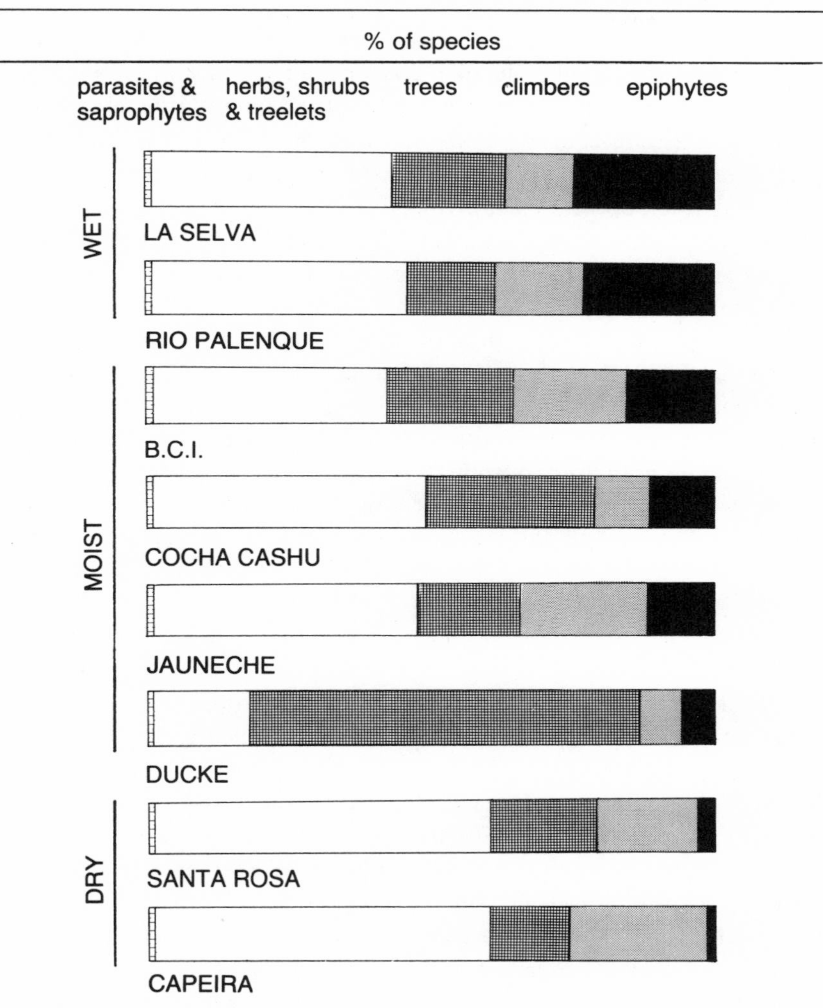

FIGURE 9.1. Percentage of species in local florulas belonging to different habit groups. La Selva, BCI, Cocha Cashu, and Ducke Reserve compared to the additional sites tabulated in Gentry and Dodson (1987). Habit groups reassigned following criteria of Gentry and Dodson (herbaceous vines included with lianas, and saprophytes and parasites tabulated separately).

been biased by a special interest in trees on the part of INPA personnel). Even though additional collecting may turn up significantly more herbs, epiphytes, or climbers, the conclusion that the Manaus area flora is overwhelmingly arborescent is not likely to change. If so, this cannot be a general property of Amazonian plant communities, since the Manu flora has exactly the opposite bias, being 50% composed of herbs and shrubs. At the opposite extreme, the La Selva flora is extremely rich in epiphytes, which constitute one-fourth of its species. La Selva has relatively few lianas, with hemiepiphytic climbers apparently to some extent replacing free-climbing ones, just as they do in the Choco (Gentry 1986b). The richness of the La Selva flora as compared to that of BCI is due largely to the proliferation of epiphytes and to some extent herbs and shrubs, especially in the large genera discussed above. The relative improverishment of the Manaus area flora, despite its richness of trees, results largely from a lack of these epiphytic and understory taxa, presumably as a consequence of its poor soil and strong dry season (see Gentry and Emmons 1987).

Tree Plot Data

Although at the local florula level Amazonia does not appear to be consistently richer than Central America, it is clearly much richer in trees. Several plots at La Selva, ranging from 2 to 4 ha, have about 100 tree species greater than 10 cm DBH (88–118) (Hartshorn 1983), and at BCI the 1-ha suplots of the Hubbell-Foster 50-ha plot average about 93 species greater than 10 cm DBH (76–116) (Foster and Hubbell, this volume); Thorington et al. (1982) found 60 species greater than 20 cm DBH in 1 ha and 112 in 5 ha. In upper Amazonia and the Choco there are between 155 and 283 species of trees greater than 10 cm DBH in 1 ha or 165–300 species greater than 10 cm DBH, including lianas (Gentry 1987, 1988). The Manaus area, in spite of its strong dry season and poor soil, shares the same high tree species diversity as Amazonian Peru, with 179 species greater than 15 cm DBH in a 1-ha plot (Prance et al. 1976). Thus, there seems to be a clear dichotomy between Amazonia and Central America in richness of tree species.

Central American and Amazonian tree communities also have differences in their floristic makeups, although the variations in diversity are more striking than the differences in familial representation. Leguminosae are the most diverse family in all lowland Neotropical tree plots, and at La Selva, at least, there are about as many species in the 1-ha plots. Moraceae, Annonaceae, Lauraceae, and Sapotaceae are also among the ten most diverse families in all tree plots (except Knight's 1975 greater than 18 cm DBH subset on BCI) but with generally fewer species in Central America. Myristicaceae, Meliaceae, Euphorbiaceae, and Palmae are among the top ten tree families in all or nearly all Peruvian sites (Gentry 1988b) but not at Manaus; except for Myristicaceae, they are also among the ten most diverse tree families in La Selva and BCI tree plots. These taxa also tend to be less diverse on poorer soil sites in Peru (Tambopata sandy, Mishana, Cabeza de Mono in table 9.3), just as on the poor soil at Manaus. At Manaus a rather different set of families specializing on poor or medium-poor soils is speciose—Chrysobalanaceae, Burseraceae, Lecythidaceae, and Vochysiaceae—along with Leguminosae, Moraceae, Lauraceae, and Sapotaceae, of course. Only twenty families are ever among the ten most diverse in any Amazonian tree plot. In general, these same families are the dominant trees of Central American forests (although Tiliaceae and Sterculiaceae may also be) but mostly with many fewer species.

To generalize from the four field stations, Central American forests are clearly much poorer in tree species than Amazonian ones, although the flora of a wet Central American forest like La Selva may be richer than many Amazonian sites due to more epiphytes and understory herbs and shrubs. In previous papers (Gentry 1982, 1986a,b; Gentry and Dodson 1987), I have discussed the prevalence of epiphytes and herbs in cloud forest environments as stemming from explosive speciation in foothill areas along the northern Andes and in southern Central America. The diversity of the La Selva flora would seem to stem in large part from this phenomenon.

Yet I have previously suggested that habitat specialization and the resultant β-diversity are major determinants of high tree-species diversity in Amazonia (Gentry 1986a, 1988b). In addition to strong selection for habitat specialization, the existence of a complex habitat mosaic provides ample opportunity for spillover from one community into another, thus increasing α-diversity. It is likely that many of the sympatric congeners that replace one another in different patches of forest on the same substrate may be ecological equivalents (Gentry 1988b). Familial composition and overall diversity of the trees in Amazonian forests seems predictable, but their exact species composition much less so. Central America has relatively uniformly fertile soils and lacks the rich array of marginal habitats and the rich communities of plants specialized for them. Thus a lower intracommunity diversity might also be expected in Central America as a result of the lesser habitat diversity (Ricklefs 1987).

Preparation of this chapter resulted in several useful insights about weaknesses in and differences among the floristic data bases at the four sites. Perhaps the greatest weakness in the data is the inadequacy of plant collections from South American sites. This is most acute in Central Amazonia, where not one reasonably complete floristic list or local florula is available. Even at Manu National Park, which has the longest species list of any of the four sites, the plant-collecting effort has been conducted sporadically and by few botanists working over relatively little time and with minuscule financial support. The numbers of new species and new records for the La Selva flora that have come out of the intensive fieldwork there in recent years demonstrate once again the importance of intensive year-round collecting programs if reasonable floristic knowledge of sites as rich in species as these field stations is to be achieved. Another glaring deficiency in the data base that comes from these comparisons is the lack of a major field station in the aseasonal part of upper Amazonia and/or the Choco region, where plant species diversity may well be higher than at any of the sites discussed here.

Finally, the importance of standardized sampling is re-emphasized. The Manaus tree plot data, for example, have a 15-cm DBH cut-off rather than the 10-cm DBH that is gaining acceptance as the norm for this kind of study. The BCI tree plot data do not include large lianas, whereas the others do. Previously available BCI tree data used 20 cm minimum DBH (Thorington et al. 1982) or an 18-cm DBH tree size class (Knight 1975). Knight's small diameter subset for BCI went down to 2 cm DBH but did not include lianas, whereas that at Cocha Cashu goes down to 2.5 cm and does include lianas (Gentry 1985). The floras themselves are not entirely comparable either, since many differences are the result of differential inclusion of aquatic areas and weedy habitats at the four sites.

In spite of these problems, the floristic data do show interesting similarities and differences among the sites. Some differences, like the prevalence of epiphytes at La Selva relative to the other sites, are not unexpected. Others, like the prevalence of shrubs at Cocha Cashu and the overwhelming predominance of trees belonging to rather few relatively large genera in the Ducke flora, are more surprising.

REFERENCES

Croat, T. 1978. *Flora of Barro Colorado Island*. Stanford Univ. Press, Stanford.

Dodson, C. and A. Gentry. 1978. Flora of the Rio Palenque Science Center. Selbyana 4: 1–623.

Foster, R. 1985. Plantas de Cocha Cashu. In M. A. Ríos (ed.), *Reporte Manu*.

Centro de Datos para la Conservación, Univ. Agraria, La Molina, Peru, pp. 5/1–5/20.

———. (This volume). Floristic composition of the Manu floodplain forest.

Foster, R. B. and N. V. L. Brokaw. 1982. Structure and history of the vegetation on Barro Colorado Island. In E. G. Leigh, Jr., A. S. Rand, and D. M. Windsor, (eds.), *The Ecology of a Tropical Forest: Structural Rhythms and Long-Term Changes*. Smithsonian Inst. Press, Washington, D.C., pp. 67–81.

Foster, R. B. and S. P. Hubbell. (This volume). The floristic composition of the Barro Colorado Island forest.

Gentry, A. H. 1982. Neotropical floristic diversity: phytogeographical connections between Central and South America, Pleistocene climatic fluctuations, or an accident of the Andean orogeny? Ann. Mo. Bot. Gard. 69: 557–593.

———. 1985. Algunos resultados preliminares de estudios botanicos en el Parque Nacional del Manu. In M. A. Ríos (ed.), *Reporte Manu*. Centro de Datos para la Conservación, Univ. Agraria, La Molina, Peru, pp. 2/1–2/22.

———. 1986a. Endemism in tropical vs. temperate plant communities. In M. E. Soulé(ed.), *Conservation Biology: The Science of Scarcity and Diversity*. Sinauer, Sunderland, Mass., pp. 153–181.

———. 1986b. Species richness and floristic composition of Choco region plant communities. Caldasia 15: 71–91.

———. 1988a. Tree species richness of upper Amazonian forests. Proc. Nat. Acad. Sci. 85: 156–159.

———. 1988b. Changes in plant community diversity and floristic composition on geographical and environmental gradients. Ann. Mo. Bot. Gard. 75: 1–34.

Gentry, A. H. and C. Dodson. 1987. Diversity and phytogeography of Neotropical epiphytes. Ann. Mo. Bot. Gard. 74: 205–233.

Gentry, A. H. and L. H. Emmons. 1987. Geographical variation in fertility and composition of the understory of Neotropical forests. Biotropica 19: 216–227.

Hammel, B. (This volume). The distribution of diversity among families, genera, and habit types in the La Selva flora.

Hartshorn, G. 1983. Plants: introduction. In D. Janzen (ed.), *Costa Rican Natural History*. Univ. of Chicago Press, Chicago, pp. 118–157.

Knight, D. H. 1975. A phytosociological analysis of species-rich tropical forest on Barro Colorado Island, Panama. Ecol. Monogr. 45: 259–284.

Prance, G. T. (This volume). The floristic composition of the forests of central Amazonian Brazil.

Prance, G. T., W. A. Rodrigues, and M. F. da Silva. 1976. Inventario florestal de um hectare de mata de terra firme, km 30 da Estrada Manaus-Itacoatiara. Acta Amazônica 6: 9–35.

Rickleffs, R. E. 1987. Community diversity: relative roles of local and regional processes. Science 235: 167–171.

Thorington, R. W., Jr., B. Tannenbaum, A. Tarak, and R. Rudran. 1982. Distribution of trees on Barro Colorado Island: a five-hectare sample. In E. G. Leigh, Jr., A. S. Rand, and D. M. Windsor (eds.), *The Ecology of a Tropical Forest: Seasonal Rhythms and Long-Term Changes*. Smithsonian Inst. Press, Washington, D.C., pp. 83–94.

PART III

Birds

10

Birds of La Selva Biological Station: Habitat Use, Trophic Composition, and Migrants

JOHN G. BLAKE, F. GARY STILES, AND BETTE A. LOISELLE

La Selva Biological Station has been the site of many bird studies during the past three decades (for example, Slud 1960; Orians 1969; Stiles 1975, 1978a,b, 1980a; Stiles and Wolf 1974, 1979; Denslow and Moermond 1982; Moermond and Denslow 1983, 1985; Levey et al. 1984; Levey 1987a,b, 1988; Moermond et al. 1986; Loiselle 1987a,b, 1988; Loiselle and Blake 1990). During this period, 410 species of birds in 19 orders and 52 families (following AOU 1983 and supplements) have been recorded from La Selva and its immediate environs, an increase of 24% from Slud's (1960) list of 331 species. La Selva owes its diverse avifauna to its climate, geographic position, and habitat diversity (see Levey and Stiles 1991). Although there have been fewer recent additions to the La Selva list (21 since 1977), more species will likely be added with continued observations (for example, *Cypseloides cryptus*, *C. cherriei*).

The species list includes 47 species recorded fewer than five times (accidentals). Most accidentals (for example, *Stercorarius pomarinus*) do not occur regularly at La Selva and are not included in subsequent analyses. The 47 accidentals include 11 altitudinal migrants. Some altitu-

We thank J. R. Karr, D. J. Levey, and T. C. Moermond for many helpful discussions and comments; D. and D. Clark for improving life at La Selva; J. A. Leon for facilitating paperwork; and F. Cortés S. and J. Dobles Z. of the Servicio de Parques Nacionales for permission to work in Parque Nacional Braulio Carrillo. The following provided support for these and related studies: Jessie Smith Noyes Foundation (BAL), National Geographic Society (JGB and BAL), Organization for Tropical Studies (BAL and JGB), Joseph Henry Fund (BAL), Paul A. Stewart Award (Wilson Ornithological Society) (BAL), Northeastern Bird Banding Association (BAL), Guyer Fellowship, University of Wisconsin (JGB), Vicerrectoría de Investigación, U. de Costa Rica (FGS), CONICIT (FGS), and Chapman Fund of the American Museum of Natural History (FGS).

dinal migrants probably occur at La Selva more frequently than records indicate (for example, *Pipra pipra*), depending on conditions (fruit supplies, rainfall) at higher elevations in adjoining Braulio Carrillo National Park (BCNP). Long-term studies at La Selva are needed to detect such occurrences.

La Selva was largely surrounded by forest in 1960 but is now bordered by pasture on most of three sides. La Selva thus forms the tip of a forest corridor that connects lowland forest to more extensive montane forest in BCNP (see fig. 1.1). Preservation of this forested transect ensures that La Selva will not become a completely isolated patch of forest. Consequently, although about twenty species have declined in abundance at La Selva over the past thirty years (for example, *Spizastur melanoleucus*, *Ara ambigua*, *Myiobius sulphureipygius*, *Thamnistes anabatinus*; Levey and Stiles, in press), it is now less likely that many species will be lost permanently, as has happened at some reserves (Leck 1979; Willis 1979; Karr 1982). Nonetheless, as more forest bordering La Selva and the corridor is cleared, further declines are possible, and continued observations at La Selva and higher elevations are necessary (Stiles 1985a,b; Loiselle 1987b).

In contrast to forest species, increases in abundance have been noted for a number of species associated with second-growth habitats (for example, *Aratinga finschii*, *Tapera naevia*, *Thraupis episcopus*; Levey and Stiles, in press). Addition of the Vargas and Las Vegas properties (see fig. 1.1) increased the area and diversity of second-growth habitats within La Selva. These second-growth areas support a diverse assemblage of resident birds, including several rare species, such as *Gymnocichla nudiceps* and *Aphanotriccus capitalis*, and are also used seasonally by many latitudinal migrants. Abundant fruit crops attract some forest species, particularly when fruit crops are low in the forest (Blake and Loiselle, unpubl. data). Thus, successional habitats may be a necessary component of the habitat requirements of some tropical forest species (see Terborgh 1983; Martin and Karr 1986). Naturally occurring changes in vegetation structure or changes that occur as a result of experimental manipulations likely will influence the distribution and abundance of many species, resulting in further alterations to the La Selva avifauna. Continued monitoring of populations in younger habitats will be necessary to detect such changes.

In this chapter we review the current species list, focusing on species composition with respect to habitat use, diet, and migratory status. We also compare bird communities in second-growth and forest habitats, based on mist-net studies. This chapter can be viewed as a companion to Levey and Stiles (in press), which focuses more on geographical affinities of the La Selva avifauna, breeding periods, molt cycles, and flocking behavior; such topics are not covered here.

General Composition of the Avifauna

The species list (see appendix 14.1) was compiled from Stiles (1977, 1983b) and Levey and Stiles (1991). Species were assigned to a single category in each of several ecological groups (preferred habitat, migratory pattern, diet) for the following analyses. Thus, analyses of species distribution patterns, such as by habitat and diet, necessarily ignore the fact that most species occur in a variety of habitats and use a variety of food types. Results should thus not be viewed as a precise determination of, for example, the number of species found in different habitats or using different foraging substrates. Additional habitat categories for many species are shown in the appendix where appropriate.

Of the 363 species included in following analyses, 256 are known or strongly believed to breed at La Selva (Levey and Stiles 1991). Breeding occurs in all months, with a broad peak from March through June (163 to 211 species breeding per month) and a low from October through December (fewer than 20 species breeding per month) (Levey and Stiles 1991). Most species have a three-to-six month breeding season, but some extend for longer periods.

Distribution among Habitats

Habitat selection patterns are an important determinant of bird community composition (Cody 1985), and diversity of habitats in a region strongly influences the number of species that can occur there (Karr 1980, 1982). La Selva supports a variety of habitats, including primary forest, forest edges and gaps, rivers and streams, riverine scrub, anthropogenic second-growth, old plantations, and surrounding pastures (table 10.1).

Most species primarily are associated with forest habitats (70% of breeding, 55% of nonbreeding). However, aquatic and aerial habitats are occupied mostly by nonbreeding species. Within primary forest, more breeding species are found in forest interior or canopy than are typically associated with forest gaps or edges. In contrast, nonbreeding species tend to be concentrated in forest edge habitats (31% versus 18% of breeding species) and are uncommon within the forest interior or canopy.

TABLE 10.1 Species commonly associated with different habitats at La Selva

Habitat	Breed[1] N (%)	Nonbreed[2] N (%)	Migrants[3] Latitudinal N (%)	Migrants[3] Altitudinal N (%)
Air and water				
air[4]	7 (3)	12 (11)	14 (17)	
large rivers	9 (4)	13 (12)	9 (11)	
ponds, marsh	3 (1)	2 (2)	2 (2)	
Nonforest				
pastures, cultivation	15 (6)	5 (5)	3 (4)	1 (4)
young second-growth	40 (16)	14 (13)	14 (17)	
plantations (e.g., cacao)	4 (2)	2 (2)	2 (2)	
Forest				
old second-growth[5]	12 (5)	11 (10)	10 (12)	
forest streams and swamp	12 (5)	3 (3)	2 (2)	
primary forest—interior	54 (21)	6 (6)	1 (1)	8 (32)
—canopy	54 (21)	6 (6)		10 (40)
gaps and edge	46 (18)	33 (31)	28 (33)	6 (24)
TOTAL	256	107	85	25

Notes: [1]Breed: species that breed or very likely breed at La Selva

[2]Nonbreed: species not known or thought to breed at La Selva. Most species are migrants; some species (e.g., *Cochlearius cochlearius*) occur at La Selva with some regularity, breed on the Caribbean slope of Costa Rica, but are not known to breed at La Selva.

[3]Some species classified as migrants also are represented by breeding individuals at La Selva (e.g., *Thalurania colombica, Legatus leucophaius*)

[4]Species that spend most of the day soaring or flying over the canopy

[5]With well-developed canopy

Many species, particularly those associated with forest canopies and edge habitats, range over both types of habitats—the foliage-air interface of Stiles (1983a). Some common species of forest habitats are: forest interor—*Tinamus major*, *Micrastur ruficollis*, *Glyphorynchus spirurus*, *Pipra mentalis*, and several antbirds; forest canopy—*Columba nigrirostris*, several parrots, *Florisuga mellivora*, *Querula purpurata*, toucans, and several flycatchers; and forest gaps and edges—*Xiphorhynchus guttatus*, *Manacus candei*, *Thryothorus nigricapillus*, and *Caryothraustes poliogaster*.

Trophic Composition

Ten diet categories are used, with assignments based on personal observations, fecal samples from mist-netted birds (Loiselle and Blake, 1990, unpubl. data), and information in the literature (Slud 1964; Ridgely 1976; Hilty and Brown 1986; Stiles and Skutch 1989). Most species depend either on insects (173 species, 48%) or on fruit or fruit and insects (109 species, 30%) (table 10.2).

Noticeable differences exist between breeding and nonbreeding species in relative importance of different food types. For example, species that forage primarily on large insects are more important among breeding (22%) than nonbreeding (4%) species, whereas the reverse is true for species that forage on small insects (26% and 43%, respectively; Stiles 1983a). Insectivores as a whole, however, are equally well represented among both breeding (48%) and nonbreeding (47%) species. In contrast, species that forage almost entirely on fruit are twice as important among breeding (12%) than nonbreeding (6%) species at La Selva.

Few differences exist between forest and nonforest species in relative distribution of species among diet groups. Nectarivores tend to be more important in forest habitats and granivores in nonforest habitats.

Of 237 species associated primarily with forest habitats (that is, forest habitat listed first in appendix 14.1), a slightly larger proportion of species forage primarily on ground or in forest understory (120 species, 51%) than in the canopy (99 species, 42%) (table 10.3). (We include in the canopy those birds that typically follow the foliage-air interface down along forest edges or into gaps.) Insectivores and frugivores (including both frugivores and frugivore-insectivores) are represented by similar numbers of ground-foraging species, such as *Sclerurus guatemalensis*, *Formicarius analis*, and *Hylopezus perspicillata* among insectivores and *Crax rubra*, *Geotrygon veraguensis*, and *Tinamus major* among frugivores.

TABLE 10.2 Distribution of species among primary diet categories

Diet	Breed	Nonbreed	Migrants		Habitat[1]	
			Latitudinal	Altitudinal	Forest	Nonforest
Small insects[2]	67	46	46		69	27
Large insects, small vertebrates (herps)	56	4	5		44	12
Fruit, small insects	37	23	17	6	43	17
Fruit, large insects, small vertebrates	11	1		1	12	
Fruit, fruit and seeds	31	6	2	7	28	9
Nectar and insects	14	7	1	8	18	3
Omnivore	6				4	2
Grass seeds, vegetation	7	3	2		1	7
Small to large vertebrates	24	17	11	3	18	3
Carrion	3		1			
TOTAL	256	107	85	25	237	80

Notes: [1] Categories for forest and nonforest habitats in table 10.1; aquatic habitats and air not included
[2] "Insects" also may include other arthropods

TABLE 10.3 Distribution of forest birds among trophic/foraging groups and strata

Guild	Water	Aquatic vegetation and rocks	Ground	Understory		Canopy[1] and outer edge of foliage		Air above canopy
				Shrub	Tree	Trees	Edges	
Insectivore								
"surface"	2	4	10					
live foliage				17	11	16	5	
dead foliage				2	1			
lianas, vines				2	3			
epiphytes, moss					1	1		
trunks, branches				2	5	10		
army ant follower			1	3	2			
air				3	4	1		7
Fruit and insects			6	13	9	16	11	
Fruit, fruit and seeds			4	3	1	20		
Nectar and insects				10		6	2	
Omnivore						4[2]		
Vertebrates	4		2	3	2	6	1	
Grass seeds		1						
TOTAL	6	5	23	58	39	80	19	7

Notes: [1]Canopy includes canopy layer of old second-growth habitats
[2]Forage primarily along bark (2 species) or in epiphytes (2)

Species that forage for insects on live foliage are the most diverse insectivore group (table 10.3) and are slightly more diverse in lower strata than in the canopy. Among insectivores, antbirds and furnariids tend to dominate in the understory, whereas flycatchers and oscines are more common in the canopy (Stiles 1983a). By contrast, frugivorous species are more heavily concentrated in the canopy (47 species) than in the understory (26 species; 36, if ground-foragers are included).

Migratory Patterns

Migrants comprise about 30% of the La Selva avifauna (table 10.1). Latitudinal migrants (species that breed north of Costa Rica, generally in North America—temperate migrants; and species that breed in Costa Rica and winter south—intra-tropical migrants) are associated predominantly with nonforest (51%) and forest-edge habitats (33%). Many migrant species, however, frequently occur in forest canopy, and during migration, migrants can be found in almost any habitat. *Hylocichla mustelina*, for example, occurs primarily within forest interior as a winter resident but is also found in early successional habitats during migration. Many latitudinal migrants often (but not always) prefer tropical second-growth habitats rather than moist lowland forests (Willis 1966; Karr 1976; Chipley 1977; Hutto 1980; Stiles 1980b; Martin 1985). Temperate migrants can be recorded at La Selva during any month except June and July but are most abundant during autumn; in spring many migrants move north via the coasts (Stiles 1983a). Common passage migrants include *Cathartes aura*, *Buteo swainsonii*, *Catharus ustulatus*, and *Tyrannus tyrannus*. Common winter residents include *Hylocichla mustelina*, *Dumetella carolinensis*, *Dendroica pensylvanica*, *Seiurus aurocapillus*, and *Oporornis formosus*.

Distribution of altitudinal migrants among habitats differs from that of latitudinal migrants (table 10.1; forest versus nonforest, $X^2 = 17.8$, $P < 0.001$, $df = 1$). Altitudinal migrants occur primarily in forest interior and canopy (72%) (table 10.1), underscoring the importance of the forest corridor that connects La Selva with higher elevation forests in BCNP. Frugivorous altitudinal migrants typically arrive at La Selva during October–November (as early as July for such species as *Cephalopterus glabricollis*) and begin to depart in mid- to late January. Some common species are *Mionectes olivaceus*, *Procnias tricarunculata*, and *Corapipo leucorrhoa*. Altitudinal migrations of hummingbirds are more variable in timing. Hummingbirds typically move downslope between April and July–August and return upslope beginning in August at the start of the breeding season (Stiles 1983c). Three hummingbirds that regularly reach La Selva are *Microchera albocoronata*, *Phaethornis guy*, and *Discosura conversii*. Another species, *Thalurania colombica*, breeds at La Selva, and then much of the population migrates upslope (Stiles 1985a).

Latitudinal and altitudinal migrants also differ with respect to diet (table 10.2). Temperate zone migrants are primarily insectivores (60%) with few species (such as *Tyrannus tyrannus*) heavily dependent on fruit; many other species also take fruit with some regularity (Loiselle and Blake 1990, unpubl. data). Altitudinal migrants primarily are frugivores (56%) or nectarivores (32%). Resource availability along the altitudinal transect is a primary influence on the extent and timing of altitudinal migrations (Stiles 1980a, 1985a; Loiselle 1987b). The arrival of frugivores and nectarivores at La Selva coincides with periods of high fruit and flower abundance, respectively (Stiles 1978a, 1980a; Loiselle 1987b; Levey 1988).

Populations of several species resident at La Selva throughout the year are augmented by the seasonal arrival of individuals from higher elevations. *Pipra mentalis*, for example, increases in abundance at La Selva when other altitudinal migrants are arriving (Loiselle 1987b). Direct evidence for such movements has been obtained through recaptures of banded individuals between La Selva and higher elevations (250 and 450 m) in BCNP (Loiselle and Blake, unpubl. data); recaptures include both *P. mentalis* and *Mionectes oleagineus*. Such seasonal movements are under further investigation (Loiselle and Blake; Stiles and Rosselli).

Bird Composition in Mist-Net Samples in Second-Growth and Forest

Mist-net data have been collected (JGB and BAL) from three habitats: young (five to eight years) and old (thirty to thirty-five years) second-growth and primary forest. Nets are arranged in grids covering about 5 ha in second-growth and 10 ha in forest. Sites are netted at about five-to-six-week intervals. Samples discussed here are from the "dry" seasons (approximately January–April) of 1985–87.

Capture Rates and Species Accumulation Curves

Capture rates of birds were, in all years and samples, highest in the young second-growth (YSG) and lowest in the forest (table 10.4); this pattern also held during the wet season (Blake and Loiselle, unpubl. data). (It is important to note that higher capture rates in the youngest

TABLE 10.4 Summary of dry season mist-net data from three sites at La Selva

Site/habitat	Year	MNH	Species (*N*)	Captures (*N*)	Captures/ 100 MNH
Young second-growth (YSG)	1985	1,252	79	701	56
	1986	1,233	76	542	44
	1987	1,240	72	485	39
Old second-growth (OSG)	1985	1,499	59	380	25
	1986	1,227	53	350	29
	1987	1,228	49	330	27
Forest	1985	1,540	50	344	22
	1986	2,169	52	443	20
	1987	2,460	53	348	14

Note: Mist-net hours (MNH) are based on the number of hours nets are operated times the number of nets

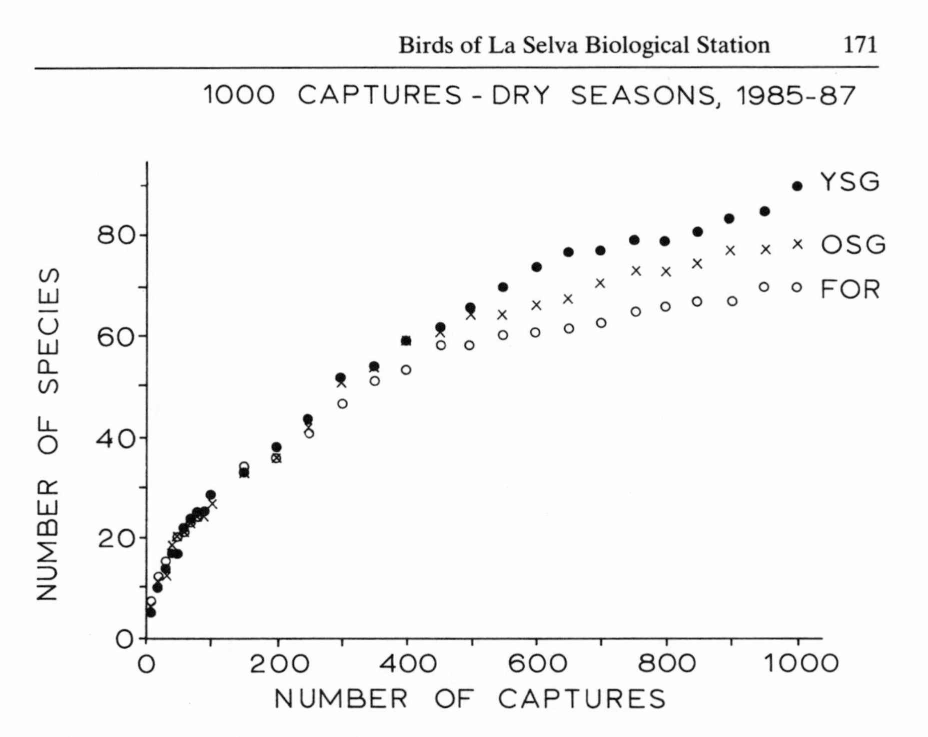

FIGURE 10.1. Species-accumulation curves based on mist-net data from young and old second-growth (YSG, OSG) and primary forest (FOR) at La Selva. A power function accounted for the greatest amount of variation in each case. *For:* ln Species = 0.47 (± 0.010) ln Captures + 1.10 (± 0.057); $r^2 = 0.988$; *OSG*: ln Species = 0.53 (± 0.012) ln Captures + 0.82 (± 0.068); $r^2 = 0.986$; *YSG*: ln Species = 0.57 (± 0.013) ln Captures + 0.64 (± 0.073); $r^2 = 0.986$. Values in parentheses = 1 standard error.

site stems partly from the fact that mist nets sample a greater proportion of vegetation in shorter [younger] habitats and thus are more likely to capture birds using the foliage-air interface). More species were also captured in young than older habitats, even when sample size was accounted for. Species-accumulation curves based on 1,000 captures from each habitat (fig. 10.1) indicate that rate of species accumulation (slope) was higher in YSG than in old second-growth (OSG) (Tukey test for multiple comparisons, $P < 0.001$) and higher in OSG than in forest ($P < 0.001$). Total species captured in our study sites (forest–85 species, OSG–94, YSG–132) represent 65%, 66%, and 73% of all species recorded (by sight, sound, or capture) on these plots. Percentages for older sites are likely an underestimate because some species, such as small canopy dwellers, have probably been overlooked.

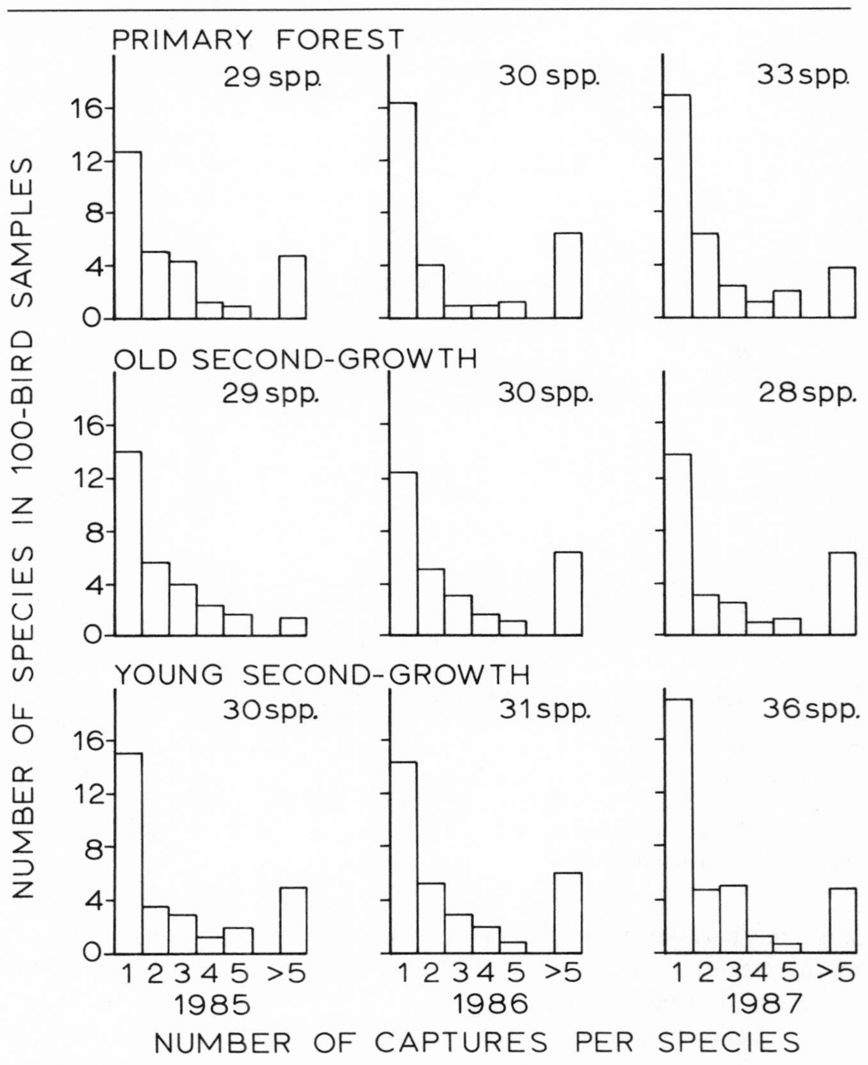

FIGURE 10.2. Numbers of species in 100-bird samples that were represented by 1–5 or more than 5 captures. Results based on three 100-bird samples per year from the dry season. Mean number of species in samples is given for each year.

Number of species represented within 100-bird samples (Karr 1980) ranged from 26 to 37 in forest, 18 to 37 in OSG, and 26 to 38 in YSG. Average number of species in dry-season samples (fig. 10.2) was, however, comparable among years and sites; no statistically significant differences were found among sites, years, or samples within years (ANOVA tests). Variation in number of species in 100 captures was due largely to changes in captures of one to a few species, notably *Catharus ustulatus*.

Relative Abundance in Mist-Net Samples

Number of species represented by a single capture in 100 capture samples accounted for 50% of all species in YSG and forest and 44% in OSG (fig. 10.2; mean of 9 samples). Species captured more than 5 times accounted for 15–19% of the totals. Rare species (1–2 captures/900 birds) accounted for a smaller proportion of total captures (N=900) from forest (28%) than OSG (40%) or YSG (45%). Conversely, the 2 most frequently captured species at each site accounted for a greater percentage of captures in forest than in younger sites (table 10.5).

Habitat preference was the most likely factor accounting for the rarity of 16 of 20 species captured only once in YSG. Many of the 16 species are more typical of forest or old second-growth (*Micrastur mirandollei*, *Trogon rufus*) or open pastures (*Columbina talpacoti*, *Crotophaga sulcirostris*). Some species rarely were captured because of size (*Phaethornis longuemareus*) or because of a general rarity at La Selva (*Vermivora pinus*). The most commonly captured species in YSG were frugivores and nectarivores.

Species captured only once in OSG were mostly outside normal habitat or foraging strata (*Columba nigrirostris*, *Monasa morphoeus*, *Ramphocelus passerinii*); 17 of 20 species fit in this category. Others were rare at La Selva in general (*Geotrygon montana*). Eight of the 10 most frequently captured species in OSG were nectarivores or frugivores, but the most frequently captured was *Glyphorynchus spirurus*, an insectivore (table 10.5).

Only 12 species were represented by a single capture in the forest sample of 900 captures. These included 5 that normally occur in the canopy (such as *Xiphorhynchus lachrymosus*, *Zimmerius vilissimus*) and 3 typically found in younger habitats (*Glaucis aenea*, *Amazilia amabilis*, *Geothlypis semiflava*). One, *Hyloctistes subulatus*, is generally rare at La Selva. Frugivores and nectarivores comprised 7 of the 10 most frequently captured species in the forest, (table 10.5) but, as in OSG, *G. spirurus*

TABLE 10.5 Ten most frequently captured species (% of 900 captures)

Forest	(%)	Old second-growth (OSG)	(%)	Young second-growth (YSG)	(%)
Glyphorynchus spirurus	14.9	*Glyphorynchus spirurus*	10.9	*Phaethornis superciliosus*	9.8
Pipra mentalis	13.8	*Mionectes oleagineus*	9.1	*Manacus candei*	8.9
Mionectes oleagineus	9.0	*Phaethornis superciliosus*	8.8	*Mionectes oleagineus*	7.1
Hylocichla mustelina	6.4	*Pipra mentalis*	7.9	*Glaucis aenea*	6.8
Phaethornis superciliosus	6.0	*Manacus candei*	7.1	*Catharus ustulatus*	5.9
Corapipo leucorrhoa	3.9	*Arremon aurantiirostris*	4.1	*Ramphocelus passerinii*	5.8
Gymnopithys leucaspis	2.8	*Threnetes ruckeri*	3.0	*Seiurus aurocapillus*	4.8
Henicorhina leucosticta	2.6	*Hylocichla mustelina*	2.7	*Saltator maximus*	4.1
Chalybura urochrysia	2.1	*Gymnopithys leucaspis*	2.4	*Pipra mentalis*	3.4
Chlorothraupis carmioli	2.1	*Catharus ustulatus*	2.4	*Threnetes ruckeri*	3.1
				Amazilia tzacatl	3.1
TOTAL	63.6		58.4		70.6

Note: Results are based on three 100-bird samples from the 1985–87 dry seasons

TABLE 10.6 Average number of species per guild, based on three 100-bird samples/year, dry seasons only

	Forest			Old second-growth (OSG)			Young second-growth (YSG)		
Guild	1985	1986	1987	1985	1986	1987	1985	1986	1987
TF	1.0	0.7	0.3	0.7	0.7		1.7	1.0	1.0
AF	2.3	2.3	1.7	2.3	2.7	3.3	2.0	3.0	3.7
TFI	0.3	0.7		1.3	1.0	1.0	1.7	2.0	1.7
AFI	4.0	7.0	6.7	4.7	6.3	6.3	8.0	9.7	9.0
TI	2.0	3.0	2.7	0.7	1.0	1.7	1.3	1.3	1.3
FI*	5.7	5.0	7.0	8.3	5.7	6.3	7.7	6.3	10.0
BI	2.7	2.0	2.7	2.3	2.0	1.7	1.3	1.3	2.0
Fly	2.3	2.3	1.7	1.7	1.7	1.0	0.3	0.7	1.0
Ant	3.7	3.0	3.3	1.7	2.7	2.0	0.3		0.3
NI	4.7	4.0	4.0	7.0	6.0	4.3	4.3	5.3	4.7
Seed			1.7	1.0	0.7	0.3	1.0	0.7	0.7
Carn	0.3		1.0	0.3		0.3	0.3		0.3

Key: TF = terrestrial frugivore; AF = arboreal frugivore; TFI = terrestrial frugivore-insectivore; AFI = arboreal frugivore-insectivore; TI = terrestrial insectivore; FI = foliage insectivore (includes foragers on dead leaves and birds that snatch insects from foliage while in flight; BI = bark insectivore; Fly = aerial or flycatching insectivore; Ant = army ant–following insectivore; NI = nectarivore-insectivore; Seed = grass seed eater; Carn = carnivore

was the most frequently captured. *Glyphorynchus* was also the most frequently captured species in Lovejoy's (1974) Brazil sites. In contrast, a manakin (*Pipra* sp.) was the most frequently captured species at the other three sites included in this volume (see Karr et al., this volume).

Trophic Structure of Mist-Net Samples

Twelve guilds were used to analyze trophic structure of mist-net samples (tables 10.6, 10.7). Arboreal frugivore-insectivores and foliage insectivores usually accounted for the greatest number of species captured in each area (table 10.6), with totals generally highest in YSG. Ant-followers were represented by more species in forest with only one species (*Gymnocichla nudiceps*) captured in YSG. Three species that forage primarily by extracting arthropods from dead leaves were captured: *Helmitheros vermivorous* and *Automolus ochrolaemus* in YSG and OSG and *Myrmotherula fulviventris* in OSG and forest. Other species that also forage on dead leaves were captured (for example, *Vermivora chrysoptera* and *Thryothorus atrogularis* in YSG; *Hylophilus ochraceiceps* in forest and OSG).

TABLE 10.7 Average number of captures per guild, all species per guild combined

	Forest			Old second-growth (OSG)			Young second-growth (YSG)		
Guild[1]	1985	1986	1987	1985	1986	1987	1985	1986	1987
TF	1.3	1.0	0.7	2.3	0.7		1.7	1.3	1.0
AF[2]	12.7	22.0	20.7	9.0	20.3	23.3	8.7	16.7	18.0
TFI[2,3]	0.3	0.7		* 4.7	3.7	4.3	6.7	3.3	2.7
AFI[3]	21.3	25.7	26.0	18.3	24.3	24.0	* 34.0	37.7	31.0
TI[3]	3.0	8.0	4.7	* 1.0	3.0	2.0	* 5.3	5.7	4.3
FI	9.7	9.0	9.7	13.0	9.7	15.3	9.3	10.3	13.7
BI[3]	19.0	14.7	16.7	* 14.7	10.0	11.7	* 1.3	1.7	4.7
Fly[3]	4.3	2.7	4.0	* 2.0	2.0	1.0	0.7	0.7	1.7
Ant[3]	12.7	4.7	6.3	5.7	9.7	4.7	* 0.3		0.7
NI[2,3]	15.0	11.7	7.7	* 28.0	16.0	12.3	30.0	21.3	21.3
Seed			2.3	1.0	0.7	1.0	1.7	1.3	0.7
Carn	0.7		1.3	0.3		0.3	0.3		0.3

Source: Based on three 100-bird samples/year, dry seasons only. Differences in captures/guild were tested with two-way ANOVA, 9 samples/site. Significance of year and site effects are indicated; no interactions were significant. A Tukey multiple comparison test (one-way ANOVA) was used to test for differences among sites when two-way ANOVA indicated a significant site effect; significant (at least $P < 0.05$) differences between sites are indicated (*). Seed eaters and carnivores were not compared.

Notes: [1] See table 10.6 for guilds
[2] Significant difference (at least $P < 0.01$) in capture rates among years
[3] Significant difference (at least $P < 0.01$) in capture rates among sites

Arboreal frugivore-insectivores constituted the most abundant guild in all but one sample, particularly in YSG (table 10.7). Nectarivores were most abundant in YSG and least common in forest; a reverse pattern held for bark insectivores and flycatchers. Arboreal frugivores (primarily manakins) accounted for substantial proportions of total captures at all sites.

Variation in trophic structure was apparent both among sites and years (table 10.7). Samples from YSG clearly separated from older sites (fig. 10.3) when analyses were based on species or captures. OSG and forest sites showed greater overlap, particularly in species composition of guilds (fig. 10.3a). Samples from 1986 and 1987 showed a greater similarity within sites, particularly when analyses were based on captures (fig. 10.3b). Separation of 1985 samples from later samples indicated substantial annual variation in trophic composition of mist-net samples (see also table 10.7).

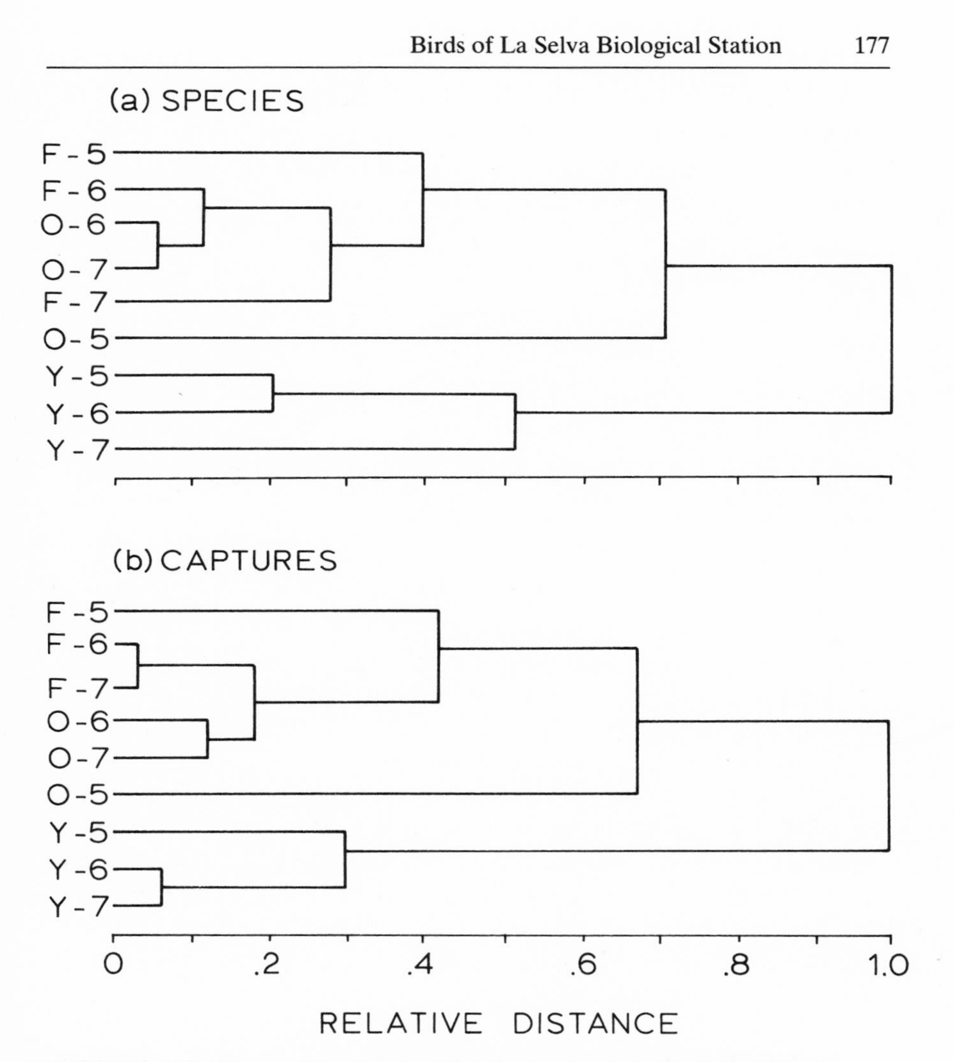

FIGURE 10.3. Cluster diagrams (weighted pair groups method; Euclidean distance measures) based (*a*) on number of species and (*b*) on number of captures in different guilds (see tables 10.6, 10.7). F = forest; O = old second-growth; Y = young second-growth; 5, 6, 7 = 1985, 1986, 1987.

Migrants in Mist-Net Samples

Preference of temperate migrants for second-growth or canopy habitats (foliage-air interface) at La Selva (Stiles 1983a) was reflected in the mist-net data. Numbers of species and captures of temperate migrants were generally highest in YSG and lowest in forest, where most migrants

TABLE 10.8 Numbers of species/captures of temperate (Temp) and altitudinal (Alt) migrants in 100-bird samples from forest, old second-growth, and young second-growth during early (sample 1), middle (2), and late (3) dry seasons

		Forest		Old second-growth (OSG)		Young second-growth (YSG)	
Year	Sample	Temp	Alt	Temp	Alt	Temp	Alt
1985	1	3/13	4/12	3/7	3/7	7/16	0
	2	2/10	3/8	4/5	2/3	6/15	0
	3	2/7	1/1	5/13	1/3	5/31	0
1986	1	1/5	4/21	3/7	3/19	7/24	2/5
	2	1/5	2/16	3/7	5/11	6/16	3/6
	3	1/9	4/5	5/18	2/2	7/35	0
1987	1	2/7	6/12	4/10	3/3	8/16	3/3
	2	2/8	5/8	2/7	2/8	5/15	3/5
	3	4/11	2/2	8/18	0	7/25	1/1

occur in the canopy (table 10.8; see also Martin 1985). The late dry season (spring) passage of migrants through La Selva was reflected in generally higher captures during later samples, particularly in YSG.

Distribution of altitudinal migrants among sites and samples was generally a reverse of the pattern observed for temperate migrants (table 10.8). Few altitudinal migrants were captured in YSG, and a substantially greater number were captured in forest. Departure of altitudinal migrants from La Selva was reflected in declining captures in forest and OSG from first to last samples in each year. Substantial annual variation in abundance of altitudinal migrants occurs; captures in 1986 were high relative to 1985 and 1987.

La Selva Biological Station supports a diverse and dynamic avifauna. Long-term changes in the avifauna have occurred and will continue to occur as habitats change locally and regionally—that is, as loss of forests continues in Central America. Long-term monitoring of bird populations at La Selva will be necessary to detect and evaluate such changes. Seasonal fluctuations in bird communities occur as migrants enter and leave La Selva and surrounding areas. Both annual variation in the extent and duration of altitudinal migrations and changes in population levels of residents contribute to the dynamic nature of the avifauna. Local movements of birds among habitats within La Selva occur regularly, often in response to changes in abundance of resources. Such movements create considerable spatial and temporal variability in assemblage characteristics at a given site.

La Selva is fortunate in having a protected, forested connection to foothill and montane forests. Such forests are seasonally required habitat for a variety of birds that occur at La Selva and thus comprise integral components of the La Selva system. Continued presence of the connection will have important consequences for preservation of many species, including regular altitudinal migrants and species that may visit La Selva only irregularly.

Diversity of habitats present at La Selva is great and is instrumental in promoting such a varied avifauna. In addition to primary forest, successional habitats of many types occur. These areas are used by many temperate migrants and, periodically, by forest residents as foraging sites. Successional habitats also support a wide variety of resident species, some rare. Maintenance of second-growth areas at La Selva will probably be necessary to ensure continued existence at La Selva of a variety of species.

In spite of considerable past and present research, there is much room for more study. Many groups of birds, for example, most insectivores, and most species have not been studied in detail. Birds are an integral and important component of tropical forests—witness their role as primary seed-dispersers for most trees and shrubs (Stiles 1985c)—and deserve more consideration in long-term studies at La Selva and elsewhere.

REFERENCES

American Ornithologists' Union. 1983. *Check-List of North American Birds*, 6th ed. American Ornithologists' Union, Washington, D.C.

Chipley, R. M. 1977. The impact of wintering migrant wood warblers on resident insectivorous passerines in a subtropical Colombian oak woods. Living Bird 15: 119–141.

Cody, M. L. (ed.). 1985. *Habitat Selection in Birds*. Academic Press, New York.

Denslow, J. S. and T. C. Moermond. 1982. The effect of accessibility on rates of fruit removal from Neotropical shrubs: an experimental study. Oecologia 54: 170–176.

Hilty, S. L. and W. L. Brown. 1986. *A Guide to the Birds of Colombia*. Princeton Univ. Press, Princeton, N.J.

Hutto, R. L. 1980. Winter habitat distribution of migratory land birds in western Mexico with special reference to small, foliage-gleaning insectivores. In A. Keast and E. S. Morton (eds.), *Migrant Birds in the Neotropics: Ecology, Behavior, Distribution, and Conservation*. Smithsonian Inst. Press, Washington, D.C., pp. 181–204.

Karr, J. R. 1976. On the relative abundance of migrants from the north temperate zone in tropical habitats. Wilson Bull. 88: 433–458.

———. 1980. Geographical variation in the avifaunas of tropical forest undergrowth. Auk 97: 283–298.

———. 1982. Avian extinction on Barro Colorado Island, Panama: a reassessment. Amer. Nat. 119: 220–239.

Karr, J. R., S. Robinson, J. G. Blake, and R. O. Bierregaard. (This volume). Birds of four Neotropical forests.

Leck, C. F. 1979. Avian extinctions in an isolated tropical wet-forest preserve, Ecuador. Auk 96: 343–352.

Levey, D. J. 1987a. Seed size and fruit handling techniques of avian frugivores. Amer. Nat. 129: 471–485.

———. 1987b. Sugar-tasting ability and fruit selection in tropical fruit-eating birds. Auk 104: 173–179.

———. 1988. Spatial and temporal variation in Costa Rican fruit and fruit-eating bird abundance. Ecol. Monogr. 58: 251–269.

Levey, D. J., T. C. Moermond, and J. S. Denslow. 1984. Fruit choice in Neotropical birds: the effect of distance between fruits on preference patterns. Ecol. 65: 844–850.

Levey, D. J. and F. G. Stiles. 1991. Birds of La Selva. In L. A. McDade, K. S. Bawa, H. A. Hespenheide, and G. S. Hartshorn (eds.), *La Selva: Ecology and Natural History of a Neotropical Rainforest*. Univ. of Chicago Press, Chicago.

Loiselle, B. A. 1987a. Migrant abundance in a Costa Rican lowland forest canopy. J. Trop. Ecol. 3: 163–168.

———. 1987b. Birds and plants in a Neotropical rain forest: seasonality and interactions. Ph.D. diss., Univ. of Wisconsin, Madison.

———. 1988. Bird abundance and seasonality in a Costa Rican lowland forest canopy. Condor 90: 761–772.

Loiselle, B. A. and J. G. Blake. 1990. Diets of understory fruit-eating birds in Costa Rica. In M. L. Morrison, C. J. Ralph, and J. Verner (eds.), *Food Exploitation by Terrestrial Birds*, Stud. Avian Biol. 13.

Lovejoy, T. 1974. Bird diversity and abundance in Amazonian forest communities. Living Bird 13: 127–191.

Martin, T. E. 1985. Selection of second-growth woodlands by frugivorous migrating birds in Panama: an effect of fruit size and plant density? J. Trop. Ecol. 1: 157–170.

Martin, T. E. and J. R. Karr. 1986. Temporal dynamics of Neotropical birds with special reference to frugivores in second-growth woods. Wilson Bull. 98: 38–60.

Moermond, T. C. and J. S. Denslow. 1983. Fruit choice in Neotropical birds: effects of fruit type and accessibility on selectivity. J. Anim. Ecol. 52: 407–420.

———. 1985. Neotropical frugivores: patterns of behavior, morphology, and nutrition with consequences for fruit selection. In P. A. Buckley, M. S. Foster, E. S. Morton, R. S. Ridgely, and N. G. Smith (eds.), *Neotropical Ornithology*. Ornithol. Monogr. 36. AOU, Washington, D.C., pp. 865–897.

Moermond, T. C., J. S. Denslow, D. J. Levey, and E. Santana C. 1986. The influence of morphology on fruit choice in Neotropical birds. In A. Estrada and T. H. Fleming (eds.), *Frugivores and Seed Dispersal in the Tropics*. W. Junk, Dordrecht, pp. 137–146.

Orians, G. H. 1969. The number of bird species in some tropical forests. Ecol. 50: 783–801.

Ridgely, R. S. 1976. *A Guide to the Birds of Panama*. Princeton Univ. Press, Princeton, N.J.

Slud, P. 1960. The birds of finca "La Selva," Costa Rica: a tropical wet forest locality. Bull. Amer. Mus. Nat. Hist. 121: 49–148.

———. 1964. The birds of Costa Rica. Bull. Amer. Mus. Nat. Hist. 125: 1–430.

Stiles, F. G. 1975. Ecology, flowering phenology, and hummingbird pollination of some Costa Rican *Heliconia* species. Ecol. 56: 285–301.

———. 1977. Checklist of the birds of La Selva and vicinity. MS, Organization for Tropical Studies, Costa Rica.

———. 1978a. Temporal organization of flowering among the hummingbird foodplants of a tropical wet forest. Biotropica 10: 194–210.

———. 1978b. Possible specialization for hummingbird-hunting in the tiny hawk. Auk 95: 550–553.

———. 1980a. The annual cycle in a tropical wet forest hummingbird community. Ibis 122: 322–343.

———. 1980b. Evolutionary implications of habitat relations between permanent and winter resident land-birds in Costa Rica. In A. Keast and E. S. Morton (eds.), *Migrant Birds in the Neotropics: Ecology, Behavior, Distribution, and Conservation*. Smithsonian Inst. Press, Washington, D.C., pp. 421–436.

———. 1983a. Birds: introduction. In D. H. Janzen (ed.), *Costa Rican Natural History*. Univ. of Chicago Press, Chicago, pp. 502–530.

———. 1983b. Checklist of birds. In D. H. Janzen (ed.), *Costa Rican Natural History*. Univ. of Chicago Press, Chicago, pp. 530–544.

———. 1983c. Cambios altitudinales y estacionales en la avifauna de la vertiente Atlantica de Costa Rica. I Symposio de Ornitologia Neotropical (IX Claz Peru, Oct. 1983): 95–103.

———. 1985a. Seasonal patterns and coevolution in the hummingbird-flower community of a Costa Rican subtropical forest. In P. A. Buckley, M. S. Foster, E. S. Morton, R. S. Ridgely, and N. G. Smith (eds.), *Neotropical Ornithology*. Ornithol. Monogr. 36; AOU, Washington, D.C., pp. 757–785.

———. 1985b. Conservation of forest birds in Costa Rica: problems and perspectives. In A. W. Diamond and T. Lovejoy (eds.), *Conservation of Tropical Forest Birds*. ICBP Tech. Publ. 4, Cambridge, pp. 141–168.

———. 1985c. On the role of birds in the dynamics of Neotropical birds. In A. W. Diamond and T. Lovejoy (eds.), *Conservation of Tropical Forest Birds*. ICBP Tech. Publ. 4, Cambridge, pp. 49–59.

Stiles, F. G. and A. F. Skutch. 1989. *A Guide to the Birds of Costa Rica*. Cornell Univ. Press, Ithaca, N.Y.

Stiles, F. G. and L. L. Wolf. 1974. A possible circannual molt rhythm in a tropical hummingbird. Amer. Nat. 108: 341–354.

———. 1979. *The Ecology and Evolution of a Lek Mating System in the Long-tailed Hermit Hummingbird*. AOU Monogr. 27, Washington, D.C.

Terborgh, J. 1983. *Five New World Primates: A Study in Comparative Ecology*. Princeton Univ. Press, Princeton, N.J.

Willis, E. O. 1966. The role of migrant birds at swarms of army ants. Living Bird 5: 187–231.

———. 1979. The composition of avian communities in remanescent woodlots in southern Brazil. Pap. Avul. Zool. 33: 1–25.

11

The Avifauna of Barro Colorado Island and the Pipeline Road, Panama

JAMES R. KARR

Barro Colorado Island (BCI) is arguably the most thoroughly studied tropical forest in the world. Observations of birds on BCI began in the 1920s with such noted scholars as F. M. Chapman, A. O. Gross, J. Van Tyne, and L. Griscom. Because BCI is a recently isolated land-bridge island, however, its avifauna has changed over the past seventy years. This analysis thus includes an area of Soberanía National Park referred to as the Pipeline Road (PR). PR is adjacent to BCI on the east side of the Panama Canal (fig. 11.1). The avifauna of the PR was little known before the mid-1960s, when observations by many individuals, including members of a very active Panama Audubon Society, added to knowledge of the species composition and biology of its avifauna. The area considered here is about 16,000 ha: 1,500 ha on BCI and the remainder in the PR area.

Any effort to describe the avifauna of a geographical area must state the principle geographic characteristics of that area. BCI is a land-bridge island isolated from the nearby mainland early in the twentieth century. The PR is an elongate mainland area east of BCI with approximate dimensions of 8 km by 18 km. Both BCI and the PR are composed of gently rolling to moderately steep topography ranging from 30 m to 200 m in

This chapter is dedicated with my appreciation to the many individuals, both professional and amateur, who have contributed to knowledge of the birds of central Panama. In addition, I acknowledge the logistical support of the Smithsonian Tropical Research Institute and the many field assistants and volunteers that have helped with the fieldwork reported here. Finally, the financial support of the National Science Foundation (DEB #82-06672) and Earthwatch/Center for Field Research have been essential to the continuation of this study.

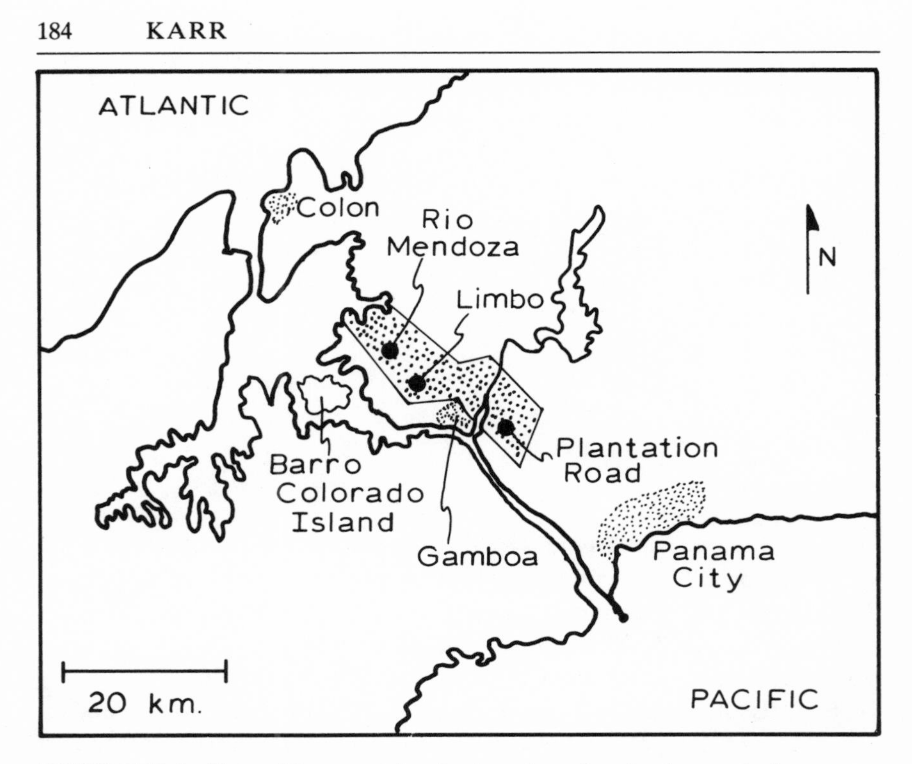

FIGURE 11.1. Central Panama, showing location of study plots and other locations mentioned in text.

elevation. Although both areas were once broadly connected to foothill areas to the east and west, those connections were severed in the past few decades by a rapidly increasing human population and resulting habitat destruction. Central Panama experiences a distinctly seasonal climate, with a dry season from late December to late April. Annual rainfall at BCI averages about 2,600 mm per year but is somewhat less (2,100 mm) at Gamboa at the south end of the PR. At the north end of the PR, rainfall is about 3,000 mm per year. As a result, the severity of the dry season varies markedly over the PR-BCI area. Finally, the rainfall gradient across the isthmus has produced a biota that is a mixture of faunal and floral elements common to wet Atlantic and dry Pacific coastal lowlands in central Panama.

The Avifauna

The first comprehensive compilation of birds known from BCI was prepared by Eisenmann (1952), and that list was updated by Willis and

Eisenmann (1979). The first such compilation of birds known from the PR was compiled by the Panama Audubon Society, and I have updated that list in recent years. The number of species recorded from the combined areas is 444 (see appendix 14.1). Of these, 69 are recorded from the PR but not at BCI and 51 from BCI but not the PR. Five species have disappeared in recent decades not only from BCI and PR but from central Panama altogether (four species of *Ara* and *Daptrius americanus*). At least 66 species have become extinct from BCI in this century (Terborgh 1974; Willis and Eisenmann 1979; Karr 1982a). About two-thirds disappeared as early successional habitats matured (that is, due to loss of habitat), but 30% of the extirpations are of forest birds, even though extensive areas of their habitat remain on BCI. Some may never have been present as breeding residents but may have been vagrants. Most of these and about 40 other forest species not found on BCI are present at the PR. The additional 40 may never have occurred on BCI, or perhaps they disappeared before they were noted (Karr 1982a). Avifaunal differences between BCI and the PR have been described previously (Karr 1982a), so they will not be discussed here.

Of the 444 species recorded from the two areas, 73 are considered vagrants—that is, they have been seen only a few times. Some are seabirds dispersing across the isthmus, whereas others are typical of Atlantic coastal forest or highland forest not present within the confines of the area considered here. Of the 73 vagrants, 18 have been recorded only at PR, 25 from both areas, and the largest number, 30, only from BCI. Field observations in the early to middle years of this century were concentrated on BCI, so the accumulation of vagrant observations is higher there.

Most nonvagrant species (282; 76%) in central Panama exhibit no regular, long-distance migration. Sixty-three (17%) are migrants from North America, and 5 (1%) migrate between Central and South America. Only 2 seem to show regular altitudinal or seasonal movements, the latter in association with rainfall patterns. A final group of 18 species, mostly large raptors and parrots, travel over very large areas on a daily basis.

Excluding the vagrants from the combined PR-BCI list leaves 370 species. Of these, 278 (75%) are known or thought to breed in the area. Each nonvagrant species was classified according to its dominant habitat, such that each species is counted only once. Fifty-five species (15%) are associated with lakes or rivers. Sixty-four species (17%) are primarily associated with aerial and second-growth habitats. The remaining 251 species (68%) are birds associated with forest, including small forest

TABLE 11.1 Number of species in each of several ecological groupings for the forest birds of central Panama

Stratum		Substrate		Diet	
Ground	28	Ground	27	Small insects	74
Shrub	56	Foliage (live)	164	Small insects and fruits	47
Understory	44	Foliage (dead)	5	Large insects	39
Canopy	117	Army ants	9	Large insects and fruits	19
Water	6	Aerial	25	Large insects, fruits and small vertebrates	6
		Water	6	Fruits	32
		Trunks and branches	14	Vertebrate predator (general)	8
		Twigs	1	Birds	1
				Mammals	2
				Herps	3
				Fish	4
				Carrion	1
				Nectar and arthropods	14
				Snails	1

streams, old second-growth, forest interior or canopy, and forest edges and gaps.

Each of the forest species was classified according to several ecological criteria (table 11.1). The classifications used and the classes in each were as follows: FORAGING STRATUM—ground; shrub; understory; canopy; and aquatic. DIET—small insects; small insects and fruits; large insects; large insects and fruit; large insects, small vertebrates, and fruits; fruits or fruits and seeds; general vertebrate predator; birds; mammals; herps; fish; carrion; nectar and arthropods. FORAGING SUBSTRATE—ground; foliage live (including fruits and flowers); foliage dead; aerial; water; trunks and branches; and twigs. In addition, species that specialize at following army ants are grouped. Of the 251 species of forest birds, 11 are nocturnal.

These categories are somewhat arbitrary. Birds, for example, do not forage with respect to distances above the ground or by measuring the length of insects. Foraging height varies as a function of vegetation structure, and birds diverge from their normal haunts under unusual circumstances. For example, I occasionally catch *Ornithion brunneicapillum* in ground-level mist nets, typically when adults accompany newly fledged young. As a guideline, shrub species are birds that typically forage from 0.5 to 5.0 m above the ground, whereas understory species concentrate in the area from 5 to 20 m. Canopy species feed at or near the foliage-air interface above 20 m, except in treefall or edge areas, where they may descend to much lower heights. These conventions follow Karr (1971), Willis (1980), and Stiles (1983).

Species that feed on small insects (74 species; 29% of forest birds) or small insects and fruits (47; 19%) are the two largest groups (table 11.1). Large insects (39; 16%) and fruits (32; 13%) are the next largest groups, followed by a combination of large insects and fruits (19; 8%). Fish are consumed by 4 species. Nectar is the dominant dietary item for 14 species; all supplement their diets with small insects.

Tropical forest birds tend to have narrowly defined foraging strata, indicating greater specialization than is found in temperate forest birds (Terborgh and Weske 1969; Karr 1971). The stratum with the largest number of species is the canopy, followed by shrub and understory (table 11.1). Within each stratum several substrates may be exploited by foraging birds. Sixty-six percent of species obtain their food by gleaning from foliage (including fruit and flowers), with ground and aerial feeding next at 11% and 9% of species, respectively.

The distribution of birds across strata varies among foraging guilds

TABLE 11.2 Number of species by selected stratal and diet categories for the forest birds of central Panama, excluding aquatic and above-canopy species

	Stratum			
Diet	Ground	Shrub	Understory	Canopy
Small insects	11	19	20	21
Large insects	6	9	12	12
Fruits and small insects	2	12	4	29
Fruits and large insects	0	1	3	15
Fruits, large insects, and small vertebrates	1	0	0	5
Fruits and/or seeds	7	4	0	21
Nectar and arthropods	0	11	0	3
Terrestrial vertebrates	0	0	5	10

(table 11.2). Both small and large insectivores have the fewest species at the ground but relatively little variation in percent of species occupying each stratum (15–31%). In contrast, the four groups that include fruits or fruits and insects in their diet are concentrated more in the canopy (62–83% of species in canopy) and have variable distributions among the four primary strata. Nectarivores are concentrated in the shrub layer and species that feed on terrestrial vertebrates are mostly found in the canopy.

Mist-Net Samples of Understory Birds

Birds were sampled using 12-m, four-shelf mist nets of 36-mm black nylon mesh. Sampling is efficient at capturing most forest undergrowth species, especially passerines, except those that exceed 100 gm and those that spend most of their time walking on the ground (Karr 1979, 1981). Mist nets provide an opportunity to accumulate information on species active in forest undergrowth in a relatively short period. Furthermore, both sample size (100 captures) and sample effort (number of mist-net hours) can be standardized. Twelve to twenty nets were deployed over an area of 2–4 ha. Each study plot was netted from three to eight days until 100 captures were accumulated. For the analysis presented here, I use from three to twenty replicate samples at each site. Analysis of samples collected by operation of mist nets in the undergrowth of forest does not provide a sample of the entire avifauna, but it does provide a sample of the birds using a segment of the tropical forest. Observer biases (familiarity with vocalizations and ability to identify birds glimpsed only briefly) that plague most census methods are avoided. Although density estimates cannot be made, population estimates are possible with Jolly-Seber models (Karr et al., in prep.). Use of mist nets has the advantage of producing a repeatable sample of forest undergrowth birds. These samples are admittedly subject to the major variation in activity of the various species, some of which are active only in the level of the nets and others that travel in and out of the area sampled (0 to 3 m high). In some cases, seasonal variation in foraging height may produce temporal variation in capture rates within a species. Similarly, interspecific variation in capture rates result from differences in the mobility, spacing systems, and feeding ecology of bird species (Karr 1971; Remsen and Parker 1983). Biases inherent in sampling with mist nets are discussed more fully in Karr (1979, 1981), Karr, Schemske, and Brokaw (1982), Remsen and Parker (1983), and Greenberg and Gradwohl (1986).

Six study plots were located in Soberanía National Park and one on BCI. All were selected as homogeneous plots with as little variation in vegetation structure among plots as could be found (Karr and Freemark 1983). Variation among the plots derives from diverse macro- and microclimatic conditions and from different histories of disturbance of vegetation (Karr 1981; Karr and Freemark 1983). Four plots are adjacent to Limbo Camp within a 2 km^2 area of moist lowland forest along the PR about 8 km northwest of Gamboa. Those plots (from dry to wet: Ridge, Road, Limbo, and Valley) were selected to represent the range of understory moisture conditions characteristic of the 2 km^2 area centered at Limbo Camp (Karr and Freemark 1983). Habitat differences among plots stemmed from local topography, presence of streams, and treefalls. All plots contain forty-to-four-hundred-year-old forest and receive about 2,600 mm of rain per year. The BCI study plot has similar climate, topography, and vegetation structure (Karr 1982a).

Two other mainland sites were sampled in 1983 and 1984. A site at Plantation Road (14 km southeast of the PR study plots) experiences a more marked dry season, an annual rainfall of 2,100 mm, and is relatively flat with a semipermanent stream. Río Mendosa is the next stream north of the Limbo Camp and occupies a relatively deep valley. It is a larger stream than is associated with any of the other study plots. The area probably receives in excess of 2,800 mm of rainfall each year and certainly experiences more regular cloud cover (Karr, pers. obs.). Finally, a plot on BCI, about 10 km west of the PR plots, was sampled ten times since 1980

(Karr 1982a). The forest is about two hundred years old (Foster and Brokaw 1982). Among the four mainland plots noted above, the area most resembles the Limbo plot, though it lacks a permanent stream.

Results

Species accumulation curves. Data from the four Limbo area plots have been accumulated since at least 1979, and all show essentially identical rates of species accumulation—80 species at 1,500 captures (fig. 11.2). Although the identity and number of species caught varies among years at individual plots (Karr 1981), the accumulation of species when the four plots are combined (Limbo mosaic) is similar to the accumulation curves for the individual plots. In addition, the number of species seems to increase consistently almost indefinitely as samples accumulate

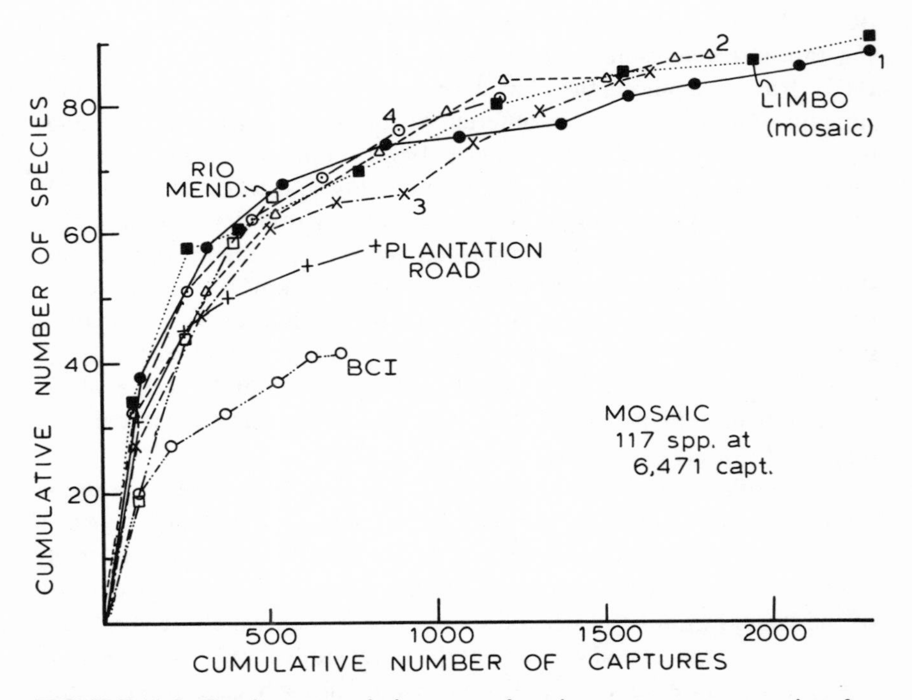

FIGURE 11.2. Species-accumulation curves for mist-net captures at a series of forest sites in central Panama. Sites near Limbo Camp from dry to wet: Ridge (*2*), Road (*3*), Limbo (*1*), and Valley (*4*); Río Mendosa; Plantation Road; Barro Colorado Island (BCI).

over time. In just over 6,400 captures at the Limbo study plot, 117 species have been caught and new species are still recorded each year. Although fewer data are available for the Río Mendosa plot, it lies within the bounds of the Limbo area plots.

Species accumulation curves are lower at both the Plantation Road and, especially, the BCI plot. The smaller number of species at the Plantation Road site is characteristic of drier sites throughout Central America (Stiles 1983; Karr, pers. obs.). Although its rainfall is similar to that of Limbo, the BCI study plot has species accumulation rates well below those of the Limbo area plots and the drier forest at Plantation Road. The lack of undergrowth species on BCI has been ascribed variously to island effects (Willis 1974; Terborgh 1974), predation (Loiselle and Hoppes 1983), and a lack of moist refuges during dry periods (Karr 1982a,b). The real explanation is probably more complex than is implied by any single theory.

Relative abundances. Accurate assessment of the relative abundances of the birds of the tropical forest is difficult. Like any sampling procedure, mist nets yield a biased estimate of relative abundances, with overestimates for mobil species and underestimates for more sedentary species. As a result, mist-net capture data cannot be used as surrogates of density or population size. Comparison of relative abundances described here are used only to compare samples collected during different time periods. Two samples taken about ten years apart at Limbo (fig. 11.3) had similar distributions of relative abundance (and accumulation of species as well; Karr, Schemske, and Brokaw 1982). Indeed, among the 9 most frequently captured species in 1977–78, 6 were also in the top 9 species in 1968–69. *Pipra mentalis* was the most frequently captured species in both samples, at 19.6% and 21.9% of captures.

Most recently, 1,000 capture samples were compared from the four study plots in the vicinity of Limbo Camp. Only March (dry season) samples are included, one set 1979–81, and the second 1984–86. *Pipra mentalis* was the most frequently captured species in both samples at 10.8% and 12.1%, respectively. *Gymnopithys leucaspis* and *Mionectes oleagineus* were next most common. Seven of the 10 most frequently captured species were found in both samples (1979–81 versus 1984–86). Seventy-eight species and 699 individuals were represented in the 1979–81 samples, and 84 species and 754 individuals were represented in the 1984–86 samples. All relative abundance curves are characterized by a rapid decline in capture proportions with a long tail of rarely captured species. A sample of

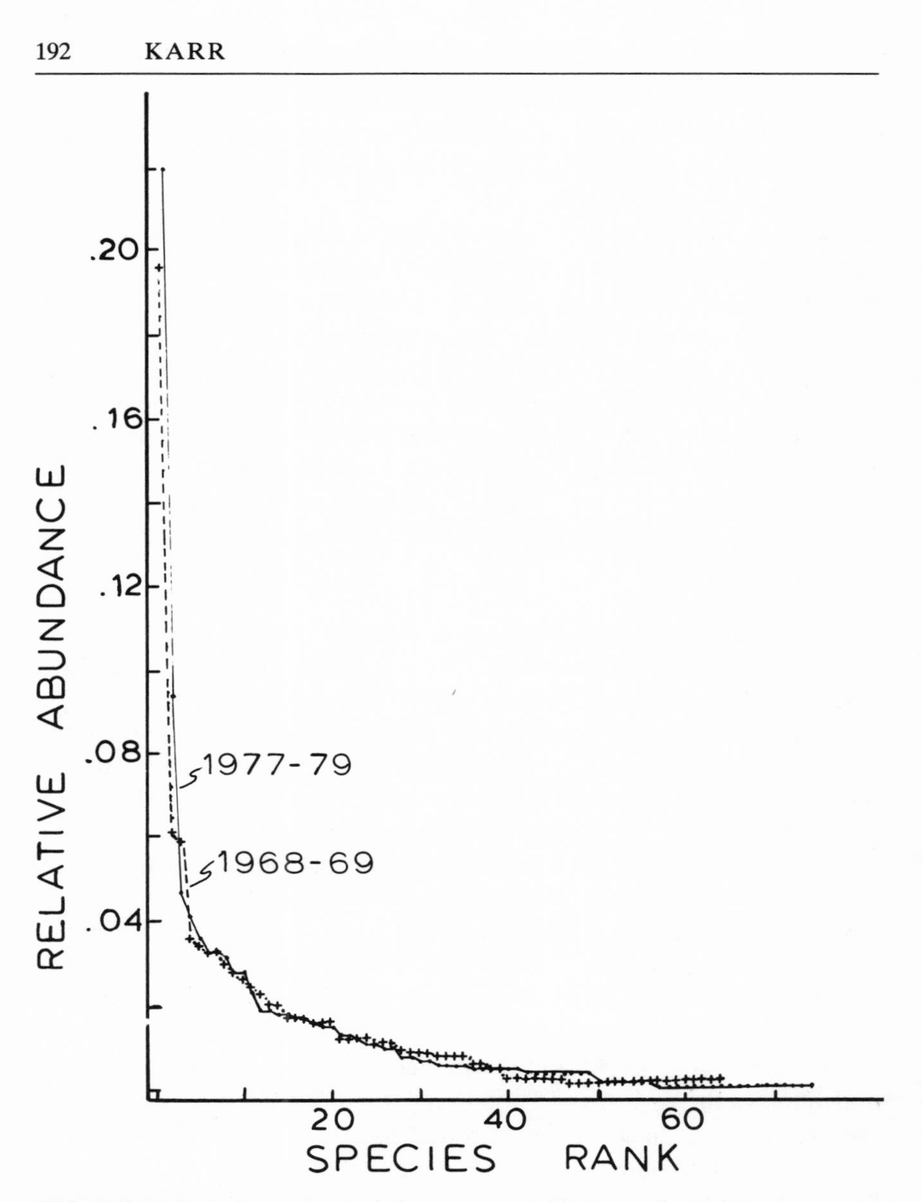

FIGURE 11.3. Proportion of captures per species from most abundant to least abundant species in two samples from Limbo Study Plot. *1968–69*: 65 species, 643 captures; *1977–78:* 74 species, 826 captures. (From Karr, Schemske, and Brokaw 1982.)

1,000 captures typically contains about 18 species represented by only one capture and an additional 50 speciess that are captured from two to twenty times each. Typically, only a dozen species make up more than 2% of the sample (that is, more than twenty captures).

Trophic structure. Temperate forest bird communities are dominated by insectivorous species, whereas tropical forest avifaunas contain species that consume a wider variety of food resources (Karr 1975). Mist nets provide a sample that permits trophic analysis (table 11.3). Among a series of forest samples in Panama, insectivores are the most diverse guild, whereas the frugivores are represented by the largest number of captures (reflecting both their abundance and high mobility). Professional ant-followers are most variable in the representation among samples, presumably because of the high mobility of the army ants upon which these species depend. The four sites near Limbo have essentially the same guild signatures. Species richness is a bit lower at Plantation Road than for Limbo area plots, primarily due to smaller numbers of ant-following species and, to a lesser extent, other insectivore groups, especially flycatchers and ground gleaners. Virtually all of the guilds tend to have reduced numbers of species in the BCI fauna except, perhaps, for the frugivores.

Capture rates. The rate of capture of birds in nets can be expressed as captures per net day or per 100 net hours. Here I use captures per 100 net hours. Previous analyses have shown that for a given site capture rates vary with time of day (high during early morning, lower in late afternoon and lowest at midday), season (higher in dry than wet season), and among years (Karr 1981, 1985). The causes of these patterns are complex and variable in both space and time (Karr, in prep.). Some changes can be correlated with changing microclimatic conditions (Karr and Freemark 1983), whereas others seem to be related to availability of food resources (ant followers, some frugivores). Shifts reflect subtle and not so subtle variation in habitat use by the birds of forest undergrowth.

Total capture rates (all species) vary, for example, between the Ridge and Limbo plots. The drier Ridge plot has higher capture rates in the wet than dry season, while the pattern at Limbo is the reverse (Karr and Brawn, in prep.). When all four Limbo area plots are combined, local variation is damped, and it is then possible to examine variation at a different (mosaic) scale. Total capture rates are consistently but not significantly ($P > .05$) higher in the dry (17.2 ± 3.8) than wet (14.1 ± 3.2)

TABLE 11.3 Guild signatures for study plot in central Panama based on 100-capture sample at each site

Plot	Sample	Frugivore Terrestrial	Frugivore Arboreal	Frugivore Insectivore Terrestrial	Frugivore Insectivore Arboreal	Insectivore Nectarivore	Insectivores Bark	Insectivores Ground	Insectivores Foliage	Insectivores Flycatcher	Ant Follower	Other
		Number of species (individuals)										
Limbo	Dry 1979	1 (1)	3 (11)	0 (0)	3 (10)	1 (2)	3 (4)	9 (11)	8 (11)	3 (12)	6 (25)	1 (1)
	Wet 1979	1 (4)	4 (24)	1 (1)	4 (8)	2 (4)	2 (2)	5 (7)	7 (19)	2 (4)	6 (10)	1 (1)
	Dry 1980	1 (1)	2 (20)	0 (0)	3 (12)	4 (12)	1 (1)	2 (2)	9 (14)	3 (6)	5 (18)	0 (0)
	Wet 1980	1 (1)	4 (21)	0 (0)	3 (6)	0 (0)	3 (5)	5 (10)	7 (16)	1 (1)	6 (18)	0 (0)
BCI	Dry 1983	1 (1)	2 (16)	0 (0)	4 (12)	1 (1)	2 (8)	2 (8)	5 (17)	1 (5)	3 (12)	0 (0)
	Dry 1985	1 (1)	2 (17)	0 (0)	2 (6)	1 (5)	2 (4)	2 (9)	7 (17)	1 (5)	3 (12)	1 (2)
Plantation Road												
	Dry 1983	0 (0)	3 (23)	0 (0)	4 (10)	2 (18)	2 (3)	6 (10)	8 (18)	2 (4)	2 (5)	0 (0)
	Wet 1983	2 (7)	3 (36)	0(0)	2 (3)	3 (5)	0 (0)	4 (11)	4 (13)	2 (2)	3 (5)	0 (0)

season (eight years of samples). However, analysis of the 20 most common species shows that 16 of the 20 species have capture rates that differ significantly between the dry and wet seasons. Some, like most flycatchers, are typically captured at higher rates in the dry season, whereas others (*Geotrygon montana*) seem to be more common in the wet season.

Capture rates also vary among years. The most striking example comes from the recent El Niño drought of 1983, when capture rates at three study plots with sufficient data for analysis exhibited the most extreme capture rates ever recorded at those sites. El Niño capture rates at the relatively dry Ridge and Road study plots were the lowest ever recorded in eight years of sampling (22% and 35% below average, respectively). In contrast, the wetter Limbo plot had the highest capture rates ever recorded for a dry season sample at the site (52% above the mean).

Survival rate. A major advantage of the use of mist nets is that their routine long-term use allows banding of individuals and tracking of their movement patterns and longevity. Mark-recapture data are now available for over 2,700 individuals of 25 species, and demographic parameters have been estimated. Those data suggest that the long-assumed high survival rates of tropical forest birds may not be true. Some species admittedly have high survival rates (greater than 70% per year), but others have survivorship as low as 30% per year (Karr et al. 1990).

Other Studies

The long history of observations of birds in central Panama goes back to the mid-nineteenth century during the days of the construction and early operation of the Panama Railroad, especially the collections of James McLeannan. Interest in the biology of central Panama increased during the construction of the Panama Canal. That stimulated efforts to define a long-term research area in central Panama, and that goal was realized when Barro Colorado Island was designated a reserve in 1923. Starting with the early observations of Chapman on BCI and Sturges in the wider canal area and continuing to modern times, birds have been the focus of much observation and research. Skutch made early observations, and Wetmore and Eisenmann made regular trips to Panama with many hours in the field in central Panama. More recently, many others have added to our knowledge of the avifauna through a range of studies that address many questions and use differing methodologies. These have included but are not limited to the following:

1. Willis (1967–80; also Willis and Eisenmann 1979; Willis and Oniki 1972) on the ecology, population dynamics, behavior of the ant-following birds, and early analyses of extinctions on BCI
2. Johnson (1953, 1954), Moynihan (1960, 1962), Wiley (1971), and Gradwohl and Greenberg (1980) on the dynamics of mixed feeding flocks
3. Greenberg and Gradwohl on territoriality and breeding (1982b), foraging ecology (1980, 1981a, 1982a), ecology of migrants (1981c, 1984, 1986), and the ecology of canopy birds (1981b)
4. Howe (1982) on the feeding and seed-dispersal systems of several canopy trees on BCI
5. Worthington (1982) on the frugivorous manakins; she showed, for example, that in the season of fruit shortage manakins did not breed, spent more time foraging, and ate more insects
6. Morton (1971, 1973, 1975, 1977) on bird vocalizations, frugivory, and migrants; Morton initiated research on wrens that has been followed by Farabaugh and more recently pioneering work by Rachel Levin on dueting by *Thryothorus nigricapillus*
7. Leck (1972) on frugivorous and nectarivorous birds on BCI

REFERENCES

Eisenmann, E. 1952. Annotated list of birds of Barro Colorado Island, Panama Canal Zone. Smithsonian Misc. Coll. 117(4): 1–62.

Foster, R. B. and N. V. L. Brokaw. 1982. Structure and history of the vegetation of Barro Colorado Island. In E. G. Leigh, Jr., A. S. Rand, and D. M. Windsor (eds.), *The Ecology of a Tropical Forest: Seasonal Rhythms and Long-Term Changes*. Smithsonian Inst. Press, Washington, D.C., pp. 67–81.

Gradwohl, J. and R. Greenberg. 1980. The formation of antwren flocks on Barro Colorado Island, Panama. Auk 97: 385–395.

———. 1982a. The effect of a single species of avian predator on the arthropods of aerial leaf litter. Ecol. 63: 581–583.

———. 1982b. The breeding season of antwrens on Barro Colorado island. In E. G. Leigh, Jr., A. S. Rand, and D. M. Windsor (eds.), *The Ecology of a Tropical Forest: Seasonal Rhythms and Long-Term Changes*. Smithsonian Inst. Press, Washington, D.C., pp. 345–351.

Greenberg, R. 1981a. Dissimilar bill shapes in New World tropical vs. temperate forest foliage gleaning birds. Oecologia 49: 143–147.

———. 1981b. The abundance and seasonality of forest canopy birds on Barro Colorado Island, Panama. Biotropica 13: 241–251.

———. 1981c. Frugivory in some migrant tropical forest wood warblers. Biotropica 13: 215–223.

———. 1984. The winter exploitation systems of bay-breasted and chestnut-sided warblers in Panama. Univ. Calif. Publ. Zool. 116: 1–107.

———. 1986. Competition in migrant birds in the nonbreeding season. Current Ornithol. 3: 281–307.

Greenberg, R. and J. Gradwohl. 1980. Leaf surface specialization of birds and arthropods in Panamanian forest. Oecologia 46: 115–124.

———. 1986. Constant density and stable territoriality in some tropical insectivorous birds. Oecologia 69: 618–625.

Howe, H. F. 1982. Fruit production and animal activity in two tropical trees. In E. G. Leigh, Jr., A. S. Rand, and D. M. Windsor (eds.), *The Ecology of a Tropical Forest: Seasonal Rhythms and Long-Term Changes*. Smithsonian Inst. Press, Washington, D.C., pp. 189–200.

Johnson, R. A. 1953. Breeding notes on two Panamanian antbirds. Auk 70: 494–496.

———. 1954. The behavior of birds attending army ant raids on Barro Colorado Island, Panama Canal Zone. Proc. Linnean Soc. N.Y. 65: 41–70.

Karr, J. R. 1971. Structure of avian communities in selected Panama and Illinois habitats. Ecol. Monogr. 41: 207–233.

———. 1975. Production, energy pathways, and community diversity in forest birds. In F. B. Golley and E. Medina (eds.), *Tropical Ecological Systems: Trends in Terrestrial and Aquatic Research*. Springer-Verlag, New York, pp. 161–176.

———. 1979. On the use of mist nets in the study of bird communities. Inland Bird Banding 51: 1–10.

———. 1981. Surveying birds with mist nets. In C. J. Ralph and J. M. Scott (eds.), *Estimating Numbers of Terrestrial Birds*. Stud. Avian Biol. 6: 62–67.

———. 1982a. Avian extinction on Barro Colorado Island, Panama: a reassessment. Amer. Nat. 119: 220–239.

———. 1982b. Population variability and extinction in a tropical land-bridge island. Ecol. 63: 1975–1978.

———. 1985. Birds of Panama: biogeography and ecological dynamics. In W. G. D'Arcy and M. D. Correa A. (eds.), *The Botany and Natural History of Panama: La Botanica e historia natural de Panamá*. Monogr. Syst. Bot. 10: 77–93. Mo. Bot. Gard, St. Louis.

Karr, J. R. and K. E. Freemark. 1983. Habitat selection and environmental gradients: dynamics in the "stable" tropics. Ecol. 64: 1481–1494.

Karr, J. R., J. D. Nichols, M. Klimkiewicz, and J. D. Brawn. 1990. Survival rates of birds of tropical and temperate forest: will the dogma survive? Amer. Nat., *in press*.

Karr, J. R., D. W. Schemske, and N. Brokaw. 1982. Temporal variation in the undergrowth bird community of a tropical forest. In E. G. Leigh, Jr., A. S. Rand, and D. M. Windsor (eds.), *The Ecology of a Tropical Forest: Seasonal Rhythms and Long-Term Changes*. Smithsonian Inst. Press, Washington, D.C., pp. 441–453.

Leck, C. F. 1982. Seasonal changes in feeding pressures of fruit and nectar-eating birds in Panama. Condor 74: 54–60.

Loiselle, B. A. and W. G. Hoppes. 1983. Nest predation in insular and mainland lowland forest in Panama. Condor 85: 93–95.

Morton, E. S. 1971. Food and migration habits of the eastern kingbird in Panama. Auk 88: 925–926.

———. 1973. On the evolutionary advantages of fruit eating in tropical birds. Amer. Nat. 107: 8–22.

———. 1975. Ecological sources of selection in avian sounds. Amer. Nat. 109: 17–34.

———. 1977. Intratropical migration in the yellow-green vireo and piratic flycatcher. Auk 94: 97–106.

Moynihan, M. H. 1960. Some adaptations which help to promote gregariousness. Proc. Int. Ornithol. Congr. 12: 523–541.

———. 1962. The organization and probable evolution of some mixed species flocks of Neotropical birds. Smithsonian Misc. Coll. 143(7): 1–140.

Remsen, J. V., Jr. and T. A. Parker, III. 1983. Contribution of river created habitats to bird species richness in Amazonia. Biotropica 15: 223–231.

Stiles, F. G. 1983. Birds: introduction. In D. H. Janzen (ed.), *Costa Rican Natural History*. Univ. of Chicago Press, Chicago, pp. 502–530.

Terborgh, J. W. 1974. Preservation of natural diversity: the problem of extinction prone species. BioScience 24: 715–722.

Terborgh, J. W. and J. S. Weske. 1969. Colonization of secondary habitats by Peruvian birds. Ecol. 50: 765–782.

Wiley, R. H. 1971. Cooperative relationships in mixed flocks of antwrens (Formicariidae). Auk 88: 881–892.

Willis, E. O. 1967. The behavior of bicolored antbirds. Univ. Calif. Publ. Zool. 79: 1–132.

———. 1972a. The behavior of plain-brown woodcreepers, *Dendroncincla fuliginosa*. Wilson Bull. 88: 377–420.

———. 1972b. The behavior of spotted antbirds. Ornithol. Monogr. 10: 1–162.

———. 1973. The behavior of ocellated antbirds. Smithsonian Contrib. Zool. 144: 1–57.

———. 1974. Populations and local extinction of birds on Barro Colorado Island, Panama. Ecol. Monogr. 44: 153–169.

———. 1980. Ecological roles of migratory and resident birds on Barro Colorado Island, Panama. In A. Keast and E. S. Morton (eds.), *Migrant Birds in the Neotropics: Ecology, Behavior, Distribution, and Conservation*. Smithsonian Inst. Press, Washington, D.C., pp. 205–225.

Willis, E. O. and E. Eisenmann. 1979. A revised list of birds of Barro Colorado Island, Panama. Smithsonian Contrib. Zool. 291: 1–31.

Willis, E. O. and Y. Oniki. 1972. Ecology and nesting behavior of the chestnut-backed antbird, *Myrmeciza exsul*. Condor 74: 87–101.

Worthington, A. 1982. Population sizes and breeding rhythms of two species of manakins in relation to food supply. In E. G. Leigh, Jr., A. S. Rand, and D. M. Windsor (eds.), *The Ecology of a Tropical Forest: Seasonal Rhythms and Long-Term Changes*. Smithsonian Inst. Press, Washington, D.C., pp. 213–225.

12

Bird Communities of the Cocha Cashu Biological Station in Amazonian Peru

SCOTT K. ROBINSON AND JOHN TERBORGH

The forests of western Amazonia contain the richest and least disturbed bird communities in the world. Several sites in eastern Ecuador and Peru support over 500 species in areas of less than 50 km^2 (Terborgh 1985). Two of these sites, the Cocha Cashu Biological Station and the Tambopata Preserve, are in the Department of Madre de Dios in southeastern Peru, within one of the centers of avian endemism identified by Haffer (1985). Ornithologists at both research stations have built their species lists to over 550 in a friendly competition for the world record among comparably sized areas. Historically, however, western Amazonian bird communities have been among the last to be described. Until very recently, too little was known about the distributions, vocalizations, taxonomy, and habitats of most Amazonian birds to attempt anything more than descriptions of the natural history of specific groups and the development of species lists. Indeed, new species continue to be discovered from lowland habitats (for example, Pierpont and Fitzpatrick 1983; Terborgh et al. 1984), and other species known only from museum skins have been "rediscovered" (Parker 1982). Similarly, until recently there has been a general ignorance of what constituted major habitats for birds (Parker et al. 1982; Remsen and Parker 1983; Terborgh 1985).

As Terborgh (1985) has argued, however, we now know enough about the basic natural history of most species to begin community-level studies. In this chapter we present an overview of the work on bird community organization at the Cocha Cashu Biological Station in Manu National Park, Peru. The bird community of this lowland site has been under intensive study since 1973, when mist-net sampling began. Since then, four dissertation projects have been completed and another five begun on various families or groups of birds. In 1980, we began an intensive, large-scale census of the bird communities of the major habitats near the biological station. We now have data on population densities, habitat selection, and territory sizes for most of the approximately 435 resident terrestrial species that occur in the Cocha Cashu area. We present data on the species composition, guild structure, and organization of bird communities in the major habitat types near Cocha Cashu. To facilitate comparisons with other sites in this volume, we will emphasize mature forest habitats and mist-net data.

Habitats

The vegetation, climate, and topography of the Cocha Cashu Biological Station are described elsewhere in this volume and in Terborgh (1983). Terborgh et al. (1984) and Terborgh (1985) described the major habitats of the Manu River. Here we provide only brief summaries to aid in the interpretation of the data presented in this chapter. *Aquatic habitats* include both the Manu River and the marsh-bordered oxbow lakes along its floodplain. These habitats contain many piscivorous species, such as herons, kingfishers, and cormorants (Willard 1985). *Beaches* occur along the inside of meander loops during the dry season, when weedy annuals grow on the upper reaches. *Tessaria/Cane* thickets represent the first phase of primary succession along the inside of meander loops. The youngest layer is a virtual monoculture of *Tessaria integrifolia*, a treelike Compositae. The next, or *Cane*, stage is dominated by a tall (up to 10 m) cane, *Gynerium sagittatum*, which also forms a virtual monoculture until overtopped by *Cecropia*, *Erythrina*, and *Ficus*. *Transition Forest* includes the middle stages of riverine succession, beginning with a *Ficus*-dominated stage that gradually becomes more diverse over time. *High Ground Forest*, the name we have given the next stage of floodplain succession, grows on rich soils that are flooded only occasionally during the rainy season. This floristically diverse habitat is characterized by its closed canopy and many tall emergent trees, some of which reach 60 m (Foster et al. 1986). Most sections of High Ground Forest include seasonal swamps, which represent the last stages of lake bed succession. As floodplain forest becomes very old, it undergoes substantial changes, which create two distinct habitats. In some poorly drained areas, monocultures of *Mauritia* palms (locally called *Aguajales*) form. The second habitat, which we call *Open Forest*, is characterized by frequent treefalls, a dense understory of bamboo and *Heliconia*, and wide spacing between canopy trees. *Upland Forests* grow on terraces well above flood level. They are dissected by steep ravines where landslides create large gaps in the canopy. Upland Forests also contain extensive patches of bamboo. *Lake Margin* vegetation consists of the shrubby second growth that replaces marshes in lake bed succession.

Methods

Faunal list. The faunal list from Cocha Cashu provided in appendix 14.1 updates the list published by Terborgh et al. (1984) and includes all species recorded within a 15-km radius of Cocha Cashu from 1973 to 1989. Most species have either been captured in one of the eighteen mist-net lines or observed by an experienced ornithologist. No areas of second growth caused by human activity occur within this area. In spite of intensive ongoing ornithological studies, only two to six species have been added to the list each year since 1982. At least twenty species, however, occur in comparable habitats elsewhere in southeastern Peru but have not yet been found in the Cocha Cashu area (T. Parker, p.c.; V. Remsen, p.c.). The species list, therefore, should not be considered complete.

Mist-net samples. Every year since 1973 we have operated mist-net lines through various habitats. Each net line consists of 12-m ATX mist nets strung end-to-end in straight lines of 50–500 m through a habitat. Nets are opened for four days from dawn until dark, a period of roughly eleven to twelve hours. All birds are measured and banded with numbered metal rings and, for some species, unique combinations of color bands.

Song censuses. From 1981 to 1987 we conducted song censuses in five intensive study plots ranging in size from 60 to 99 ha. Each plot contained an extensive trail system designed so that trails were no more than 200 m apart. Following the methods developed by Kendeigh (1944), observers walked fixed routes along these trails and recorded the locations of every bird heard or observed. In particular, all simultaneous registrations of two or more singing males were noted. Censuses were conducted at dawn and dusk during the late dry and early wet seasons (August–November), when most species breed. Each route was censused at least twenty-five times.

From these data, we made summary maps of all registrations for each species; these were used to estimate territory boundaries and population densities. For many families, we supplemented song censuses with data from observations of color-marked birds. This practice was followed especially with flock-living species, which Munn (1985) studied in detail.

The methods used to census parrots, manakins, and various other nonterritorial species will be described elsewhere (Terborgh et al., in press). We have not yet censused Aguajales or Open Forest habitats and therefore restrict our data analysis to the habitats we have censused, including Beaches, *Tessaria*/Cane, Transition Forest, High Ground Forest, and Upland Forest.

Results

Faunal Composition

Appendix 14.1 lists the 550 or so bird species recorded in the vicinity of Cocha Cashu. Of these, 467 occur in terrestrial habitats, which for our purposes include forest streams, lake margin shrubs, and wooded swamps but not open water or the airspace overhead. As of 1987, 32 of the terrestrial species have been recorded only as "vagrants"—that is, wandering individuals of species not known to occur regularly in any Cocha Cashu habitat. Most vagrants normally occur at higher elevations or in more open areas where forests have been cleared by humans. Below, we include only the 435 species known to occur regularly in one or more of the terrestrial habitats we have censused.

The number of resident species recorded in each habitat for which we have complete census data increases with successional age (table 12.1). Open beaches, which have only limited vegetation, have the fewest species, and most of these are terrestrial insectivores and granivores. Lake Margin shrubs and *Tessaria*/Cane thickets have a surprisingly large number of species, considering the structural simplicity of the habitats. Species richness increases abruptly in Transition Forest, partly because this habitat is a combination of several successional stages. Species diversity reaches its maximum in High Ground Forest and Upland Forest, where over half of all terrestrial species occur.

Less than a quarter (102) of the 435 regular terrestrial species were recorded in only one of the habitats censused (table 12.1). Upland Forest had the highest number of exclusive species (32), of which 12 were restricted to bamboo thickets. All of these bamboo specialists, however,

TABLE 12.1 Species richness of Cocha Cashu habitats for which both mist-net and song census data are available

	Resident[1] species (*N*)	Exclusive[2] (%)	Migrants[3] (%)	Additional[4] Species (*N*)
Beach	26	38.5	19.2	13
Tessaria/cane	61	27.9	29.5	48
Transition forest	212	9.4	7.1	70
High ground forest	239	3.3	3.3	29
Upland forest	245	13.1	2.0	28
Lake margin	83	18.1	7.2	26

Notes: [1]Breed or forage regularly in the habitat, including migrants that defend territories
[2]Found only in the habitat among those censused (not including Open forest and Aguajales)
[3]Includes austral, intratropical, and Nearctic migrants
[4]Recorded in that habitat on an irregular basis either in mist nets or during song censuses

also occur in sections of floodplain forest where bamboo thickets form. High Ground Forest had the fewest exclusive species (8); most species that occurred in this habitat were also found in Transition and/or Upland Forest. If High Ground Forest and Upland Forest are combined in a "mature forest" category, they have 104 exclusive species, roughly a quarter of all terrestrial species. Beaches and *Tessaria*/Cane, the most structurally distinct habitats, have the highest percentages of exclusive species. Of the 435 species, 80 are restricted to early successional vegetation, including Lake Margins, Beaches, *Tessaria*/Cane, the younger sections of Transition Forest, and extensive treefalls. Aguajales and Open Forest, not yet intensively censused, contain at least 5 species that do not regularly occur elsewhere. Open Forest habitats also have populations of many of the species restricted to early successional habitats among the major census plots.

Early successional habitats appear to have a higher proportion of nonresident species. Migrants make up a higher percentage of the species in early successional habitats than in mature forest (Robinson et al. 1988). Similarly, *Tessaria*/Cane and Transition Forest have the largest number of "additional" species—that is, those recorded in a habitat that do not breed or otherwise occur there regularly. As detailed below, early successional bird communities consist of a small core of resident species,

TABLE 12.2 Guild breakdown of forest (including Transition, High Ground, and Upland) birds

Stratum	Trophic level	Species (*N*)	Percent
Ground	Insectivore	24	7.3
	Omnivore	16	4.9
Shrub	Insectivore	34	10.3
	Nectarivore	8	2.4
	Frugivore	7	2.1
	Omnivore	2	0.6
	Ant-follower	7	2.1
Understory	Insectivore	38	11.5
	Fruit	7	2.1
	Omnivore	5	1.5
Bark	Insectivore	25	7.5
Canopy	Insectivore	35	10.6
	Nectarivore	6	1.8
	Omnivore	44	13.4
	Fruit/mast	44	13.4
Nocturnal	Insectivore	7	2.1
Carnivore	Fish	3	0.9
	Birds	6	1.8
	Mammals	4	1.2
	Miscellaneous predator (insects, frogs, lizards, birds)	5	1.5
	Herps	1	0.3
	Carrion	2	0.6
TOTAL		330	

many of which are extremely common, and a large and variable number of wanderers from other habitats.

The guild breakdown of forest bird communities (table 12.2) shows the predominance of insectivores (163 species). Even though fruit is eaten by 125 species, only 58 eat fruit almost exclusively; the other 67 species supplement their diet with arthropods and/or nectar and are classed as omnivores in table 12.2. The bark-foraging guild is extremely well represented with 25 species, most of which are woodcreepers (Dendrocolaptidae). Understory and shrub communities are dominated by insectivores (94 of 122 species), whereas canopy communities are dominated by frugivores and omnivores (88 of 135 species) (see also Terborgh 1980). Terrestrial species include both insectivores and omnivores, as well as one of the few terrestrial frugivores in the world, the pale-winged trumpeter (*Psophia leucoptera*).

TABLE 12.3 Net lines used for analyses for Cocha Cashu communities

Habitat	Net line	Dates mo/yr	Nets
Tessaria/cane	South beach	8/83, 11/85	14
	Trans-Manu beach	7/73, 8/74, 8/75, 9/76	8
		8/86	14
Transition forest	Trail 7	7/74, 8/75, 9/80	13(*1974*), 41(*1975*), 21(*1980*)
	Trail 20	10/86	27
High ground forest	Trail 5 (swamp)	9/81	32
	Trail 8 (stream)	8/76	12
	Trail 10 west	8/81	26
	Trail 10 east	11/82	19
	Trail 7 north	8/81	15
	Trail 4	9/80	12
	Trail 3	8/73, 7/74, 11/80, 7/81, 9/82	36
Upland forest	Ridge top	8/74, 8/75, 7/76	38
	Bamboo	10/85	31
	Ravine	10/86	35

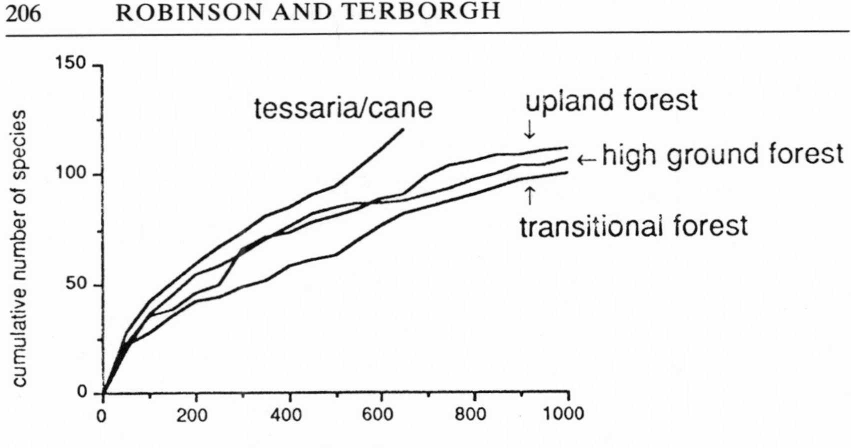

FIGURE 12.1. Species-accumulation curves for mist-net captures in four successional habitats. The *Tessaria*/cane curve includes all captures from the net lines that have been run in this habitat. The Upland forest curve includes data from the ridgetop (8/74, 8/75), ravine (10/86), and bamboo (10/85) net lines. The High ground forest curve includes the swamp (9/81), trail 3 (8/81), stream (7/76), and trail 10 (9/81, 11/82) net lines. The Transition forest curve includes data from the trail 20 (10/86) and trail 7 (8/74, 8/75, 8/80) net lines.

Ecological Patterns from Mist-Net Data

For each habitat, we have data from at least two mist net lines, some of which were run for several years (table 12.3). These data can be analyzed in many ways, a few of which are given here.

Species-accumulation curves. In general, the species-accumulation curves (fig. 12.1) for most habitats were relatively similar in spite of the considerable differences in diversity (table 12.1) and habitat structure. Nets in Transition Forest accumulated species more slowly than in other habitats, largely because of the overwhelming dominance of a few species of hummingbirds (see below). By the 1,000th capture, however, this habitat had only 5–8% fewer species than High Ground Forest. *Tessaria*/Cane had by far the highest species accumulation rate, largely because of the abundance of forest birds that wander through this habitat and because the mist nets sampled virtually the entire community of this relatively low (1–8 m) vegetation. Of the 120 species netted in this habitat, at least 48 were not known to breed or occur regularly in *Tessaria* or Cane. Unlike the other habitats, the species accumulation curves in *Tessaria*/Cane also

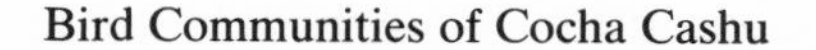

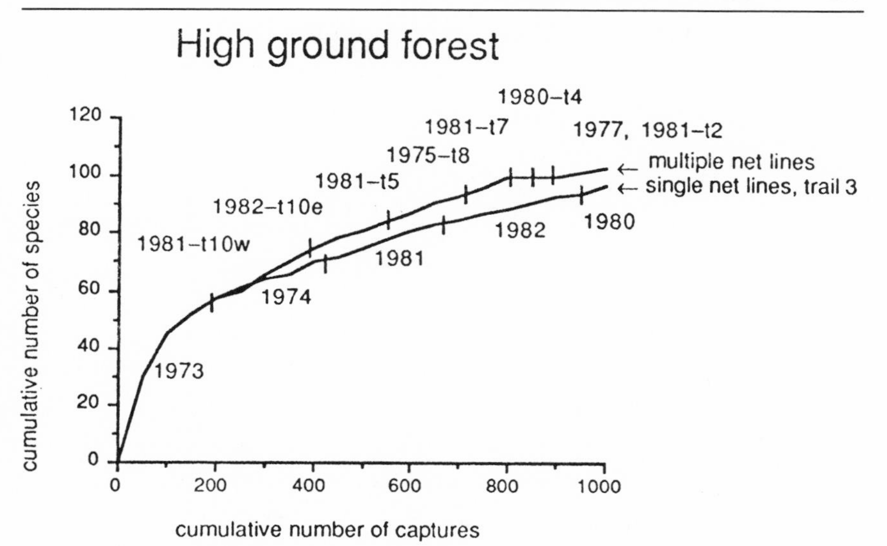

FIGURE 12.2. Comparison of species-accumulation curves for multiple and single net lines in High ground forest (t = Trail). The curve for the single net line includes data from four years on trail 3, an area of relatively uniform forest. The curve for multiple net lines includes data from the edge of a seasonal swamp (trail 5), a permanent stream (trail 8), and an area with several major treefall gaps (trail 10 east [e] and west [w]), as well as from uniform forest interior (trails 3, 4, and 7).

showed no clear tendency to level off by the 400th or 500th capture, a finding that suggests an extremely variable species composition over time. In contrast, only 6 of 107 species in High Ground Forest showed no evidence of breeding or regular nonbreeding occurrence. Of the 45 species captured by Munn in canopy nets 30–50 m above the ground, 22 had never been captured in ground-level nets inside mature forest. In contrast, only 11 of the 81 species known to occur regularly in *Tessaria*/Cane have never been captured in this habitat. Therefore, comparisons of species accumulation curves among structurally different habitats in fig. 12.1 reveal little about overall bird species diversity.

The curves shown in fig. 12.1 are based on data from two to five net lines in each habitat. The Upland Forest data, for example, include the ridgetop, bamboo, and ravine net lines (table 12.3); the High Ground Forest curve includes data from Trail 3 (1981), Trail 10 (1981, 1982), Trail 4 (1980), Trail 5 (1981), and Trail 8, which borders a permanent stream. These net lines were chosen to represent the major topographic and

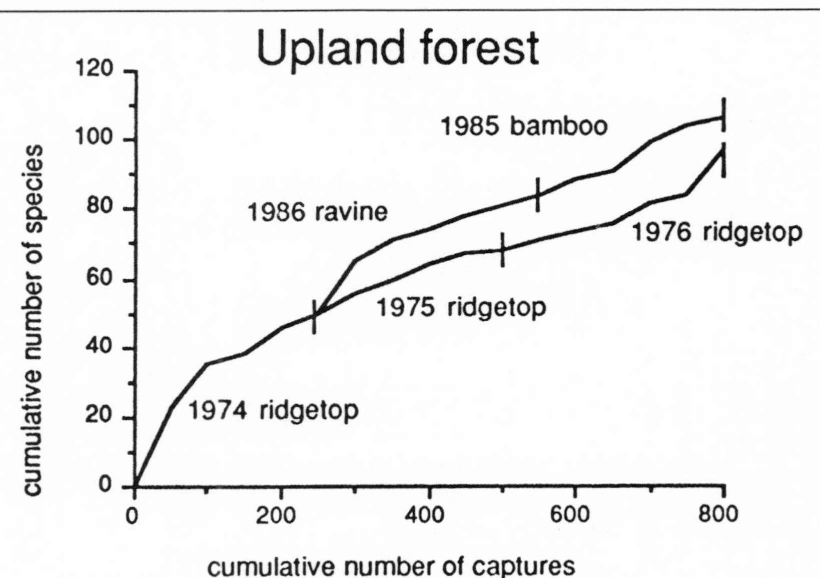

FIGURE 12.3. Species-accumulation curves of multiple and single net lines in Upland forest. The curve for the single net line includes three consecutive years of data from the ridgetop net line. The curve for multiple net lines include data from the ridgetop (8/74), ravine (10/86), and bamboo (10/85) net lines.

vegetation formations within each habitat. What do these "extra" net lines contribute to the species accumulation curves? To answer this question, we compared curves from the same net line run in different years with those from all of the different net lines in High Ground (fig. 12.2) and Upland Forest (fig. 12.3). In High Ground Forest, species accumulated at a slightly higher rate when data from multiple net lines were included. By the 1,000th capture, however, the difference between the two curves was only six species (fig. 12.2). The difference among the species accumulation curves for multiple and single net lines in the Upland Forest is more dramatic; by the 800th capture, the two curves differed by nineteen species. This suggests that High Ground Forest is a more homogeneous bird community than Upland Forest, where many species occur only in ravines or bamboo thickets.

Trophic structure of standard samples. The guild structure of standard samples differs considerably among and within habitats (table 12.4). Nectarivores dominate Transition Forests, due largely to the abundance of dense *Heliconia* thickets favored by several hummingbird genera

TABLE 12.4 Guild breakdown of standard 100-capture samples in different net lines

Date (mo/yr)	Habitat	Captures (species)								
		Insectivores								
		Ground	Foliage	Bark	Ants	Fruit	Seeds	Nectar	Omnivore	Predator
8/83	*Tessaria*/cane	7 (4)	32 (21)	5 (2)	1 (1)	5 (3)	15 (2)	4 (2)	25 (6)	
11/85	*Tessaria*/cane	2 (2)	39 (19)	4 (1)		3 (1)	7 (2)	23 (4)	18 (8)	1 (1)
9/80	Transition forest	2 (1)	22 (11)	5 (4)		6 (2)		58 (7)	5 (3)	
10/86	Transition forest	4 (2)	9 (6)	2 (2)	8 (1)	12 (5)		56 (8)	7 (2)	2 (1)
7/81	High ground forest	8 (6)	44 (10)	8 (7)	15 (3)	15 (5)		4 (3)	5 (2)	
8/81	High ground forest	2 (1)	55 (20)	9 (6)	5 (2)	16 (1)		7 (4)	5 (4)	1 (1)
8/82	High ground forest	10 (6)	31 (13)	9 (6)	16 (3)	16 (4)		13 (6)	2 (2)	2 (1)
10/85	Upland bamboo	8 (5)	41 (16)	9 (4)	5 (2)	10 (6)		18 (6)	9 (6)	
8/74	Upland ridgetop	12 (6)	26 (12)	18 (4)	3 (1)	20 (5)		6 (4)	5 (3)	
10/86	Upland ravine	13 (6)	38 (17)	11 (4)	2 (1)	7 (4)		8 (4)	21 (8)	
9/82	Canopy		30 (16)	3 (2)		16 (7)		11 (6)	40 (14)	

(*Phaethornis*, *Threnetes*, *Glaucis*, *Campylopterus*). Omnivores occur most commonly in early successional habitats and in the canopy. Ant followers occurred in all mature forest understory net lines in variable numbers that probably reflected local army ant activity when the nets were open. Insectivores were well represented in all samples except those in Transition Forest, where hummingbirds overwhelmed the samples. Bark-foraging insectivores occurred at roughly the same frequencies in all mature forest samples except the 1974 Upland Ridgetop sample, where a woodcreeper, *Glyphorynchus spirurus*, was the most common species. The diversity of foliage-gleaning insectivores varied greatly, even within habitats, but was generally highest in *Tessaria*/Cane thickets, where foliage-dwelling insects abound (unpubl. data). Terrestrial insectivores also varied within and among habitats but were somewhat more common in mature habitats, most of which seldom flood during the wet season. Canopy nets in High Ground Forest run by Munn were notable for the high relative abundance and diversity of omnivores, which tend to be more common in the canopy than in the understory (Terborgh 1980).

Capture rates. By far the highest capture rates per net day were in the *Tessaria*-dominated net lines, where each net averaged more than five captures per day (table 12.5). Remsen (p.c.) found similarly high capture rates in *Tessaria* habitats along the Amazon River. These rates reflect both the abundance of birds in this habitat and that mist nets sample most of the community. Capture rates were also high in Transition Forest, almost entirely because of the abundance of hummingbirds, which accounted for over 50% of all captures there (see table 12.4). High Ground Forest capture rates varied between net lines with some indication that capture rates may have been slightly higher during the late dry-early wet season than during the dry season. Capture rates on the Trail 3 High Ground Forest line varied from lows of 1.20 (July 1978) and 1.33 (June 1977) captures per net day to a high of 1.88 (September 1982). Upland Forest capture rates also varied among the ravine (3.2 captures/net day), bamboo (2.5), and ridge top (1.8 in 1974 and 1975, 2.0 in 1976) net lines. Of the net lines with at least 1,000 net hours, the number of species captured ranged from 49 in the Upland Ridgetop (August 1974) to 68 in the Upland Ravine (October 1986), 66 in Upland Bamboo (October 1985), and 65 in the Upland Ridge Top (October 1976) net lines. The most rapid accumulation of species, however, occurred in *Tessaria*/Cane, where over 120 species were captured in 2,140 net hours.

TABLE 12.5 Summary statistics of capture rates for selected net lines

Date (mo/yr)	Habitat	Net hours	Captures	Captures/ net day[1]	Individuals	Individuals/ net day[1]	Species
8/83	*Tessaria*/cane	325	194	7.2	168	6.2	58
11/85	*Tessaria*/cane	230	154	8.0	145	7.6	46
8/86	*Tessaria*/cane	144	73	6.1	70	5.8	39
9/80	Transition forest	1,000	273	3.3	240	2.9	51
10/86	Transition forest	722	260	4.3	220	3.7	51
8/81	High ground forest	1,064	189	2.1	145	1.6	55
11/82	High ground forest	593	152	3.1	122	2.5	44
7/81	High ground forest	1,467	219	1.8	171	1.4	55
10/85	Upland bamboo	1,223	252	2.5	193	1.9	66
10/86	Upland ravine	1,165	306	3.2	243	2.5	68
8/74	Upland ridgetop	1,672	250	1.8	176	1.3	49
8/75	Upland ridgetop	1,672	251	1.8	204	1.5	55
10/76	Upland ridgetop	1,782	294	2.0	201	1.4	65

Note: [1] 1 net day = 12 net hours

Ecological Patterns from Song Census Data

In an attempt to overcome some of the limitations of mist-net censuses, we have undertaken song censuses of the major habitats of Cocha Cashu (Terborgh et al., in press). Some of the key results are summarized here.

Territory size. Birds of mature forest tend to have large territories. Territory sizes of insectivores in High Ground Forest, for example, range from 3 ha to well over 100 ha. Most species have territories of 5–50 ha, including all of the species that live in permanent understory and canopy flocks (Munn 1985). Birds of early successional habitats, however, often have very small territories, often less than 0.25 ha.

Patchiness. Few species occupy all of the study plots. Even in the relatively uniform High Ground Forest plot, half of all species occupied less than 50% of the area. Canopy flocks, for example, were absent from two large (10–20 ha) sections of the plot (Munn 1985). Similarly, bird community composition varied considerably among *Tessaria*/Cane thickets on different bends of the river.

Species richness. High Ground Forest and Upland Forest plots had over 230 resident species, fewer than half of which were captured in mist nets. To date, these are the most diverse communities censused, with over five times the number of species found in temperate zone forests.

Rarity. The census data provide the first complete picture of the extent to which tropical species occur at low population densities in mature forest. In High Ground Forest and Upland Forest, over half of the species had population densities of fewer than three breeding pairs/100 ha. Very few species, however, were rare everywhere we censused; most species that were rare in one habitat were much more common at a different successional stage.

Biomass. In the undisturbed Cocha Cashu bird community, a few species dominated the overall biomass of approximately 187 kg/100 ha. Fewer than 10% of all species accounted for over 60% of the total biomass. Most of these species were large frugivores (including granivores), such as Cracids, Tinamous, and Trumpeters (Psophiidae), which are the first species to be hunted to extinction after an area has been settled. Insectivores, which accounted for nearly half of all species in mature forest communities, made up less than 15% of the total biomass.

Lacunae. Censuses of apparently uniform High Ground Forest plots revealed large (10–15 ha) lacunae where species richness declined by 40–50% for no obvious reason. These data suggest that large plots are necessary for community comparisons.

Summary of Bird Community Studies at Manu

The many community-oriented studies of various groups of birds at the Cocha Cashu Biological Station are briefly summarized below.

1. Multi-species flocks. Munn and Terborgh (1979) and Munn (1985) have described the complex and often mutualistic interactions among the many species that live year-round in understory and canopy flocks. Understory flocks defend 4–6-ha territories and canopy flocks defend 12–20-ha areas.
2. Fitzpatrick (1980b, 1981, 1985) has studied the relationships among morphology, foraging ecology, and systematics of flycatchers (Tyrannidae) at Cocha Cashu and elsewhere in South America.
3. Willard (1985) conducted the first detailed study of the organization of the diverse fish-eating bird community of oxbow lakes.
4. Pierpont (1986) studied the foraging ecology, territorial behavior, and interspecific interactions of the woodcreepers (Dendrocolaptidae), several of which exhibit interspecific territoriality.
5. Robinson (1985a,b, 1986) studied the foraging ecology and the role of nest predation in determining habitat selection and population dynamics of caciques, oropendolas, and other Icterinae in the Cocha Cashu area. These large passerines can be abundant locally, but populations fluctuate greatly in response to recent predation events at colonies.
6. Janson (1983) studied the role of fruit color and morphology in dispersal by birds and mammals.
7. Robinson et al. (1988) and Fitzpatrick (1980a) showed that Nearctic migrants are confined largely to early successional habitats in the Cocha Cashu area.
8. Terborgh and Robinson (in prep.) have shown that many con-

geners that segregate among habitats exhibit strong interspecific aggression and may be defending interspecific territories. Similarly, many patterns of relative abundance within syntopic congeners may be influenced by interspecific aggression.

9. Munn has studied the ecology of the nineteen species of parrots that occur around Cashu, with special emphasis on the macaws (*Ara* species).
10. M. Foster has studied the population dynamics of lek breeding birds (*Pipra* species, *Tyranneutes*, *Lipaugus*) over a 700-ha section of floodplain forest.

Ongoing studies include the ecology of trumpeters (Psophiidae) (P. Sherman and J. Price), Cracidae (R. Gutierrez), hoatzins (B. Torres), lake-margin breeders (P. Eason, K. Petren, J. Leak, D. Velarde), toucans (B. Baca), and raptors (S. Robinson).

The extraordinary diversity of the Manu bird community results from a combination of many elements. Historical factors such as the location of hypothesized Pleistocene refugia (Haffer 1985) may account for some of the additional richness of western Amazonian communities. The diversity of habitats created largely by river action also plays a crucial role (Remsen and Parker 1983). The Manu landscape is tremendously complex, with a mixture of riverine and lacustrine successional habitats, mature forest, open forest, and permanent swamps. Even upland habitats, which are not directly exposed to the river, have a variety of distinct habitats, including ravines, dry ridgetops, and bamboo. Most species in the Cocha Cashu area occur commonly in one or more habitats and less commonly in several others. Interspecific interactions can result in remarkably precise habitat segregation within genera. In the genus *Xiphorhynchus*, for example, the aggressively dominant *X. spixii* occupies mature forest, *X. ocellatus* is confined to transitional habitats, *X. obsoletus* is confined to seasonal swamps, and *X. picus* occurs only along the margins of lakes and rivers (Pierpont 1986). In a less diverse landscape, only one or two of these species would be present.

Habitat diversity alone does not account for the remarkable variety of the Manu community because each habitat also contains extraordinarily diverse bird communities. Mature High Ground and Upland Forest study plots of 80–100 ha have over 230 resident of breeding species, far more than have been recorded for comparably sized areas anywhere else in the world. Even small patches of structurally simple *Tessaria*-

dominated habitats have more species (table 12.1) than most mature temperate forest habitats (Terborgh 1985). The great diversity and richness of mature forests are also associated with low population densities and patchy distributions of all but a few species. Species accumulation curves continue to increase steadily even after 800 mist-net captures (fig. 12.3). The rarity and patchiness of most species may enable so many species to coexist in the same habitat.

The low population densities and patchy distributions of most species amply demonstrate the need to preserve large, contiguous areas that include the full range of successional habitats. Even species of median abundance require large preserves to maintain adequate populations. Nomadic species that wander seasonally, such as macaws and fruitcrows, need even larger preserves. Our census results also show the importance of such large birds as tinamous and various cracids, which are invariably the first birds hunted to extinction following settlement. Reserves must be large enough to provide a refuge from hunting, as well as to preserve intact habitats. The huge (1.6 million ha) Manu National Park may be just large enough to maintain adequate populations of large raptors, parrots, and frugivores.

REFERENCES

Fitzpatrick, J. W. 1980a. Wintering of North American tyrant flycatchers in the Neotropics. In A. Keast and E. S. Morton (eds.), *Migrant Birds in the Neotropics: Ecology, Behavior, Distribution, and Conservation*. Smithsonian Inst. Press, Washington, D.C., pp. 67–78.

———. 1980b. Foraging behavior of Neotropical tyrant flycatchers. Condor 82: 43–57.

———. 1981. Search strategies of tyrant flycatchers. Anim. Behav. 29: 810–821.

———. 1985. Form, foraging behavior, and adaptive radiation in the Tyrannidae. In P. A. Buckley, M. S. Foster, E. S. Morton, R. S. Ridgely, and F. G. Buckley (eds.), *Neotropical Ornithology*. Ornithol. Monogr. 36, AOU, Washington, D.C., pp. 447–470.

Foster, R. B., J. Arce B., and T. S. Wachter. 1986. Dispersal and sequential plant communities in Amazonian Peru floodplain. In A. Estrada and T. H. Fleming (eds.), *Frugivores and Seed Dispersal*. W. Junk, Dordrecht, pp. 357–370.

Haffer, J. 1985. Avian zoogeography of the Neotropical lowlands. In P. A. Buckley, M. S. Foster, E. S. Morton, R. S. Ridgely, and F. G. Buckley (eds.), *Neotropical Ornithology*. Ornithol. Monogr. 36, AOU, Washington, D.C., pp. 113–145.

Janson, C. 1983. Adaptation of fruit morphology to dispersal agents in a Neotropical forest. Science 219: 187–188.

Kendeigh, S. C. 1944. Measurement of bird population. Ecol. Monogr. 53: 183–208.

Munn, C. A. 1985. Permanent canopy and understory flocks in Amazonia: species composition and population density. In P. A. Buckley, M. S. Foster, E. S. Morton, R. S. Ridgely, and F. G. Buckley (eds.), *Neotropical Ornithology*. Ornithol. Monogr. 36, AOU, Washington, D.C., pp. 683–712.

Munn, C. A. and J. Terborgh. 1979. Multi-species territoriality in Neotropical foraging flocks. Condor 81: 338–347.

Parker, T. A., III. 1982. Observations of some unusual rainforest and marsh birds in southeastern Peru. Wilson Bull. 94: 477–493.

Parker, T. A., III, S. A. Parker, and M. A. Plenge. 1982. *An Annotated Checklist of Peruvian Birds*. Buteo Books, Vermillion, S.D.

Pierpont, N. 1986. Interspecific aggression and the ecology of woodcreepers (Aves: Dendrocolaptidae). Ph.D. diss., Princeton Univ.

Pierpont, N. and J. W. Fitzpatrick. 1983. Specific status and behavior of *Cymbilaimus sanctaemariae*, the bamboo antshrike, from southwestern Amazonia. Auk 100: 645–652.

Remsen, J. V., Jr. and T. A. Parker, III. 1983. Contribution of river-created habitats to bird species richness in Amazonia. Biotropica 15: 223–231.

Robinson, S. K. 1985a. The yellow-rumped cacique and its associated nest pirates. In P. A. Buckley, M. S. Foster, E. S. Morton, R. S. Ridgely, and F. G. Buckley (eds.), *Neotropical Ornithology*. Ornithol. Monogr. 36, AOU, Washington, D.C., pp. 898–907.

———. 1985b. Coloniality in the yellow-rumped cacique as a defense against nest predators. Auk 102: 506–519.

———. 1986. Three-speed foraging during the breeding cycle of yellow-rumped caciques (Icterinae: *Cacicus cela*). Ecol. 67: 394–405.

Robinson, S. K., J. W. Terborgh, and J. W. Fitzpatrick. 1988. Habitat selection and relative abundance of migrants in southeastern Peru. Proc. Int. Ornithol. Congr. 19: 2298–2307.

Terborgh, J. 1980. Vertical stratification of a Neotropical forest bird community. Acta. XVII Congr. Int. Ornithol.: 1005–1012.

———. 1983. *Five New World Primates: A Study in Comparative Ecology*. Princeton Univ. Press, Princeton, N.J.

———. 1985. Habitat selection in Amazonian birds. In M. L. Cody (ed.), *Habitat Selection in Birds*. Academic Press, New York, pp. 311–338.

Terborgh, J. W., J. W. Fitzpatrick, and L. Emmons. 1984. Annotated checklist of bird and mammal species of Cocha Cashu Biological Station, Manu National Park, Peru. Fieldiana (Zool.), n.s., 21: 1–29.

Terborgh, J., S. K. Robinson, T. A. Parker III, C. A. Munn, and N. Pierpont. In press. Structure and organization of an Amazonian forest bird community. *Ecol. Monogr.*

Willard, D. E. 1985. Comparative feeding ecology of twenty-two tropical piscivores. In P. A. Buckley, M. S. Foster, E. S. Morton, R. S. Ridgely, and F. G. Buckley (eds.), *Neotropical Ornithology*. Ornithol. Monogr. 36, AOU, Washington, D.C., pp. 788–797.

13

Species Composition and Trophic Organization of the Understory Bird Community in a Central Amazonian Terra Firme Forest

RICHARD O. BIERREGAARD, JR.

The avifauna of Neotropical forests is the world's richest in species (Pearson 1977; Terborgh et al. 1984) and includes species complexes with highly specialized behavioral and social adaptations, such as mixed-species insectivorous flocks (Munn and Terborgh 1979; Munn 1985; Powell 1985) and obligate army ant–following birds (Willis and Oniki 1978).

Although superficially homogeneous, the tropical forests that cover 7 $\times 10^6$ km^2 of the Amazon basin differ greatly in their vegetative structure, taxonomic composition, edaphic conditions, and rainfall (for Brazil see Projeto Radam, vols. 1–15 [1972–78]). Soil types and fertility are related to long- and short-term geological history (Beurlen 1970; Sombroek 1984); owing to the relatively recent incorporation of material derived from the volcanic Andes, soils in the western basin tend to be richer than those in the east, which derived from the preweathered crystalline shields to the north and south (Sombroek 1984). Forests seasonally flooded by sediment-rich "white water" originating in the Andes (*várzea* forests, as they are called in Brazil), have more fertile soils than adjacent terra firme

Many volunteers have most energetically and capably run the net lines and assisted with data management since the beginning of the project. The University of Amazonas provided computer time. The current manuscript has benefited from discussions with and suggestions made by B. Zimmerman, D. Stotz, J. Bates, B. Klein, T. Lovejoy, L. Harper, M. Wong, J. Karr, and two anonymous reviewers.

This study was supported by the World Wildlife Fund, the Instituto Nacional de Pesquisas da Amazônia (INPA), the Instituto Brasileiro de Desenvolvimento Florestal (IBDF), a grant from the National Park Service, Cooperative Agreement CX-0001-0-0041. This is publication number 45 in the Minimum Critical Size of Ecosystems Project (Projeto Dinâmica Biológica de Fragmentos Florestais) Technical Series.

forests (Sombroek 1984). The study areas described here are situated on these ancient, nutrient-poor soils in the Alter do Chao formation (Santos 1984).

In this chapter I describe a mist-netting and banding program in the terra firme forests some 80 km north of Manaus, Amazonas, Brazil (2°20′S × 60°W), and present some initial generalizations about the avian understory community and its trophic structure as deduced from relative abundances and capture and recapture rates in mist nets. Because my intention is to provide a data set for comparison with other studies of tropical avifaunas, the discussion is primarily descriptive. Quantitative data are presented from mist-netting in the understory. A complete list of species seen both in the canopy of the undisturbed forest as well as in the adjacent pasture and second-growth forests was presented by Stotz and Bierregaard (in press) and is available in Karr et al. (this volume). The mist-net data are the quantitative backbone of the avifaunal surveys of the Minimum Critical Size of Ecosystems Project, being developed jointly by World Wildlife Fund and the National Institute for Research in Amazônia (INPA) (Lewin 1984; Lovejoy et al. 1983, 1984, 1986).

Study Area

The study areas are on three 15,000-ha cattle ranches in an agricultural research and development district (fig. 4.1). Ranchers are felling virgin forest to create cattle pasture; the study areas thus lie in a mosaic of pasture, second-growth, isolated forest fragments, and virgin forest (fig. 13.1). The data reported here were collected in virgin forest running unbroken for hundreds to thousands of km in most directions. The area has been accessible by road for only the past thirteen years. During this time, some hunting has occurred, but any effects seem to be minimal. The presence of many large predators, such as *Felis onca* (jaguar), *F. concolor* (puma), *Harpia harpyja* (harpy eagle) and *Morphnus guianensis* (crested eagle), as well as large guans and curassows (Cracidae), suggests that the effects of human predation have been slight.

With its roots in extremely nutrient-poor, sandy or clayey, alic, yellow latosols (Camargo 1979; Chauvel 1982), also termed xanthic ferralsols (FAO/UNESCO 1974), the canopy reaches an average height of about 30–35 m with occasional emergents up to 55 m. The dense canopy permits little light to penetrate to the relatively open understory, which is characterized by many stemless palms (Guillaumet and Kahn 1983).

FIGURE 13.1 Isolated 10-ha and 1-ha forest reserves of the MSCE Project near Manaus, Brazil. Another pair of reserves is visible in the distance. *Photo:* © R. Bierregaard/WWF.

The thirty-year average for annual rainfall in Manaus is 2,186 mm, with a pronounced rainy season from November to May. More than 300 mm of rain fall in both March and April, whereas fewer than 80 mm fall each month from July through September (Anon. 1978).

Methods

Community composition of understory birds was investigated in an intensive mist-netting and mark-recapture program from October 1979 through February 1987. All nets were NEBBA-type ATX (36-mm mesh, 12 m x 2 m, tethered with four trammels). Eight, sixteen, or thirty nets were deployed in unbroken transects of 100 m, 200 m, and 400 m, respectively. Nets were separated from one another by less than 1 m. Net lines were clustered in six areas of virgin forest, the farthest clusters being some 35 km apart. Within each cluster the net lines were 200–2,000 m apart. Data were collected from over fifty-five net lines, some of which were in operation for as long as six years.

Each net line of eight and sixteen nets was opened on average once every forty days for one day. Nets were opened in the predawn darkness (0500–0600 hours) and checked approximately every hour until 1,400 hours, when net closing began. Net lines with thirty nets were run from dawn to dusk on five consecutive days.

Netting was also carried out in isolated forest fragments, and observations of birds of the canopy, second-growth, and pastures were recorded. Through these ancillary studies a fairly complete description of the avifauna of the terra firme forest site north of Manaus is available (Stotz and Bierregaard, in press). Changes in avian community structure associated with isolation are reported elsewhere (Bierregaard and Lovejoy, in press-a,b). Relative abundances were calculated based on the number of birds banded and the number of captures for each species.

To describe the trophic organization of the avian community, species were assigned to guilds based on principal food consumed, strata in the forest in which foraging takes place, and the substrate upon which prey is taken. The assignment of species to guilds was based on observations by myself or other participants in the Minimum Critical Size of Ecosystems (MCSE) project, or those made by Willis (1977) nearby in Reserva Ducke. Data on habitat, strata occupied, foraging substrate, and diet are presented for individual species in Karr et al. (this volume). To facilitate a comparison with a two-year banding study in Pasoh Forest in Malaysia (Wong 1986), netted species were also classified into the ten categories used by Wong (1986), with the addition of a new category for "obligate," or "professional" (Willis and Oniki 1978), army ant followers. Percentages for each category were calculated from captures, number of individuals, and biomass (number of individuals times average weight for the species).

Species-encounter curves (number of species versus number of captures) were calculated from all net lines combined, as well as from thirty-five net lines that had more than 200 captures. These were compared to data from an additional twenty net lines lumped into three clusters, each sampling an area of 500–700 ha.

Seasonality was tested by a Chi-square for the thirty-seven species for which we had over 200 captures. Capture data from all years were combined for each species into two five-month seasons. The rainy season was defined as December–April and the dry season as June–October. The transitional months of May and November were excluded from the analyses. Expected values were calculated by multiplying the number of captures of all species for each five-month season by each species' relative abundance in the sample with all years and months (excluding May and November) combined.

Results

A linear regression of capture rates against the number of days since the first sample shows a significant ($p < .001$) negative correlation, although the slope is slight:

$$C = .228 - .06 \cdot D$$

where C is captures per net-hour and D is the number of days since the first sample (see fig. 13.2). Net lines were occasionally skipped one or more times from the normal rotation, serendipitously offering an opportunity to test the effect of sampling frequency on learned net-avoidance. No significant relation was found in a regression between capture rate and the number of days since the given net line was run.

The percentage of birds recaptured on a given sample day approaches an asymptote of roughly 43% after about one year (fig. 13.2; see Bierregaard and Lovejoy [in press-a] for details). The number of species seen or netted on the three ranches is 352 (Stotz and Bierregaard, in press; see appendix 14.1). Specimens of about 90 species have been collected and are maintained in a small collection at INPA. Recordings of over 180 species known from the study area have been archived at the

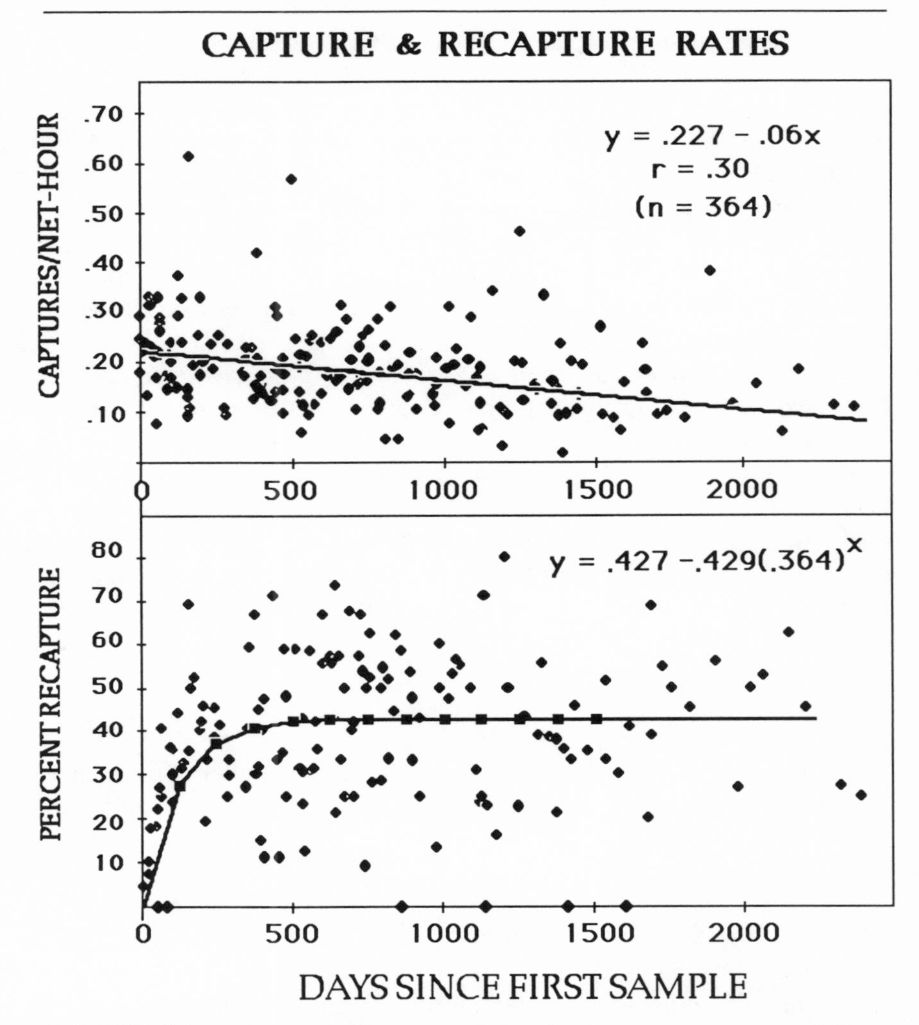

FIGURE 13.2. Capture rates (captures per net-hour) and recapture rates (same-day recaptures included) as a function of the number of days since first sampling on a net line with an average interval between samples of forty days.

Library of Natural Sounds at Cornell University, and all available photographs have been donated to VIREO at the Academy of Natural Sciences.

After seven years of banding in virgin forest, 24,957 captures of 14,026 individuals of 143 species (41% of the list for the area) have been recorded in about 136,000 net-hours. On a given net line (100-m or 200-m transect), an average of 37 and 49 species were caught in the first 100

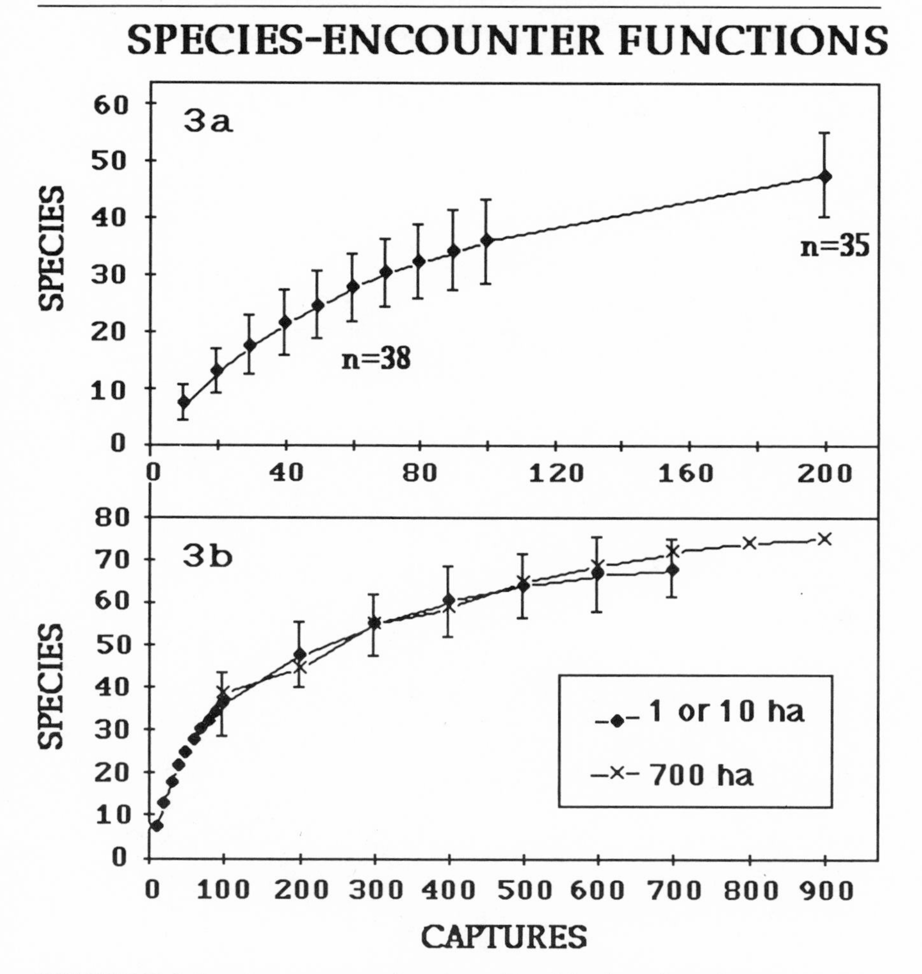

FIGURE 13.3 Species-encounter functions. *Top:* Number of species versus number of captures for 100-m or 200-m net lines sampling 1-ha or 10-ha reserves with 95% confidence limits. *Bottom:* Number of species encountered versus number of captures for 1-ha and 10-ha reserves compared to data from 3 clusters of 5–7 net lines sampling areas of 500–700 ha.

and 200 captures, respectively (fig. 13.3a). Species-encounter functions, based on five to seven net lines sampling an area of approximately 500–700 ha, were virtually identical to those from single net lines for the first 500 captures and began to diverge from the species-encounter function for net lines sampling smaller areas only after 600 captures (fig. 13.3b). The difference was not tested statistically.

TABLE 13.1 Twenty most common species in mist-netting sample

Species	Feeding guild[1]	Individuals (%)	Captures (%)	Capture rates[2]	
				Individuals[3]	All captures[4]
Pithys albifrons	AA	7.14	10.17	7.31	18.53
Hylophylax poecilinota	I	5.46	7.99	5.59	14.55
Glyphorhynchus spirurus	I, MSF	4.81	4.91	4.92	8.95
Pipra pipra	F, I	4.72	3.65	4.83	6.66
Gymnopithys rufigula	AA	4.18	5.67	4.28	10.34
Mionectes macconnelli	MSF, F	3.02	2.15	3.09	3.91
Thamnomanes ardesiacus	MSF	2.98	3.59	3.05	6.54
T. caesius	MSF	2.95	2.37	3.02	4.32
Turdus albicollis	F, I	2.60	2.78	2.66	5.07
Xiphorhynchus pardalotus	MSF	2.33	3.47	2.39	3.85
Myrmotherula longipennis	MSF	2.15	1.96	2.20	3.56
Dendrocincla merula	AA	2.11	2.50	2.16	4.56
Myrmotherula gutturalis	MSF	2.08	2.13	2.13	3.88
Percnostola rufifrons	I, AA	2.00	2.00	2.04	3.64
Myiobius barbatus	I, MSF	1.92	1.62	1.96	2.95
Hypocnemis cantator	I	1.90	1.66	1.95	3.02
Microbates collaris	I	1.86	2.04	1.91	3.72
Automolus infuscatus	MSF	1.82	2.02	1.86	3.67
Geotrygon montana	F	1.81	1.12	1.85	2.04
Schiffornis turdinus	F, I	1.78	2.55	1.82	4.65

Notes: [1] AA=army-ant follower; F = frugivore; I = primarily insectivorous, forages independent from mixed-species flocks; MSF = regular participant in insectivorous mixed-species flocks

[2] Per 1,000 net hours

[3] First captures only

[4] Recaptures included

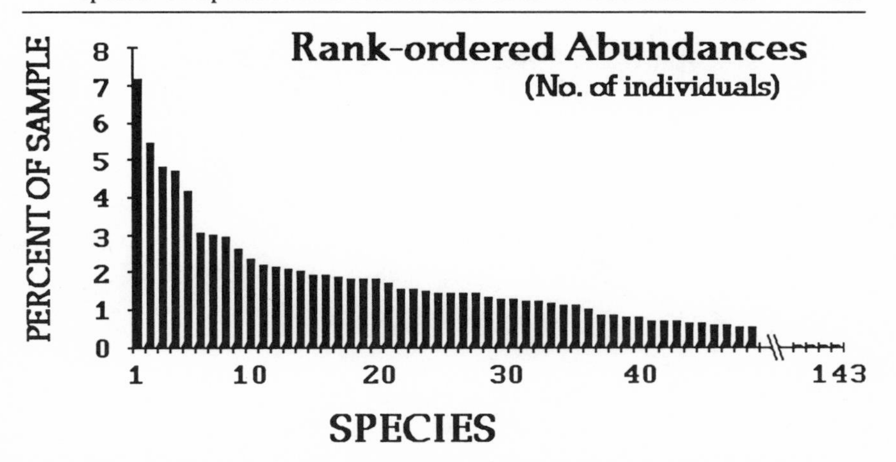

FIGURE 13.4. Rank-ordered abundance distribution of 14,026 individuals of 143 species netted in virgin forest net-lines.

The 20 most common species in the sample represent 65% of all captures and 59.6% of all individuals banded (table 13.1). No one species dominates the rank-abundance distribution, and a very long "tail" of rare species is apparent (fig. 13.4). The most abundant bird in the sample, *Pithys albifrons* (white-plumed antbird), represents only 7.1% of individuals banded and 10.1% of all captures. Some 127 species (90% of the sample) meet Karr's (1971) definition of rarity (less than 2% of the sample), whereas 106 species represented less than 1% of all individuals banded. In fact, of the 143 species captured, 10 (7%) were represented by only one individual. Of the 106 species representing less than 1% of the sample, 49 (34% of all netted species) are probably underrepresented in the netting sample because they behave in such a way as to rarely encounter the nets or are too large to be caught in them: 9 species predominantly walk rather than fly through the forest; 13 are relatively large (> 80 g) for the mesh size of our nets; 16 are canopy species that rarely descend to the forest floor; 9 are species of pasture or second-growth, which rarely venture into primary forest; and 2 are specialists in rare forest habitats (old, dense treefalls and large streams).

The birds in the understory are predominantly insectivorous. Obligate army ant followers, mixed-species flock members, and other species classified as primarily insectivorous comprise 85.4% of captures and 80.4% of individuals but only 69.4% of biomass (calculated from the number of individuals banded) (table 13.2). Including all species that prey on insects, both primarily and secondarily (see appendix 14.1), the group represents 92.8% of all individuals in the sample.

TABLE 13.2 Relative importance of feeding guilds in the understory of undisturbed forest near Manaus

Feeding category	Number of species[1] (%)	Captures[2] (%)	Individuals[3] (%)	Biomass[4] (%)
Frugivores	12.59	1.54	2.42	10.88
Frugivore/insectivore	10.49	9.22	11.21	10.98
Frugivore/carnivore	2.10	0.03	0.06	0.42
Subtotal/Primarily fruit	27.28	10.79	13.69	22.28
Army ant–follower	2.10	18.35	13.43	14.90
Insectivore	44.06	35.00	34.12	30.24
Insectivore/frugivore	2.10	2.56	1.80	2.36
Insectivore/carnivore	0.70	0.24	0.35	1.58
Mixed-flock insectivores	14.08	27.13	27.13	18.87
Mixed-flock insectivore/frugivore	0.70	2.15	3.02	1.47
Subtotal/Primarily insects	61.64	85.43	80.43	69.42
Nectarivore	5.59	2.89	4.59	0.88
Subtotal/Nectarivore	5.59	2.89	4.59	0.88
Piscivore	0.70	0.05	0.07	0.14
Piscivore/insectivore	0.70	0.04	0.04	0.02
Large arthropods/small vertebrates	0.70	0.01	0.01	0.05
Small vertebrates/insects	3.50	0.78	1.12	7.21
Subtotal/Miscellaneous	5.60	0.88	1.24	7.42
TOTALS	100.10	99.99	99.95	100.00

Notes: [1] Percentage of 143 species mist-netted and assigned to prey category
[2] Percentage of all captures per prey category
[3] Percentage of all birds banded per prey category
[4] Individuals banded × mean weight per species at study site (Bierregaard 1988)

The three species of obligate army ant followers include the most abundant and the fifth most common species in the netting sample and represent 18.4% of captures and 13.4% of individuals banded (table 13.2). The mixed-species, insectivorous flocks are perhaps the most conspicuous group of understory birds in the Amazon basin. These flocks usually contain one pair each of between 8 and 13 core species that defend congruent territories to which they are faithful over a period of years (see Powell 1985 and Munn 1985 for reviews). In the Manaus area, regular flock participants account for 29.3% of all captures and 30.7% of individuals banded (table 13.2).

Powell (p.c.) reports that two species that regularly forage with insectivorous mixed-species flocks, *Myrmotherula gutturalis* (spot-throated antwren) and *Automolus infuscatus* (olive-backed foliage-gleaner), frequently (more than 50% of all foraging observations) forage in clusters of dead leaves (see Remsen and Parker 1984). Only 22% of Powell's foraging observations (p.c.) of *Philydor erythrocercus* (rufous-rumped foliage-gleaner), a species considered a dead-leaf foraging specialist by Remsen and Parker (1984), were in dead leaves. Whereas Remsen and Parker (1984) list *Xiphorhynchus guttatus* (buff-throated woodcreeper) as a regular dead-leaf forager, Powell (p.c.) reports that *X. pardalotus* (chestnut-rumped woodcreeper) fills this niche in our study area (*X. guttatus* is not present at the Manaus study sites). These dead-leaf foragers represent 6.9% of all birds banded and 7% of all captures in the sample.

Three species in the mixed-species flocks, *Xenops minutus* (plain xenops), *Campylorhamphus trochilirostris* (red-billed scythebill), and *Piculus flavigula* (yellow-throated woodpecker), regularly peck and probe on hanging dead branches (Powell, p.c.), a foraging technique similar to the dead-leaf foraging discussed by Remsen and Parker (1984). This group represents 1.42% of the individuals banded and 0.96% of all captures.

Primarily frugivorous species represent only 10.8% of captures and 13.7% of all individuals. Because of the large size of some frugivorous species, the group represents 22.3% by biomass of all birds banded (table 13.1). After correcting for biomass, the most important species in all guilds in the sample is *Geotrygon montana* (ruddy quail-dove).

Nectar feeders (hummingbirds) are rare in the understory, accounting for only 2.9% of captures and 4.6% of all individuals (table 13.2). Due to their very small size (most weigh less than 10 g), they represent only 0.9% of the sample when corrected for biomass.

Migrants are extremely rare in the netting sample, in terms of both number of species as well as total captures. *Catharus fuscescens* (veery) and *C. minimus* (gray-checked thrush), the only migrants netted, account for 0.07% and 0.05% of all individuals banded, respectively, and none has been recaptured. Over 90% of their captures occur in the northern late winter or early spring, with the remaining captures all in late autumn.

The migrants recorded but not captured are birds of the canopy, second-growth, or open areas not sampled in the netting program. *Dendroica striata* (blackpoll warbler) and *D. fusca* (blackburnian warbler) occasionally participate in canopy mixed-species insectivorous flocks but are rare in them. *Contopus borealis* (olive-sided flycatcher) and *C. virens* (eastern wood-pewee) occur in old second-growth and primary forest. *Buteo platypterus* (broad-winged hawk) is rarely seen on the ranches, probably because there is little of its preferred habitat of old second-growth or disturbed primary forest there. In the disturbed areas around the city of Manaus, *B. platypterus* is not uncommon during the northern winter.

In the pastures adjacent to the forest study tracts, *Progne subis* (purple martin), recorded in the study after Stotz and Bierregaard (in press) compiled the area list, and *Hirunda rustica* (barn swallow) are seen during the northern winter. Nine species of migrant charadriform shorebirds have been recorded around the small ponds in the ranch pastures (see appendix 15.1).

Austral migrants include *Empidinomus aurantioatrocristatus* (crowned slaty-flycatcher), *Tyrannus savana* (southern fork-tailed flycatcher), *Myiodynastes maculatus* (streaked flycatcher), *Pyrocephalus rubinus* (vermillion flycatcher), and *Vireo olivaceus chivi* (red-eyed vireo).

The null hypothesis that relative abundance in the capture sample would be equal for dry and rainy seasons was rejected ($p < .05$) for only 10 of 37 species tested. The ground-feeding, frugivorous *Geotrygon montana* and both hummingbirds for which we had 200 or more captures, *Phaethornis superciliosus* (long-tailed hermit) and *P. bourcieri* (straight-billed hermit), were captured disproportionately in the rainy season. Among the obligate army ant followers, *Pithys albifrons* and *Dendrocincla merula* (white-chinned woodcreeper) were more likely to be caught in the dry season than wet, whereas the other ant follower, *Gymnopithys rufigula* (rufous-throated antbird), showed no seasonal difference.

The other species for which the null hypothesis of equal capture probabilities in rainy and dry season was rejected include *Deconychura stictolaema* (Spot-throated woodcreeper) and *Hylophilus ochraceiceps* (tawny-crested greenlet), both members of understory, mixed-species flocks; *Formicarius colma* (rufous-capped antthrush), a leaf-litter insectivore; *Galbula albirostris* (yellow-billed jacamar), a sallying insectivore; and *Mionectes macconnelli* (MacConnell's flycatcher), a frugivore-insectivore. All these species were captured disproportionately in the dry season.

Populations of *Geotrygon montana* fluctuate from year to year, as well as seasonally. In certain years, and occasionally for several years running, the species was very rare in the nets and then suddenly appeared in large numbers in the following rainy season.

Discussion

The decrease in capture rate with repeated sampling (fig. 13.2) suggests that some birds are learning to avoid nets or that netting and banding are increasing mortality of the population being sampled and that this mortality is not being compensated for by local reproduction or the immigration of nonterritorial individuals from nearby forest. Powell (p.c.) reports that all color-banded *Myrmotherula* species (antwrens) marked in his first field season were replaced by new individuals in the following year. This suggests that, though our research activities may indeed be increasing mortality, local movements compensate for this unnatural loss from the population.

The gentle slope of the curve relating capture rate to time since first sample suggests that either the learning of net avoidance is incomplete or that not all species are learning to avoid capture. Whereas there was no correlation between the amount of time since a net line was last run and capture rate when intersample periods of three weeks or more were considered, daily sampling on a given net line for five consecutive days results in very low capture rates by the fifth sample day (pers. obs.). This suggests that net-avoidance is forgotten somewhat without frequent reinforcement and that the sampling frequency (three-week minimum between samples) used in this study is productive for this community.

Even after repeated sampling on a given net line, more than half the birds in the nets on a given capture day are new to that net line. The new birds consist of young, recently fledged in the area, "rare" species (see below), individuals that hold no territory and are looking for a vacant territory, and neighboring territory holders that trespass, apparently exploring their neighbors' territories (Powell, p.c.). Bierregaard and Lovejoy (in press-a) show how isolation of a forest patch by deforestation results in a marked decrease in the percentage of new captures, suggesting that wandering members of a floating population are accounting for a significant portion of the avian activity in the forest understory.

The long tail of "rare" species in the sample is typical of avian communities in tropical forests (Karr 1971; Lovejoy 1974; Pearson 1977; Wong 1986). These "rare" species include species that do in fact occur at low densities, as well as relatively common birds with low probability of either encountering nets (canopy species) or of remaining in the nets once they hit them because of their size (such as cracids and tinamous) or because they usually walk rather than fly through the forest (such as *Grallaria* species [antpittas]).

Although some relatively common species are thus being underrepresented in the netting data, other common species are overrepresented. Species that travel over large areas to forage or visit isolated leks, such as understory frugivores (Graves et al. 1983), nectar feeders, and army ant followers (Willis and Oniki 1978), are more likely to encounter a net than species that feed in smaller territories; thus, the former should be captured more frequently than the latter. And because they spend much of their time perched, species that sally to flying insects, such as jacamars (Galbulidae), will be caught less often than species that move constantly through the undergrowth gleaning insects off foliage or bark, such as some antbirds (Formicariidae) and woodcreepers (Dendrocolaptidae).

The rank-abundance distribution differs somewhat from those reported from studies in western Amazonia and in Central America, where the most abundant species is often a small, frugivorous manakin (Pipridae) and represents 15–30% of all individuals in mist-net samples (Karr 1971; Terborgh and Weske 1969; Fitzpatrick, p.c.). At Manaus, the most abundant manakin, *Pipra pipra* (white-crowned manakin), is the third most abundant species in the nets, representing 4.7% of all individuals.

When samples of 1,000 captures were compared between the four forest sites studied in this volume by Karr et al. (this volume), the scarcity of frugivores in the Manaus forests was obvious. Among the ten most common species at each site, frugivores represented 17.2%, 15.2%, and 14.8% of the 1,000-capture samples from La Selva in Costa Rica, Pipeline Road in Panama, and Manu National Park in Peru, respectively, and only 9.1% of the Manaus sample (see table 14.6 in Karr et al., this volume).

Nectar feeders show a similar pattern. Species among the ten most common at each site representing 8.1%, 5.9%, and 3.9% of the 1,000-capture samples from La Selva, Pipeline Road, and Manu, respectively, whereas none was among the ten most commonly captured species in the comparable Manaus sample.

The relative scarcity of understory frugivores and nectar feeders is doubtless related to the remarkably low productivity of understory fruiting and flowering shrubs in the study area. Gentry and Emmons (1987) reported that transect counts of understory fruits and flowers yielded values one or two orders of magnitude less than comparable surveys in the western Amazon. Whether the scarcity of these species extends to the canopy cannot be addressed because I have no quantitative data from canopy netting.

The Manaus guild signatures (fig. 13.5) are broadly similar to those presented by Wong (1986) from Pasoh Forest in Malaysia, when number of individuals/guild and biomass/guild are compared. The two notable exceptions are the presence in Manaus of the obligate army ant–following guild and the relative scarcity in Manaus of aerial sallying insectivores. That the guilds seem similar in individuals and biomass, whereas the number of species/guild seems to show greater divergence between the two avifaunas, suggests ecological convergence in the face of taxonomically different ancestral stocks. The relative paucity of aerially sallying insectivores in Manaus may be explained by a difference in resource availability between this Amazonian forest and that in Malaysia. Or the differences may be an artifact of differences in our classification of species into "aerially sallying" and "aerially gleaning" guilds.

Because most birds in the forest understory are nonmigratory, it is not surprising that most species show little or no seasonality in their abundances in the nets. Most of those that do may be changing some aspect of their foraging behavior rather than coming and going from the forest around the study area. Hummingbirds, for example, may find more flowers available in the canopy than in the understory in the dry season, explaining their relative rarity in the nets during this time. It is puzzling why *Pithys albifrons* should show strong seasonal fluctuations, whereas the confamilial *Gymnopithys rufigula* shows none. Both birds are of roughly similar size (20.36 versus 29.14 g [Bierregaard 1988]) and prey almost entirely on insects fleeing from army ants. The increase in relative abundances in the nets for the seven other species that showed some seasonality might be explained by the presence of recently fledged young, since most species in the area breed in the dry season (pers. obs.).

The changes in relative abundance of *Geotrygon montana* in the netting sample reflect a real change in population levels in the forest. The

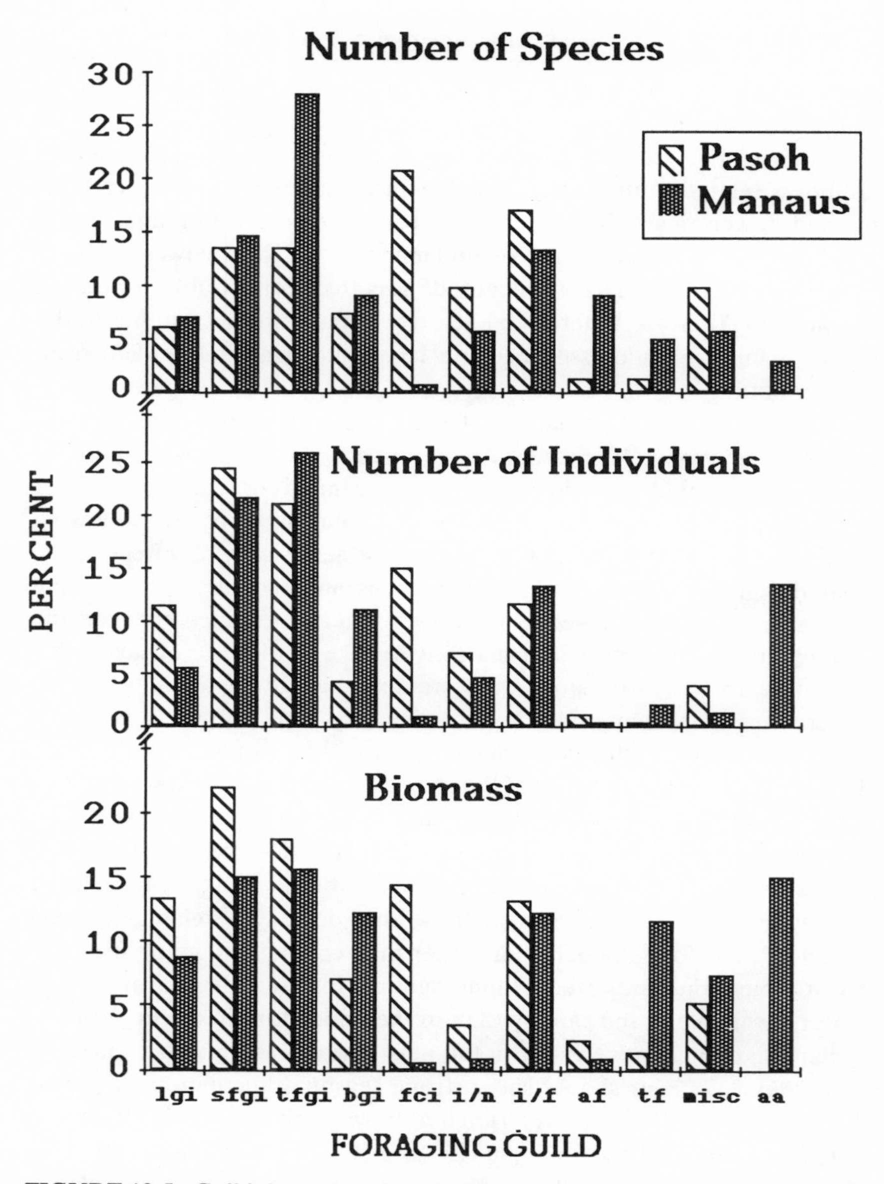

FIGURE 13.5. Guild signatures of the understory avifauna near Manaus, Brazil, and Pasoh, Malaysia (Wong 1986). Guild abbreviations, using Wong's (1986) terminology, are: lgi = leaf-litter–gleaning insectivore; sfgi = shrub foliage–gleaning insectivore; tfgi = tree foliage-gleaning insectivore; bgi = bark-gleaning insectivore (includes pecking, probing, etc.); fci = flycatching (aerial) insectivore; i/n = insectivore/nectar feeder; i/f = insectivore/frugivore; af = arboreal frugivore; tf = terrestrial frugivore; aa = army-ant follower; misc = various invertebrates and vertebrates (includes raptors).

species does not avoid nets and is conspicuous in the forest. During periods when it is uncommon in the nets (both seasonally and annually) it is also rarely seen or heard in the forest. Because it feeds on the ground, it cannot be foraging above net level. Either the population of the species, undergoes drastic changes in mortality or fecundity or both, or large-scale, regional population movements are occurring. The abrupt increase in its abundance in the nets suggests that these fluctuations are the result of regional movements as populations of *G. montana* track some resource through space. This is the only resident species in the study area whose range includes western Mexico and the Lesser Antilles, suggesting that populations are highly mobile and could therefore be expected to exhibit pronounced annual fluctuations at a given locale.

The changes in *G. montana* population levels in the study area suggest that they are moving to a forest type that is qualitatively different in its fruiting phenology from that of the study area. But where could this be? The várzea, or river-edge forest, is flooded when *G. montana* are absent from our study sites and exposed while the birds are breeding in the study area. There is no altitudinal gradient through which to move. About 80 km north of the study areas, soils begin shifting to more nutrient-rich podsols (Anon. 1978) that might offer richer seed and fruit crops in years when our study areas are relatively unproductive. No phenological data from the region to the north are available, so the question will probably remain unanswered until we get year-round data from elsewhere in Amazonia.

Robinson et al. (in press) reported that most migrants from the northern hemisphere that overwinter in South America do so along the eastern slope of the Andes. Pearson (1980) reported that 12.7–10.7% of the species recorded at three lowland sites in Amazonian Peru were migrant species. The comparable percentage from the study areas in Manaus (5.9%) is about half those reported by Pearson (1980).

The two thrushes that pass through the Manaus area either have an asymmetrical migration route or different behavior on the southbound and northbound legs of their migration. Crawford and Stevenson (1984) have shown that *C. fuscescens* is far more abundant in Florida during its fall migration than in spring, whereas *C. minimus* is encountered with equal frequencies in both seasons. It appears then, that even before reaching South America, migrating *C. fuscescens* use seasonally different routes to and from their wintering areas. Perhaps in South America the southward migration of these two species occurs east or west of the Manaus region, whereas the northward migration passes through the central basin. Alternatively, the birds may pass over the forest on the southern migration, depending on fat resources accumulated on the breeding grounds, and migrate northward through the forest, foraging as they go.

REFERENCES

Anon. 1978. Projeto Radambrasil; Folha SA20 Manaus. Ministério de Minas e Energia: Depto. Nacional de Produção Mineral, Rio de Janeiro.

Beurlen, K. 1970. *Geologie von Brasilien: Beiträge zur regionalen Geologie der Erde*. Gebr. Borntraeger, Berlin.

Bierregaard, R. O., Jr. 1988. Morphological data from understory birds in terra firme forest in the central Amazonian basin. Rev. Brasil. Biol. 48: 169–172.

Bierregaard, R. O., Jr., and T. E. Lovejoy. In press-a. Birds in Amazonian forest fragments: effects of insularization. Proc. XIX Int. Ornithol. Congr.

———. In press-b. Effects of isolation of understory bird communities in Amazonian forest fragments. Acta Amazônica.

Chauvel, A. 1982. Os latossolos amarelos, álicos, argilosos dentro dos ecossistemas das bacias experimentais do INPA e da região vizinha. Acta Amazônica 12: 47–60.

Crawford, R. L. and H. M. Stevenson. 1984. Patterns of spring and fall migration in northwest Florida. J. Field Ornithol. 55: 196–203.

FAO/UNESCO. 1971. *Soil Map of the World at 1:5,000,000.* Vol. 1: *Legend.* UNESCO, Paris.

Gentry, A. H. and L. H. Emmons. 1987. Geographic variation in fertility, phenology, and composition of the understory of Neotropical forests. Biotropica 19: 216–227.

Graves, G. R., M. B. Robbins, and J. V. Remsen, Jr. 1983. Age and sexual difference in spatial distribution and mobility in manakins (Pipridae); inferences from mist-netting. J. Field Ornithol. 54: 407–412.

Guillaumet, J. and F. Kahn. 1983. Structure et dynamisme de la forêt. Acta Amazônica 12: 61–77.

Karr, J. R., 1971. Structure of avian communities in selected Panama and Illinois habitats. Ecol. Monogr. 41: 207–231.

Karr, J. R., S. Robinson, J. G. Blake, and R. O. Bierregaard, Jr. (This volume). Birds of four Neotropical forests.

Lewin, R. 1984. Parks: how big is big enough? Science 225: 611–612.

Lovejoy, T. E. 1974. Bird diversity and abundance in Amazon forest communities. Living Bird 13: 127–191.

Lovejoy, T. E., R. O. Bierregaard, Jr., J. M. Rankin, and H. O. R. Schubart. 1983. Ecological dynamics of forest fragments. In S. L. Sutton, T. C. Whitmore, and A. C. Chaddwick (eds.), *Tropical Rain Forest: Ecology and Management*. Blackwell, Oxford, pp. 377–384.

Lovejoy, T. E., J. M. Rankin, R. O. Bierregaard, Jr., K. S. Brown, Jr., L. H. Emmons, and M. van der Voort. 1984. Ecosystem decay of Amazon forest remnants. In M. H. Nitecki (ed.), *Extinctions*. Univ. of Chicago Press, Chicago, pp. 295–325.

Lovejoy, T. E., R. O. Bierregaard, Jr., A. B. Rylands, J. R. Malcolm, C. E. Quintela, L. H. Harper, K. S. Brown, Jr., A. H. Powell, G. V. N. Powell, H. O. R. Schubart, and M. B. Hays. 1986. Edge and other effects of isolation of Amazon forest fragments. In M. E. Soulé (ed.), *Conservation Biology: The Science of Scarcity and Diversity*. Sinauer, Sunderland, Mass., pp. 257–285.

Munn, C. A. 1985. Permanent canopy and understory flocks in Amazônia: species composition and population density. In P. A. Buckley, M. S. Foster, E. S. Morton, R. S. Ridgely, and F. G. Buckley (eds.), *Neotropical Ornithology*. Ornithol. Monogr. 36, AOU, Washington, D.C., pp. 683–712.

Munn, C. A. and J. W. Terborgh. 1979. Multi-species territoriality in Neotropical foraging flocks. Condor 81: 338–347.

Pearson, D. L. 1977. A pantropical comparison of bird community structure on six lowland rain forest sites. Condor 79: 232–244.

———. 1980. Bird migration in Amazonian Ecuador, Peru, and Bolivia. In A. Keast and E. S. Morton (eds.), *Migrant Birds in the Neotropics: Ecology, Behavior, Distribution, and Conservation*. Smithsonian Inst. Press, Washington, D.C., pp. 273–283.

Powell, G. V. N. 1985. Sociobiology and adaptive significance of interspecific foraging flocks in the Neotropics. In P. A. Buckley, M. S. Foster, E. S. Morton, R. S. Ridgely, and F. G. Buckley (eds.), *Neotropical Ornithology*. Ornithol. Monogr. 36, AOU, Washington, D.C., pp. 713–732.

Projeto Radam (or Radam Brasil). 1972–1978. Levantamento de Recursos Naturais. Vols. 1–15. Ministério de Minas e Energia: Depto. Nacional de Produção Mineral, Rio de Janeiro.

Remsen, J. V., Jr. and T. A. Parker III. 1984. Arboreal dead-leaf-searching birds of the Neotropics. Condor 86: 36–41.

Robinson, S. K., J. Terborgh, and J. W. Fitzpatrick. 1988. Habitat selection and relative abundance of Nearctic migrants in southeastern Peru. Acta XIX Congr. Int. Ornithol. Vol. 2. Univ. of Ottawa Press, Ottawa, pp. 2298–2307.

Santos, J. O. S. 1984. A parte setentrional do cráton Amazônico (Escudo das Güianas) e a bacia Amazônica. In C. Schobbenhaus, D. A. Campos, G. R. Derze, and H. E. Asmus (eds.), *Geologia do Brasil: texto explicativo do mapa geológico do Brasil e da área oceânica adjacente incluindo depósitos minerais, escala 1:2,500,000*. Depto. Nacional da Produção Mineral, Brasília, pp. 57–91.

Sombroek, W. G. 1984. Soils of the Amazon region. In H. Sioli (ed.), *The Amazon: Limnology and Landscape Ecology of a Mighty Tropical River and Its Basin*. W. Junk, Dordrecht, pp. 521–535.

Stotz, D. F. and R. O. Bierregaard, Jr. In press. The birds of the fazendas Porto Alegre, Dimona and Esteio North of Manaus, Amazonas, Brazil. Rev. Brasil. Biol.

Terborgh, J. and J. S. Weske. 1969. Colonization of secondary habitats by Peruvian birds. Ecol. 50: 765–782.

Terborgh, J. W., J. W. Fitzpatrick, and L. Emmons. 1984. An annotated checklist of bird and mammal species of Cocha Cashu Biological Station, Manu National Park, Peru. Fieldiana (Zool.), n.s., 21: 1–29.

Willis, E. O. 1977. Lista preliminar das aves da parte noroeste e áreas vizinhas da Reserva Ducke. Rev. Brasil. Biol. 37: 585–601.

Willis, E. O. and Y. Oniki. 1978. Birds and army ants. Ann. Rev. Ecol. Syst. 9: 243–263.

Wong, M. 1986. Trophic organization of understory birds in a Malaysian dipterocarp forest. Auk 103: 100–116.

14

Birds of Four Neotropical Forests

JAMES R. KARR, SCOTT K. ROBINSON, JOHN G. BLAKE,
AND RICHARD O. BIERREGAARD, JR.

As a group, birds are especially well known at all four tropical sites discussed in this volume. In addition, natural history information permits comparative analysis of taxonomic composition, number of species, and a variety of ecological attributes of the avifaunas. We discuss the birds of four Neotropical forest sites: La Selva, Costa Rica (CR); Barro Colorado Island and Pipeline Road, Panama (PA); Manaus, Brazil (BR); and Manu, Peru (PE). Two subareas in PA are considered: Barro Colorado Island (BCI), an island isolated in a freshwater lake following construction of the Panama Canal, and Pipeline Road (PR), a forested area on the nearby mainland that is part of Soberanía National Park.

Research on the birds of BCI began in the early twentieth century, with emphasis on developing a list of species present and general natural history. The pioneering work of Slud at CR was a more complete and systematic compilation of species present at La Selva and their natural histories. Research at PE began only fifteen years ago, with emphasis on community studies and detailed studies of the ecology of certain species and groups, whereas research at BR began most recently as part of a study of the role of forest fragmentation on the extinction of many taxa, especially birds.

Here we review the similarities and differences among the four avifaunas and discuss potential explanations of those patterns. First, we compare birds at the faunal level. Then we focus on selected topics at the level of forest birds, often with emphasis on upland forest, the only major forest type present at all four sites. Subjects such as taxonomic and ecological diversity are emphasized. We document high species richness at several scales, the predominance of rare species at all sites, similarities among sites for the most common species, the trophic organization of the undergrowth fauna, and the relative territory size for similar species at the four sites.

The first step to take in any such comparison is to standardize samples. Generally we discuss only the birds of upland forest, and we use mist-net samples with a standard sample effort (number of captures—Karr 1979, 1981). Although the pattern of distribution of nets in forest varied among sites, no evidence exists to suggest that results are affected by those differences. Finally, for analysis of data from mist-netting of birds, we use only dry season data to avoid the confounding effects of seasonal variation.

Environmental and Geographic Context

Differences among the avifaunas are no doubt associated with five factors (table 14.1): topographic complexity, presence of bodies of water, regional landscape, climatic pattern, and history of human influence. CR has a 3–5-km wide connection to highland areas, whereas BR and PE are isolated from highland influence. PA was connected to mid-elevation forest until the 1950s, when a new highway was built across the isthmus. Since the late 1960s, clearing of forest along that road and adjacent secondary roads has isolated PA from nearby mid-elevation sites.

Water obviously attracts many bird species. Three of four sites (CR, PA, and PE) have medium to large rivers and associated small tributary streams. Oxbow lakes provide an important avian habitat at PE but do not occur at other areas. PA is associated with a large freshwater lake produced by construction of the Panama Canal and is close to both the Atlantic and Pacific oceans. No rivers are near the study areas in BR.

The regional landscape at the two South American sites (BR, PE) is extensive forest, whereas the Central American sites are forest patches adjacent to extensive tracts of cleared land. Substantial urban and industrial areas exist near the PA study area.

Climates among the areas are superficially similar. All have average temperatures above 24°C and annual rainfall above 2,000 mm. La Selva is the wettest (more than 4,000 mm), and CR has no marked dry season. PA, PE, and BR have a four-to-five month dry season during which less than 15% of annual rainfall occurs. PE is somewhat cooler than the other areas, with temperatures each year as low as 14–16°C and in some years dropping to 8°C. The history of human influences varies strikingly among the areas. PE remains in essentially pristine condition, with no human

TABLE 14.1 Comparison of the environmental and geographic context of the four study regions

Factor	Costa Rica	Panama	Brazil	Peru
Topographic Complexity	Lowlands with connection to highlands	Lowlands with recently broken connection to foothills	Vast lowland area	Vast lowland area
Presence of Water	Two medium rivers, several small streams, and swamp forest	Near two oceans, large river, and artificial lake	Small streams only	Large river; oxbow lakes; seasonal swamps; extensive floodplain
Regional Landscape	Remnant forest in a matrix of pasture and farmland	Island isolated in artificial lake and forest on adjacent mainland	Lowland forest extending indefinitely in all directions	Lowland forest broken by the dynamics of a major river channel
Climatic Pattern	Annual rainfall 4,000 mm; no marked dry season	Annual rainfall of 2,600 mm with considerable local variation 2,100–3,000; dry Dec.–Apr.; mean temp stable at about 27°C	Rainfall 2,200 mm; dry July–Sept.; monthly mean temperature stable at about 27°C	Rainfall 2,080 mm; dry May–Sept.; monthly mean temperature 22–25°C; annual low 14–16°C
History of Human Influence	Massive forest destruction regionally; hunting presence high	Regional forest destruction; creation of large lake; hunting presence high	Regional clearing of forest less than one decade old	Low density and low impact of natives

impact on vegetation and with populations of birds and mammals that have not been hunted for at least twenty years. Habitat destruction in BR is only a decade old but is substantial in the study region. Both PA and CR have experienced major habitat destruction and hunting pressure for many decades but are now protected. Finally, the biogeographic context of the areas differs with two (BR and PE) in the extensive forest of the Amazon basin and two (CR and PA) in the narrow extension of Neotropical forest into CA where dry and wet forest intergrade.

The Avifaunas

Although our primary focus here is on birds of forest, we briefly discuss the entire avifaunas known from each site (table 14.2). More species have been recorded at the PE site (554) than at any other site. The Central American sites are intermediate and BR is the lowest with 351. All regions (except BR) have about 50 species (9–14% of all species) that are vagrants in the area. The complete list for BR includes few nonforest species because of a lack of large rivers and second-growth habitats but also because research at the BR site emphasizes forest. As a result, no comprehensive regional faunal list and no vagrant list has been complied for BR. Subtraction of vagrants, second-growth, aquatic, and above-canopy species reduces the list to 244–332 species. A species was counted as a forest bird if it regularly occurs in forest habitat, even though it may be more abundant outside forest. This differs somewhat from the approach used in some of the chapters.

The Central American sites support fewer forest birds than South American sites (table 14.2). The larger number of species in PA is due to the higher richness of Neotropical groups in an area that is geographically closer to the richer South American faunas. The South American sites support over 300 species of forest birds, 15–25% higher than the Central American sites. Taxonomic diversity of forest birds at the generic level is up to 20% lower in Central than South America, but the faunal diversities of the four areas are indistinguishable at the familial level (table 14.2).

The same basic food resources are consumed at all four sites (table 14.3), with several major food resources accounting for most of the differences in species richness between South and Central America: insectivores, frugivores, and, to a lesser extent, carnivores. The only reverse in the trend is the high number of insectivore-nectarivores in CR, a pattern probably due to the proximity of high elevation habitats or, perhaps, the abundance of second-growth areas. For both insectivores and omni-

TABLE 14.2 Number of birds known from four Neotropical areas

	Costa Rica	Panama	Brazil	Peru
Total species	410	443	351	554
Nonvagrant species	363	381	?	508
Aquatic species*	28	57	16	51
Forest species	244	251	300	332
Forest families	36	39	38	37
Forest genera	205	179	208	230

Note: * Some aquatic species also counted as vagrants, especially in Panama

TABLE 14.3 Number of forest species assigned to broad diet groups

Diet	Costa Rica	Panama	Brazil	Peru
Insectivores	111	123	157	163
Omnivores	66	66	66	69
Frugivores	29	31	40	58
Insectivores-nectarivores	18	12	12	14
Carnivores, including piscivores	15	18	21	26
Other	5	1	4	2
TOTAL	244	251	300	332

vores, the relative importance of species that feed on small insects or small fruits (as opposed to large) is consistent among the four forests: 60–63% for insectivores and 66–73% for omnivores.

Another major faunal difference among the areas is the richer north temperate zone migrant fauna in Central America (CR—80 species; PA—66) than in South America (BR—20; PE—40). PE has about 20 additional species that are austral migrants. Among forest species, more are north temperate/tropical migrants in Central America (CR—37; PA—34) than in South America (BR—10; PE—7).

Avifauna Sampled with Mist Nets

Species Richness

We consider first the accumulation of species as a function of number of mist-net captures (fig. 14.1). The rate of accumulation of species is clearly highest in PE and somewhat lower but consistent among the other sites. The slowest rate of accumulation of species, only slightly above 50% of those for CR, PA/PR, and BR, occurs on PA/BCI.

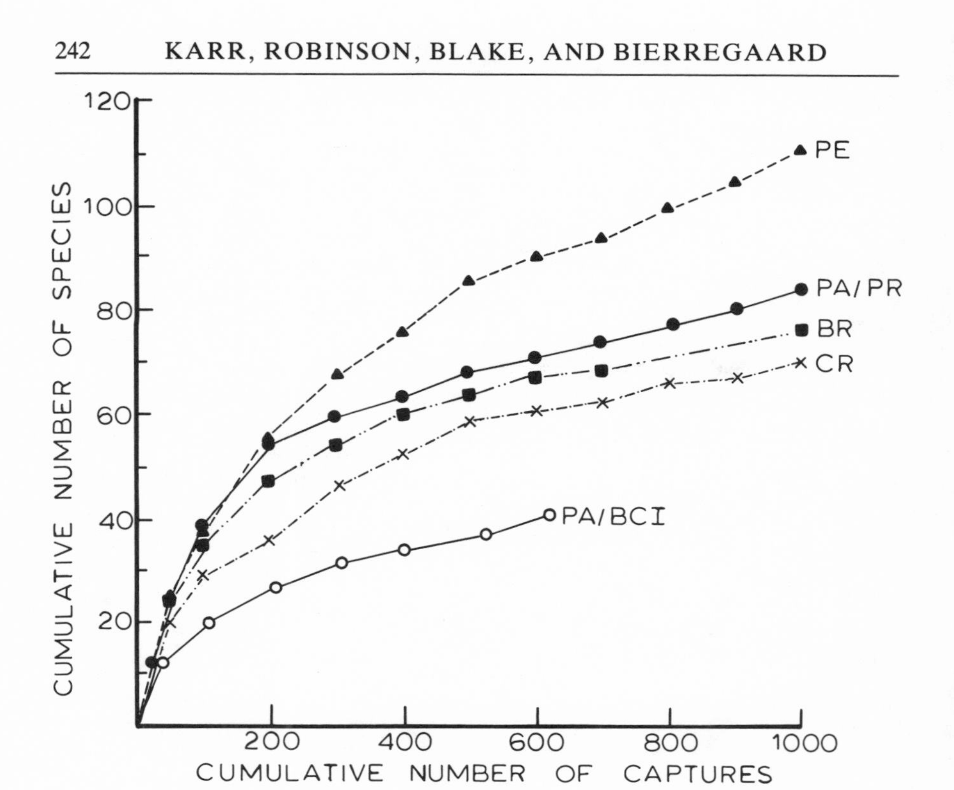

FIGURE 14.1 Species accumulation curves for undergrowth mist-net samples at a series of five Neotropical upland forest sites: PE = Manu, Peru; PA/PR = Pipeline Road, Panama; BR = Manaus, Brazil; CR = La Selva, Costa Rica; PA/BCI = Barro Colorado Island, Panama.

Another approach is to use jackknife procedures (Burnham and Overton 1979) to obtain a robust estimate of species richness from a set of 100-capture, dry-season samples (table 14.4). CR, PA/PR, and BR are similar at about 55 species. PA/BCI and PE are somewhat lower, the former due to loss of species since the island was isolated in Gatun Lake early this century. Somewhat surprisingly, given the results of the accumulation curve analysis, PE's estimates of species richness are well below those for the other tropical mainland sites. A high number of species represented by only one or two captures in the PE sample is probably responsible for this, suggesting a limitation to jackknife procedure. The value for a site on BCI is especially low, as was the accumulation curve (fig. 14.1). Variability of species richness estimates is greater among samples in Central America than in South America for reasons that are not

TABLE 14.4 Jackknife estimates of species richness based on 100-capture samples of mist-netted birds

Study region	*N*	Mean	SD
La Selva, Costa Rica	9	56	20.2
Pipeline Road, Panama	9	54	12.3
BCI, Panama	4	28	6.6
Manaus, Brazil	15	55	9.8
Manu, Peru	5	41	5.5

TABLE 14.5 Summary of selected attributes of 1,000-capture samples from upland forest

	Costa Rica	Panama	Brazil	Peru
Total number of species	70	84	76	111
Number of species represented by:				
21 or more captures	10	14	17	13
11–20 captures	14	10	15	19
2–10 captures	33	43	34	44
1 capture	13	17	10	35
Number of captures in top ten species	637	548	412	398

obvious to us at this time, although variation in abundances of altitudinal and latitudinal migrants may be a factor.

Patterns of Abundance

Mist-net samples (1,000 captures) illustrate that the differences in species richness among the forest avifaunas is true of understory as well as for the entire forest avifaunas. In addition, two other general points emerge: rare species predominate (table 14.5) and the same or ecologically similar species are most frequently captured at all sites (table 14.6). The species richness of these 1,000-capture samples is highest in PE with 111 species and lowest in CR with only 70. The predominance of rare species in three samples is clear (table 14.5). An extraordinary proportion (78–88%) of the species captured each contributed less than 2% of the captures, and many species are represented by only one capture at each site.

TABLE 14.6 Number of captures (with rank) for the ten most frequently captured species in 1,000-capture samples

Food habits group/species	Costa Rica	Panama	Brazil	Peru
Nectarivores				
Phaethornis superciliosus	60 (5)	59 (4)	+	39 (3)
Chalybura urochrysia	21 (9)			
Frugivores				
Pipra mentalis	134 (2)	121 (1)		
P. pipra	+		62 (1)	
P. fascicauda				94 (1)
P. coronata		31 (9)		31 (5)
Corapipo leucorrhoa (altitudinal migrant)	38 (6)			
Geotrygon montana	+	+	29 (8)	23 (8)
Insectivore-frugivores				
Mionectes oleagineus	94 (3)	72 (2)		+
Hylocichla mustelina (migrant)	63 (4)	+		
Turdus albicollis			28 (10)	+
Ant-followers				
Gymnopithys leucaspis	29 (7)	78 (2)		
G. rufigula			54 (3)	
Hylophylas naevioides	+	44 (5)		
H. poecilonota			58 (2)	24 (6)
Pithys albifrons			46 (4)	
Dendrocincla fuliginosa	+	39 (7)	+	+

Bark gleaners				
Glyphorynchus spirurus	152 (1)	+	46 (4)	84 (2)
Flycatcher/foliage gleaner				
Platyrinchus coronatus	+	42 (6)	+	+
Thamnomanes caesius			30 (6)	
Foliage gleaner				
Thamnomanes ardesiacus			30 (6)	+
Myrmotherula longipennis			29 (8)	+
M. axillaris	+	+	+	22 (10)
Myrmoborus myotherinus				34 (4)
Automolus ochrolaemus	+	+	+	23 (8)
Henicorhina leucosticta	25 (8)	+		
Ground gleaner				
Cyphorhinus phaeocephalus	+	32 (8)		
Microcerculus marginatus	+	+		24 (6)
Formicarius analis	21 (9)	+	+	+
Sclerurus guatemalensis	+	30 (10)		

Note: + = present in area but not in top ten species

At the other end of the abundance spectrum, the ten most frequently captured species are very similar ecologically. A manakin (*Pipra*) is the most frequently captured species at three areas (table 14.6). In the exceptional area (CR), the second most common species is a *Pipra* (table 14.5). Both Central American sites have three species of frugivores in the top ten species. The ruddy quail-dove (*Geotrygon montana*), present at all sites, was among the top ten species at both South American sites. *Glyphorynchus spirurus*, a small woodcreeper, is among the top five species at three of the four areas. Ant followers are especially abundant at PA (three species) and BR (two species) but are less common in CR and PE. Upland forest (never flooded) in PE yielded only one relatively rare species, in contrast to high ground forest (occasionally flooded) in PE, which has many more species and individuals (S. Robinson, unpubl. data). The only area with a north temperate migrant in the top ten species is CR, where 63 captures of wood thrush (*Hylocichla mustelina*) were recorded. The long-tailed hermit (*Phaethornis superciliosus*) is in the top five species in all areas but BR; this species was represented by about 60 captures in CR and PA. No hummingbird was recorded in 1,000 captures in BR, and *P. superciliosus* was present at lower densities in Peru than in Central America. Ground-gleaners were present in all areas in the ten most captured species. The black-faced antthrush (*Formicarius analis*), present at all four areas, was among the most abundant species only in CR. Sallying flycatchers are not well represented at any area. Small insectivores in such genera as *Thamnomanes*, *Myrmotherula*, *Myrmoborus*, and *Automolus* are especially well represented in the ten most captured species in South America but are unrepresented in the Central American samples. Only one dead-leaf forager, the buff-throated foliage-gleaner (*Automolus ochrolaemus*) at PE, occurs in the ten most abundant species.

Trophic Structure

The complex of food resources available to and exploited by birds defines the trophic structure of the community. Mist-net samples allow comparisons of the trophic structure of the undergrowth avifauna. (Trophic structure at the forest avifauna level is discussed in the chapters on specific sites.) Eleven foraging guilds were defined for analysis following the methods of Karr (1980). The following guilds represent a combination of food type and foraging location: two frugivore groups (terrestrial and undergrowth); five insectivore groups (terrestrial [leaf litter], bark, foliage, sally [flycatch], and ant followers); two groups of mixed feeders (insectivore-nectarivore and undergrowth insectivore-frugivores); and finally miscellaneous, which includes such rarely captured species as raptors and piscivores. Species richness within these ecological groups tends to be similar among the areas for frugivores and bark-gleaning insectivores. The species richness of most other guilds, however, tends to be higher in PE than in either of the Central American areas. Only among the ant followers do the Central American sites have higher species richness. Because almost nothing is known about the availability of any of these food resources among the sites, these results can at best be used to generate hypotheses about the differences and similarities in food resource availability as well as differences in the abundances of nonavian competitors among the sites.

We analyzed the guild structure for number of captures by guild (Karr 1980) with Sorenson's distance measure and a group average cluster procedure. To examine how much guild signatures vary within a site over time, we included two to five 100-capture samples from each area. Samples from both PR (mainland) and BCI (island) in Panama were included (fig. 14.2). Similarity among samples reflected geographic proximity. Generally, samples from a site were more similar to each other than to samples from any other site. Sites that are close together clustered until Central and South American groups were formed. Of particular interest is the clumping of the BCI and PR samples, suggesting that the reduced species richness at BCI occurs without selective loss of certain foraging guilds common to the forest undergrowth (Karr 1982).

Territory Sizes and Abundances

Limited data on a few species of undergrowth insectivores suggest that territory size varies among the geographic areas. On BCI, Greenberg and Gradwohl (1985) demonstrated territory sizes of about 1 ha for the small antwrens, and mark-recapture studies on the nearby mainland (Karr, unpubl. data) corroborate this conclusion. In contrast, work in PE suggests that the same or similar species have territory sizes in the range of 4–5 ha. Territory sizes of antwrens are 8–10 ha at BR (Powell 1985). Perhaps these territory size patterns are due to variation in soil fertility, a suggestion concordant with known patterns of soil fertility and the composition of understory floras (Gentry and Emmons 1987; Stark et al., in prep.). Another explanation is that higher faunal richness requires lar-

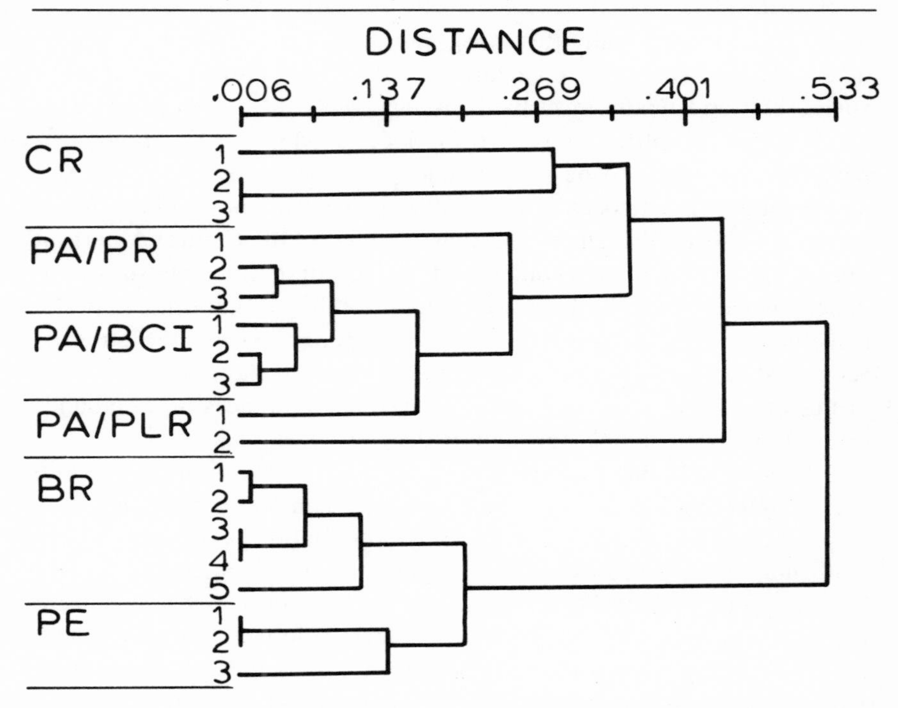

FIGURE 14.2. Dendrogram of guild signatures using number of captures in each feeding guild for 100-capture samples. Letters indicate sample location: CR = La Selva, Costa Rica; PA/PR = Pipeline Road, Panama; PA/BCI = Barro Colorado Island, Panama; PA/PLR = Plantation Road near Summit Gardens, Panama; BR = Manaus, Brazil; PE = Manu, Peru. Numbers indicate different samples at the same site.

ger territories because each species is constrained to a narrower foraging zone; this fits for a comparison of PE and PA but is not compatible when BR is added.

Mixed-species flocks found in the understory and middle levels of these forests illustrate this difference. These flocks contain many fewer core species in Central America (2–4) than in Amazonia, where they commonly contain 8–13 core species. The higher species richness of the South American flocks may constrain foraging zones for the flock species, and thus they require larger territories. Or perhaps the larger flocks are more effective at detecting predators, allowing flock members to forage with the advantage of a collective vigilance.

Discussion

Avifaunal richness among these four sites derives from varying influence of nearby nonforest habitat and from the greater richness of forest avifaunas in South America. Even within forest, some increase may be attributed to availability of new microhabitats, such as bamboo thickets, but the higher species richness within similar patches of forest in South America is also clear. Among nonforest habitats, aquatic and riparian areas are especially important. Many species are present in PA because of the extensive regional availability of aquatic environments, whereas lack of those habitats in BR results in inclusion of few aquatic species in the BR list. The number of north temperate migrants is smaller in Amazonia than in Central America, probably because of the differing migration distances. The trophic structures of the four forest faunas are superficially similar, both at the faunal level and at the level of guild signatures based on undergrowth mist-net samples.

Throughout this discussion we have described these avifaunas as relatively static entities; this is an oversimplification from several perspectives. Habitat disturbance and colonization is common to all areas from the meandering of river channels in PE, PA, and CR to the extensive modification of habitats by humans throughout this century in CR, PA, and, more recently, BR. Haffer (1969) argued that the richness of the Neotropical avifauna results from altered distributions of forest and nonforest habitats driven by climatic change. As habitat destruction from human influence increases, historical influences such as habitat expansion and shrinkage, isolation by geographic barriers, and the speciation events that result from these will be rarer.

Other factors affect the avifaunas of each of three forests. Shifting availability of habitats in North America may alter the species composition and abundances of migrants from north temperate areas. On a local scale, regional movements of birds are important. The dominant dynamic at CR is altitudinal movements of many groups, especially frugivores and nectarivores. Local movements in PA include some altitudinal changes (the yellow-eared toucanet [*Selenidera spectabilis*], for example, arrives from foothills early in the dry season) but are dominated by movements from dry to wet forest during dry periods. In addition, considerable local movement has been documented as a response to changing microclimatic conditions within a few square kilometers (Karr and Freemark 1983). All of these movements are tied to seasonality, either local effects of changing rainfall regimes or the changing temperate zone seasons. As noted by several recent authors, seasonal patterns are not consistent among species (Bierregaard, this volume; Blake et al., this volume; Karr et al. 1982; Karr and Freemark 1983). The role of seasonal dynamics is least understood at PE, where little fieldwork has been conducted outside the dry season.

Perhaps least is known about the long-term population dynamics of tropical forest birds. General observations clearly indicate that many species vary in abundances over time. *Geotrygon montana* is one of the most variable species in PA (J. Karr, pers. obs.) and in South America (Bierregaard, this volume; S. Robinson, pers. obs.), perhaps because of its tendency to aggregate under mast-fruiting plants with seeds that fall to the forest floor (S. Robinson, pers. obs.).

Although much has been learned about Neotropical birds in the past decade, our knowledge remains rudimentary. Even the nest is unknown for such familiar species as the Ocellated antbird (*Phaenostictus mcleannani*), a common ant follower in PA. Understanding the ecology and evolution of these avifaunas depends on insightful accumulation of data on individual species and groups of species. Integration of such data will improve our knowledge of community patterns and will stimulate yet another round of research projects.

REFERENCES

Burnham, K. P. and W. S. Overton. 1979. Robust estimation of population size when capture probabilities vary among animals. Ecol. 60: 927–936.

Gentry, A. H. and L. H. Emmons. 1987. Geographical variation in fertility, phenology, and composition of the understory of Neotropical forests. Biotropica 19: 216–227.

Greenberg, R. and J. Gradwohl. 1985. A comparative study of the social organization of antwrens on Barro Colorado Island, Panama. In P. A. Buckley, M. S. Foster, E. S. Morton, R. S. Ridgely, and F. G. Buckley (eds.), *Neotropical Ornithology*. Ornithol. Monogr. 36, AOU, Washington, D.C., pp. 845–855.

Haffer, J. 1969. Speciation in Amazonian forest birds. Science 165: 131–137.

Karr, J. R. 1979. On the use of mist nets in the study of bird communities. Inland Bird Banding 51: 1–10.

———. 1980. Geographical variation in the avifaunas of tropical forest undergrowth. Auk 97: 283–298.

———. 1981 Surveying birds with mist nets. In C. J. Ralph and J. M. Scott (eds.), *Estimating Numbers of Terrestrial Birds*. Stud. Avian Biol. 6: 62–67.

———. 1982. Avian extinction on Barro Colorado Island, Panama: a reassessment. Amer. Nat. 119: 220–239.

Karr, J. R., D. W. Schemske, and N. Brokaw. 1982. Temporal variation in the undergrowth bird community of a tropical forest. In E. G. Leigh, Jr., A. S. Rand, and D. M. Windsor (eds.), *The Ecology of a Tropical Forest: Seasonal Rhythms and Long-Term Changes*. Smithsonian Inst. Press, Washington, D.C., pp. 441–453.

Karr, J. R. and K. E. Freemark. 1983. Habitat selection and environmental gradients: dynamics in the "stable" tropics. Ecol. 64: 1481–1494.

Powell, G. V. N. 1985. Sociobiology and adaptive significance of interspecific foraging flocks in the Neotropics. In P. A. Buckley, M. S. Foster, E. S. Morton, R. S. Ridgely, and F. G. Buckley (eds.), *Neotropical Ornithology*. Ornithol. Monogr. 36, AOU, Washington, D.C., pp. 713–732.

Stark, N., A. H. Gentry, and H. Zuuring. Prediction of tropical plant species diversity from soil and precipitation. Biotropica (submitted).

Appendix 14.1. List of species recorded for each of the four study regions, including their status or habitat preference at each site, mass (grams), and ecological classification for each species.

SPECIES[a]	CR[b]	PA[c]	BR[d]	PE[e]	MASS[f]	GUILD[g,h]
TINAMIDAE	3	2	4	9		
Tinamus tao				FU	X,X,X,2000	G-FR-G
Tinamus major	F	UT	U	TFU	1100,1160,X,1170	G-FR-G
Tinamus guttatus			U	U	X,X,X,695	G-FR-G
Crypturellus cinereus				TUF	X,X,X,465	G-FR-G
Crypturellus soui	ST	ST	T	TUO	245,250,X,205	G-FR-G
Crypturellus undulatus				ZTFO	X,X,X,540	G-FR-G
Crypturellus bartletti				F	X,X,X,241	G-FR-G
Crypturellus variegatus			U	FU	X,X,384,352	G-FR-G
Crypturellus boucardi	F				500,X,X,X	G-FR-G
Crypturellus atrocapillus				O	X,X,X,453	G-FR-G
PODICIPEDIDAE	1	2	1	1		
Tachybaptus dominicus	R	L	L	L	150,125,X,X	
Podilymbus podiceps		L			450,340,X,X	
PELECANIDAE		1				
Pelecanus occidentalis		V			3000,3450,X,X	
PHALACROCORACIDAE	1	1				
Phalacrocorax olivaceous	R	L			1100,1150,X,X	
SULIDAE		1				
Sula leucogaster		V			1100,1000,X,X	
ANHINGIDAE	1	1	1	1		
Anhinga anhinga	R	L	L	LR	1200,1250,X,X	
FREGATIDAE	1	1				
Fregata magnificens	V	V			1200-1700,1400,X,X	
ARDEIDAE	13	12	6	12		
Ardea herodias	R	L			2500,2400,X,X	
Ardea cocoi			RL	LR	X,1600,X,X	
Casmerodius albus	R	L	L	LR	950,875,X,X	
Egretta thula	R	L		R	375,360,X,X	
Egretta caerulea	R	L			325,320,X,X	
Egretta tricolor	R	L			350,375,X,X	
Butorides striatus=virescens	RM	L		L	210,175,X,X	
Agamia agami	Q	U		TFUL	550,500,X,609	W-FI-W
Bubulcus ibis	P	L	S	LR	350,340,X,X	
Pilherodius pileatus			U	LR	X,555,X,X	W-FI-W
Nycticorax nycticorax	R	L		L	800,750,X,X	
Nycticorax violaceus	R				625,680,X,X	
Tigrisoma lineatum	Q	LU	U	TFU	850,840,X,X	W-FI-W
Ixobrychus involucris				L	X,X,X,73	
Ixobrychus exilis	M	L		L	80,85,X,X	
Cochlearius cochlearius	R	L	U	LF	600,550,X,X	W-FI-W
CICONIIDAE	1	1	1	2		
Mycteria americana	V	V	S	R	2500,2400,X,X	
Jabiru mycteria				R	6500,6400,X,X	
THRESKIORNITHIDAE	2	1		2		
Mesembrinibis cayennensis	Q	V		L	650,750,X,670	W-LI-W
Ajaia ajaja	R			R	1400,1500,X,X	
ANATIDAE	3	9		4		
Dendrocygna viduata		L			650,700,X,X	
Dendrocygna bicolor		L		L	750,680,X,X	
Dendrocygna autummalis	RM	L			800,550,X,X	
Neochen jubata				R	X,X,X,X	
Anas discors	M	L			400,350,X,X	
Anas americana		L			700,545,X,X	
Aythya affinis		L			800,600,X,X	

Aythya collaris		L			700,680,X,X	
Cairina moschata	QR	L		RL	3000,2700,X,X	W-SE-W
Oxyura dominica		L		L	375,365,X,X	
CATHARTIDAE	3	3	4	4		
Sarcoramphus papa	AF	UT	US	ALL	3500,3200,3125,X	G-CA-G
Coragyps atratus	A	S	US	RZL	1800,1800,1375,X	
Cathartes aura	A	ALL	US	PR	1400,1300,X,X	G-CA-G
Cathartes melambrotos			US	ALL	X,1200,X,1130	G-CA-G
ACCIPITRIDAE	28	28	17	28		
Pandion haliaetus	R	L		LR	1500,1515-1837,X,X	
Gampsonyx swainsonii			S	P	X,X,X,X	
Elanus leucurus		S			X,280,X,X	
Elanus caeruleus	P				350,X,X,X	
Elanoides forficatus	A	A	US	A	480,450,X,400	A-SI-F
Leptodon cayanensis	G	U		TFU	440,500,X,550	C-PR-B
Chondrohierax uncinatus	G	U		T	275,280,X,X	C-SN-B
Harpagus bidentatus	FG	U	U	TFU	180,185,X,258	C-LI-B
Ictinia plumbea	A	A	US	A	280,247,X,240	A-SI-F
Ictinia mississippiensis	A	A		V	280,275,X,X	A-SI-F
Rostrhamus sociabilis				LP	375,355,X,X	
Helicolestes hamatus				P	X,420,X,X	
Accipiter bicolor	F		U	ZTFU	260-400,250-454,X,211	C-BI-A
Accipiter superciliosus	GT	U	U	TFU	75-130,70-120,80,117	C-BI-A
Accipiter poliogaster				U	X,X,X,X	C-BI-A
Buteo swainsoni	A	ALL		R	1000,908-1069,X,X	
Buteo platypterus	AG	ALL	T	R	450,360-412,X,X	U-PR-F
Buteo brachyurus	A	V		TO	480,460-530,X,490	G-BI-G
Buteo magnirostris	S	V	T	ZR	290,226-272,X,311	G-LI-G
Buteo nitidus	G	V	S	V	425,420F,X,X	
Buteo albonotatus		V		RLO	750,565,X,565	
Buteo jamaicensis	A				900,875-1307,X,X	
Buteo albicaudatus		S			950,865-1101,X,X	
Leucopternis albicollis	F	U	US		725,652-820,X,X	C-HE-F
Leucopternis semiplumbea	F	U			269,278,X,X	U-HE-F
Leucopternis plumbea		U			X,483,X,X	U-HE-F
Leucopternis schistacea				TF	X,X,X,355	C-HE-F
Leucopternis kuhli				U	X,X,X,340	C-HE-F
Leucopternis melanops			U		X,X,X,X	C-HE-F
Leucopternis princeps	A				1000,X,X,X	U-HE-F
Buteogallus meridionalis		S			X,775-840,X,X	
Buteogallus nigricollis			L		X,725-900,X,X	
Buteogallus anthracinus	Q	L			800,793-1199,X,X	G-LI-G
Buteogallus urubitinga	G	V	US	LRFT	1100,900,X,1450	G-PR-G
Harpyhaliaetus solitarius	A			V	3000,X,X,X	G-PR-G
Morphnus guianensis	A	TU	U	TFU	X,1750,X,1750	C-MA-F
Harpia harpyja		TU	U	TFU	4500-7500,4250,X,4500	C-MA-F
Spizastur melanoleucus	FA			RLOF	850,780-525,X,850	G-PR-F
Spizaetus ornatus	FA	U	U	TFUO	1200,1004-1607,X,950	C-PR-F
Spizaetus tyrannus	FA	U	U	TFUO	1000,904-1105,X,1025	C-PR-F
Circus cyaneus	P	V			450,472-570,X,X	
Geranospiza caerulescens	GF	U		V	350,325-430,X,X	U-PR-B
FALCONIDAE	8	10	9	7		
Micrastur semitorquatus	GF	U	U	ULFT	575-710,575-750,X,550	C-PR-F
Micrastur ruficollis	FT	U	U	TF	165-200,161-196,200-230	U-PR-F
Micrastur gilvicollis			U	FU	X,X,211,220	U-PR-F
Micrastur mirandollei	FGT	U	U		420-550,450,X,X	U-PR-F
Daptrius americanus	G	U	UTS	PFUTO	550,497,X,583	C-SO-F
Daptrius ater			UT	LRPO	X,X,X,370	C-OM-F
Herpetotheres cachinnans	GFC			TF	600,675-800,X,650	C-HE-B
Milvago chimachima		V	S		330,375,X,X	
Polyborus plancus		V	S		1000,834,X,X	
Falco peregrinus	A	V			610-950,640-825,X,X	
Falco rufigularis	RG	S	UTS	RLTUF	140-200,150,X,155	A-LI-A
Falco femoralis		V			220-330,285,X,X	
Falco sparverius	P	V			115,86-113,X,X	

CRACIDAE	3	3	5	4		
Ortalis motmot			T	RLZT	X,X,X,402	C-FR-F
Ortalis cinereiceps	SG	ST			500,536,X,X	C-FR-F
Penelope jacquacu			U	TFUO	X,X,X,1325	C-FR-F
Penelope marail			U		X,X,X,X	G-FR-G
Penelope purpurascens	FGC	TU			1700,1800,X,X	C-FR-F
Aburria pipile			U	TFUO	X,1200,X,X	C-FR-F
Crax rubra	F	U			4000,3800,X,X	G-FR-G
Crax alector			U		X,X,X,X	G-FR-G
Crax mitu				TFULO	X,X,X,3061	G-FR-G
PHASIANIDAE	3	2	1	2		
Odonthoporus gujanensis		U	U	Z	300,X,X,310	G-SO-G
Odonthoporus stellatus				TFU	X,X,X,325	G-SO-G
Odontophorus guttatus	F				300,300,X,X	G-SO-G
Odontophorus erythrops	F				280,X,X,X	G-SO-G
Rhynchortyx cinctus	F	U			150,165,X,X	G-SO-G
OPISTHOCOMIDAE				1		
Opisthocomus hoazin				L	X,X,X,855	
ARAMIDAE	1			1		
Aramus guarauna	R			LP	1100,1028-863,X,X	
PSOPHIIDAE			1	1		
Psophia leucoptera				TFU	X,X,X,1240	G-LO-G
Psophia crepitans			U		X,X,X,X	G-LO-G
RALLIDAE	6	6	4	7		
Rallus nigricans				LP	X,X,X,X	
Rallus maculatus				V	X,X,X,X	
Aramides cajanea	QTF	TU	UW	TFULP	460,425-385,X,515	G-SI-G
Aramides calopterus			UW		X,X,X,X	G-SI-G
Porzana flaviventer		W			25,25,X,X	
Laterallus albigularis	MS	L			42,47,X,X	
Laterallus melanophaius			S	LP	X,X,X,51	
Laterallus viridis			S		X,X,X,X	
Laterallus exilis	PM				33,32,X,X	
Gallinula chloropus	M	L		L	365,312,X,X	
Porphyrula martinica	M	L		L	235,215,X,205	
Porphyrula flavirostris				L	X,X,X,X	
Fulica americana		L			650,565-460,X,X	
Amaurolimnas concolor	Q				95,X,X,X	G-SI-G
HELIORNITHIDAE	1	1	1	1		
Heliornis fulica	QR	L	LR	L	115,135,X,X	W-SI-W
EURYPYGIDAE	1	1	1	1		
Eurypyga helias	QR	U	RLU	RLFU	255,210,X,220	W-SI-W
JACANIDAE	1	1	1	1		
Jacana jacana		L	S	L	95,115,X,X	
Jacana spinosa	M				95,79-112,X,X	
CHARADRIIDAE	1	1	2	3		
Pluvialis dominica			RL	R	150,125,X,X	
Charadrius collaris		V	RLS	R	35,27,X,X	
Charadrius vociferus	PM				95,84,X,X	
Hoploxypterus cayanus				R	X,X,X,X	
SCOLOPACIDAE	3	4	9	12		
Tringa solitaria	M	L	RL	RL	45,55,X,X	
Tringa flavipes		L	RL	RL	85,78-91,X,X	
Tringa melanoleuca	R		RL	RL	160,221,X,X	
Actitis macularia	R	L	RL	RL	40,33,X,X	
Calidris alba				V	55,47,X,X	
Calidris fuscicollis			RL	R	45,X,X,X	
Calidris minutilla			RL		21,21,X,X	
Calidris melanotos			RL	RL	75,X,X,X	
Calidris bairdii				R	39,35,X,X	
Micropalama himantopus			RL	RL	X,60,X,X	
Tryngites subruficollis				R	60,47,X,X	
Bartramia longicauda				R	X,95,X,X	
Limnodromus griseus				V	95,102,X,X	
Gallinago gallinago		L	RL		105,108,X,X	

PHALAROPODIDAE				1		
Phalaropus tricolor				RL	40,X,X,X	
RECURVIROSTRIDAE				1		
Himantopus mexicanus				V	X,162,XX	
LARIDAE	2	13		2		
Stercorarius parasiticus		V			475,500,X,X	
Stercorarius pomarinus	V				700,495,X,X	
Larus delawarensis		V			450,502,X,X	
Larus argentatus		V			900,830,X,X	
Larus atricilla		L			275,297,X,X	
Larus pipixcan		V			250,260,X,X	
Phaetusa simplex				RL	X,220-250,X,X	
Sterna superciliaris				RL	X,X,X,X	
Sterna hirundo		L			110,105,X,X	
Sterna fuscata	V	V			180,185,X,X	
Sterna albifrons		V			X,47,X,X	
Sterna (Thalasseus) maxima		L			450,470,X,X	
Sterna (Thalasseus) sandvicensis		L			190,270,X,X	
Chlidonias niger		L			50,38,X,X	
Gelochelidon nilotica		V			180,180-155,X,X	
Anous stolidus		V			160,165,X,X	
RYNCHOPIDAE				1		
Rynchops niger				RL	350-260,351-235,X	
COLUMBIDAE	11	11	6	8		
Columba cayennensis	P	T		RLT	250,210,X,259	C-FR-F
Columba speciosa	FGC	U			320,259,X,X	C-FR-F
Columba subvinacea			U	T	170,165,128,126	C-FR-F
Columba nigrirostris	FGCT	U			140,160,X,X	C-FR-F
Columba plumbea			U	TFUO	X,X,X,207	C-FR-F
Columba flavirostris	PS				230,X,X,X	
Zenaida macroura	P				100,99,X,X	
Columbina talpacoti	PS	V	S	Z	53,47,X,47	
Columbina passerina			S		40,X,X,X	
Columbina picui				V	X,X,X,48	C-FR-F
Columbina minuta		V			33,40,X,X	
Claravis pretiosa	SP	S		Z	69,69,X,X	
Leptotila rufaxilla				LTO	155,X,X,173	G-FR-G
Leptotila verreauxi	S	S	T		165,150,X,X	G-FR-G
Leptotila cassinii	GTC	U			144,155,X,X	G-FR-G
Geotrygon montana	FT	TU	U	TFU	123,128,113,115	G-FR-G
Geotrygon veraguensis	FT	U			153,155,X,X	G-FR-G
Geotrygon violacea		U			150,102,X,X	G-FR-G
PSITTACIDAE	8	10	14	20		
Ara ambigua	F	U			1300,X,X,X	C-FR-F
Ara ararauna			U	TFPUO	X,1130,X,1125	C-FR-F
Ara macao		U	U	TFUPO	900,1000,X,1015	C-FR-F
Ara chloroptera		U	U	TFUP	X,1050,X,1250	C-FR-F
Ara severa		U		TFUPO	X,360,X,430	C-FR-F
Ara manilata			PU	PO	X,X,X,370	C-FR-F
Ara couloni				PTO	X,X,X,251	C-FR-F
Aratinga leucophthalmus				TFU	X,X,X,191	C-FR-F
Aratinga weddellii				LPTO	X,X,X,108	C-FR-F
Aratinga finschi	P				150,X,X,X	
Aratinga nana	FGTC				95,X,X,X	C-FR-F
Pyrrhura picta				TFO	X,X,X,67	C-FR-F
Pyrrhura rupicola				TFU	X,X,X,76	C-FR-F
Forpus sclateri				TFUO	X,X,X,27	C-FR-F
Forpus passerinus			U		X,X,X,X	C-FR-F
Forpus sp. nov.				U	X,X,X,X	C-FR-F
Brotogeris jugularis	SP	TU			65,63,X,X	C-FR-F
Brotogeris cyanoptera				TFUO	X,X,X,67	C-FR-F
Brotogeris chrysopterus			U		X,X,X,X	C-FR-F
Brotogeris sancthithomae				LTFO	X,X,X,64	C-FR-F
Touit huetii				V	X,X,X,X	
Touit purpurata			U		X,X,X,X	C-FR-F

Pionites leucogaster				TFUO	X,X,X,168	C-FR-F
Pionites melanocephalus			U		X,X,X,X	C-FR-F
Pionopsitta haematotis	FGC	U			165,145,X,X	C-FR-F
Pionopsitta caica			U		X,X,X,X	C-FR-F
Pionopsitta barrabandi				FUO	X,X,X,184	C-FR-F
Pionus menstruus		U	U	TFO	220,235,X,293	C-FR-F
Pionus fuscus			U		X,X,X,X	C-FR-F
Pionus senilis	FGC				220,213,X,X	C-FR-F
Amazona ochrocephala		V		TFO	X,405,X,510	C-FR-F
Amazona farinosa	FGC	U	U	FUO	600,750-625,X,798	C-FR-F
Amazona autumnalis	FGC	U	U		420,416,X,X	C-FR-F
Deroptyus accipitrinus			U		X,X,X,X	C-FR-F
CUCULIDAE	6	9	5	11		
Coccyzus erythrophthalmus	S	V		ZT	43,49,X,43	U-LI-F
Coccyzus americanus	S	U		TUF	50,60,X,48	U-LI-F
Coccyzus melacoryphus				ZTO	X,X,X,48	U-LI-F
Coccyzus euleri			UT		X,X,X,X	U-LI-F
Piaya cayana	SFG	TU	T	TFUO	107,111,X,104	C-LI-F
Piaya minuta		S		LP	X,X,X,40	U-LI-F
Piaya melanogaster			U	FU	X,X,X,107	C-LI-F
Crotophaga major		L		LP	X,157-140,X,165	
Crotophaga ani		S	ST	LP	115-95,117-104,X,105	G-LI-G
Crotophaga sulcirostris	PS				80-70,69,X,X	
Tapera naevia	P	S			55,51,X,X	
Dromococcyx phasianellus		U		L	96,86,X,X	U-LI-F
Dromococcyx pavoninus			UT	TUO	X,X,X,56	U-LI-F
Neomorphus geoffroyi	F	U		TFU	350,340,X,340	G-LI-R
TYTONIDAE	1		1			
Tyto alba	PS		ST		425,439-516,X,X	G-PR-G
STRIGIDAE	6	8	5	8		
Otus choliba		S		V	160,143,X,X	
Otus watsonii			U	TFU	X,X,148,144	U-LI-F
Otus guatemalae	F	TU			150,100,X,X	U-LI-F
Lophostrix cristata	FT	U	U	FU	400,510,X,510	C-LI-F
Pulsatrix perspicillata	FCG	U	UT	FU	750,850,X,795	C-PR-F
Glaucidium minutissimum	FGC	U	U	TFUO	60,61,55,61	U-LI-F
Glaucidium brasilianum				LZT	70,70,X,67	U-LI-F
Ciccaba huhula				FU	X,X,X,370	C-LI-F
Ciccaba virgata	GTC	U	U	TFUO	275,230-307,X,318	C-LI-F
Ciccaba nigrolineata	F	U			350,417-500,X,X	C-LI-F
Rhinoptynx clamator		V			440,320-425,X,X	
STEATORNITHIDAE		1				
Steatornis caripensis		V			430,404,X,X	
NYCTIBIIDAE	2	2	4	3		
Nyctibius griseus	GP	ST	UT	LRUO	230,185,X,175	C-LI-A
Nyctibius bracteatus			U	FU(?)	X,X,X,X	C-LI-A
Nyctibius aethereus			U		X,X,X,X	C-LI-A
Nyctibius grandis	F	U	U	FUTO	600,585,X,575	C-LI-A
CAPRIMULGIDAE	4	5	6	8		
Lucoralis semitorquatus	GF	U	S	FUO	75,75,X,87	A-SI-A
Chordeiles acutipennis		V	S		45,60,X,X	
Chordeiles minor	A		S	A	65,62,X,X	
Chordeiles rupestris				RL	X,X,X,X	
Nyctidromus albicollis	SP	ST	PT	ZRLT	67,53,55,64	A-SI-A
Nyctiphrynus ocellatus			U	FU	36,X,X,43	A-SI-A
Caprimulgus rufus		S			95,95,X,X	
Caprimulgus nigrescens			TU		X,X,X,X	A-SI-A
Caprimulgus carolinensis	GF	V			110,110,X,X	A-SI-A
Caprimulgus sericocaudatus				V		
Hydropsalis climacocerca				ZRL	X,X,X,48	
Hydropsalis brasiliana				V	X,X,X,54	

APODIDAE	6	5	4	6		
Streptoprocne zonaris	A	A		A	98,80,X,111	
Chaetura cinereiventris	A			A	17,14,X,19	
Chaetura pelagica	A	V			21,X,X,X	
Chaetura spinicauda		A	US		18,15,X,X	
Chaetura brachyura		A	US	A	X,19,X,X	
Chaetura egregia				A	X,X,X,24	
Panyptila cayennensis	A	A	PS	A	18,20,X,X	
Panyptila sanctihieronymi	A				50,X,X,X	
Cypseloides niger	A				35,X,X,X	
Tachornis squamata			P	A	X,X,X,X	
TROCHILIDAE	25	21	11	18		
Glaucis hirsuta		ST		ZTFUO	X,7,X,6	S-NI-F
Glaucis aenea	S				5,X,X,X	
Threnetes ruckeri	GTF	TU			6,6,X,X	S-NI-F
Threnetes leucurus				ZTFUO	X,X,X,5	S-NI-F
Phaethornis superciliosus	GFTS	TU	U	ZTFUO	6,6,6,5	S-NI-F
Phaethornis bourcieri			U		X,X,4,X	S-NI-F
Phaethornis hispidus				ZTFUO	X,X,X,5	S-NI-F
Phaethornis longuemareus	FGTS	STU			3,3,X,X	S-NI-F
Phaethornis stuarti				ZTFUO	X,X,X,3	S-NI-F
Phaethornis guy	FG	V			6,6,X,X	S-NI-F
Eutoxeres condamini=aquila	FGT	V		V	12,11,X,10	S-NI-F
Campylopterus largipennis			U	ZTFUO	X,X,9,8	S-NI-F
Phaeochroa cuvierii	G	ST			9,9,X,X	S-NI-F
Florisuga mellivora	FGTC	TU	U	TFUO	7,7,6,7	C-NI-F
Colibri delphinae	G				7,7,X,X	C-NI-F
Anthracothorax nigricollis		S		L	X,7,X,X	
Anthracothorax sp? virdigu			ST		X,X,X,X	C-NI-F
Anthrocothorax prevostii	S				8,7,X,X	
Klais guimeti	GFTC	V			3,3,X,X	S-NI-F
Lophornis delattrei		S			3,X,X,X	
Lophornis helenae	FGC				3,X,X,X	C-NI-F
Lophornis chalybea				UF	X,X,X,3	C-NI-F
Discosura conversii	GFC				3,X,X,X	C-NI-F
Discosura longicauda			U		X,X,X,X	C-NI-F
Popelairia popelairii				UF	X,X,X,3	C-NI-F
Thalurania furcata			U	ZTFUO	X,4,X,5	S-NI-F
Thalurania colombica	FGTC	STU			5,X,X,X	S-NI-F
Chlorostilbon canivetii		ST			3,X,X,X	S-NI-F
Damophila julie		TS			X,4,X,3	S-NI-F
Lepidopyga coeruleogularis		ST			4,4,X,X	S-NI-F
Hylocharis cyanus				ZTFUO	X,X,X,5	C-NI-F
Hylocharis sapphirina			U		X,X,X,X	C-NI-F
Hylocharis eliciae	T				3,4,X,X	S-NI-F
Chrysuronia oenone				TFUO	X,X,X,4	C-NI-F
Amazilia amabilis	GTC	T			4,4,X,X	S-NI-F
Amazilia versicolor			ST		X,X,X,X	S-NI-F
Amazilia edward		S			5,5,X,X	
Amazilia tzacatl	SCG	S			5,5,X,X	S-NI-F
Amazilia lactea				TF	X,X,X,5	C-NI-F
Amazilia rutila	T				5,X,X,X	U-NI-F
Amazilia cyanura	S				5,X,X,X	
Microchera albocoronata	FGCT				3,X,X,X	C-NI-F
Polyplancta aurescens				TFO	X,X,X,6	S-NI-F
Chalybura buffonii		S			X,6,X,X	
Chalybura urochrysia	FGTC				6,7,X,X	S-NI-F
Heliodoxa jacula	GF				8,9,X,X	S-NI-F
Topaza pella			U		X,X,10,X	C-NI-F
Heliothryx aurita			U	FU	X,X,5,6	C-NI-F
Heliothryx barroti	FGTC	TU			6,6,X,X	C-NI-F
Heliomaster longirostris	FGC	S		T	7,7,X,X	C-NI-F
Archilochus colubris	S				3,4,X,X	

TROGONIDAE	5	5	4	5		
Trogon massena	FGTC	TU			145,140,X,X	C-LO-F
Trogon melanurus		TU	U	TFUO	X,115,92,122	C-LO-F
Trogon viridis		TU	U	TFU	X,80,49,X	C-LO-F
Trogon collaris	FG			TFUO	70,64,X,59	C-LO-F
Trogon curucui				TFUO	X,X,X,61	C-LO-F
Trogon rufus	FGT	TU	U		51,52,50,X	U-LO-F
Trogon violaceus	TGC	TU	U	TFU	56,57,54,44	U-LO-F
Trogon clathratus	F				130,X,X,X	C-LO-F
ALCEDINIDAE	6	6	3	5		
Ceryle torquata	R	L	RL	LR	290,327,X,310	
Ceryle alcyon	R	L			150,148,X,X	
Chloroceryle amazona	R	L		LR	110,125,X,120	
Chloroceryle americana	RQ	R		LR	37,39,X,32	W-FI-W
Chloroceryle inda	Q	U	U	LTFU	60,59,51,54	W-FI-W
Chloroceryle aenea	Q	U	U	LTFU	17,16,14,14	W-FI-W
MOMOTIDAE	3	3	1	3		
Electron platyrhynchum	FQGT	U		TFUO	60,62,X,65	C-LO-F
Baryphthengus martii				U	X,X,X,154	U-LO-F
Baryphthengus ruficapillus	FQGT	U			176,162,X,X	U-LO-F
Momotus momota	GT	ST	U	TF	120,112-102,135-111	S-LO-F
GALBULIDAE	2	1	4	4		
Jacamerops aurea	F	U	UT	FU	63,65,61,79	U-LI-F
Galbula cyanescens				RLTFUO	X,X,X,24	U-LI-A
Galbula albirostris			U		X,X,18,X	U-LI-A
Galbula leucogastra			TU		X,X,X,X	U-LI-A
Galbula dea			U		X,X,X,X	C-LI-A
Galbula ruficauda	GCQ				25,26,X,X	C-LI-A
Galbalcyrhynchus purusianus				LTO	X,X,X,50	U-LI-A
Brachygalba albogularis				U	X,X,X,X	U-LI-A
BUCCONIDAE	4	6	7	10		
Bucco macrorhynchos	FG	ST	U	TFUO	105,96,54,120	C-LI-F
Bucco pectoralis		U			X,68,X,X	C-LI-F
Bucco tectus	FGC	TU	U	F	27,33,X,X	C-LI-F
Bucco macrodactylus				ZTBFO	X,X,X,25	U-LI-F
Bucco capensis			U	TF	X,X,54,53	U-LI-F
Bucco tamatia			U		X,X,36,X	U-LI-F
Nystalus striolatus				TFU	X,X,X,50	C-LI-F
Malacoptila semicincta				FU	X,X,X,44	S-LI-F
Malacoptila fusca			U		X,X,44,X	S-LI-F
Malacoptila panamensis	FGTC	U			42,44,X,X	S-LI-F
Nonnula ruticapilla			UT	TUBFO	X,X,20,22	S-LI-F
Nonnula frontalis		U			X,15,X,X	S-LI-F
Monasa nigrifrons				ZTFUO	X,X,X,85	C-LI-A
Monasa atra			UT		X,X,89,X	C-LI-A
Monasa morphoeus	FGTC	V		FU	111,110,X,70	C-LI-A
Chelidoptera tenebrosa				LRO	X,X,X,47	
CAPITONIDAE		1	1	4		
Capito niger			U	TFU	X,X,X,64	C-LO-F
Capito maculicoronatus		TU			X,X,X,X	C-LO-F
Eubucco bourcierii				TFU	35,34,X,X	C-LO-F
Eubucco richardsoni				TFU	X,X,X,35	C-LO-F
Eubucco tucinkae				ZTL	X,X,X,41	C-LO-F
RAMPHASTIDAE	5	4	4	8		
Aulacorhynchus prasinus	G			T	161,X,X,132	C-LO-F
Pteroglossus torquatus	FGTC	TU			220,226,X,X	C-LO-F
Pteroglossus castanotis				TFO	X,X,X,310	C-LO-F
Pteroglossus inscriptus				TO	X,X,X,126	C-LO-F
Pteroglossus viridis			U		X,X,140,X	C-LO-F
Pteroglossus flavirostris				TFU	X,X,X,135	C-LO-F
Pteroglossus beauharnaesii				TF	X,X,X,235	C-LO-F
Selenidera reinwardtii				TU	X,X,X,150	C-LO-F
Selenidera culik			U		X,X,149,X	C-LO-F
Selenidera spectabilis	FG	V			220,229,X,X	C-LO-F

Ramphastos culminatus=vitellinus			U	TFUO	X,X,X,369	C-LO-F
Ramphastos tucanus=cuvieri			U	TFUO	X,X,X,734	C-LO-F
Ramphastos swainsonii	FGTC	U			750,X,X,X	C-LO-F
Ramphastos sulfuratus	FGTC	U			500,399,X,X	C-LO-F
PICIDAE	7	8	11	16		
Picumnus rufiventris				ZTFUB	X,X,X,21	S-SI-B
Picumnus exilis			U		X,X,X,X	S-SI-B
Picumnus borbae				V	X,X,X,9	
Chrysoptilus punctigula				L	X,60,X,79	
Piculus chrysochloros		V	U	TFU	X,88,X,88	C-SI-B
Piculus flavigula			U		X,X,34,X	C-SI-B
Piculus leucolaemus	FGC			TFU	50,X,X,69	C-SI-B
Piculus callopterus		V			X,X,X,X	
Celeus loricatus	FGC	U			83,74,X,X	C-SO-B
Celeus elegans			U	T	X,X,138,136	C-SO-B
Celeus grammicus			U	TF	X,X,72,79	C-SO-B
Celeus flavus				TFUO	X,X,X,102	C-SO-B
Celeus undataus			U		X,X,X,X	C-SO-B
Celeus spectabilis				B	X,X,X,111	C-SO-B
Celeus torquatus			U	FU	X,X,X,134	C-SO-B
Celeus castaneus	FGC				103,90,X,X	C-SO-B
Dryocopus lineatus	GT	TU	ST	LRT	197,180,X,209	C-SO-B
Melanerpes cruentatus			TS	TFUO	X,X,X,59	C-SO-B
Melanerpes rubricapillus		S			55,53,X,X	
Melanerpes pucherani	FGTC	U			63,54,X,X	C-SO-B
Veniliornis passerinus				ZTO	X,X,X,35	U-SI-B
Veniliornis cassini			U		X,X,34,X	U-SI-B
Veniliornis affinis				ZTFU	X,X,X,36	U-SI-B
Veniliornis fumigatus	GT				34,31,X,X	U-SI-B
Campephilus guatemalensis	FGTC				231,240,X,X	C-LO-B
Campephilus melanoleucos		U		FUOT	X,225,X,X	C-LO-B
Campephilus rubricollis			U	FU	X,X,216,220	C-LO-B
Campephilus haematogaster		U			X,224,X,X	C-LO-B
DENDROCOLAPTIDAE	9	9	13	18		
Dendrocincla fuliginosa	FGT	U	U	TFU	42,41,40,31	S-LI-R
Dendrocincla merula			U	TFU	X,X,53,46	S-LI-R
Dendrocincla homochroa	F	TU			44,45-37,X	S-LI-R
Deconychura longicauda	F	U	U	TFU	24,24,29,28	U-LI-B
Deconychura stictolaema			U		X,X,17,X	U-LI-B
Sittasomus griseicapillus			U	TFUO	14,14,14,16	U-SI-B
Glyphorynchus spirurus	FGTC	U	U	FU	16,15,14,14	S-SI-B
Nasica longirostris				TFO	X,X,X,92	C-LI-B
Dendrexetastes rufigula			T	TFUO	X,X,70,70	C-LI-F
Hylexetastes stresemanni				V	X,X,X,111	
Hylexetastes perrotti			U		X,X,115,X	U-LI-B
Xiphocolaptes promeropirhynchus				FU	140,140,X,136	C-LI-B
Dendrocolaptes certhia	FGTC	U	U	TFU	71,68,66,73	U-LI-R
Dendrocolaptes picumnus			U	TFU	65,88,79,80	S-LI-R
Xiphorhynchus picus				LR	X,40,X,41	
Xiphorhynchus pardalotus			U		X,X,38,X	C-SI-B
Xiphorhynchus obsoletus				FW	X,X,X,39	U-SI-B
Xiphorhynchus ocellatus				TUO	X,X,X,32	U-SI-B
Xiphorhynchus spixii				FU	X,X,X,40	U-SI-B
Xiphorhynchus guttatus	GCT	STU		TFUO	47,47,X,65	S-LI-B
Xiphorhynchus lachrymosus	FGC	U			61,51,X,X	C-LI-B
Xiphorhynchus erythropygius	FG				46,48,X,X	U-LI-B
Lepidocolaptes souleyetii	GTCS	V			28,25,X,X	U-SI-B
Lepidocolaptes albolineatus			U	TFU	X,X,19,33	U-SI-B
Campylorhamphus trochilirostris		V	U	ZTB	X,37,34,38	S-SI-B
FURNARIIDAE	5	5	11	29		
Furnarius leucopus				RLTF	X,X,X,44	G-SI-G
Synallaxis cabanisi				V	X,X,X,21	
Synallaxis brachyura	S				17,18,X,X	
Synallaxis albigularis				ZL	X,X,X,17	
Synallaxis gujanensis				ZL	X,X,X,19	

Synallaxis rutilans			U	U	X,X,17,18	S-SI-F
Cranioleuca gutturata				FU	X,X,X,14	S-SI-D
Metopothrix aurantiacus				FW	X,X,X,11	U-SI-F
Thripophaga fusciceps				ZFBO	X,X,X,25	U-SI-F
Berlepschia rikeri				P	X,X,X,35	
Hyloctistes subulatus	F			FUO	32,35,X,29	U-LI-D
Ancistrops strigilatus				FU	X,X,X,37	U-LI-D
Simoxenops ucayalae				B	X,X,X,51	
Philydor erythrocercus		V	U	TFU	X,29,23,25	U-LI-D
Philydor pyrrhodes			U	TFU	X,X,31,29	U-LI-D
Philydor rufus				ZT	34,31,X,34	U-LI-D
Philydor erythropterus				TFUO	X,X,X,30	C-LI-D
Philydor ruficaudatus				ZTF	X,X,X,29	U-LI-D
Automolus infuscatus			U	FU	X,X,31,38	S-LI-D
Automolus dorsalis				B	X,X,X,30	S-LI-D
Automolus rubiginosus			U	U	52,X,37,X	S-LI-D
Automolus ochrolaemus	TGF	U	U	ZTFU	44,40,35,34	S-LI-D
Automolus rufipileatus				ZT	X,X,X,37	S-LI-D
Automolus melanopezus				U	X,X,X,31	S-LI-D
Xenops sp. (milleri/tenuirostris)			U	TFU	X,X,11,X	U-SI-T
Xenops rutilans				TFU	13,12,X,13	U-SI-T
Xenops minutus	GTC	U	U	TFU	12,11,12,12	U-SI-T
Sclerurus albigularis				U	38,35,X,X	G-SI-G
Sclerurus mexicanus		U	U	UF	28,26,25,X	G-SI-G
Sclerurus rufigularis			U		X,X,21,X	G-SI-G
Sclerurus caudacutus			U	TFU	X,X,39,36	G-SI-G
Sclerurus guatemalensis	F	U			35,34,X,X	G-SI-G
FORMICARIIDAE	20	23	34	56		
Cymbilaimus lineatus	GTC	ST	U	TFU	34,37,35,40	U-LI-F
Cymbilaimus sanctaemariae				B	X,X,X,29	U-LI-F
Frederickena unduligera				TFU	X,X,X,85	S-LI-F
Frederickena viridis			U		X,X,68,X	S-LI-F
Taraba major	S	T		ZTO	70,68,X,60	S-LI-F
Thamnophilus doliatus	S	S		Z	31,28,X,28	
Thamnophilus aethiops				TFU	X,X,X,27	S-LI-F
Thamnophilus murinus			U		X,X,18,X	S-LI-F
Thamnophilus schistaceus				TFU	X,X,X,21	U-LI-F
Thamnophilus punctatus	FGT	STU			24,22,X,X	U-LI-F
Thamnophilus amazonicus				V	X,X,X,19	
Thamnistes anabatinus	FG	U			21,21,X,X	C-LI-F
Dysithamnus puncticeps		U			17,15,X,X	U-SI-F
Dysithamnus mentalis	F				15,14,X,X	U-SI-F
Dysithamnus striaticeps	FG				18,X,X,X	U-SI-F
Pygiptila stellaris				TFUO	X,X,X,25	U-LI-F
Neoctantes niger				UBO	X,X,X,32	S-SI-B
Thamnomanes ardesiacus			U	TFU	X,X,18,18	S-SI-F
Thamnomanes schistogynus				ZTFUO	X,X,X,17	S-LI-F
Thamnomanes caesius			U		X,X,18,X	S-LI-F
Myrmotherula brachyura		TU	U	TFUO	X,7,X,8	U-SI-F
Myrmotherula sclateri				TFU	X,X,X,8	C-SI-F
Myrmotherula obscura				U	X,X,X,8	C-SI-F
Myrmotherula surinamensis		U	U	L	X,10,X,8	S-SI-F
Myrmotherula hauxwelli				FU	X,X,X,11	U-SI-F
Myrmotherula guttata			U		X,X,10,X	
Myrmotherula gutturalis			U		X,X,9,X	
Myrmotherula leucophthalma				FU	X,X,X,10	S-SI-D
Myrmotherula fulviventris	FGT	U			11,10,X,X	U-SI-D
Myrmotherula ornata				B	X,X,X,10	S-SI-D
Myrmotherula axillaris	FT	U	U	ZTFU	8,8,8,8	U-SI-F
Myrmotherula longipennis			U	TF	X,X,8,9	S-SI-F
Myrmotherula iheringi				TFUB	X,X,X,8	S-SI-F
Myrmotherula menetriesii			U	TFU	X,X,8,9	S-SI-F
Dichrozona cincta				FU	X,X,X,16	G-SI-G
Herpsilochmus rufimarginatus				V	X,X,X,12	
Herpsilochmus dorsimaculatus			U		X,X,X,X	C-SI-F

Microrhopias quixensis	GTC	U		TFUB	8,8,X,11	U-SI-F
Drymophila devillei				B	X,X,X,10	S-SI-D
Terenura humeralis				TFU	X,X,X,8	C-SI-F
Terenura spodioptila			U		X,X,X,X	C-SI-F
Cercomacra tyrannina	GS	U	ST		18,17,18,X	S-SI-F
Cercomacra cinerascens			U	TFU	X,X,X,18	U-SI-F
Cercomacra nigrescens				Z	X,X,X,18	
Cercomacra sp. nov.				B	X,X,X,19	S-SI-F
Cercomacra serva				FUO	X,X,X,17	S-SI-F
Myrmoborus leucophrys				ZTFUO	X,X,X,21	S-SI-F
Myrmoborus myotherinus				FUT	X,X,X,20	S-SI-F
Hypocnemis cantator			U	TB	X,X,11,13	S-SI-F
Hypocnemoides maculicauda				LFW	X,X,X,13	S-SI-F
Hypocnemoides melanopogon			U		X,X,X,X	S-SI-F
Percnostola lophotes				ZTBO	X,X,X,28	S-LI-F
Percnostola rufifrons			U		X,X,29,X	S-LI-F
Percnostola leucostigma			U		X,X,24,X	S-LI-F
Sclateria naevia			U	LFWO	X,X,22,22	S-SI-G
Gymnocichla nudiceps	SG	V			34,31,X,X	S-LI-R
Myrmeciza ferruginea			U		X,X,20,X	G-LI-G
Myrmeciza longipes		S			X,28,X,X	
Myrmeciza hemimelaena				ZTFUO	X,X,X,16	S-LI-F
Myrmeciza exsul	FGT	U			29,27,X,X	G-LI-G
Myrmeciza laemosticta		U			25,X,X,X	G-LI-G
Myrmeciza hyperythra				TFUO	X,X,X,41	S-LI-F
Myrmeciza goeldii				ZTFUO	X,X,X,42	S-LI-F
Myrmeciza fortis				FU	X,X,X,46	S-LI-R
Myrmeciza atrothorax			U	ZBO	X,X,19,18	S-LI-F
Myrmeciza immaculata	F				40,X,X,X	S-LI-R
Gymnopithys salvini				FUO	X,X,X,25	S-LI-R
Gymnopithys leucaspis	FG	U			30,30,X,X	S-LI-R
Gymnopithys rufigula			U		X,X,29,X	S-LI-R
Rhegmatorhina melanosticta				TFU	X,X,X,32	S-LI-R
Phaenostictus mcleannani	F	U			50,51,X,X	S-LI-R
Hylophylax naevia			U	TFU	X,X,13,13	S-SI-F
Hylophylax poecilonota			U	FU	X,X,17,18	S-LI-R
Hylophylax naevioides	FG	U			17,17,X,X	S-SI-R
Phlegopsis nigromaculata				TFU	X,X,X,45	S-LI-R
Myrmornis torquata		U	U		X,48,45,X	G-LI-G
Pithys albifrons			U		X,X,20,X	S-LI-R
Chamaeza nobilis				FU	X,X,X,130	G-LI-G
Formicarius colma			U	FU	X,X,46,49	G-LI-G
Formicarius analis	FG	U	U	TFU	61,57,62,58	G-LI-G
Formicarius rufifrons				T	X,X,X,X	G-LI-G
Grallaria eludens				F	X,X,X,X	G-LI-G
Grallaria varia			U		X,X,122,X	G-LI-G
Hylopezus berlepschi				TFUO	X,X,X,48	G-SI-G
Hylopezus macularius			U		X,X,42,X	G-SI-G
Hylopezus perspicillatus	F	U			48,42,X,X	G-SI-G
Hylopezus fulviventris	TG				44,X,X,X	G-SI-G
Myrmothera campanisona			U	TFUO	X,X,48,47	G-LI-G
Pittasoma michleri		U			110,109,X,X	G-LI-R
Conopophaga peruviana				FU	X,X,X,23	G-SI-G
Conopophaga aurita			U		X,X,24,X	G-SI-G
RHINOCRYPTIDAE				1		
Liosceles thoracicus				F	X,X,X,81	G-SI-G
COTINGIDAE	6	4	5	8		
Cephalopterus glabricollis	F				450-320,X,X,X	C-FR-F
Iodopleura isabellae				FT	X,X,X,X	C-SO-F
Iodopleura fusca			T		X,X,X,X	C-FR-F
Lipaugus vociferans			U	TFU	X,74,81	C-LO-F
Lipaugus unirufus	FG	U			75,86,X,X	C-FR-F
Porphyrolaema porphyrolaema				FU	X,X,X,60	C-FR-F
Cotinga maynana				TFUO	X,X,X,69	C-FR-F
Cotinga cayana			U	TUO	X,X,X,53	C-FR-F

Cotinga cotinga			U		X,X,X,X	C-FR-F
Cotinga amabilis	V				72,X,X,X	C-FR-F
Cotinga nattererii		TU			X,X,X,X	C-FR-F
Xipholena punicea			U		X,X,X,X	C-FR-F
Carpodectes nitidus	FGC				105,X,X,X	C-FR-F
Conioptilon mcilhennyi				ZTO	X,X,X,90	C-LO-F
Gymnoderus foetidus				TU	X,X,X,275	C-FR-F
Querula purpurata	FGC	U		TFU	115-100,104,X,125	C-LO-F
Procnias tricarunculata	FCS	V			220-145,X,X,X	C-FR-F
PIPRIDAE	6	6	9	8		
Schiffornis major				TFW	X,X,X,31	S-SI-F
Schiffornis turdinus	F	U	U		35,33,34,X	S-SI-F
Piprites chloris			U	TFU	X,X,17,20	S-SI-F
Piprites griseiceps	FG				16,X,X,X	S-SI-F
Sapayoa aenigma		U			X,21,X,X	U-SI-F
Tyranneutes stolzmanni				FU	X,X,X,9	U-SO-F
Tyranneutes virescens			U		X,X,9,X	U-SO-F
Neopelma chrysocephalum			U	V	X,X,X,15	S-FR-F
Machaeropterus pyrocephalus				BW	X,X,X,10	S-FR-F
Pipra coronata		U		TFU	12,10,X,9	S-FR-F
Pipra serena			U		X,X,11,X	S-FR-F
Pipra fasciicauda				TFU	X,X,X,17	S-FR-F
Pipra chloromeros				FU	X,X,X,17	S-FR-F
Pipra mentalis	FGT	STU			15,15,X,X	S-FR-F
Pipra erythrocephala			U		X,11,12,X	S-FR-F
Pipra pipra	F		U		13,11,12,X	S-FR-F
Neopipo cinnamomea			U		X,X,7,X	S-SO-F
Chiroxiphia lanceolata		S			19,18,X,X	
Corapipo leucorrhoa	FGT				12,X,X,X	S-FR-F
Corapipo gutturalis			U		X,X,8,X	S-FR-F
Manacus vitellinus		ST			X,17,X,X	S-FR-F
Manacus candei	GTSC				18,X,X,X	S-FR-F
TYRANNIDAE	51	61	49	84		
TYRANNINAE	46	57	42	77		
Zimmerius gracilipes			T	TFU	X,X,X,9	C-SO-F
Zimmerius vilissimus	FCTGS	STU			9,9,X,X	C-SO-F
Ornithion inerme				FU	X,X,X,7	C-SI-F
Ornithion brunneicapillum	FGC	U			7,7,X,X	C-SI-F
Camptostoma obsoletum		S		Z	8,8,X,9	
Phaeomyias murina				Z	X,10,X,10	
Sublegatus modestus				Z	14,X,X,12	
Sublegatus arenarum		V			X,X,X,X	
Phyllomyias griseiceps			T		X,X,X,X	C-SI-F
Tyrannulus elatus		STU	UT	TFUO	8,8,X,8	C-SO-F
Myiopagis gaimardii		U	UT	TFU	X,14,X,12	C-SI-F
Myiopagis viridicata		S		ZT	13,14,X,11	C-SI-F
Myiopagis caniceps		U	U	?	X,X,X,11	C-SI-F
Elaenia spectabilis				LZ	X,X,X,24	C-SO-F
Elaenia parvirostris			TS	RZT	X,X,X,13	C-SO-F
Elaenia strepera				RZ	X,X,X,17	
Elaenia gigas				RZ	X,X,X,32	
Elaenia cristata				ZT	X,X,X,X	C-SO-F
Elaenia chiriquensis		S			18,15,X,X	
Elaenia flavogaster	SP	S			25,24,X,X	
Inezia inornata				Z	X,X,X,5	
Capsiempis flaveola	SP	V			8,8,X,X	
Phylloscartes flavovirens		V			X,X,X,X	
Mionectes olivaceus	FT	U		FU	14,14,X,15	S-SO-F
Mionectes oleagineus	FGTC	U		ZTFU	13,X,X,11	S-SO-F
Mionectes macconnelli			U	TUB	X,X,12,14	S-SO-F
Leptopogon amaurocephalus	CTS			TFBU	12,12,X,11	S-SI-F
Corythopis torquata			U	FU	X,X,15,17	S-SI-F
Myiornis ecaudatus			TS	FU	X,X,X,6	C-SI-F
Myiornis atricapillus	FGC	U			5,X,X,X	C-SI-F
Oncostoma olivaceum		TU			X,7,X,X	S-SI-F

Oncostoma cinereigulare	TG				7,7,X,X	S-SI-F
Lophotriccus eulophotes				B	X,X,X,7	U-SI-F
Lophotriccus vitiosus			U		X,X,X,X	U-SI-F
Lophotriccus galeatus			U		X,X,8,X	U-SI-F
Hemitriccus flammulatus				B	X,X,X,9	U-SI-F
Hemitriccus minor			U		X,X,9,X	U-SI-F
Hemitriccus zosterops				FU	X,X,X,8	U-SI-F
Hemitriccus iohannis				T	X,X,X,X	U-SI-F
Poecilotriccus albifacies				B	X,X,X,8	U-SI-F
Todirostrum latirostre				Z	X,X,X,9	
Todirostrum maculatum				Z	X,X,X,8	
Todirostrum chrysocrotaphum			T	TFUO	X,X,X,7	C-SI-F
Todirostrum cinereum	S	S			7,7,X,X	
Todirostrum sylvia	S	S			7,7,X,X	
Todirostrum nigriceps	FGC				6,X,X,X	C-SI-F
Rhynchocyclus olivaceus		U	U		23,22,20,X	U-SI-F
Rhynchocyclus brevirostris	FC				23,24,X,X	U-SI-F
Ramphotrigon fuscicauda				TFU	X,X,X,18	U-SI-F
Ramphotrigon ruficauda			U	FU	X,X,19,19	U-SI-F
Ramphotrigon megacephala				B	X,X,X,20	U-SI-F
Tolmomyias assimilis	FGC	TU	U	TFU	14,X,14,17	C-SI-F
Tolmomyias poliocephalus			U	ZTFU	X,X,X,12	C-SI-F
Tolmomyias flaviventris				ZT	X,X,X,14	C-SI-F
Tolmomyias sulphurescens	T	ST			15,12,X,X	U-SI-F
Cnipodectes subbrunneus		U			X,23,X,X	C-SI-F
Platyrinchus coronatus	F	U	U	FU	9,9,9,10	S-SI-F
Platyrinchus saturatus			U		X,X,10,X	S-SI-F
Platyrinchus platyrinchos			U	FU	X,X,12,12	S-SI-F
Onychorhynchus coronatus			U	FUO	X,X,15,14	U-SI-A
Onychorhynchus mexicanus	QGCT	U			20,19,X,X	U-SI-A
Terenotriccus erythrurus	FGTC	U	U	TFUO	8,7,7,7	S-SI-A
Myiobius barbatus			U	U	X,X,11,X	S-SI-A
Myiobius atricaudus		S			10,10,X,X	S-SI-A
Myiobius sulphureipygius	FG	U			13,12,X,X	S-SI-A
Myiophobus fasciatus		S		Z	10,10,X,10	
Contopus borealis	GC	U	T	V	32,33,X,31	C-SI-A
Contopus virens	GC	ST	UT	ZTO	13,14,X,14	U-SI-A
Contopus sordidulus	GC	ST			14,13,X,X	U-SI-A
Contopus cinereus	SP	S			13,12,X,X	
Empidonax alnorum	SCG	S		ZL	13,12,X,X	S-SI-A
Empidonax traillii	SCG				13,X,X,X	S-SI-A
Empidonax euleri				ZTFU	X,X,X,11	S-SI-A
Empidonax flaviventris	TFG	U			12,11,X,X	S-SI-A
Empidonax virescens	TCG	U			12,12,X,X	S-SI-A
Empidonax minimus	CT				10,10,X,X	S-SI-A
Empidonax albigularis	S				X,X,X,X	
Aphanotriccus capitalis	CSTG				11,X,X,X	S-SI-F
Cnemotriccus fuscatus				Z	X,X,X,X	
Pyrocephalus rubinus			ST	LZ	X,13,X,14	C-SI-A
Ochthoeca littoralis				RZ	X,X,X,13	
Muscisaxicola fluviatilis				R	X,X,X,14	
Fluvicola pica				L	X,12,X,X	
Colonia colonus	GCS	TU		V	15,16,X,X	A-SI-A
Satrapa icterophrys				R	X,X,X,X	
Attila bolivianus				ZTFUO	X,X,X,45	U-LI-F
Attila spadiceus	GFTC	U	TU	TFU	38,38,32,35	U-LI-F
Casiornis rufa				V	X,X,X,X	
Rhytipterna simplex			U	TFU	X,X,35,36	C-LI-F
Rhytipterna holerythra	FGC	U			38,38,X,X	U-LI-F
Laniocera hypopyrrha			U	FU	X,X,49,51	C-LO-F
Laniocera rufescens	FG	U			48,49,X,X	U-LI-F
Sirystes sibilator		U	TU	FU	X,32,X,38	C-SI-A
Myiarchus tuberculifer	TGC	U	TU	TFUO	20,20,X,X	C-SI-F
Myiarchus swainsoni				TO	X,X,X,X	C-SI-F
Myiarchus ferox			T	Z	X,31,X,28	C-SI-A

Myiarchus panamensis		S			32,32,X,X	
Myiarchus tyrannulus				T	34,X,X,X	C-SI-A
Myiarchus crinitus	TGF	STU			34,34,X,X	C-SO-F
Pitangus lictor		LR		LT	X,26,X,28	C-OM-W
Pitangus sulphuratus	SP	S	T	RLTO	59,59,X,55	C-OM-F
Megarhynchus pitangua	GTCF	ST		RLTO	63,72,X,66	C-OM-F
Conopias parva	GFC	U	U		27,22,X,X	C-SI-F
Conopias sp.				O	X,X,X,X	C-SI-F
Myiozetetes similis	SRP	S		RLT	29,30,X,33	S-SO-F
Myiozetetes granadensis	SGT	L		RLTO	31,29,X,32	S-SO-F
Myiozetetes cayanensis		L	TS		X,27,X,X	S-SO-F
Myiozetetes luteiventris				O	X,X,X,23	S-SO-F
Myiodynastes luteiventris	GC	S		ZTLO	52,47,X,48	C-SO-F
Myiodynastes maculatus	G	STU	T	TO	45,46,X,51	C-SO-F
Legatus leucophaius	GPC	STU	TS	LTFUO	25,26,X,23	C-FR-F
Empidonomus aurantioatrocristatus			TU	TF	X,X,X,X	C-SO-F
Empidonomus varius			TS	V	X,X,X,X	C-SO-F
Tyrannus dominicensis		V			44,50,X,X	
Tyrannus melancholicus	SP	ST	S	RL	40,41,X,47	C-SO-A
Tyrannus tyrannus	GS	TU		ZTL	41,38,X,41	C-SO-F
Tyrannopsis sulphurea			U	A	X,X,X,X	
Muscivora tyrannus (savana)		V	ST	V	X,28,X,X	A-SI-A
TITYRINAE	5	4	7	7		
Pachyramphus castaneus				T	X,X,X,X	C-SO-F
Pachyramphus polychopterus	TC	T		ZT	22,19,X,22	U-SO-F
Pachyramphus aglaiae	G				33,X,X,X	
Pachyramphus marginatus			U	FU	X,X,19,18	C-SI-F
Pachyramphus cinnamomeus	GCS	T			23,20,X,X	C-SI-F
Pachyramphus minor			TU	TFU	X,X,X,37	C-SI-F
Pachyramphus surinamus			TU		X,X,X,X	C-SI-F
Tityra cayana			U	TFU	X,X,X,66	C-LO-F
Tityra inquisitor	GFTC	U		T	50,41,X,45	C-LO-F
Tityra semifasciata	GFTC	U		TFU	79,80,X,88	C-LO-F
Phoenicircus carnifex			U		X,X,X,X	U-SO-F
Haematoderus militaris			U		X,X,X,X	C-FR-F
Perissocephala tricolor			U		X,X,X,X	U-FR-F
HIRUNDINIDAE	7	11	4	12		
Tachycineta albiventer				RL	X,X,X,18	
Tachycineta albilinea	R	L			14,14,X,X	
Tachycineta bicolor		A		RL	20,21,X,X	
Progne tapera		A		RL	X,X,X,X	
Progne chalybea	PA	A	T	RL	40,39,X,X	
Progne subis		A		RL	45,52,X,X	
Neochelidon tibialis		U	T	F	X,9,X,X	A-SI-A
Notiochelidon cyanoleuca		V		L	10,X,X,15	
Atticora fasciata				LR	X,X,X,X	
Stelgidopteryx ruficollis	RP	L	TS	LR	15,14,X,X	
Stelgidopteryx serripennis	RP				16,X,X,X	
Riparia riparia	AP	L		LR	13,14,X,X	
Hirundo rustica	AP	L	TS	LR	17,18,X,X	
Hirundo pyrrhonota	AP				20,X,X,X	
Petrochelidon pyrrhonota		L		L	X,20,X,X	
CORVIDAE	1	1		1		
Cyanocorax violaceus				ZT	X,X,X,262	C-LO-F
Cyanocorax affinis		T			205,210,X,X	C-LO-F
Cyanocorax morio	S				235,X,X,X	
TROGLODYTIDAE	9	11	5	6		
Campylorhynchus zonatus	GCT				29,36,X,X	C-LI-F
Campylorhynchus turdinus				TFU	X,X,X,28	C-SI-F
Campylorhynchus albobrunneus		T			X,X,X,X	C-SI-F
Thryothorus leucotis		S	T		X,19,15,X	S-SI-F
Thryothorus coraya			T		X,X,17,X	S-SI-F
Thryothorus modestus	S	S			25,19,X,X	
Thryothorus rufalbus		S			25,26,X,X	
Thryothorus atrogularis	TSG				23,X,X,X	S-SI-F

Thryothorus thoracicus	CTG				16,X,X,X	S-SI-F
Thryothorus nigricapillus	GRT	T			25,22,X,X	S-SI-F
Thryothorus fasciatoventris		T			30,23,X,X	S-SI-F
Thryothorus rutilus		S			16,15,X,X	
Thryothorus genibarbis				TBZU	X,X,X,19	S-SI-D
Troglodytes aedon=musculus	P	S	ST	R	12,15,12,X	S-SI-F
Henicorhina leucosticta	FG	U			17,17,X,X	S-SI-F
Microcerculus marginatus		U		FU	X,19,X,18	G-SI-G
Microcerculus philomela	F				18,X,X,X	G-SI-G
Microcerculus bambla			U	V	X,X,17,X	G-SI-G
Cyphorhinus phaeocephalus	F	U			24,25,X,X	G-SI-G
Cyphorhinus arada				TFU	X,X,20,30	G-SI-G
Cyphorhinus marginatis			U		X,X,X,X	G-SI-G
MIMIDAE	1	2		1		
Donacobius atricapillus				LP	X,35,X,X	
Dumetella carolinensis	S	S			39,39,X,X	
Mimus gilvus		V			X,X,X,X	
MUSCICAPIDAE	11	9	6	9		
TURDINAE	8	6	3	7		
Catharus ustulatus	TSCF	STU		ZTFU	30,30,X,27	S-SO-F
Catharus minimus	TCS	TU	U		30,27,30,X	S-SO-F
Catharus fuscescens	TGC	TU	U		32,31,28,X	S-SO-F
Hylocichla mustelina	FGT	T			48,46,X,X	S-SO-F
Turdus assimilis	F	V			72,68,X,X	S-SO-F
Turdus grayi	SP	ST			71,74,X,X	S-SO-F
Turdus nigriceps				V	X,X,X,49	
Turdus amaurochalinus				Z	X,X,X,59	
Turdus ignobilis				ZTL	X,X,X,58	S-SO-F
Turdus lawrencii				FUW	X,X,X,68	S-SO-F
Turdus obsoletus (=hauxwelli)	GT			TFU	74,X,X,72	U-SO-F
Turdus albicollis			U	FU	X,X,50,52	S-SO-F
Myadestes melanops	GTF				31,32,X,X	U-FR-F
SYLVIINAE	3	3	3	2		
Polioptila plumbea	FGC	TU		V	7,7,X,X	C-SI-F
Polioptila guianensis			U		X,X,X,X	C-SI-F
Ramphocaenus rufiventris		ST			X,10,X,X	S-SI-F
Ramphocaenus melanurus	TG		U	ZTBO	11,X,9,12	S-SI-F
Microbates cinereiventris	FG	U			11,12,X,X	S-SI-F
Microbates collaris			U		X,X,11,X	S-SI-F
VIREONIDAE	9	9	8	7		
VIREONINAE	8	7	6	5		
Vireo olivaceus	GTCF	ST	U	ZTFUO	18,17,X,15	C-SO-F
Vireo flavoviridis	GTC	U		ZTFUO	16,17,X,X	S-SO-F
Vireo gilvus	G				12,X,X,X	C-SI-F
Vireo philadelphicus	GCT				12,11,X,X	S-SO-F
Vireo flavifrons	GCTF	TU			17,18,X,X	C-SI-F
Vireo griseus	T,S				12,X,X,X	C-SI-F
Hylophilus decurtatus	GCFT	TU			9,10,X,X	C-SI-F
Hylophilus flavipes		V			13,12,X,X	
Hylophilus semicinereus			U		X,X,X,X	C-SI-F
Hylophilus aurantiifrons		S			X,10,X,X	
Hylophilus ochraceiceps	FG	TU	U	TFU	11,12,10,11	U-SI-F
Hylophilus thoracicus				U	X,X,X,14	C-SI-F
Hylophilus hypoxanthus				TFUO	X,X,X,17	C-SI-F
Hylophilus pectoralis?			U		X,X,X,X	C-SI-F
Hylophilus muscicapinus			U		X,X,11,X	C-SI-F
Hylophilus brunneiceps			U		X,X,X,X	C-SI-F
VIREOLANIINAE	1	1	1	1		
Vireolanius leucotis			U	TF	X,X,X,26	C-SI-F
Vireolanius pulchellus	FGC	TU			30,25,X,X	C-SI-F
CYCLARHININAE		1	1	1		
Cyclarhis gujanensis		V	T	LO	31,X,X,27	C-SI-F

EMBERIZIDAE	87	83	44	72		
PARULINAE	30	27	3	4		
Mniotilta varia	GTC	T			10,11,X,X	U-SI-B
Protonotaria citrea	SG	V			13,13,X,X	S-SI-F
Helmitheros vermivorus	TS	T			14,13,X,X	S-SI-D
Vermivora chrysoptera	GSC	ST			9,9,X,X	U-SI-D
Vermivora pinus	S	ST			8,9,X,X	U-SI-D
Vermivora peregrina	GSC	T			9,9,X,X	C-NI-F
Dendroica aestiva		S			X,10,X,X	
Dendroica magnolia	TSC	T			9,9,X,X	U-SI-F
Dendroica petechia	SP	L			9,X,X,X	
Dendroica tigrina	GS	V			11,11,X,X	
Dendroica virens	GT	V			9,9,X,X	C-SI-F
Dendroica cerulea	GFC	V			8,9,X,X	C-SI-F
Dendroica fusca	GTCF	T	T		9,10,X,X	C-SI-F
Dendroica pensylvanica	GSTF	T			9,10,X,X	C-SI-F
Dendroica castanea	GTCF	T			11,10,X,X	C-SO-F
Dendroica striata		V	TU		12,9,X,X	S-SI-F
Dendroica coronata	SG				13,11,X,X	
Seiurus aurocapillus	TGS	T			19,19,X,X	G-SI-G
Seiurus noveboracensis	QRS	T			16,17,X,X	G-SI-G
Seiurus motacilla	Q	TU			19,19,X,X	G-SI-G
Oporornis formosus	TGF	TU			14,14,X,X	G-SI-G
Oporornis philadelphia	S	V			12,11,X,X	
Oporornis agilis				Z	14,X,X,13	
Oporornis tolmiei	S				12,11,X,X	
Wilsonia citrina	GS	V			10,9,X,X	
Wilsonia canadensis	GCTF	T			11,10,X,X	S-SI-F
Wilsonia pusilla	GST				7,8,X,X	S-SI-F
Setophaga ruticilla	GFTC	T			9,8,X,X	U-SI-F
Geothlypis aequinoctialis				ZL	17,X,X,X	
Geothlypis trichas	S	V			10,10,X,X	
Geothlypis semiflava	SP				15,X,X,X	
Geothlypis poliocephala	PS				16,14,X,X	
Basileuterus chrysogaster				V	X,X,X,11	
Basileuterus delatrii		S			X,11,X,X	
Phaeothlypis rivularis			U		X,X,13,X	S-SI-F
Phaeothlypis fulvicauda	QR	U		FU	15,14,X,15	W-SI-W
Icteria virens	S				25,27,X,X	
COEREBINAE	1	1	1			
Coereba flaveola	S	ST	TU		9,9,9,X	S-NI-F
THRAUPINAE	25	27	25	42		
Schistochlamys melanopis				V	X,X,X,25	?,?,?
Conothraupis speculigera				UZ	X,X,X,23	U-FR-F
Rhodinocicla rosea		S			50,48,X,X	S-SI-F
Mitrospingus oleagineus=cassinii	GTC	TU			42,37,X,X	S-SO-F
Chlorothraupis carmioli	FG	U			36,34,X,38	U-SO-F
Heterospingus rubrifrons		TU			30,38,X,X	S-FR-F
Lamprospiza melanoleuca			UT	FU	X,X,X,39	C-SO-F
Cissopis leveriana				LT	X,X,X,66	U-SO-F
Thlypopsis sordida				Z	X,X,X,17	
Hemithraupis guira				FU	X,X,X,13	C-SO-F
Hemithraupis flavicollis			U	FU	X,14,11,17	C-SO-F
Nemosia pileata				Z	X,X,X,16	
Eucometis penicillata		U		F	31,30,X,X	U-SI-R
Lanio versicolor				TFU	X,X,X,19	C-LI-F
Lanio leucothorax	FG				40,36,X,X	C-LI-F
Lanio fulvus			U		X,X,25,X	C-LI-F
Tachyphonus rufiventer				TFU	X,X,X,17	C-SI-F
Tachyphonus luctuosus	FGT	TU		TFUO	16,15,X,13	C-SO-F
Tachyphonus rufus	SG	T			32,35,X,X	C-SO-F
Tachyphonus delatrii	FG	U			20,18,X,X	U-SO-F
Tachyphonus cristatus			U		X,X,18,X	C-SO-F
Tachyphonus surinamus			U		X,X,21,X	U-SO-F
Habia rubica				TFU	38,36,X,33	S-SO-F

Habia fuscicauda	GSC	ST			40,39,X,X	S-SO-F
Piranga rubra	GCT	ST		T	29,29,X,X	C-SO-F
Piranga olivacea	GCF	ST		T	28,30,X,19	C-SO-F
Cyanicterus cyanicterus			U		X,X,34,X	C-SI-F
Ramphocelus nigrogularis				LT	X,X,X,26	S-SO-F
Ramphocelus carbo			ST	LRZO	X,X,24,27	S-SO-F
Ramphocelus dimidiatus		S			X,30,X,X	
Ramphocelus icteronotus		T			X,34,X,X	S-SO-F
Ramphocelus sanguinolenta	SP				38,X,X,X	
Ramphocelus passerinii	SGC				30,33,X,X	S-SO-F
Thraupis episcopus	SP	T	ST	LRP	30,30,X,40	C-SO-F
Thraupis palmarum	STP	T	ST	LRUP	38,38,35,36	C-SO-F
Euphonia chlorotica				V	X,X,X,11	
Euphonia laniirostris		ST		LRT	15,X,X,16	C-FR-F
Euphonia anneae	GF				14,X,X,X	S-FR-F
Euphonia chrysopasta			TU	TFUO	X,X,X,15	C-FR-F
Euphonia minuta	GFC	U	UT	TFU	10,11,X,10	C-FR-F
Euphomia cayannensis			U		X,X,16,X	C-FR-F
Euphonia xanthogaster				TFUO	X,17,X,14	C-FR-F
Euphonia rufiventris				TFUO	X,X,X,15	C-SO-F
Euphonia luteicapilla	CS	ST			13,12,X,X	C-FR-F
Euphonia fulvicrissa		U			X,11,X,X	C-FR-F
Euphonia gouldi	FGC				12,13,X,X	C-FR-F
Euphonia plumbea			T		X,X,9,X	C-FR-F
Chlorophonia cyanea				V	X,X,X,13	
Tangara mexicana			T	TFUO	X,X,X,20	C-SO-F
Tangara chilensis			U	TFUO	X,X,15,22	C-FR-F
Tangara shrankii				TFUO	X,X,X,20	C-SO-F
Tangara xanthogastra				U	X,X,X,14	C-SO-F
Tangara gyrola	G	U	U	U	23,22,X,X	C-SO-F
Tangara nigrocincta				TFUO	X,X,X,17	C-SO-F
Tangara velia			U	TFU	X,X,X,21	C-SO-F
Tangara callophrys				TFUO	X,X,X,23	C-SO-F
Tangara larvata	STC	TU			18,19,X,X	C-SO-F
Tangara inornata	TSC	TU			22,18,X,X	C-SO-F
Tangara punctata			U		X,X,14,X	C-FR-F
Tangara varia			U		X,X,10,X	C-SO-F
Tangara icterocephala	GSC				23,22,X,X	C-SO-F
Dacnis lineata			U	TFU	X,X,X,12	C-SO-F
Dacnis flaviventer				TUO	X,X,X,14	C-SO-F
Dacnis cayana	GFCS	T	U	TFU	14,13,13,14	C-SO-F
Dacnis venusta	GFC	U			16,16,X,X	C-SO-F
Chlorophanes spiza	GFC	TU	U	TFUO	19,18,16,18	C-SO-F
Cyanerpes caeruleus			U	TFUO	X,13,X,16	C-SO-F
Cyanerpes nitidus			U		X,X,X,9	C-SO-F
Cyanerpes cyaneus		TU		V	14,13,X,X	C-SO-F
Cyanerpes lucidus	GFC	TU			11,12,X,X	C-SO-F
Tersina viridis				RLTU	X,30,X,28	C-SO-F
Conirostrum speciosum			TU	Z	X,X,X,7	C-SO-F
Conirostrum bicolor					X,X,X,X	C-SO-F
CARDINALINAE	11	8	3	5		
Pitylus grossus	FGT	U	U	UO	47,43,44,45	C-SO-F
Pheucticus ludovicianus	GSTC	TU			45,45,X,X	C-SO-F
Pheuticus tibialis	G				70,X,X,X	C-SO-F
Cyanocompsa cyanoides	GTC	TU	TU	ZFTUO	31,32,X,X	S-SO-F
Passerina cyanea	S	V			14,14,X,X	
Spiza americana		V			28,27,X,X	
Saltator maximus	STC	T		TLFUO	47,47,X,46	S-SO-F
Saltator coerulescens	S			ZL	52,X,X,56	
Saltator albicollis	S	S			40,40,X,X	
Saltator atriceps	GTCS	V			79,X,95,X,X	C-SO-F
Guiraca caerulea	S				31,X,X,X	
Caryothraustes humeralis				U	X,X,X,28	S-SO-F
Caryothraustes poliogaster	GTC				37,42,X,X	S-SO-F
Caryothraustes canadensis			U		X,34,33,X	C-FR-F

EMBERIZINAE	8	8	5	10		
Ammodramus aurifrons			S	R	X,13,15,18	
Volatinia jacarina	PS	V	S		9,9,X,X	
Sporophila americana				B	X,X,X,X	S-SE-F
Sporophila lineola				R	X,X,X,X	
Sporophila nigricollis		V		R	9,9,X,X	
Sporophila caerulescens				RLZ	X,X,X,10	
Sporophila castaneiventris			S	RZ	X,X,X,8	
Sporophila aurita	SP	V			11,10,X,X	
Sporophila torqueola	PS				10,8,X,X	
Sporophila minuta		V			8,9,X,X	
Sporophila schistacea		V		V	13,11,X,X	
Oryzoborus angolensis			S	L	X,X,13,X	
Oryzoborus funereus	SP	S			12,12,X,X	
Oryzoburus nuttingi	SPM				24,X,X,X	
Arremon taciturnus			TU	TFUO	X,X,22,28	G-SO-G
Arremon aurantiirostris	GFT	TU			34,31,X,X	G-SO-G
Arremonops conirostris	SP	S			37,41,X,X	
Paroaria gularis				RZ	X,X,X,31	
Tiaris olivacea	SP				10,9,X,X	
ICTERINAE	12	12	7	11		
Psarocolius oseryi				TFU	X,X,X,235-132	C-OM-F
Psarocolius decumanus		U		TFU	X,X,X,350-154	C-OM-F
Psarocolius angustifrons				LZTFU	X,X,X,430-203	C-OM-F
Psarocolius yuracares				TFU	X,X,X,510-250	C-OM-F
Psarocolius wagleri	GFTC	TU			225-125,214-113,X,X	C-OM-F
Psarocolius montezuma	GFTC				520-230,430-222,X,X	C-OM-F
Psarocolius viridis			U		X,X,X,X	C-OM-F
Cacicus cela		ST		LTFUO	X,113-68,X,107-69	C-OM-F
Cacicus uropygialis	FGCT	TU			68-53,57,X,X	C-LO-F
Cacicus solitarius				LT	X,X,X,84	C-OM-F
Cacicus holosericeus	TG	SU			70-60,71-55,X,X	S-SI-F
Cacicus haemorrhous			TU		X,X,X,X	C-OM-F
Cassidix mexicanus	P	V			230-125,191-101,X,X	
Icterus cayanensis			U	TFU	X,X,X,43	U-OM-D
Icterus chrysocephalus			U		X,X,X,X	C-SO-F
Icterus icterus				LT	X,X,X,52	
Icterus spurius	CST	STU			20,25,X,X	C-SO-F
Icterus mesomelas	T	L			70,54,X,X	U-SO-F
Icterus chrysater		T			X,43,X,X	U-SO-F
Icterus dominicensis	SCT				32,X,X,X	U-SO-F
Icterus galbula	GSTC	T			31,34,X,X	C-SO-F
Scaphidura oryzivora	FGC	U	S	RLT	200-140,225-150,X,154	G-OM-G
Agelaius xanthophthalmus				LP	X,X,X,X	
Dolichonyx oryzivorus		V		ZL	25,32,X,X	
Molothrus aeneus	P				68-56,66-58,X,X	
Molothrus bonarensis			S		X,40-31,X,X	
Sturnella magna	P				85,98-73,X,X	
Sturnella militaris			S		48-41,46-36,X,X	
PASSERIDAE	1					
Passer domesticus	V				26,X,X,X	G-OM-G

a. A few recent records have been added to this tabulation but manuscripts for the four areas have not been revised to incorporate those records.

b. Costa Rica habitat designations: F=Forest; T=Old second growth; S=Young second growth; R=River; A=Aerial; M=Pond, marsh; Q=Forest stream, swamp; P=Pasture, cultivated area; G=Forest edge and gaps; C=Plantations - Peyibaye, cacao; V=Vagrant.

c. Panama habitat designations: U=Upland forest; T=Transition, old second growth; S=Young second growth; L=Lake; A=All; W=Swamp, marsh; V=Vagrant.

d. Brazil habitat designations: U=Upland forest; T=Old second growth; S=Young second growth; L=Lake; R=River; W=Swamp, marsh; P= Pastizal (palm savanna, marsh); V=Vagrant.

e. Peru habitat designation: L=Lake; R=River; A=Aerial; Z=Zabolo (cane, Tessaria); T=Transition Forest; S=Shrub upland; F=Mature bottomland; U=Upland forest; W=Swamp; P=Pastizal; O=Open Forest gap; V=Vagrant.

f. Mass (grams) - Weights are provided for each region if known. If two weights are provided, the format is male-female. When available for a species we use our own data in this tabulation. We have used published tabulations for the countries in question when data are not available from our research sites.

g. Guild designation based on these attributes; Foraging Strata - Diet - Substrata using the following codes: STRATA: G=Ground; S=Shrub; U=Understory; C=Canopy; A=Above Canopy; W=Water. DIET: SI=Small insects; SO=Small insects and fruit; LI=Large insects, small verts.; LO=Large insects, fruit and small verts.; FR=Fruits or fruits + seeds; SE=Grass seeds; PR=Carn - Verts, large insects; BI=Birds; MA=Mammals; HE=Herps; FI=Fish; CA-Carrion; OM=Omnivore; SN=Snails; NI=Nectar, small insects/spiders. SUBSTRATA: G=Ground; F=Foliage live include fruits + flowers; D=Foliage - dead; R=Army ants; A=Air; W=Water; B=Branches, trunk; T=Twigs.

h. Some species vary in their ecologies among the study regions (e.g. antfollower in some areas but not others). No effort was made to distinguish those differences in this tabulation.

PART IV

Mammals

15

Mammals of La Selva, Costa Rica

DON E. WILSON

The mammal fauna of La Selva Biological Station has been studied for fewer than twenty years, yet it represents one of the better known faunas in Central America. The site is located almost in the center of the Central American Caribbean Lowland Forest, and the mammals are typical of a fauna found from tropical Mexico to northern South America.

Knowledge of the mammals of La Selva began to accumulate during the 1960s, when the property was still owned by Leslie Holdridge. Collections were made at La Selva or in the general vicinity of Puerto Viejo by Richard Casebeer, Ronald Linsky, and Andrew Starrett of the University of Southern California and Craig Nelson of the University of Kansas (Casebeer et al. 1963; Starrett and Casebeer 1968).

Bats. Shortly after Finca La Selva was purchased by the Organization for Tropical Studies (OTS) in 1968, I began netting bats there along with many students in OTS courses, among them Richard LaVal and David Armstrong. Michael Mares visited in 1971. I continued my study of bats during OTS courses through 1974, and the number of species of bats known from the site increased to 32 (Armstrong 1969; Gardner et al. 1970; Mares and Wilson 1971). During 1973 and 1974, LaVal sampled the bat fauna intensively using both mist nets and harp traps. His efforts dramatically increased the number of reported species to 63 (fig. 15.1), several of which were new to Costa Rica (LaVal 1977; LaVal and Fitch 1977).

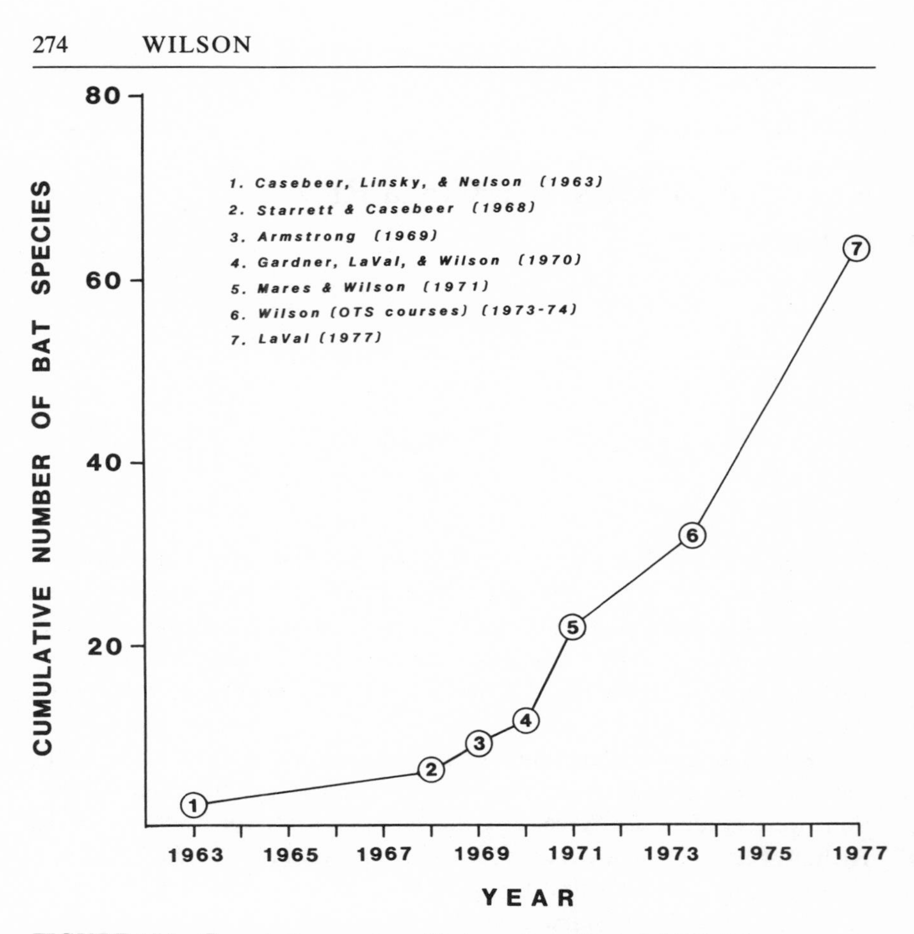

FIGURE 15.1. Cumulative number of bat species collected, 1963–77.

Small mammals. Theodore Fleming studied the rodent fauna in 1970 and 1971. The number of identified species of small mammals, including marsupials, from La Selva increased greatly during this period (Fleming 1973), with Fleming collecting 14 species of rodents and 5 of marsupials. OTS courses rarely include trapping studies, so documenting rare species of small mammals from the site has proceeded slowly.

Large mammals. As the number of trained observers increased, large mammals were observed more frequently. Information on all mammal species has continued to accumulate with OTS's continued sponsorship of courses at the site and with the increased number of resident

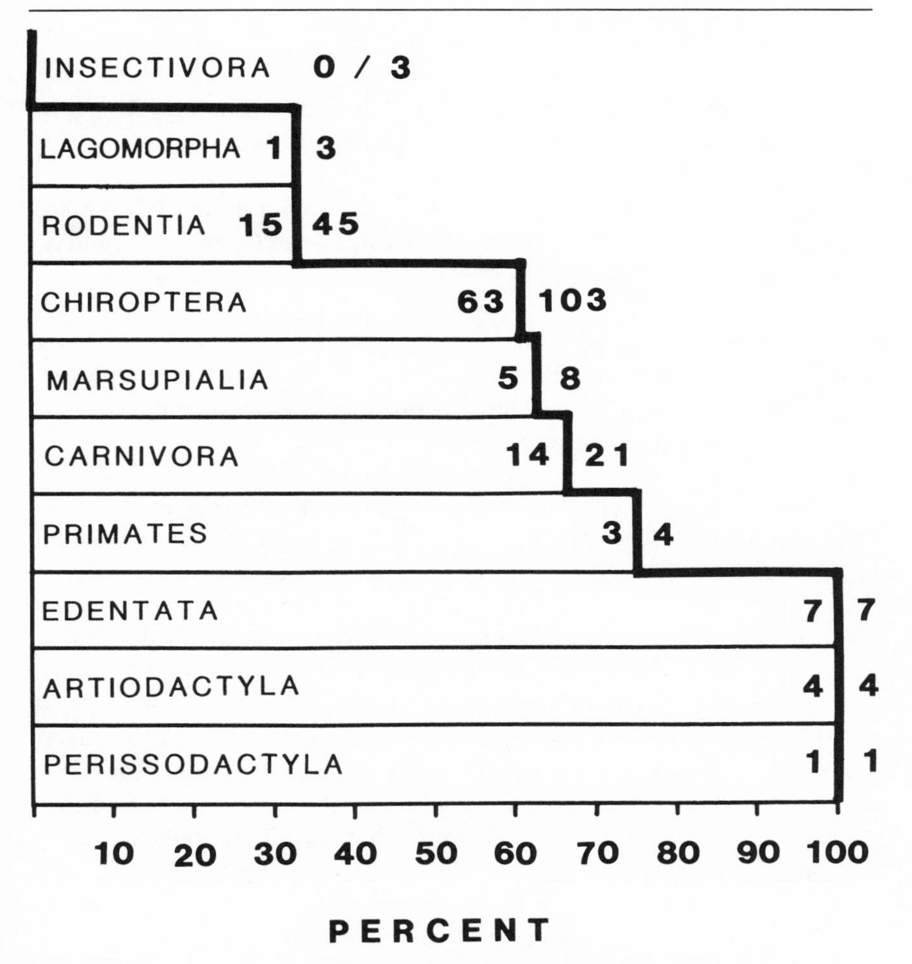

FIGURE 15.2. Percentage of the Costa Rican mammal fauna known from La Selva. To the left of the heavy line are the number of species known from La Selva; to the right, the number known from the country as a whole.

researchers. Mammal sightings are recorded in a log book, and considerable information has accumulated over the years. Our knowledge of the entire mammal fauna of Costa Rica has likewise increased tremendously since Goodwin's (1946) faunal study. The most recent estimate (Wilson 1983) of the country's terrestrial mammalian fauna is 199 species, 113 of which occur at La Selva (fig. 15.2).

Current Status

Insectivores. Insectivores occur only at high elevations in Costa Rica, so the 3 known species are not found at La Selva.

Lagomorphs. Of the 2 rabbit species that do not occur at La Selva, *Sylvilagus floridanus* is restricted to the dry forests of Guanacaste, and *S. dicei* is a highland form.

Rodents. The rodent fauna of La Selva, representing only one-third of that recorded for Costa Rica, is probably not yet completely known. Taxonomic problems in this order may preclude complete comparison of the numbers with the other groups.

Bats. Bats make up at least 60% of the mammal species both at La Selva and in Costa Rica as a whole (fig. 15.2). This estimate is probably a bit low; at least 16 species not yet recorded at La Selva should occur there, based on their overall distributions and habitat requirements. During a month of intensive collecting in the Zone Protectora, adjacent to and upslope from La Selva, I added 11 species not known from La Selva proper.

This emphasizes the point that even at a location that has received considerable attention, continual inventory is critically needed. Too often we devote attention to compelling ecological questions only to find that the systematic status of our organisms is still in flux. Taxonomic revisions of groups like bats are hampered by inadequate samples from throughout a species's range.

The problem of encountering fugitive species is common to a variety of tropical organisms. Figure 15.3 uses the data set compiled by LaVal and Fitch (1977) during a year of intensive bat sampling. Of the 57 species they encountered at La Selva, 20 were represented by 3 or fewer individuals. Of those 20, only a single individual was collected for 15. An additional 10 species were known from fewer than 10 specimens each. This means that over half of the fauna was represented by fewer than 10 individuals each in their sample. At the other end of the scale, 2 species accounted for 762 of the 1,865 individuals captured.

Marsupials. Marsupials are well represented in the La Selva fauna (fig. 15.2). However, 2 of the 3 species known from Costa Rica but not yet from La Selva, will probably be shown to occur there. The third spe-

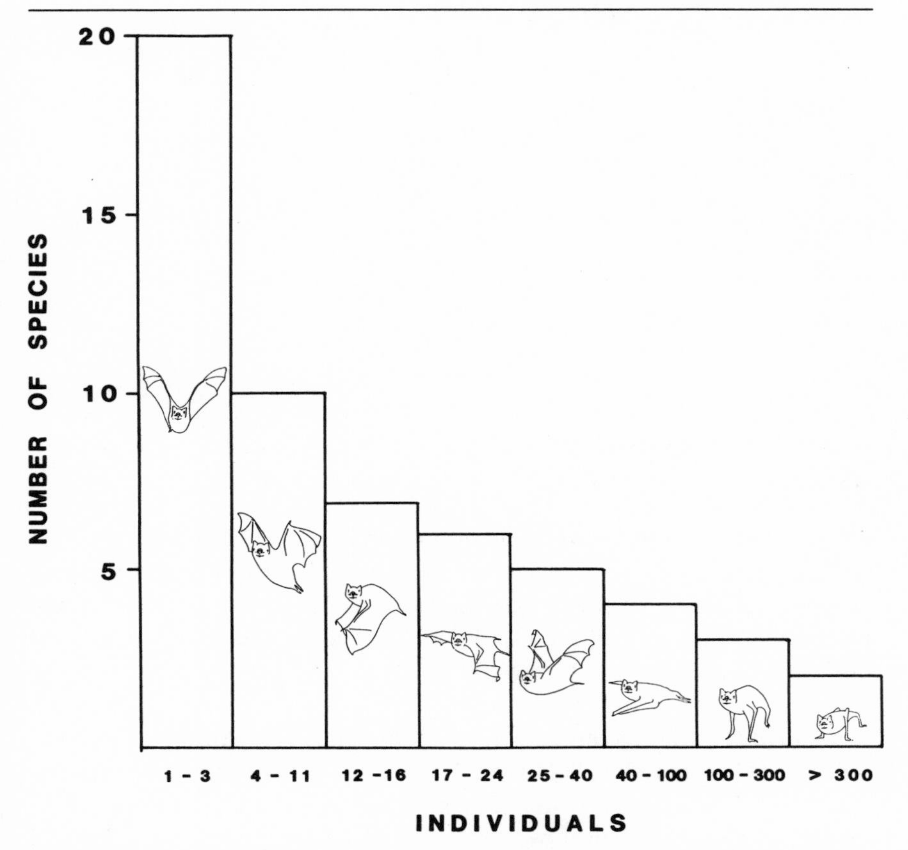

FIGURE 15.3. Number of individuals per species of bats collected by LaVal and Fitch (1977).

cies missing is *Didelphis virginiana*, a northern species that reaches only Guanacaste Province.

Carnivores. Two-thirds of Costa Rica's carnivores are found at La Selva. Although poorly represented in collections, these animals are more easily observed and documented than are the smaller forms. The unrecorded species are almost entirely dry forest species, such as the coyote (*Canis latrans*) and gray fox (*Urocyon cinereoargenteus*), which are found only on the Pacific side of the country.

Primates. Three of the four primates occurring in Costa Rica are found at La Selva and seem to be doing well. Their numbers have in-

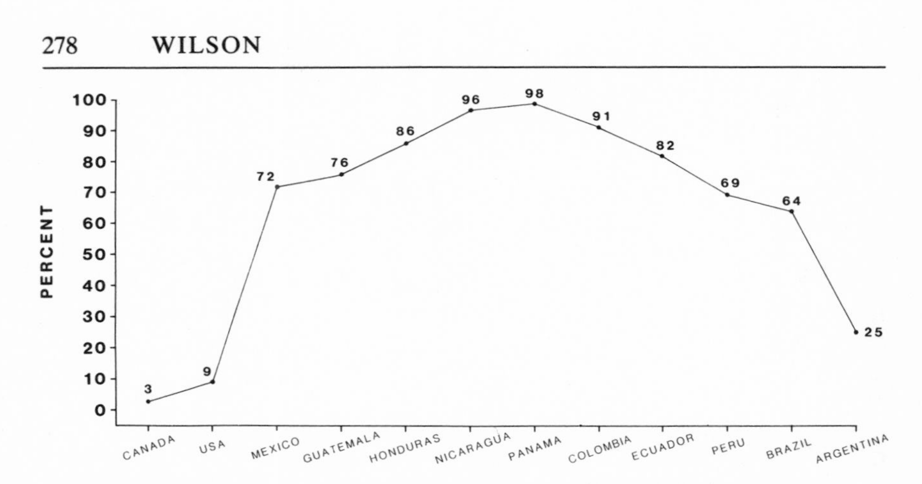

FIGURE 15.4. Percentage of the La Selva mammal fauna that extends into countries north and south of Costa Rica.

creased with the protection from hunting afforded by OTS during the past twenty years. The only species not represented is the squirrel monkey (*Saimiri oerstedii*), which ranges into Costa Rica from the south through the Pacific wet forests of the Osa Peninsula and reaches its distributional limit slightly to the north of there.

Edentates. Sloths, anteaters, and armadillos are well represented at La Selva. Undoubtedly all 7 Costa Rican species occurred there in historical times. The giant anteater (*Myrmecophaga tridactyla*) has not been seen at La Selva in the past two decades and is almost certainly extirpated from the area. This may be the only large mammal missing from the site.

Ungulates. All of the original hoofed stock apparently are still extant at La Selva. The one questionable species is the white-lipped peccary (*Tayassu pecari*), whose occurrence has not been verified unequivocally in recent years. The collared peccary (*Dicotyles tajacu*) is common, and both species of deer (*Mazama americana* and *Odocoileus virginianus*) have probably increased in numbers with protection from hunting.

The La Selva mammal fauna seems to be an excellent example of the mammals of Costa Rica. It is particularly representative of the wet-forest fauna of the Caribbean versant. Whether this is a wide-ranging fauna or a localized association can be tested by looking at the geographic distributions of the species known to occur at La Selva and by tracing their limits northward and southward (fig. 15.4, tables 15.1–3). These distributions

TABLE 15.1 Proportion of La Selva mammal fauna as a percentage of Costa Rican mammal fauna

	La Selva (*N*)	Costa Rica (*N*)	(%)
Marsupialia	5	8	62
Insectivora	0	3	0
Chiroptera	63	103	61
Primates	3	4	75
Edentata	7	7	100
Lagomorpha	1	3	33
Rodentia	15	45	33
Carnivora	14	21	67
Artiodactyla	4	4	100
Perissodactyla	1	1	100
TOTAL	113	199	56

suggest that La Selva sits in the center of a well-defined and wide-ranging mammalian fauna that occupies Central America from southern Mexico to northern South America.

Four species of La Selva mammals extend their range northward as far as Canada: long-tailed weasels (*Mustela frenata*), raccoons (*Procyon lotor*), mountain lions (*Felis concolor*), and white-tailed deer (*Odocoileus virginianus*). The mountain lion is the most widely distributed species of mammal in the Western Hemisphere (other than *Homo sapiens*) and ranges southward into Argentina.

Several species reach the United States, including the nine-banded armadillo (*Dasypus novemcinctus*), collared peccary (*Dicotyles tajacu*), and such carnivores as coatis (*Nasua nasua*) and all of the cats except the margay (*Felis wiedii*). These are all essentially tropical species that marginally range into the subtropics. The main northward limit of the Middle American fauna is in Mexico, where 72% of the La Selva fauna also occurs. These species occur in southern Mexico below the plateau, and most extend north through the tropical and subtropical zones along both coasts.

Because La Selva is located in the northern part of Costa Rica, not far from the Nicaraguan border, the number of species that also occur in Nicaragua is high (96%). Perhaps more surprisingly, 98% also occur in Panama, the nearest neighbor to the south. This probably indicates that the fauna in general is centered a bit further south and is representative of the Middle American humid tropical forest fauna, which has its southern limit in western Ecuador.

TABLE 15.2 La Selva mammalian fauna that occur north of Costa Rica

	Nicaragua (%)	Honduras (%)	Guatemala (%)	Mexico (%)	USA (%)	Canada (%)
MARSUPIALIA	100	100	100	100		
CHIROPTERA	96	85	76	71		
Emballonuridae	100	100	100	100		
Noctilionidae	100	100	50	50		
Mormoopidae	100	100	100	100		
Phyllostomidae	97	85	78	73		
Phyllostominae	94	83	72	67		
Glossophaginae	100	83	83	67		
Carolliinae	100	100	67	67		
Stenodermatinae	100	85	85	85		
Desmodontinae	100	100	100	100		
Furipteridae						
Thyropteridae	100	100	100	100		
Vespertilionidae	100	100	85	85		
Molossidae	100	100	50	50		
PRIMATES	100	100	67	67		
EDENTATA	100	85	71	42	14	
LAGOMORPHA	100	100	100	100		
RODENTIA	86	73	46	46		
CARNIVORA	100	92	92	92	57	21
ARTIODACTYLA	100	100	100	100	50	25
PERISSODACTYLA	100	100	100	100		
TOTAL	96	86	76	72	9	3

TABLE 15.3 La Selva mammalian fauna that occur south of Costa Rica

	Panama (%)	Colombia (%)	Ecuador (%)	Peru (%)	Brazil (%)	Argentina (%)
MARSUPIALIA	100	80	80	60	60	60
CHIROPTERA	96	93	85	79	73	19
Emballonuridae	100	100	100	100	100	
Noctilionidae	100	100	100	100	100	
Mormoopidae	100	100	100	100	100	
Phyllostomidae	100	95	83	76	66	14
Phyllostominae	100	100	88	88	88	11
Glossophaginae	100	83	67	50	16	16
Carolliinae	100	100	100	100	67	
Stenodermatinae	100	92	78	64	57	14
Desmodontinae	100	100	100	100	100	100
Furipteridae	100	100	100	100	100	
Thyropteridae	100	100	100	100	100	
Vespertilionidae	85	85	85	85	85	57
Molossidae	50	50	50			
PRIMATES	100	100	67			
EDENTATA	100	100	85	85	71	42
LAGOMORPHA	100	100	100	100	100	100
RODENTIA	100	73	53	20	20	6
CARNIVORA	100	92	92	85	78	42
ARTIODACTYLA	100	100	100	100	100	75
PERISSODACTYLA	100	100	100			
TOTAL	98	91	82	69	64	25

The actual southern limit for many species seems to be Brazil, where 64% of the La Selva species also occur. Argentina has 25% of La Selva species, a proportion of the fauna that extends into subtropical regions to the south in the same marginal sense that others reach the United States in the north.

In summary, the mammal fauna at La Selva is representative of a wet tropical forest region that extends from southern Mexico to northern South America. It is not a localized fauna, and endemism at the species level is nonexistent. Studies conducted on mammals at La Selva therefore are likely to have broad application throughout the Neotropical region.

REFERENCES

Armstrong, D. M. 1969. Noteworthy records of bats from Costa Rica. J. Mamm. 50: 808–810.

Casebeer, R. S., R. B. Linsky, and C. E. Nelson. 1963. The phyllostomid bats, *Ectophylla alba* and *Vampyrum spectrum*, in Costa Rica. J. Mamm. 44: 186–189.

Fleming, T. H. 1973. The number of rodent species in two Costa Rican forests. J. Mamm. 54: 518–521.

Gardner, A. L., R. K. LaVal, and D. E. Wilson. 1970. The distributional status of some Costa Rican bats. J. Mamm. 51: 712–729.

Goodwin, G. G. 1946. Mammals of Costa Rica. Bull. Amer. Mus. Nat. Hist. 87: 271–474.

LaVal, R. K. 1977. Notes on some Costa Rican bats. Brenesia Dept. Hist. Nat. Mus. Nac. Costa Rica, 10/11: 77–83.

LaVal, R. K. and H. S. Fitch. 1977. Structure, movements, and reproduction in three Costa Rican bat communities. Occ. Pap. Mus. Nat. Hist. Univ. Kans. 69: 1–28.

Mares, M. A. and D. E. Wilson. 1971. Bat reproduction during the Costa Rican dry season. BioScience 21: 471–477.

Starrett, A. and R. S. Casebeer. 1968. Records of bats from Costa Rica. Contrib. Sci. Los Angeles Co. Mus. 148: 1–21.

Wilson, D. E. 1983. Checklist of mammals. In D. H. Janzen (ed.), *Costa Rican Natural History*. Univ. of Chicago Press, Chicago, pp. 443–447.

APPENDIX 15.1. Checklist of the Mammals of La Selva

MARSUPLIALIA	
Didelphidae	
Caluromys derbianus	Woolly opossum
Chironectes minimus	Water opossum
Marmosa mexicana	Mexican mouse-opossum
Didelphis marsupialis	Southern opossum
Philander opossum	Gray four-eyed opossum
CHIROPTERA	
Emballonuridae	
Centronycteris maximiliani	Thomas' bat
Cormura brevirostris	Wagner's sac-winged bat
Cyttarops alecto	Short-eared bat
Peropteryx kappleri	Greater doglike bat
Rhynchonycteris naso	Brazilian long-nosed bat
Saccopteryx bilineata	Greater white-lined bat
S. leptura	Lesser white-lined bat
Noctilionidae	
Noctilio albiventris	Lesser bulldog bat
N. leporinus	Greater bulldog bat
Mormoopidae	
Pteronotus parnelli	Parnell's mustached bat
Phyllostomidae	
Macrophyllum macrophyllum	Long-legged bat
Micronycteris brachyotis	Dobson's large-eared bat
M. daviesi	Davies' large-eared bat
M. hirsuta	Hairy large-eared bat
M. megalotis	Brazilian large-eared bat
M. minuta	Gervais' large-eared bat
M. nicefori	Niceforo's large-eared bat
M. schmidtorum	Schmidt's large eared bat
Mimon cozumelae	Cozumel spear-nosed bat
M. crenulatum	Striped spear-nosed bat
Phylloderma stenops	Northern spear-nosed bat
Phyllostomus discolor	Pale spear-nosed bat
P. hastatus	Spear-nosed bat
Tonatia bidens	Spix's round-eared bat
T. brasiliense	Pygmy round-eared bat
T. silvicola	D'Orbigny's round-eared bat
Trachops cirrhosus	Fringe-lipped bat
Vampyrum spectrum	False vampire bat
Glossophaginae	
Choeroniscus godmani	Godman's bat
Glossophaga commissarisi	Commissaris' long-tongued bat
G. soricina	Pallas' long-tongued bat
Hylonycteris underwoodi	Underwood's long-tongued bat
Lichonycteris obscura	Brown long-nosed bat
Lonchophylla robusta	Panama long-tongued bat

Carolliinae	
Carollia brevicauda	Silky short-tailed bat
C. castanea	Allen's short-tailed bat
C. perspicillata	Seba's short-tailed bat
Stenodermatinae	
Artibeus jamaicensis	Jamaican fruit-eating bat
A. lituratus	Big fruit-eating bat
A. phaeotis	Pygmy fruit-eating bat
A. watsoni	Thomas' fruit-eating bat
Chiroderma villosum	Shaggy-haired bat
Ectophylla alba	Honduran white bat
Sturnira ludovici	Anthony's bat
Uroderma bilobatum	Tent-making bat
Vampyressa nymphaea	Big yellow-eared bat
V. pusilla	Little yellow-eared bat
Vampyrodes caraccioli	San Pablo bat
Vampyrops helleri	Heller's broad-nosed bat
Desmodontinae	
Desmodus rotundus	Vampire bat
Furipteridae	
Furipterus horrens	Eastern smoky bat
Thyropteridae	
Thyroptera tricolor	Spix's disk-winged bat
Vespertilionidae	
Eptesicus brasiliensis	Brazilian brown bat
E. furinalis	Argentine brown bat
Myotis albescens	Silver-tipped bat
M. elegans	Elegant myotis
M. nigricans	Black myotis
M. riparius	Riparian myotis
Rhogeessa tumida	Central American yellow bat
Molossidae	
Molossus bondae	Bonda mastiff bat
M. sinaloae	Allen's mastiff bat
PRIMATES	
Cebidae	
Alouatta palliata	Mantled howler monkey
Ateles geoffroyi	Geoffroy's spider monkey
Cebus capucinus	White-faced capuchin
EDENTATA	
Bradypodidae	
Bradypus variegatus	Three-toed sloth
Choloepidae	
Choloepus hoffmanni	Two-toed sloth
Dasypodidae	
Cabassous centralis	Five-toed armadillo
Dasypus novemcinctus	Nine-banded armadillo

Taxon	Common name
Myrmecophagidae	
Cyclopes didactylus	Silky anteater
Myrmecophaga tridactyla	Giant anteater
Tamandua mexicana	Northern tamandua
LAGOMORPHA	
Leporidae	
Sylvilagus brasiliensis	Forest rabbit
RODENTIA	
Sciuridae	
Microsciurus alfari	Alfaro's pygmy squirrel
Sciurus granatensis	Red-tailed squirrel
S. varieqatoides	Variegated squirrel
Heteromyidae	
Heteromys desmarestianus	Desmarest's spiny pocket mouse
Muridae	
Nyctomy's sumichrasti	Sumichrast's vesper rat
Oryzomys alfari	Alfaro's rice rat
O. bombycinus	Long-whiskered rice rat
O. caliginosus	Dusky rice rat
O. fulvescens	Pygmy rice rat
Tylomys watsoni	Watson's climbing rat
Erethizontidae	
Coendou mexicanus	Prehensile-tailed porcupine
Agoutidae	
Agouti paca	Paca
Dasyproctidae	
Dasyprocta punctata	Agouti
Echimyidae	
Hoplomys gymnurus	Armored rat
Proechimys semispinosus	Tomes' spiny rat
CARNIVORA	
Mustelidae	
Conepatus semistriatus	Striped hog-nosed skunk
Eira barbara	Tayra
Galictis vittata	Grison
Lutra longicaudis	Southern river otter
Mustela frenata	Long-tailed weasel
Procyonidae	
Bassaricyon gabbii	Olingo
Nasua nasua	Coati
Potos flavus	Kinkajou
Procyon lotor	Raccoon
Felidae	
Felis concolor	Puma
F. onca	Jaguar
F. pardalis	Ocelot
F. wiedii	Margay
F. yagouaroundi	Jaguarundi

ARTIODACTYLA	
Tayassuidae	
Dicotyles tajacu	Collared peccary
Tayassu pecari	White-lipped peccary
Cervidae	
Mazama americana	Red brocket deer
Odocoileus virginianus	White-tailed deer
PERISSODACTYLA	
Tapiridae	
Tapirus bairdii	Baird's tapir

16

Neotropical Mammal Densities: How Unusual Is the Community on Barro Colorado Island, Panama?

WILLIAM E. GLANZ

Barro Colorado Island (BCI), Panama, is well known among tropical biologists for its abundant and easily observable mammals. Within fifteen years after the island was protected as a wildlife reserve, both Allee and Allee (1925) and Chapman (1929) remarked on the abundance of mammals there. Recently some biologists have suggested that mammals may be unusually plentiful on BCI, in comparison to other Neotropical sites. Terborgh and Winter (1980) have commented: "Completely protected for more than 50 years, Barro Colorado has become a veritable zoo without cages. Medium and large mammals are unwary and remarkably abundant. These include the collared peccary [*Tayassu tajacu*], agouti [*Dasyprocta punctata*], coati mundi [*Nasua nasua*], three-toed and two-toed sloths [*Bradypus variegatus* and *Choloepus hoffmanni*], tamandua [*Tamandua mexicana*], armadillo [*Dasypus novemcinctus*], and howler monkey [*Alouatta palliata*]. Terborgh has travelled extensively in the Neotropics, and nowhere, even in areas protected from hunting, has he found any of these mammals to be as abundant as on Barro Colorado." In addition, Eisenberg et al. (1979) have noted: "Vegetational succession, local extinctions, and the loss of large predators on BCI [have] produced a somewhat artificial faunal assemblage." In this chapter, I review several theories previously advanced to explain the abundance and possible faunal imbalance of mammals on BCI. I then discuss the species composition and densities of non-flying mammals on Barro Colorado to supplement and update previous analyses by Enders (1935), Eisenberg and Thorington (1973), and Glanz (1982). These data are compared with those from other sites in the Neotropics, to evaluate whether the BCI mammal community is unusual in its composition and densities.

Before considering possible mechanisms of faunal disruption on Barro Colorado, it is useful to consider some factors that may confound the analysis. First, the Central American mammal fauna is less diverse than that of the Amazon Basin, particularly in species richness of marsupials (Streilein 1982), primates (Eisenberg 1979), and hystricognath rodents (Mares and Ojeda 1982). This pattern may reflect the smaller area of Central America, its past separation from the South American continent, and its present isolation by the Andes from the Amazon Basin. Central Panama has fewer mammal species than eastern or western Panama (Handley 1966), probably because substantial mountain ranges, with their high habitat diversity, are geographically distant.

Next, central Panama has had prolonged exposure to European humans and their impacts, and these influences have probably been greater than at many other Neotropical areas. European settlement dates from 1519, and human impacts increased greatly with the completion of the trans-isthmian railroad in 1853 and the beginning of canal construction in 1882. Thus, when BCI was isolated by the rising waters of Gatun Lake in 1914, much of it had been hunted, farmed, or disturbed by human activities during the preceding century (Foster and Brokaw 1982). Some mammalian species, therefore, may already have been reduced in numbers or locally extinct.

Finally, current ecological factors may produce high extinction rates on BCI (Willis 1974; Terborgh 1974; Karr 1982). These include its small size (15 km^2), which limits total population size, its isolation (by water gaps of at least 300 m) and resultant low recolonization rates, and its relatively homogeneous habitat of mature forest (approximately half is "old forest," about four hundred years in age, and half is "young forest," at least seventy years old, which is approaching "old forest" in structure; see Foster and Brokaw 1982), which limits opportunities for species that require certain localized or successional microhabitats.

Hypotheses of Faunal Imbalance on Barro Colorado Island

Allee and Allee (1925) and Chapman (1929) both noted that local observers attributed the abundance of mammals on Barro Colorado to the island acting as a Noah's Ark, drawing numerous mammals from surrounding drowned habitat. Chapman (1929) and Enders (1939) dismissed this theory, citing the small ratio of flooded to nonflooded habitat if surrounding "mainland" areas are considered and the low likelihood that excessive densities in 1914 would persist into the 1920s and 1930s, when

TABLE 16.1 Mammalian extinctions on Barro Colorado Island

Species	Common name	Weight (kg)	Status
Tapirus bairdii	Tapir	300.0	RI
Panthera onca	Jaguar	36.0	E1
Felis concolor	Puma	29.0	E
Tayassu pecari	White-lipped peccary	35.0	E
Ateles geoffroyi	Spider monkey	7.0	RI
Felis wiedii	Margay	6.0	E1
Procyon lotor	Raccoon	5.0	E?
P. cancrivora	Crab-eating raccoon	5.0	E?
Bassaricyon gabbii	Olingo	2.0	E1
Sigmodon hispidus	Cotton rat	0.1	E
Microsciurus alfari	Pygmy squirrel	0.06	E
Zygodontomys brevicauda	Cane rat	0.05	E?
Oryzomys fulvescens	Pygmy rice rat	0.04	E

Sources: Body weights from Eisenberg and Thorington (1973), Eisenberg (1981), Terborgh (1983), and Emmons (1987). Extinctions based on Enders (1935, 1939), Chapman (1929), Eisenberg and Thorington (1973), and app. 16.1.
Status Key:
E = probably extinct
E? = possibly extinct, no recent records
E1 = possibly extinct as a population, one recent sighting
RI = formerly extinct, reintroduced and established

they were observing high mammalian densities. This controversy, however, is worth noting because it implies that high mammalian abundance was characteristic of BCI very early in its history.

Most recent hypotheses of faunal imbalance focus on current ecological aspects of the island and their consequences. Of the known or likely mammalian extinctions on BCI (table 16.1), several large species would probably be limited by its small area. These include the jaguar (*Panthera onca*) and the puma (*Felis concolor*), which exist at low densities and require large home ranges (Emmons 1987; Schaller and Crawshaw 1980), and the white-lipped peccary (*Tayassu pecari*), which occurs in large social groups that range widely (Kiltie and Terborgh 1983). Most of the small mammal extinctions listed may be related to habitat, because three species require grassland or other clearings (*Oryzomys fulvescens*, *Sigmodon hispidus*, and *Zygodontomys brevicauda*) and the fourth (*Microsciurus alfari*) inhabits thick, successional vegetation (Enders 1935).

Several extinctions of large species, however, may be more related to human influence than to natural factors. Spider monkeys were absent from BCI when it was formed, probably because their bold behavior and tasty flesh made them desirable game during the canal construction era. Jaguars may also have been absent from Barro Colorado; Enders (1935) reported possible jaguar signs, but none of these descriptions (primarily large-diameter feces and scrapes) provide definitive evidence for jaguar rather than puma (for other characters, see Emmons 1987; Rabinowitz and Nottingham 1986; Schaller and Crawshaw 1978). Enders (1939) implied that increased poaching on BCI during 1932–37 may have caused the virtual extinction of puma and tapir on the island. Protection from poaching has improved greatly since 1950 (Glanz 1982); both spider monkeys and tapirs were reintroduced by 1960 and have persisted.

Most recent comments on mammalian density changes on Barro Colorado have invoked the absence of jaguars and pumas as the primary cause (Eisenberg et al. 1979; Terborgh and Winter 1980; DeSteven and Putz 1984; Emmons 1984, 1987). Extinction of these large cats has presumably permitted some terrestrial prey species, such as agoutis, pacas (*Agouti paca*), and peccaries, to increase in numbers. Terborgh and Winter (1980) and Emmons (1984) have also argued that extinctions of certain avian predators have had similar effects on some arboreal mammals. Raptors reported attacking arboreal mammals at a pristine forest site, Cocha Cashu, Peru (Terborgh 1983), have been compared with those on BCI (table 16.2). The two sites were formerly very similar in raptor species composition, and extinctions at BCI include only the largest species, the harpy eagle (*Harpia harpyja*), and possibly the next largest species, the crested eagle (*Morphnus guianensis*).

In contrast to predation, other hypotheses for high apparent densities of mammals on Barro Colorado have received much less attention. Eisenberg et al. (1979) mentioned the advanced state of forest succession on BCI as favoring higher densities of canopy species, in comparison to Venezuelan sites with younger forests or broken canopies. Eisenberg (1979) related forest maturation on BCI to its high densities of howler monkeys. Similarly, less second-growth forest could help explain declines in the relative importance of white-faced monkeys (*Cebus capucinus*) and tamarins (*Saguinus oedipus*). Eisenberg (1979) also presented evidence for apparent density compensation (MacArthur 1972), with howlers reaching high densities at other sites besides Barro Colorado, where *Cebus* or spider monkeys (*Ateles*) are rare or absent. Emmons (1984) also noted correlations between habitat characteristics and dominance of *Alouatta* within primate communities and considered such relationships as an alternative hypothesis to theories based on extinctions of large predators.

TABLE 16.2 Raptors at Cocha Cashu and Barro Colorado Island that are potential predators of arboreal mammals

Species	Size (kg)	Status at Cocha Cashu	Status at Barro Colorado	Diet
Harpy eagle *Harpia harpyja*	4–5	Rare	Extinct by 1950; recently seen nearby	Sloths, primates, other mammals
Crested eagle *Morphnus guianensis*	1.7	Rare	Occasional; one possible recent sighting	Monkeys, opossums, rodents, large reptiles
Black hawk-eagle *Spizaetus tyrannus*	0.9–1.1	Rare	Regular	Opossums, squirrels
Ornate hawk-eagle *Spizaetus ornatus*	0.8–1.6	Uncommon	Rare	Large birds, small primates, kinkajous, squirrels
Black-and-white hawk-eagle *Spizastur melanoleucus*	0.7–0.8	Rare	No counterpart	Birds, mammals
Slate-colored hawk *Leucopternis schistacea*	1.0	Uncommon	Larger congener *L. albicollis* uncommon	Snakes, frogs, rodents, small primates
Bicolored hawk *Accipiter bicolor*	0.2–0.5	Uncommon	No counterpart	Birds, small primates
Collared forest-falcon *Micrastur semitorquatus*	0.5	Rare	Rare	Birds, rodents
Spectacled owl *Pulsatrix perspicillata*	0.6–1.0	Uncommon	Fairly common	Rodents

Sources: Status from Terborgh et al. (1984) and Willis and Eisenmann (1979); Diet based on Terborgh (1983) and references therein, Smith (1970), Bierregaard (1984), Lyon and Kuhnigk (1985), and Glanz (unpubl. obs.)

Smythe (1978) and Glanz (1982) noted that Barro Colorado Island is very well protected from poaching for a Neotropical forest site; low densities elsewhere could be related to poaching or unregulated hunting by humans. Freese et al. (1982) and Emmons (1984) found evidence that high hunting pressure depressed mammal densities at some Amazonian sites but did not apply this idea to BCI. Glanz (1982) also suggested that even occasional hunting pressure could affect mammalian wariness and hence detectability, producing higher apparent densities at a well-protected site than at a hunted site with similar actual densities. Finally, Glanz (1982) has suggested that BCI may be a very productive site, in terms of seasonally critical food resources, in comparison to most other Neotropical forests. Emmons (1984) used site productivity in differentiating mammal density patterns at Amazonian sites but regarded Barro Colorado as comparable to her white-water sites in Ecuador and Peru. Terborgh (1983) compared annual fruit drop at Cocha Cashu, Peru, with that measured by Smythe (1970) on BCI and found the two sites to be similar in gross production (kg/ha) of fruit.

Species Richness of Nonflying Mammals on Barro Colorado Island

The number of nonflying mammal species reported from Barro Colorado (see appendix 16.1) have been compared with two Costa Rican (from Wilson 1983) and two Peruvian sites (Patton et al. 1982; Terborgh et al. 1984) (table 16.3). Species totals for each of the comparative sites exceed that of BCI, despite fewer years of biological study at these other sites. The small size of Barro Colorado, its isolation, and the subsequent extinctions are likely causes of this difference. All Central American sites

TABLE 16.3 Species richness of nonflying mammals at five Neotropical sites

Site	Number of species recorded
Barro Colorado Island, Panama	39 (49 originally − 13 extinctions + 3 additions = 39)
La Selva, Costa Rica	43
Osa Peninsula, Costa Rica	59
Rio Cenepa, Peru	62
Cocha Cashu, Peru	68 (hypothetical species omitted)

Sources: Appendix 16.1, Terborgh et al. (1984), Patton et al. (1982), and Wilson (1983)

have fewer confirmed nonflying species than South American sites; furthermore, the Rio Cenepa, Peru, total is undoubtedly an underestimate, since it represents only four years of fieldwork. Thus, biogeographic differences in regional diversity also seem to contribute to the low species richness of BCI. An updated species list for BCI is provided in appendix 16.1.

Methods for Estimating Densities of Neotropical Mammals

Techniques for determining mammalian densities vary greatly among taxonomic groups. Visual counts are possible for large, diurnal species, such as howler monkeys, but are difficult for cryptic or nocturnal species. Mark-recapture density estimates are possible for small, nocturnal rodents and marsupials, but not all species are easily live-trapped. Many of the published mark-recapture estimates, moreover, involve different techniques with different assumptions, making separate studies difficult to evaluate and compare. To assess densities in a taxonomically diverse community is thus very difficult; most recent surveys of tropical mammal communities have employed less precise but more broadly applicable techniques than single-species studies. Visual censuses from trail transects have been widely used on BCI (Eisenberg and Thorington 1973; Glanz 1982), in Guatemala (Cant 1977; Walker and Cant 1977), in Venezuela (Eisenberg et al. 1979), in Peru (Emmons 1984, 1987), and in a variety of Neotropical primate communities (Eisenberg 1979; Freese et al. 1982). Techniques are described in Robinette et al. (1974), Eisenberg (1979), Glanz (1982), and National Research Council (1982). Results can be expressed as sightings/hour, as sightings/km, or as animals/unit area if the width of the strip being sampled can be estimated. These techniques can be calibrated by conducting such strip censuses at sites where more precise single-species density estimates are available. See Malcolm (this volume) for more on these techniques.

Density Variations among Years

If mammal densities vary greatly among years, comparisons of sites may lead to very different conclusions depending on which years are chosen for analysis. I reviewed censuses and anecdotal data from Barro Colorado Island prior to 1977 in Glanz (1982) and noted some long-term trends. These included a progressive decline in *Cebus* monkey numbers since the 1960s, a concurrent increase in squirrel (*Sciurus granatensis*)

TABLE 16.4 Changes in sighting rates of mammals on Barro Colorado Island

Species	Census type	1977–78	1982	1983–84	1985–86
Marsupialia					
Didelphis marsupialis	n	0.67	0.31	0.33	—
Edentata					
Tamandua mexicana	d	0.08	0.05	0.10	0.07
Dasypus novemcinctus	n	0.55	0.63	0.04	—
Primates					
*Alouatta palliata**	d	0.56	0.81	0.84	0.70
*Cebus capucinus**	d	0.10	0.21	0.33	0.21
Lagomorpha					
Sylvilagus brasiliensis	n	0.31	0.09	0.08	—
Rodentia					
Sciurus granatensis	d	2.51	1.20	1.05	0.96
Oryzomys spp.	n	0.33	0.03	0.08	—
Proechimys semispinosus	n	0.50	0.17	0.33	—
Dasyprocta punctata	d	3.26	1.35	2.26	2.15
Agouti paca	n	0.70	0.61	0.98	—
Carnivora					
*Nasua nasua**	d	0.22	0.29	0.18	0.24
Potos flavus	n	0.39	0.23	0.25	—
Artiodactyla					
*Tayassu tajacu**	d	0.09	0.13	0.36	0.26
*Tayassu tajacu**	n	0.10	0.10	0.21	—
Mazama americana	d	0.04	0.04	0.15	0.19
Mazama americana	n	0.28	0.49	0.66	—
Total diurnal sighting rate		7.04	4.22	5.38	4.88
Total nocturnal sighting rate		5.06	3.20	3.50	—
Total diurnal km censused		314	168	101	98
Total nocturnal km censused		62	57	24	—

Notes: d = diurnal
n = nocturnal
* = group sightings/km (all others individual sightings/km)
— = not censused

numbers, and apparent increases in the sighting rates of some nocturnal species, including armadillos (*Dasypus novemcinctus*), pacas (*Agouti paca*), and brocket deer (*Mazama americana*).

My sighting rates for selected species during four sample periods from 1977 to 1986 include all species with diurnal sighting rates of 0.1/km in at least one period and all nocturnal species with sighting rates of at least 0.2/km in 1977 or 1982 (table 16.4). Omitted species were generally sighted too infrequently for sighting rates to stabilize. Two primarily frugivorous groups, the marsupials and the rodents, decreased concurrently from 1977 to 1982, with drops in sighting rates ranging from slight (*Agouti*) to 50–90% (*Didelphis*, *Sciurus*, *Proechimys*, *Dasyprocta*, and *Oryzomys*). These declines may be related to poor fruit yields on BCI during 1979 and 1981 (Giacalone et al. 1989). Two primates showed contrasting patterns. *Cebus* monkeys increased threefold over the period. Howler monkeys showed a slight increase, but a change in our recording of troop "fragments" is responsible for most of the apparent 1977–82 increase. Only one mammal species, the nine-banded armadillo (*Dasypus novemcinctus*), showed a dramatic decline after 1982; sightings fell by more than 90% after the prolonged dry season of 1983. Finally, two large terrestrial species, the collared peccary (*Tayassu tajacu*) and the brocket deer (*Mazama americana*), appeared to have increased two- to fourfold over the period, although the peccary data are complicated by a concurrent decrease in average group size.

These results suggest that published density figures for some frugivores, especially for some marsupials and rodents, should be used with caution, recognizing that these groups can vary manyfold from year to year. Armadillos likewise are subject to major density changes. Hence, for such species a twofold or threefold difference in densities between sites might be insignificant, but a tenfold difference would certainly be worthy of consideration. Most other species vary less from year to year, but two- to fourfold changes were observed even in some large species. In comparing densities at Neotropical sites, I will focus on differences that exceed these levels.

Comparisons of Mammal Densities among Neotropical Sites

In the following discussion, densities of mammals on Barro Colorado are taken from my estimates in Glanz (1982), which were based primarily on visual trail censuses in 1977–78, with calibrations for some species (*Alouatta*, *Sciurus*, *Dasyprocta*, and *Proechimys*). Results are compared with estimates from five other rainforest sites in the Neotropics: Guatopo, Venezuela (Eisenberg et al. 1979), Cocha Cashu, Peru (Terborgh 1983), Cabassou, French Guiana (Charles-Dominique et al. 1981), Sierra Chame, Guatemala (Hendrichs 1977), and Tikal, Guatemala (Cant 1977; Walker and Cant 1977). All these studies used some form of visual transect census supplemented by other techniques. I omit from this comparison studies of Neotropical mammal communities in nonforest habitats (e.g., Schaller 1983, and the Masaguaral site in Eisenberg et al. 1979) or studies of just one taxonomic group (e.g., Freese et al. 1982).

The analyses below are organized by mammalian orders, but they emphasize comparisons of ecologically similar genera and species at different sites. I caution the reader, however, that the assumption of ecological similarity of different species in one genus, or even different populations within one species, may be invalid in certain cases. The Central American howler monkey *Alouatta palliata*, for example, differs significantly in diet among sites and differs greatly in social structure from the South American *A. seniculus* (Milton 1980). I compare *Alouatta* populations assuming they are approximate ecological equivalents, but further studies of primate communities may suggest more appropriate comparisons. Taxonomic nomenclature used follows Honacki et al. (1982), but future revisions of genera and species also may affect the validity of some comparisons.

Marsupialia. Densities of the common opossum (*Didelphis marsupialis*) are remarkably similar among sites (table 16.5). The mouse opossum genus *Marmosa* is more common at the younger forest sites with more brush and vines (Guatopo and Cabassou), as predicted by its habitat preferences on BCI (Enders 1935). Densities at these two sites were determined by trapping, which probably samples these small mammals more accurately than the visual trail censuses used at the other sites. Only one species, *M. robinsoni*, occurs at BCI and Guatopo, while Cocha Cashu and Cabassou each have two *Marmosa* species. The arboreal species of *Caluromys* and *Caluromysiops* occur at comparable densities at BCI, Guatopo, and Cocha Cashu. The more terrestrial genera *Philander* and *Metachirus* are absent at Guatopo, uncommon at BCI, and slightly more common at Cocha Cashu. Why these latter two marsupial groups are extremely abundant at Cabassou is unclear; perhaps sampling was more intensive, or perhaps lower primate abundance there permits density compensation by these marsupial taxa.

Edentata. Barro Colorado is notorious for its high densities of sloths (Eisenberg and Thorington 1973; Brockie 1986), estimated by Montgomery and Sunquist (1975) using range sizes and overlaps of radio-tracked animals and fecal deposits counted on a very small plot. These techniques are prone to bias and difficult to apply to other sites, but no other technique short of intensive, tree-by-tree searches seems appropriate (see

TABLE 16.5 Reported densities for marsupials and edentates at selected Neotropical sites

Genera	BCI, Panama (N/km^2)	Guatopo, Venezuela (N/km^2)	Cocha Cashu, Peru (N/km^2)	Cabassou, French Guiana (N/km^2)
Marsupialia				
Caluromys and *Caluromysiops*	20	20	10	100–200
Didelphis	47	65	55	25–50
Philander and *Metachirus*	13	—	37	100–200
Marmosa	27	137	25	45–200
Edentata				
Bradypus and *Choloepus*	633	2	<1	300–600
Tamandua	5	6	?	?
Dasypus	53	4	?	?

Sources: Barro Colorado (BCI) figures from Glanz (1982), Guatopo from Eisenberg et al. (1979), Cocha Cashu from Terborgh (1983), and Cabassou from Charles-Dominique et al. (1981)

Charles-Dominique et al. 1981). In spite of this, sloths are frequently encountered on BCI, and I revised previous estimates only slightly downward in Glanz (1982). Nevertheless, I consider these estimates (table 16.5) inadequate and await better data. (See also my discussion of long-term variations in biomass, below.)

Sloths are rare at Guatopo, Venezuela; Eisenberg et al. (1979) attribute this to the young, broken-canopy forest being poor habitat. Sloths are also scarce in the mature forest of Cocha Cashu, however, and Emmons (1984) invokes avian predation to explain this. Noting that harpy eagles in Guyana eat many sloths (Fowler and Cope 1964; Rettig 1978) and that harpy eagles are extinct on BCI with its many sloths, she implies that the presence of these raptors at Cocha Cashu may explain the rarity of sloths there. Sloths are common, however, in much of the Guiana region, in Suriname (Walsh and Gannon, 1967) and at Cabassou, French Guiana (see table 16.5), and harpy eagles are present in countable numbers in both these areas, even in secondary forests (Haverschmidt 1968; Thiollay 1985). It seems more reasonable that harpy eagles eat many sloths where sloths are common and eat more primates where

TABLE 16.6 Reported densities of primates at three Neotropical sites

Genera	BCI, Panama (N/km^2)	Guatopo, Venezuela (N/km^2)	Cocha Cashu, Peru (N/km^2)
Alouatta	87.0	16.0	20
Ateles	0.9	5.6	25
Cebus	9.3	27.0	75*
Aotus	0.9	—	24
Saguinus	2.7	—	33
Others	—	—	86

Sources: Barro Colorado (BCI) from Glanz (1982), Guatopo from Eisenberg et al. (1979), and Cocha Cashu from Terborgh (1983)
Note: *Two species

sloths are rare, and that other factors depress sloth densities in southern Peru. Terborgh (1984) notes that occasional winter storm fronts sweep that region each year, bringing temperatures as low as 8°C. Such cold is unknown in Panama and the north coast of South America. If sloths are poor thermoregulators (see McNab 1978; Montgomery and Sunquist 1978), this climatic factor could explain the scarcity of sloths at Cocha Cashu. Nutritional factors may also be involved in these contrasting densities if coastal regions, such as Panama and Suriname, receive higher inputs of nutrients like sodium, which may be limiting for folivores (U. Roze, p.c.), than such inland regions as Cocha Cashu.

Densities of other edentates (anteaters and armadillos) are poorly known in other rain forest habitats. *Tamandua* anteaters are similar in abundance at BCI and Guatopo (table 16.5). The high densities of nine-banded armadillos (*Dasypus novemcinctus*) on BCI are noteworthy, but note my above discussion of variations in densities among census periods.

Primates. Primates are less abundant in the young forest of Guatopo than in the older forests of Barro Colorado and Cocha Cashu (table 16.6), a not unexpected result if a forest's suitability to primates is correlated with the density of its canopy. Eisenberg et al. (1979) remark that the density of howler monkeys *Alouatta seniculus* at Guatopo is very similar to that of *A. palliata* on BCI in the 1930s, when its forest was young.

Cocha Cashu has more than twice the primate density of BCI (table 16.4), but these numbers are spread among 11 species (with 2 more in neighboring forests; Terborgh 1983), whereas howler monkeys dominate the BCI primate community (more than 85% of all individuals). Emmons (1984) also suggests that this difference may be related to predation by harpy eagles at Cocha Cashu but acknowledges that other habitat and zoogeographic factors may be involved. She notes that high densities of howlers have been found in other communities with few primate species (see also Heltne and Thorington 1976; Eisenberg 1979; Crockett 1987), especially in Central American and drier South American forests, in other marginal habitats and in isolated forest patches. Emmons suggests "an alternative to the lack of predators hypothesis for *Alouatta* on BCI is that its island nature or some other aspect of its ecology has made BCI a marginal habitat for frugivorous monkeys, with a consequent increase in howlers." Such a response could be a form of density compensation in response to the absence of competition from other species, but it is unclear why BCI is a marginal habitat and why other monkeys outcompete howlers elsewhere. Other monkey genera, such as *Cebus* and *Saguinus* often reach high densities in secondary forests (Eisenberg 1979); their abundance relative to *Alouatta* at many sites (Freese et al. 1982) may be related to basic habitat preferences rather than to interspecific competition.

Rodentia. Perhaps the most obvious animal in the Barro Colorado forest is a large, terrestrial rodent, the agouti *Dasyprocta punctata*. This species is a preferred game animal in much of its range, and it can be very wary. Its visibility on BCI therefore may be due in part to the absence of hunting there. Densities of agoutis are lower at all other sites studied so far (see table 16.7) but those reported at Guatopo approach levels on BCI. Density estimates for another large rodent, the nocturnal paca (*Agouti paca*), are similarly elevated on BCI (table 16.7). The abundance of agoutis and pacas on BCI suggests a common mechanism of density increase. Smythe (1978) argues that, since both rodents are preferred game animals with relatively low reproductive rates, hunting and poaching, even in "reserves" with inadequate protection, could significantly lower their densities. Although this hypothesis may apply to most accessible forested sites in the Neotropics, it seems inadequate for uninhabited sites like Cocha Cashu. Emmons (1984, 1987) presents a strong case for the role of large felid predators in depressing numbers of such large rodents. Her dietary data for jaguar and puma (Emmons 1987) and data from Schaller and Vasconcelos (1978), Schaller (1983), and Rabinowitz and Nottingham (1986) indicate that large (greater than 1 kg) rodent species (agoutis, acouchis [*Myoprocta* species], capybaras [*Hydrochaeris hydrochaeris*], and pacas) are either preferred prey of large cats or taken in proportion to their abundance. Emmons's calculations (1984,

TABLE 16.7 Reported densities for rodents and carnivores at six Neotropical sites

Genera	BCI, Panama (N/km^2)	Guatopo, Venezuela (N/km^2)	Cocha Cashu, Peru (N/km^2)	Cabassou, French Guiana (N/km^2)	Sierra Chame, Guatemala (N/km^2)	Tikal, Guatemala (N/km^2)
Rodentia						
Sciurus	180.0	25.0	25.0	common	100.0	>9
Oryzomys	133.0	234.0	180.0	50–300	?	?
Proechimys	200.0	126.0	230.0	very common	—	—
Agouti	40.0	18.0	3.5	common	30.0	?
Dasyprocta	100.0	63.0	5.2	common	30.0	8
Myoprocta	—	—	5.3	—	—	—
Hydrochaeris	<1.0	—	1.6	—	—	—
Carnivora						
Nasua	24.0	—	<1.0	P	15.0	20
Potos	20.0	—	20.0	20–30	20.0	74
Tayra	1.6	2.0	?	P	1.0	2
Felis	0.8	1.3	0.8	P	0.4	P
Panthera	—	0.02	0.4	P	0.1	P

Sources: Barro Colorado (BCI) data from Glanz (1982), Guatopo from Eisenberg et al. (1979), Cocha Cashu from Terborgh (1983) with revisions from Emmons (1987), Cabassou from Charles-Dominique et al. (1981), Sierra Chame from Handrichs (1977), and Tikal from Cant (1977)

Note: P = present at site, including rare and occasional

1987) of estimated kill rates suggest a major impact of these cats on populations of large rodents, and I agree that this hypothesis is currently the best explanation for elevated agouti and paca numbers on BCI in comparison to those at uninhabited sites.

Other rodent species on BCI show different patterns (table 16.7). The small terrestrial genera *Oryzomys* and *Proechimys* are either comparable in densities between BCI and other sites or less common on BCI. Other studies of *Proechimys* populations at mainland sites (Fleming 1971; Guillotin 1982) also found densities in the same range (40–560/km^2), but Gliwicz (1973) reported higher densities (850/km^2) on a small island in Panama near BCI. Emmons (1987) suggests that ocelots (*Felis pardalis*) have major impacts on these small rodents (weighing less than 1 kg); since ocelots are present at BCI and most mainland sites but probably absent on Gliwicz's site, the data are consistent with this hypothesis.

Published densities for squirrels, *Sciurus* species, are very high for BCI (table 16.7; see also Glanz et al. 1982; Giacalone et al., in press) and one Guatemalan site, and lower for most other sites. Reasons for these patterns are unclear. Terrestrial predators are clearly unimportant, because they would have difficulty capturing so agile a prey species. Avian predators may have an effect, but BCI has several raptor species known to attack and take squirrels (table 16.2), especially the black hawk-eagle (*Spizaetus tyrannus*) and the ornate hawk-eagle (*S. ornatus*). Most Neotropical squirrels feed primarily on hard, nutlike fruits (Glanz et al. 1982; Emmons 1984), resources few other arboreal mammals can utilize, so abundances of squirrels may be related more to site-specific production rates of these foods. Other explanations, including disease, predation of young in the nest by snakes and *Cebus* monkeys, and competition with agoutis and peccaries for fallen nutlike fruits are also possible. *Sciurus* densities have varied greatly through time on BCI (Glanz 1982, and see above), so perhaps squirrels occasionally reach similarly high densities at other Neotropical sites.

Carnivora. Two procyonid species are common on Barro Colorado, the coati (*Nasua nasua*) and the kinkajou (*Potos flavus*). Coati densities are high relative to most other sites (table 16.7), although sighting rates at Tikal, Guatemala (Cant 1978), are comparable. Coatis have benefited from human garbage, food stores, and cultivated trees in the laboratory clearing on BCI since the 1920s (Chapman 1929), but reductions in these food sources since the 1950s have not dramatically reduced their numbers (Kaufmann 1960; Eisenberg and Thorington 1973; Glanz 1982).

Terborgh and Winter (1980) attributed their abundance to absence of the large cats. Schaller and Crawshaw (1980) report a jaguar killing coatis, so this explanation is plausible, but it does not explain coati abundance at Tikal, where the larger cats are still present.

Kinkajou (*P. flavus*) numbers on BCI are similar to those at three other sites (Cocha Cashu, Cabassou, and Sierra Chame), but much lower than at Tikal. Reasons for this pattern are not obvious. Walker and Cant (1977) noted a strong association between kinkajous and the availability of an abundant, preferred fruiting tree, *Pouteria campechiana*. The lower diversity of some frugivorous groups in Guatemala (such as marsupials and primates) may also permit higher kinkajou densities.

Other Carnivora are rare on BCI. As noted above, ocelots are present and appear to occur at normal densities, whereas jaguars and pumas are absent (table 16.7). Jaguarundis (*Felis yaguaroundi*) are very rare (few good observations since 1977). Tayras (*Eira barbara*) are seen more frequently but are still uncommon, and their density is comparable to those at other Neotroical sites (table 16.7).

Ungulates. Density estimates of two ungulate genera, *Tayassu* (peccaries) and *Mazama* (brocket deer) on Barro Colorado are remarkably similar to those from Cocha Cashu (table 16.8). This is surprising, because these taxa are also suitable prey for jaguars and possibly pumas (Emmons 1987). Perhaps due to the size and speed of these animals, their populations are affected less by the large cats than are agouti and paca populations. The large litters of the peccaries may also permit a greater potential reproductive rate than is possible in agouti and paca populations. Lower ungulate densities at Guatopo may reflect occasional hunting, but no obvious factor explains the high estimates for Sierra Chame.

Mainland Sites in Panama: Testing Predation versus Human Impacts

The analysis above compares densitites at some sites that are 1,000 km or more apart. Vegetation at these sites ranges from young (Guatopo) to undisturbed by humans (Cocha Cahsu), and plant species composition undoubtedly varies greatly among sites. I have noted the differences in lowest temperatures between Cocha Cashu and the sites nearer coastlines. Considering these and other possible site differences, as well as species differences in the mammalian faunas themselves, it would seem

TABLE 16.8 Reported densities of ungulates at four Neotropical sites

<table>
<tr><th>Genera</th><th>BCI, Panama (N/km²)</th><th>Guatopo, Venezuela (N/km²)</th><th>Cocha Cashu, Peru (N/km²)</th><th>Sierra Chame, Guatemala (N/km²)</th></tr>
<tr><td>Perissodactyla</td><td></td><td></td><td></td><td></td></tr>
<tr><td>Tapirus</td><td>0.5</td><td>0.6</td><td>P</td><td>—</td></tr>
<tr><td>Artiodactyla</td><td></td><td></td><td></td><td></td></tr>
<tr><td>Tayassu</td><td>9.3</td><td>.19</td><td>5.6</td><td>10</td></tr>
<tr><td>Mazama</td><td>2.0</td><td>5.3</td><td>2.6</td><td rowspan="2">20</td></tr>
<tr><td>Odocoileus</td><td>0.7</td><td>—</td><td>—</td></tr>
</table>

Sources: BCI data from Glanz (1982), Guatopo from Eisenberg et al. (1979), Cocha Cashu from Emmons (1987) and Terborgh et al. (1984), and Sierra Chame from Hendrichs (1977)

prudent to compare sites where such variables are more similar. In a separate study (Glanz, in press), I have reported census results for several mainland sites in Panama that presumably have similar climates, floras, and faunas to those on BCI but have different predation and human impact levels. I briefly summarize my results here. In 1977–78, I censused two heavily hunted areas, the Pipeline Road region, 3–6 km east of BCI, and Paraiso, 30 km southeast of BCI. The forest at Pipeline Road was similar to that on BCI, while that at Paraiso was much drier (Glanz 1984). Howler and *Cebus* monkeys, agoutis, coatis, and squirrels were present but scarce in the Pipeline area, and total mammal sighting rates were only 0.52/km, as opposed to 7.04/km on BCI. Sighting rates were higher at Paraiso (2.51/km), but 86% of all sightings were of two squirrel species, and all more desirable game species (agoutis, peccaries, deer) were rare or absent. Clearly, unregulated hunting was having a great impact on mammals at these sites.

In 1980 two mainland peninsulas south of BCI were incorporated into the Barro Colorado National Monument and protected from hunting. By 1982, poaching on these peninsulas was infrequent, though still higher than on BCI. I censused these areas in the 1982 wet season and 1983–84 dry season (table 16.9). In contrast to the hunted sites, these protected sites had many mammals. Total sighting rates on one peninsula, Peña Blanca, were 56% of those on BCI, and those in the other (Gigante) were 42% of island rates. Statistical tests (Glanz, in press) indicate that these differences were due largely to lower agouti densities on both mainland sites than on BCI and to lower squirrel densities on Gigante. Howler monkeys were rarer on the mainland, whereas *Cebus*

TABLE 16.9 Sighting rates of diurnal mammals on Barro Colorado Island and two recently protected peninsulas on the mainland, 1982 and 1983–84

Genus	Barro Colorado	Gigante	Peña Blanca
Primates			
*Alouatta**	0.82	0.64	0.64
*Cebus**	0.25	0.52	0.46
*Saguinus**	0.03	0.14	0.14
Edenates			
Tamandua	0.07	0.02	0.07
Rodents			
Sciurus	1.14	0.23	0.61
Dasyprocta	1.69	0.34	1.11
Carnivores			
*Nasua**	0.25	0.02	0.07
Eira	0.01	0.02	0.04
Ungulates			
*Tayassu**	0.23	0.05	0
Mazama	0.08	0	0.04
Odocoileus	0.01	0.02	0.04
Total sighting rate	4.65	2.00	2.71
Total km censused	268	44	28

Note: Sighting rates given as individuals/km (or groups/km for * species), calculated over all km censused in both sample periods

monkeys and tamarins were more abundant, differences that could be related to habitat. Sighting rates of all primates combined were similar on BCI and the two mainland sites.

From these comparisons, I conclude that the general rarity of mammals noticed by casual observers at many forested sites in central Panama is due more to unregulated hunting than to biological differences between sites. Where hunting can be greatly reduced, the mammal species at such sites can return to density levels comparable to BCI. The lower agouti numbers on Gigante and Peña Blanca, however, are consistent with the predation hypothesis of faunal imbalance on BCI. That is, if large cats are still present there, they could be reducing agouti numbers below those seen on BCI. The lowered squirrel densities at Gigante remain unexplained. Perhaps productivity of hard nuts is lower there, or perhaps predators on young in the nest are more abundant.

TABLE 16.10 Biomass of frugivorous mammals (including certain omnivorous groups) at three Neotropical sites

Group	Guatopo, Venezuela (kg/km²)	Cocha Cashu, Peru (kg/km²)		Barro Colorado, Panama (kg/km²)		
		1983	1985	1973	1977	1982
Marsupials	62	60	38	110	67	35
Primates	167	650	650	421	510	583
Procyonids	18	110	40	120	112	118
Rodents: large[1]	270	168	59	454	520	339
small[2]	84	114	65	186	155	57
Peccaries	34	230	168	153	215	236
TOTAL	635	1,332	1,020	1,444	1,579	1,368

Sources: Modified from Terborgh (1983: 29). Guatopo data calculated from Eisenberg et al. (1979), Cocha Cashu 1983 from Terborgh (1983), Cocha Cashu 1985 calculated from densitities in Emmons (1987), BCI 1973 from Eisenberg and Thorington (1973), BCI 1982 from Glanz (1982), BCI 1982 from Glanz's calculated densities from that year.
Notes: [1] Includes agoutis, acouchis, capybaras, and pacas
[2] Includes rats, mice, and squirrels

Biomass Comparisons and Variations

Terborgh (1983) compared the biomass of frugivores (including partially frugivorous omnivores, such as cricetid rodents and procyonids) at three Neotropical sites (Guatopo, Cocha Cashu, and BCI, with the BCI estimates based on Eisenberg and Thorington 1973). I present here (table 16.10) both his biomass calculations with revised estimates for BCI based on my 1977–78 and 1982 census results and revised estimates for certain taxa at Cocha Cashu from Emmons (1987). The 1977–78 data for BCI are from years of high marsupial and rodent densities, and the 1982 data show much lower densities of these groups. Emmons' (1987) figures also reflect, in part, variation between years (Emmons, p.c.). Terborgh (1983) found frugivore biomass at Guatopo to be about one-half that of the two mature-forest sites, but Cocha Cashu and BCI were very similar in total biomass of the frugivores analyzed (see table 16.10 for additional support for the similarity in biomass noted by Terborgh). Although Cocha Cashu has a greater biomass of primates than BCI, it has fewer large rodents (agoutis, acouchis, pacas, and capybaras). The differences in biomass of these groups virtually compensate each other, according to Terborgh (1973), a result that could be fortuitous or could suggest competition among frugivores. These patterns are similar whether one uses the high 1977–78 figures for BCI or the lower (by 15%) values of 1982, which followed years of poor fruit crops. Note, however, that Emmons's (1987) densities lower the Cocha Cashu biomass total by about 30% but that the relative rankings of groups remains unchanged. This analysis suggests that, though some frugivorous groups (such as marsupials and rodents on BCI) may vary greatly from year to year, total biomass appears to vary proportionately less due to the less variable, long-lived groups (such as primates on BCI) and that site-specific differences among dominant groups (large rodents on BCI and primates at Cocha Cashu) persist between "good" and "bad" years.

Total mammalian biomass estimates are available for few Neotropical sites, largely because all taxa have not been studied at each site. Eisenberg and Thorington (1973) estimated total biomass of mammals on BCI at 4,431 kg/km^2. From the 1977–78 densities in Glanz (1982), I calculate a total biomass of 3,839 kg/km^2. If these densities are adjusted to the generally lower 1982 figures, the total biomass becomes 3,361 kg/km^2. For comparison, mammal biomass at Guatopo is 1,489 kg/km^2 (Eisenberg, 1979). I stress that BCI biomasses are highly dependent upon sloth density estimates, which may comprise 40–50% of the total. If the sloth figures are too high, as I have suggested above, conclusions could be significantly altered. Furthermore, I have sketchy data that suggest that sloths may vary greatly over several years. While conducting behavioral watches of squirrels, I have been able to scan 0.1-ha plots of forest canopy intensively for 15–45 minutes while the focal squirrel was resting. In 1977–78 I located sloths with relative ease, finding 8 animals in the 22 plots for which I have notes. From 1981 to 1984 I recorded checking 31 such plots and found only 1 sloth. If this 90% decrease in detection rate represents a 90% decrease in sloth biomass, total mammalian biomass on BCI for 1982 would be as low as 2,614 kg/km^2. Others have noticed die-offs of sloths on BCI before (Enders, unpubl. ms; K. Milton, p.c.; Foster 1982), but no quantitative data from such episodes have been published. Clearly, close monitoring and accurate censusing of sloths at this and other sites would be helpful.

The above comparisons of mammal numbers at different sites answer few questions and pose many others. Nevertheless, several general conclusions on whether the mammal community on Barro Colorado is unusual are worth stating. First, the densities of marsupials on BCI are similar to those at other Neotropical sites and therefore not "abnormal." The primate community is unusual only in the the same ways as other sites that have few primate species in mature forest habitat, in that howler monkey densities are high and other species of younger forest (such as tamarins) are rare. Sloth densities on BCI are high but similar to those recorded in French Guiana. Current density estimates for BCI are still speculative, however, and better ways of censusing sloths are needed. Agouti and paca numbers are also high, and the lack of predation by large cats on BCI appears to be the best available explanation for the differences in densities between that island and uninhabited sites like Cocha Cashu. Nevertheless, these large rodents are often poached at some protected sites and hunted heavily at many unprotected sites. Densities of small nocturnal rodents on BCI are comparable to other sites, but squirrel densities are relatively high. After studying squirrels on BCI for ten years, I still find this puzzling. Site differences in abundance of their preferred hard, nutlike foods may explain some differences in squirrel densities, but probably not all. Raptor abundance may also be important, but not raptor species composition (table 16.2). The impact of nest predators like *Cebus* monkeys and arboreal snakes on tropical squirrel populations needs further study. Large cats are absent from BCI, but ocelots are present. Densities of the small, omnivorous Carnivora range from normal (tayra) to high (coati). Among the ungulates, white-lipped peccaries are extinct, but other species (collared peccary, brocket deer, and tapir) occur at unsurprising densities.

These patterns thus can be used to evaluate certain of the faunal imbalance hypotheses mentioned earlier. First, the isolation and small size of BCI are probably responsible for most extinctions of large species, whereas habitat change explains most extinctions of small species. Habitat change and vegetational succession are also implicated in the success of certain species, such as howler monkeys, and declines in others, such as tamarins. Density compensation following extinctions may also be involved in the success of some species. Lack of top predators, I believe, is less important than many suggest. It explains high densities of agoutis and pacas on BCI but few other patterns. Lack of large avian predators on BCI may release sloths from one mortality agent, but the climatic extremes at Cocha Cashu are more likely to explain lower sloth densities there than on BCI. Barro Colorado is a productive site in terms of soils (Foster and Brokaw 1982) and fruit yields (Terborgh 1983), but more work is needed on the availability of specific food types and fruit abundances during lean seasons. Finally, in comparing sites throughout the Neotropics, human impacts, in terms of unregulated hunting, poaching, and wariness imposed by humans, may be more important than any biological factor in producing the scarcity of mammals one sees.

REFERENCES

Allee, W. C. and M. H. Allee. 1925. *Jungle Island*. Rand McNally, Chicago.
Bierregaard, R. O., Jr. 1984. Observations of the nesting biology of the Guiana crested eagle (*Morphnus guianensis*). Wilson Bull. 96: 1–5.
Brockie, R. E. and A. Moeed. 1986. Animal biomass in a New Zealand forest compared with other parts of the world. Oecolgia 70: 24–34.
Cant, J. G. H. 1977. A census of the agouti (*Dasyprocta punctata*) in seasonally dry forest at Tikal, Guatemala, with some comments on strip censusing. J. Mamm. 58: 688–690.
Chapman, F. M. 1929. *My Tropical Air Castle*. D. Appleton, New York.
Charles-Dominique, P., M. Atramentowicz, M. Charles-Dominique, H. Gérard, A. Hladik, C. M. Hladik, and M. F. Prévost. 1981. Les mammifères frugivores arboricoles nocturnes d'une forêt guyanaise: inter-relations plantes-animaux. Rev. Ecol. (Terre et Vie) 35: 341–435.
Crockett, C. 1985. Population studies of red howler monkeys (*Alouatta seniculus*). Nat. Geog. Res. 1: 264–273.
DeSteven, D. and F. E. Putz. 1984. Impact of mammals on early recruitment of a tropical canopy tree, *Dipteryx panamensis*, in Panama. Oikos 43: 207–216.
Eisenberg, J. F. 1979. Habitat, economy, and society: some correlations and hypotheses for Neotropical primates. In I. S. Bernstein and E. O. Smith (eds.), Primate Ecology and Human Origins. Garland Press, New York, pp. 215–262.
———. 1981. *The Mammalian Radiations*. Univ. of Chicago Press, Chicago.
Eisenberg, J. F., M. A. O'Connell, and P. V. August. 1979. Density, productivity, and distribution of mammals in two Venezuelan habitats. In J. F. Eisenberg (ed.), *Vertebrate Ecology in the Northern Neotropics*. Smithsonian Inst. Press, Washington, D.C., pp. 187–207.
Eisenberg, J. F. and R. W. Thorington, Jr. 1973. A preliminary analysis of a Neotropical mammal fauna. Biotropica 5: 150–161.
Emmons, L. H. 1984. Geographic variation in densities and diversities of non-flying mammals in Amazonia. Biotropica 16: 210–222.
———. 1987. Comparative feeding ecology of felids in a Neotropical rainforest. Behav. Ecol. Sociobiol. 20: 271–283.
———. 1988. A field study of ocelots (*Felis pardalis*) in Peru. Rev. Ecol. (Terre et Vie) 43: 133–157.
Enders, R. K. 1935. Mammalian life histories from Barro Colorado Island, Panama. Bull. Mus. Comp. Zool., Harvard 78: 383–502.
———. 1939. Changes observed in the mammal fauna of Barro Colorado Island, 1929–1937. Ecol. 20: 104–106.
Fleming, T. H. 1971. Population ecology of three species of Neotropical rodents. Misc. Publ., Mus. Zool., Univ. Mich. 143: 1–77
Foster, R. B. 1982. Famine on Barro Colorado Island. In E. G. Leigh, Jr., A. S. Rand, and D. M. Windsor (eds.), *The Ecology of a Tropical Forest: Seasonal Rhythms and Long-Term Changes*. Smithsonian Inst. Press, Washington, D.C., pp. 201–212.
Foster, R. B. and N. V. L. Brokaw. 1982. Structure and history of the vegetation of Barro Colorado Island. In E. G. Leigh, Jr., A. S. Rand, and D. M. Windsor

(eds.), *The Ecology of a Tropical Forest: Seasonal Rhythms and Long-Term Changes*. Smithsonian Inst. Press, Washington, D.C. pp. 67–81.

Fowler, J. M. and J. B. Cope. 1964. Notes on the harpy eagle in British Guiana. Auk 81: 257–273.

Freese, C. H., P. Heltne, N. Castro R., and G. Whitesides. 1982. Patterns and determinants of monkey densities in Peru and Bolivia, with notes on distributions. Int. J. Primatol. 3: 53–89.

Giacalone, J., W. E. Glanz, and E. G. Leigh, Jr. 1989. Fluctuaciones poblacionales a largo plazo de *Sciurus granatensis*. In E. G. Leigh (ed.), *Ecología de un bosque tropical*. Smithsonian Tropical Research Inst, Balboa, Panama.

Glanz, W. E. 1984. Food and habitat use by two sympatric *Sciurus* species in central Panama. J. Mamm. 65: 342–347.

———. 1982. The terrestrial mammal fauna of Barro Colorado Island: censuses and long-term changes. In E. G. Leigh, Jr., A. S. Rand, and D. M. Windsor (eds.), *The Ecology of a Tropical Forest: Seasonal Rhythms and Long-Term Changes*. Smithsonian Inst. Press, Washington, D.C., pp. 455–468.

———. In press. Mammalian densities at protected vs. hunted sites in central Panama. In J. Robinson and K. Redford (eds.), *Neotropical Wildlife: Use and Conservation*. Univ. of Chicago Press, Chicago.

Glanz, W. E., R. W. Thorington, Jr., J. Giacalone-Madden, and L. R. Heaney. 1982. Seasonal food use and demographic trends in *Sciurus granatensis*. In E. G. Leigh, Jr., A. S. Rand, and D. M. Windsor (eds.), *The Ecology of a Tropical Forest: Seasonal Rhythms and Long-Term Changes*. Smithsonian Inst. Press, Washington, D.C., pp. 239–252.

Gliwicz, J. 1973. A short characteristics of a population of *Proechimys semispinosus* (Tomes, 1860)—a rodent species of the tropical rain forest. Bull. Acad. Pol. Sci. Biol. 21: 413–418.

Guillotin, M. 1982. Place de *Proechimys cuvieri* (Rodentia, Echimyidae) dans les peuplements micromammaliens terrestres de la forêt guyanaise. Mammalia 46: 299–318.

Handley, C. O., Jr. 1966. Checklist of the mammals of Panama. In R. L. Wenzel and V. J. Tipton (eds.), *Ectoparasites of Panama*. Field Mus. Nat. Hist., Chicago, pp. 753–795.

Heltne, P. G., D. C. Turner, and N. J. Scott. 1976. Comparison of census data on *Alouatta palliata* from Costa Rica and Panama. In R. W. Thorington, Jr. and P. G. Heltne (eds.), *Neotropical Primates: Field Studies and Conservation*. Nat. Acad. Sci., Washington, D.C., pp. 10–19.

Hendrichs, H. 1977. Untersuchungen zur Säugetierfauna in einem palaeotropischen und einem neotropischen Bergregenwaldgebiet. Säugetierk. Mitteil. 25: 214–225.

Honacki, J. H., K. E. Kinman, and J. W. Koeppl (eds.). 1982. *Mammal Species of the World*. Allen Press, Lawrence, Kans.

Karr, J. R. 1982. Avian extinction on Barro Colorado Island, Panama: a reassessment. Amer. Nat. 119: 220–239.

Kaufmann, J. H. 1962. Ecology and social behavior of the coati, *Nasua narica* on Barro Colorado Island, Panama. Univ. Calif. Publ. Zool. 60: 95–222.

Kiltie, R. A. and J. Terborgh. 1983. Observations on the behavior of rain forest peccaries in Peru: why do white-lipped peccaries form herds? Z. Tierpsych. 62: 241–255.

Lyon, B. and A. Kuhnigk. 1985. Observations on nesting ornate hawk-eagles in Guatemala. Wilson Bull. 97: 141–147.

MacArthur, R. H. 1972. *Geographical Ecology*. Harper and Row, New York.

Malcolm, J. R. (This volume). Estimation of mammalian densities in continuous forest north of Manaus.

McNab, B. K. 1978. Energetics of arboreal folivores: physiological problems and ecological consequences of feeding on an ubiquitous food supply. In G. G. Montgomery (ed.), *The Ecology of Arboreal Folivores*. Smithsonian Inst. Press, Washington, D.C., pp. 153–162.

Milton, K. 1980. *The Foraging Strategy of Howler Monkeys: A Study in Primate Economics*. Columbia Univ. Press, New York.

Montgomery, G. G. and M. E. Sunquist. 1975. Impact of sloths on Neotropical forest energy flow and nutrient cycling. In F. Golley and E. Medina (eds.), *Tropical Ecological Systems*. Springer-Verlag, New York, pp. 69–98.

———. 1978. Habitat selection and use by two-toed and three-toed sloths. In G. G. Montgomery (ed.), *The Ecology of Arboreal Folivores*. Smithsonian Inst. Press, Washington, D.C., pp. 329–359.

National Research Council. 1981. *Techniques for the Study of Primate Population Ecology*. Nat. Acad. Press, Washington, D.C.

Patton, J. L., B. Berlin, and E. A. Berlin. 1982. Aboriginal perspectives of a mammal community in Amazonian Peru: knowledge and utilization patterns among Aguaruna Jivaro. In M. A. Mares and H. H. Genoways (eds.), *Mammalian Biology in South America*. Spec. Publ. 6, Pymatuning Lab. Ecol., Univ. Pittsburgh, pp. 129–142.

Rabinowitz, A. R. and B. G. Nottingham, Jr. 1986. Ecology and behaviour of the jaguar (*Pantthera once*) in Belize, Central America. J. Zool., Lond. 210: 149–159.

Rettig, N. L. 1978. Breeding behavior of the harpy eagle (*Harpia harpyja*). Auk 95: 629–643.

Schaller, G. B. 1983. Mammals and their biomass on a Brazilian ranch Arq. Zool., Mus. Zool., Univ São Paulo 31: 1–36.

Schaller, G. B. and P. G. Crawshaw, Jr. 1980. Movement patterns of jaguar. Biotropica 12: 161–168.

Schaller, G. B. and J. Vasconcelos. 1978. Jaguar predation on capybara. Z. Saugetierk. 43: 296–301.

Smith, N. G. 1970. Nesting of king vulture and black hawk-eagle in Panama. Condor 72: 247–248.

Smythe, N. 1970. Relationships between fruiting seasons and seed dispersal methods in a Neotropical forest. Amer. Nat. 104: 25–35.

———. 1978. The natural history of the Central American agouti, *Dasyproctta punctata*. Smithsonian Contrib. Zool. 257: 1–52.

Streilein, K. E. 1982. Behavior, ecology, and distribution of the South American marsupials. In M. A. Mares & H. H. Genoways (eds.), *Mammalian Biology in South America*. Spec. Publ. 6, Pymatuning Lab. Ecol., Univ. Pittsburgh, pp. 231–250.

Terborgh, J. 1983. *Five New World Primates: A Study in Comparative Ecology.* Princeton Univ. Press, Princeton, N.J.

———. 1974. Preservation of natural diversity: the problem of extinction-prone species. BioScience 24: 715–722.

Terborgh, J. and B. Winter. 1980. Some causes of extinction. In M. E. Soulé and B. A. Wilcox (eds.), *Conservation Biology: An Ecological-Evolutionary Perspective*. Sinauer, Sunderland, Mass., pp. 119–133.

Terborgh, J., J. W. Fitzpatrick, and L. Emmons. 1984. An annotated checklist of bird and mammal species of Cocha Cashu Biological Station, Manu National Park, Peru. Fieldiana (Zool.), n.s., 21: 1–29.

Thiollay, J.-M. 1985. Composition of falconiform communities along successional gradients from primary rainforest to secondary habitats. In I. Newton and R. Chancellor (eds.), *Conservation Studies on Raptors*. ICBP Tech. Publ. 5, Cambridge, pp. 181–190.

Walker, P. L. and J. G. H. Cant. 1977. A population survey of kinkajous (*Potos flavus*) in a seasonally dry tropical forest. J. Mamm. 58: 100–102.

Walsh, J. and R. Gannon. 1967. *Time Is Short and the Water Rises*. T. Nelson, Camden, N.J.

Willis, E. O. 1974. Populations and extinctions of birds on Barro Colorado Island, Panama. Ecol. Monogr. 44: 153–169.

Wilson, D. E. 1983. Checklist of mammals. In D. H. Janzen (ed.), *Costa Rican Natural History*. Univ. of Chicago Press, Chicago, pp. 443–447.

APPENDIX 16.1. Nonflying Mammals of Barro Colorado Island

This list is based primarily on "Checklist of Mammals of Barro Colorado Island," by C. Koford (1957), revised by N. Smythe, J. Hayden, and F. Bonaccorso (1973), mimeo, Smithsonian Tropical Research Institute. Annotations by Glanz. Abbreviations: (E) = probably extinct; (E?) = possibly extinct, no recent records; (E1) = probably extinct as population, one recent sighting; (AI) = accidental introduction; (RI) = formerly extinct, reintroduced and established; (RC) = recent colonization.

MARSUPIALIA	
Didelphidae	
Caluromys derbianus	Wooly opossum
Marmosa robinsoni	Brown murine opossum
Philander opossum	Four-eyed opossum
Metachirus nudicaudatus	Brown-masked opossum
Didelphis marsupialis	Common opossum
Chironectes minimus	Water opossum
PRIMATES	
Cebidae	
Aotus trivirgatus	Night monkey
Ateles geoffroy	Spider monkey (RI)
Alouatta palliata	Howler monkey
Cebus capucinus	White-faced monkey
Callithricidae	
Saguinus oedipus	Tamarin
EDENTATA	
Myrmecophagidae	
Tamandua mexicana	Vested anteater
Cyclopes didactylus	Silky anteater
Bradypodidae	
Bradypus variegatus	Three-toed sloth
Choloepidae	
Choloepus hoffmanni	Two-toed sloth
Dasypodidae	
Dasypus novemcinctus	Nine-banded armadillo
Cabassous centrails	Five-toed armadillo
LAGOMORPHA	
Leporidae	
Sylvilagus brasiliensis	Forest rabbit
RODENTIA	
Sciuridae	
Sciurus granatensis	Red-tailed squirrel
Microsciurus alfari	Pygmy squirrel (E)
Heteromyidae	
Heteromys desmarestianus	Spiny pocket mouse

Cricetidae	
Oryzomys fulvescens	Pygmy rice rat (E)
O. capito	Talamancan rice rat
O. concolor	Climbing rice rat
O. bicolor	Climbing mouse
Sigmodon hispidus	Cotton rat (E)
Zygodontomy brevicauda	Cane rat (E?)
Muridae	
Mus musculus	House mouse (AI, E?)
Rattus rattus	Black rat (AI)
Erethizontidae	
Coendou rothschiloi	Porcupine
Hydrochaeridae	
Hydrochaeris hydrochaeris	Capybara (RC)
Agoutidae	
Agouti paca	Paca
Dasyproctidae	
Dasyprocta punctata	Agouti
Echimyidae	
Proechimys semispinosus	Spiny rat
Diplomys labilis	Tree rat
CARNIVORA	
Procyonidae	
Procyon lotor	Raccoon (E?)
P. cancrivorous	Crab-eating raccoon (E?)
Nasua nasua	Coati
Potos flavus	Kinkajou
Bassaricyon gabbii	Olingo (E1)
Mustelidae	
Eira barbara	Tayra
Lutra annectens	Otter
Felidae	
Panthera onca	Jaguar (E1)
Felis concolor	Puma (E)
F. pardalis	Ocelot
F. yagouaroundi	Jaguarundi
F. wiedii	Margay (E1)
PERISSODACTYLA	
Tapiridae	
Tapirus bairdii	Tapir (RI)
ARTIODACTYLA	
Tayassuidae	
Tayassu pecari	White-lipped peccary (E)
T. tajacu	Collared peccary
Cervidae	
Mazama americana	Red brocket deer
Odocoileus virginianus	White-tailed deer

17

Ecological Structure of the Nonflying Mammal Community at Cocha Cashu Biological Station, Manu National Park, Peru

CHARLES H. JANSON AND LOUISE H. EMMONS

There are few lists of the mammal faunas of single localities in lowland moist tropical forests (e.g., Eisenberg and Thorington 1973; Pine 1973; Eisenberg et al. 1979; Glanz 1982, and this volume; Patton et al. 1979; Crespo 1982; Wilson, this volume). Unlike birds, many mammal species are cryptic and nocturnal and cannot be identified with confidence in the field. Some cannot even be identified in the hand, and construction of a nearly complete list thus requires a long and intensive collection program, with preservation of vouchers (Patton et al. 1979; Wilson, this volume). As a result, all lists of Amazonian mammal species for specific sites that might be defined as single communities are still incomplete. The only accurate data on comparative ecology for most tropical mammals are derived from intense effort studying individual species (e.g., Emmons 1981; Emmons et al. 1983; Terborgh 1983). At Cocha Cashu Biological Station (CC), Manu National Park, Peru, researchers have compiled quantitative data on population densities and feeding ecology of many of the more common or conspicuous large mammals in the community (Milton 1980; Janson et al. 1981; Kiltie 1981; Emmons 1982; Kiltie 1982; Janson 1984; Kiltie and Terborgh 1983; Terborgh 1983; Terborgh and Janson 1983; Emmons 1984; Terborgh et al. 1984; Janson 1985; Janson and Terborgh 1986; Wright 1985; Goldizen and Terborgh 1986; Janson 1986; McFarland 1986; Terborgh 1986; White 1986; Wright 1986; Emmons

We wish to thank the Dirección General Forestal y de Fauna for their permission to work in the Manu National Park. Our research was supported by grants to CHJ from the National Science Foundation (BNS-8007381), and to LHE from Wildlife Conservation International and World Wildlife Fund-U.S. This is contribution 678 from the Department of Ecology and Evolution, State University of New York, Stony Brook.

1987; Terborgh and Stern 1987; Brecht, unpubl. data). In this chapter, we synthesize these data to provide a detailed description of this Neotropical mammal community. Although many small mammals remain unstudied at CC, we hope these data will give a useful baseline for comparison with other sites.

Cocha Cashu may offer a vision of what many mammal communities in Amazonia were like before the intrusion of civilized people. Unlike most study areas, this site has remained free of systematic hunting since 1970, when researchers began to occupy the area (previously the area was probably used sporadically by Amerindian hunters). Similar highly productive floodplain areas elsewhere in Amazonia have always been the first to be colonized and feel the impact of human settlement (Meggers 1971). Yet at CC there are still high densities of large animals that have been greatly reduced elsewhere by hunting (jaguar, ocelot, giant otter, peccaries). In this paper, we focus on three factors that help explain the high diversity and pattern of abundances of mammal species at CC: high plant productivity, high habitat diversity, and freedom from hunting.

Study Site and Methods

Cocha Cashu is near the edge of Manu Park at 11°22′S, 71°22′W. The area comprises undisturbed lowland tropical forest with a pronounced three-month dry season (June–August) and annual rainfall close to 2,000 mm (Terborgh 1983; Terborgh et al. 1986). A few conspicuous plant species lose their leaves during the dry season, but most species retain their foliage. Seasonal variation in temperature is slight, with the dry season averaging about 1°C less than the rainy season; occasional strong Patagonian winter storm systems, however, can keep maximum daily temperatures below 18°C for several consecutive days in the dry season (Terborgh 1983).

Within 2 km^2 the CC study area encompasses at least six distinct successional plant associations caused by the meandering Manu River (Terborgh 1983; Foster et al. 1986). Over 1,200 plant species have been collected on floodplain soils near CC (Foster, unpubl. data), of which over half produce fleshy fruits consumed primarily by mammals and birds (Janson 1983).

Each researcher chose methods suitable to his or her study. Arboreal mammals with relatively small home ranges (such as most primate species) were followed from dawn to dusk for several days to weeks (Janson 1984; Terborgh 1983; Terborgh and Janson 1983; Wright 1985; McFarland 1986). During such follows, data on food intake were taken either systematically at fixed intervals or opportunistically whenever animals entered fruit trees. Data on ranging patterns were recorded by pacing distances and compass directions as groups moved; researchers also noted the locations and times of crossing of mapped trails in the study area. Population densities were determined by dividing the number of individuals in a group by the area of their home range, multiplied by the extent of overlap between adjacent home ranges (Janson and Terborgh 1986).

For terrestrial species with relatively small home ranges (most rodents, ocelot), data on ranging behavior were obtained by systematic trapping and tracking individuals with radio-collars (Emmons 1982, 1984, 1987). Food habits were determined in part by direct observation, in part by stomach contents of animals collected at park boundaries, and in part by analysis of scats (Emmons 1982, 1987). In a few cases, diets recorded in other studies on the same or similar species are used in table 17.1: diets of didelphids except *Caluromysiops* are from Charles-Dominique et al. (1981) and for *Oryzomys capito* from Guillotin (1982). Population densities were calculated from ranging data when possible or from repeated strip-census surveys using the King estimator (Emmons 1984; Janson and Terborgh 1986).

For species with home ranges extending well beyond the 6 km^2 of the trail system at CC (white-lipped peccaries, jaguar, puma) or vagrants into the study area (coati), ranging data are necessarily fragmentary. Ranging patterns were estimated from the temporal pattern of use of the trail system (based on footprints, visual sightings, and scats; Kiltie and Terborgh 1983; Emmons 1987). Species not resident in the study area are denoted vagrants in table 17.1. Population densities, when estimated, were based on ranging data. Diet information came from direct observations of feeding animals, stomach contents of animals collected at the park boundary, and analysis of scats.

Body masses for many species were taken from live animals captured on the study site and the data reported represent the mean of all individuals weighed (including juveniles). The data on population density and body mass for rodents, felids, and nocturnal mammals are taken from Emmons (1982, 1984, 1987), whereas those of primates and miscellaneous species are from Terborgh (1983), except for *Tapirus terrestris* body weights from Hames (1979). Although Terborgh (1983) also reports densities of rodents and nocturnal mammals (based on Emmons's unpublished results), the data given here reflect a larger data set and are thus likely to be more accurate.

TABLE 17.1 Species, ecological data, and population densities at Cocha Cashu (modified from Terborgh et al. 1984)

Species[1]	Habitats[2]	Canopy height[3]	Trophic status[4]	Activity cycle[5]	Population density[6]	Body mass (kg)	Biomass (kg/km^2)	Energy intake (kJ/day/ind.)[7]
Marsupialia								
Didelphidae								
Metachirus nudicaudatus	Fh	T	I,F	N	0.03/km,12	0.42	5.04	464.9
Micoureus cinerea	Fh	Sc,U,C	I,F,N	N	0.04/km,10	0.15	1.5	239.6
Marmosops noctivaga	Fh,Fo	U,V,T	I,F,N	N	0.04/km,15	0.08	1.2	159.8
Caluromysiops irrupta	Fh	C,Sc	F,N,I?	N	0.03/km,10	0.25	2.5	332.9
Caluromys lanatus	Fh	C,Sc	—	N	R	—	—	—
Philander opossum	Lm,Fo,Fh	T,U,V	I,F	N	0.05/km,25	0.27	6.75	349.8
Didelphis marsupialis	Fh,Fo	T,Sc,C	I,F	N	0.15/km,55	1.0	55.0	812.8
Glironia sp.	—	C	—	N	R	—	—	—
Primates								
Cebidae								
Aotus trivirgatus	Fh,Ft	F,Sc,U	F,I,L,N	N	40	0.7	28.0	571.4
Callicebus moloch	Lm,Rm,Fh	Sc,C,U	F,U,L,I	D	24	0.7	16.8	571.4
Pithecia sp. (*monachus*?)	Fh	C,Sc	F,?	D	V	1.0	—	783.4
Alouatta seniculus	Fh,Ft	C,Sc,F	F,L,U	D	30	6.0	180.0	3825.3
Cebus apella	All	U,Sc,F,C	F,I,S,N,P,V	D	40	2.6	104.0	1825.0
C. albifrons	All	Sc,F,U,C,T	F,I,S,N,V	D	35	2.4	84.0	1700.1
Saimiri sciureus	All	Sc,V,C,F	F,I,N	D	60	0.8	48.0	643.0
Ateles paniscus	Fh,Ft	C,F	F,L,N	D	25	7.0	175.0	4384.4
Lagothrix lagothricha	Fh	C,F	F,L	D	1(V?)	8.0	8.0	4934.4
Callitrichidae								
Cebuella pygmaea	Lm,Rm,Fs	U,V,Sc	G,I,N,F	D	5	0.1	0.5	102.1
Saguinus fuscicollis	Fh,Fo	V,Sc,C	F,I,N,G	D	16	0.35	5.6	309.4
S. imperator	Fh,Fo	V,U,Sc	F,I,N,G	D	12	0.4	4.8	348.2
+*Callimico goeldii*	B?	V,U?	—	D	V	0.5	—	424.2

TABLE 17.1 (*Continued*)

Species[1]	Habitats[2]	Canopy height[3]	Trophic status[4]	Activity cycle[5]	Population density[6]	Body mass (kg)	Biomass (kg/km^2)	Energy intake (kJ/day/ind.)[7]
Edentata								
Myrmecophagidae								
Myrmecophaga tridactyla	Fh,Ft	T	A	D,N	R	—	—	—
Tamandua tetradactyla	Fh,Ft	T,Sc,C	A	D,N	U	5.0	—	3255.3
Cyclopes didactylus	Ft	C	—	N?	R	—	—	—
Bradypodidae								
Bradypus variegatus	Fh	C	L	D	0.01/km	—	—	—
Megalonychidae								
* *Choloepus hoffmanni*	Fh	C	—	N	R	—	—	—
Dasypodidae								
Dasypus novemcinctus	Rm,Fh	T	—	N,D	0.03/km	—	—	—
Priodontes maximus	Fh	T	—	N,D	V	—	—	—
Lagomorpha								
Leporidae								
Sylvilagus brasiliensis	Fs,Fo	T	L	N	R-C	1.0	—	783.4
Rodentia								
Sciuridae								
* *Sciurus spadiceus*	Fh,Lm	T,U,Sc,V	S,F	D	0.45/km,25	0.6	5.0	269.5
* *S. ignitus*	Fh,Lm	U,Sc,V	S,F	D	0.04/km	0.22	—	162.0
* *S. sanborni*	Fh?	T,U?	—	D	R	—	—	—
Muridae (Cricetini)								
* *Oryzomys capito*	Fh	T	S,F,I	N	0.14/km,180 (*O. capito*, *O. nitidus*, *O. macconnelli* bracketed)	0.07	12.6	90.7
O. nitidus	Fh	T	S,I?	N				
O. macconnelli	Fh	T	S,I?	N				
* *Oligoryzomys microtis*	Fh	T,U	S,I?	N				
* *Oecomys superans*	Fh	U,Sc,C	S?	N	U	—	—	—
O. bicolor	Fh	U,Sc,C	S?	N	U	—	—	—
Oxymycterus sp.	—	—	I	N	—	—	—	—
Nectomys squamipes	Fh,Lm	T	F,I	N	R	—	—	—
* *Rhipidomys couesi*	Fh	C,Sc	—	N	R	—	—	

Erethizontidae								
Coendou prehensilis	Fh	Sc,C	L,F,S	N	R	4.5	—	748.5
Hydrochaeridae								
Hydrochaeris hydrochaeris	Lm,Rm,Z	T	L	D,N	1.6	45.0	72.0	2405.4
Dasyproctidae								
Agouti paca	Fh	T	F	N	0.08/km,3.5	8.0	28.0	1002.0
Dasyprocta variegata	Fh	T	S,F	D	5.2	4.0	20.8	705.1
Myoprocta pratti	Fh	T	S,F	D	5.3	1.5	7.95	428.8
Dinomyidae								
* *Dinomys branickii*	—	T	—	N	V	—	—	—
Echimyidae								
* *Proechimys simonsi* 2N = 32	Fh	T	M,S,F	N }	1.48/km,230	0.28	64.4	179.8
* *P. steerei* 2N = 24	Fh,Rm	T	M,S,F	N }		0.26(+)		
P. brevicauda	—	T	M,S,F	N	—	—	—	—
Echimys sp.	Fh	U,Sc	S,F	N	R	0.3	—	—
Mesomys hispidus	Fh	U,Sc	F,L,I	N	R	0.2	—	154.4
Dactylomys dactylinus	Lm,Rm,B	Sc,C,V	L	N	0.09/km	—	—	—
Carnivora								
Procyonidae								
Procyon cancrivorous	Fsm	T,W	—	N	V	—	—	—
Nasua nasua	Fh	T,Sc,C	F,I	D	0.2,V?	4.5	0.9	2965.5
Potos flavus	Fh	C,Sc	F,N	N	0.19/km,25.3	2.2	55.7	1574.1
Bassaricyon gabbii	Fh	C,Sc	F,N,I	N	0.03/km,6.8	1.0	6.8	783.4
Mustelidae								
Eira barbara	Fh,Ft,Z	T,Sc,C	F,V?	D	U	—	—	—
Galictis vittata	All	T	V	D	R	—	—	—
Lutra longicaudis	R,L	W	Fish	D,N?	V	—	—	—
Pteronura brasiliensis	L,R	W	Fish	D	<0.5	—	—	—
Felidae								
Felis concolor	Fh	T	V	N,D	0.024	29.0	0.82	15424.9
F. pardalis	All	T	V,Fish,I	N,D	0.8	0.93	0.74	734.7
F. yagouarundi	All	T	V	D,N	R	7.0	—	—
Panthera onca	All	T	V	N=D	0.035	35.0	1.18	18218.0

TABLE 17.1 (*Continued*)

Species	Habitats[2]	Canopy height[3]	Trophic status[4]	Activity cycle[5]	Population density[6]	Body mass (kg)	Biomass (kg/km²)	Energy intake (kJ/day/ind.)[7]
Perissodactyla								
Tapiridae								
Tapirus terrestris	Fs,Rm,Fh,A	T	L,F	N,D	0.5	160.0	80.0	69927.2
Artiodactyla								
Tayassuidae								
Tayassu pecari	Fh,Ft,A	T	S,L	D,N	3-V	35.0	105-<5	18218.0
T. tajacu	Fh,Fs	T	S,L	D	5.6	25.0	140.0	13526.2
Cervidae								
Mazama americana	Fh	T	L?,F	N,D	2.6	30.0	78.0	15894.7
M. gouazoubira	Fh,Ft	T	L?,F	D?	V	—	—	—

Notes: Abbreviations for each topic are listed in order of importance for a given species

[1] * = species for which skeletal material was collected either from natural kills at Cocha Cashu or by trapping at the boundary of the park. + = species seen within the park but not confirmed at Cocha Cashu.

[2] Abbreviations are as follows:

Fh = high-ground forest (*Dipteryx, Quararibea*)
Fo = forest openings, usually treefalls
Lm = lake margins
Fsm = forest stream margins
Rm = river margins
Fs = swamp forest (*Ficus trigona*)
Ft = transitional forest (*Guatteria*, *Brosimum*, *Inga*)
All = all terrestrial habitats
B = bamboo thickets
Z = *zabolo* = river-edge stands of *Gynerium* cane with *Cecropia*
L = in oxbow lake
R = in river
A = *aguajal* = permanently flooded *Mauritia* palm forest

[3] Abbreviations are as follows:
T = terrestrial
U = understory, ca. 5 meters or less
C = canopy, ca. 25 m or more
V = vines
Sc = subcanopy, ca. 5–25 meters
W = over or in water
F = in or near fruiting trees

[4] Abbreviations are as follows:
F = fruit pulp
I = invertebrates (mostly insects)
N = nectar
L = leaves
U = unripe fruit
S = seeds
P = pith and meristems
V = vertebrates (usually terrestrial)
G = gums
A = ants
M = mycorrhizal fungi
Fish = fish
? = behavior indicates category, but confirmation lacking

[5] Abbreviations are as follows:
N = nocturnal
D = diurnal

[6] Population density is given by one of several measures: a) rate of encounter/km walked during transect censuses, given as '#/km'; b) population density/km^2, given as a number with no accompanying symbol; c) crude classification of encounter rate by human observers (V = vagrant, R = rare, U = uncommon, C = common). Vagrant species are not resident in the study area, but appear on occasion. Rare species are residents, but are seen or trapped once to a few times a year even when observers are present year-round. Uncommon species are seen or trapped about once a month. Common species are recorded at least several times a month. Note that some species (such as nocturnal canopy dwellers) may be classed as rare only because they are difficult to detect.

[7] Estimates are from regressions in Nagy (1987). For marsupials, we use the regression for nonherbivorous marsupials. For rodents, we use the regression for rodents. For all other mammals, we use the regression for nonrodents.

From August 1980 through November 1981, one hundred plastic bags (0.0665 m^2 collecting area) were suspended from wire hoops 50 cm above the ground at 25-m intervals along trails in the center of the study area (Janson 1984; Terborgh et al. 1986). Each bag captured remnant fruit crops falling off canopy trees above the trap, as well as seeds naturally dispersed by arboreal fruit-eaters from distant areas. The sum of these seed sources should closely reflect the number of fruits of a given species actually produced in the forest because (1) most fruit-eating animals in the study site physically damage only a small fraction of the seeds they ingest (Janson, unpubl. data) and (2) the density of arboreal seed-eating mammals is low (see below). The plastic bags were cleaned of all contents approximately every fourteen days, and all fruits, seeds, and flowers were counted, weighed, and identified whenever possible.

Mean caloric values per wet weight of ripe fruit were estimated from data taken on 231 species of freshly collected fruit at CC (Janson, in prep.). For fruits of a known species, the wet weight of pulp per fruit was multiplied by the apparent sucrose density in the pulp (measured with a Bausch-and-Lomb temperature-corrected sucrose densitometer) times 16.76 KJ/g (appropriate for carbohydrates). Energy values derived from apparent sucrose densities correlate well with energy values based on chemical analyses of percentage soluble carbohydrates, proteins, and fats (Stiles, White, and Janson, unpubl. data). For fruits of unknown species, energy values were estimated from regressions of energy value against fruit or seed sizes of known species. The energy value (in KJ) of seeds was calculated as 18.87 times the wet weight of the seed (excluding seed coats if these represented more than 10% of the seed weight); this conversion was derived from the energy density of 16 domesticated grain and nut species (Adams 1975).

To calculate energy production per day in fruit pulp or seed, the number of seeds of a given species caught in all traps per biweekly sample period was converted to an equivalent number of whole fruits of that species. This number was then multiplied by the energy value of pulp or seed per fruit of that species to give the total energy value caught. These total energy values per species were summed across all species, and the resulting grand total energy value was divided by the total area (ha) and duration (days) of sampling for that period to produce values in KJ/ha/day. Seeds from species that ripen fruit during a restricted season were assigned to the biweekly sample period in which the seeds were caught. Seeds from species with no distinct phenological synchrony among individuals (e.g., many fig species) were divided equally among all sampling periods. The seeds of at least 110 plant species were caught in the traps during eighteen months.

Habitat-specific patterns of monthly fruit availability were derived from data on phenology, habitat association, crop size, and maturation rate for 231 fleshy-fruited species at CC (Janson, unpubl. data); for lack of adequate phenological records, 70 very rare or aseasonal species were excluded. Phenology data on 57 species were taken on 517 marked census trees by CHJ; for the remaining species, mean month of fruit maturation was estimated from opportunistic records taken by CHJ and other botanists at the site. The number of species fruiting in each month in a given habitat was weighted by the average monthly biomass production of ripe fruit per plant of those species. Thus, in each habitat, a relatively few highly productive plant species dominate the seasonal pattern of fruit production.

Relative insect abundance was estimated by collecting all insects attracted during the night to a twelve-watt ultraviolet lamp suspended 1.5 m above the forest floor 300 m from the nearest open water. The trap was set for three to four nights centered on the waning and waxing half moons (to control for nighttime illuminance). All insects caught were separated by taxonomic order, counted, measured, and weighed without drying.

Results

To date, 70 species of nonflying mammals have been identified with high certainty at CC or nearby (table 17.1). This list is almost certainly incomplete (Emmons, unpubl. data); published and museum records of the ranges of Neotropical mammals suggest that records are missing for at least 2–3 more marsupials (*Marmosa* spp.), 4–5 more rodent species, 3–4 more carnivores (felids and mustelids), 2 other armadillos, and possibly 1 primate (*Pithecia* sp.). In addition, many identifications (especially of small mammals) must remain tentative, because few of the species listed are represented by specimens from the national park or surrounding area. For this reason, we excluded any discussion of the bat fauna at Cocha Cashu, even though 24–25 species have tentatively been identified there (Terborgh et al. 1984; Emmons unpubl. data).

The estimated mammalian biomass from table 17.1 is 1,416 kg/km^2, nearly 300 kg/km^2 less than the 1,705 kg/km^2 estimated by Terborgh (1986) for nonflying mammals at CC. The difference in biomass estimates derives mostly from divergent figures for the population density of the paca (*Agouti paca*, 24/km² in Terborgh [1983] vs. 3.5/km² in Emmons [1987]), kinkajou (*Potos flavus*, 43.2/km² in Terborgh [1983] vs. our revised estimate of 25.3/km²), and tapir (*Tapirus terrestris*, 150 kg/km² in Terborgh [1986] vs. our estimate of 80 kg/km²). Although a biomass of 1,400+ kg/km² is a tentative figure (we have no data for many species in table 17.1), it is so strongly influenced by the biomass of the well-known common large mammals that it is unlikely to change appreciably with the addition of new or revised data on the population densities of small or rare species.

The mammalian biomass is unevenly divided among trophic categories. If we classify each mammal species by the major element in its diet (the first abbreviation in the trophic status column of table 17.1), the community is dominated by fruit-pulp eaters (748 kg/km²) rather than by seed-eaters (366 kg/km²), leaf-eaters (230 kg/km²), or faunivores (72 kg/km²). For this calculation, *Proechimys* spp. (Rodentia) were considered seed-eaters even though the most common items in stomach contents were fungal fruiting bodies. The figure for seed-eaters may be an overestimate, since it includes 105 kg/km² for the white-lipped peccary (*Tayassu pecari*), based on data from 1976–78, but this species has nearly vanished from CC since 1981. This trophic breakdown of biomass might change somewhat if we could divide each species in table 17.1 according to the energetic contribution of different foodstuffs to its diet; such data, however, are available for only a few species (Janson 1985).

Different mammal orders dominate different aspects of abundance at CC. The most species-rich groups of nonflying mammals are rodents (24 spp.), primates (13 spp.), carnivores (12 spp.), and marsupials (8 spp.), together accounting for more than 80% of known species. The taxa with the highest population densities are the rodents (>450 inds./km²), primates (288 inds./km²), marsupials (>127 inds./km²), and carnivores (>33 inds./km²). Finally, the orders contributing most to the total biomass are the primates (655 kg/km²), ungulates (403 kg/km², including white-lipped peccary), and rodents (211 kg/km²).

A striking characteristic of CC is the presence of relatively high densities of felids (table 17.1). These dominant terrestrial predators may affect the composition of the mammal community by their nonrandom selection of prey species. For instance, ocelots and large felids (jaguar and puma) show distinct preferences for terrestrial prey of different sizes at CC (fig. 17.1; from data in Emmons 1987). Medium-sized mammals (2–10 kg) suffer the greatest intensity of predation by felids because they are chosen with appreciable frequency by all cat species, whereas large

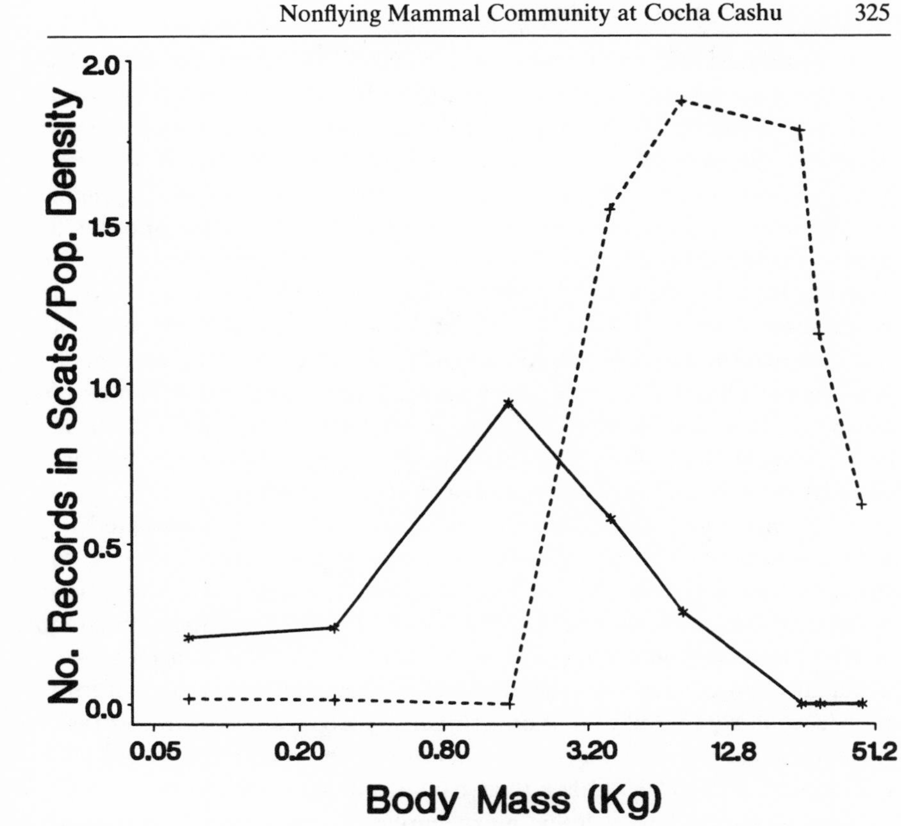

FIGURE 17.1. Relative occurrence of terrestrial prey of different sizes in the diets of felids. The *solid line* is for ocelot (*Felis pardalis*), the *dashed line* for both jaguar (*F. onca*) and puma (*F. concolor*). The prey types, in increasing order of size, are: *Oryzomys* spp., *Proechimys* spp., *Myoprocta pratti, Dasyprocta variegata, Agouti paca, Tayassu tajacu, Mazama americana*, and *Hydrochaeris hydrochaeris*.

prey are taken only by large cats and small prey only by small cats (fig. 17.1). In addition, however, small prey are eaten by many other kinds of predators (snakes, birds, and others).

In most years, the Manu River floods its banks at least once during the rainy season. In both 1977 and 1981, for example, the water level of the Manu River rose about 1.5 m above its banks. In the 1983–84 wet season, the river rose above its banks nine times. In the flat terrain of CC, such flooding inundated six of the seven major habitat types. In all three years, some river water reached the high-ground forest, but little of this habitat was flooded. Even the high-ground forest, however, is not immune from flooding. In October 1982 the Manu River flooded its banks by more than 3 m, covering virtually the entire 6 km^2 of the study area, including over 95% of the high-ground forest. Such extreme floods appear to be rare (park personnel claim that the 1982 flood was the greatest since before 1971). In addition, about 20–30% of the high-ground forest is under shallow standing water for about four months a year, reducing the habitat available for terrestrial mammals.

The forest at CC is highly productive (18.11 $\times$ 10^6 KJ/ha/year of fruit pulp, seeds, and nectar; fig. 17.2), probably because of the influx of nutrient-rich silt deposited by the annual floods of the Manu River on all low-lying habitats. The estimated mammalian standing crop biomass of 14.16 kg/ha would consume about 3.08 $\times$ 10^6 KJ/ha/year (table 17.1, from regressions in Nagy 1987). Thus, it appears that mammals annually consume at most 17% of the production of fruits and seeds at CC. This figure may be misleading, because the biomass of mammals is dominated by fruit-pulp eaters (52.8%), whereas pulp contributes only about 16% to the total energy production. In fact, the expected maximum consumption of all fruit-pulp eaters (1.82 $\times$ 10^6 KJ/ha/year) is 62% of the estimated energy production in pulp and nectar. Thus, most of the fruit pulp produced at CC is consumed by mammals (the biomass of birds that eat fruit pulp is small compared to that of mammals; Terborgh 1986). In contrast, the expected maximum energy consumption of the standing crop of seed-eating mammals (0.73 $\times$ 10^6 KJ/ha/year) is only 4.8% of the energy production in seeds. This value is more similar to consumption values for tropical folivores and leaves (1.6–4.6%; Leigh and Smythe 1978) than for temperate-zone granivores and seeds (10–50%; Odum et al. 1962).

Production of fruits is highly seasonal at CC, reaching an area-wide minimum during the dry season of June–August (fig. 17.2). The minimum observed energy production in pulp and seeds of seasonally fruiting species was 5,657.5 KJ/ha/day (September 1–15), whereas the maximum was about twenty-five–fold higher at 142,588 KJ/ha/day (February 11–25). The average energy production in pulp and seeds of aseasonally

→

FIGURE 17.2. (*A*) Seasonal variation in energy production in nectar of mammal-pollinated flowers and in pulp of fleshy fruits at Cocha Cashu. Fruits and flowers were caught in plastic bags suspended near the ground. *Solid black* indicates average production by aseasonally fruiting species, *hatched areas* show production of seasonally fruiting species, and the *clear area* is the production of nectar. (*B*) Seasonal variation in energy production in seeds. Shading means the same as in *A*, except that the *clear area* is the production of wind-dispersed seeds.

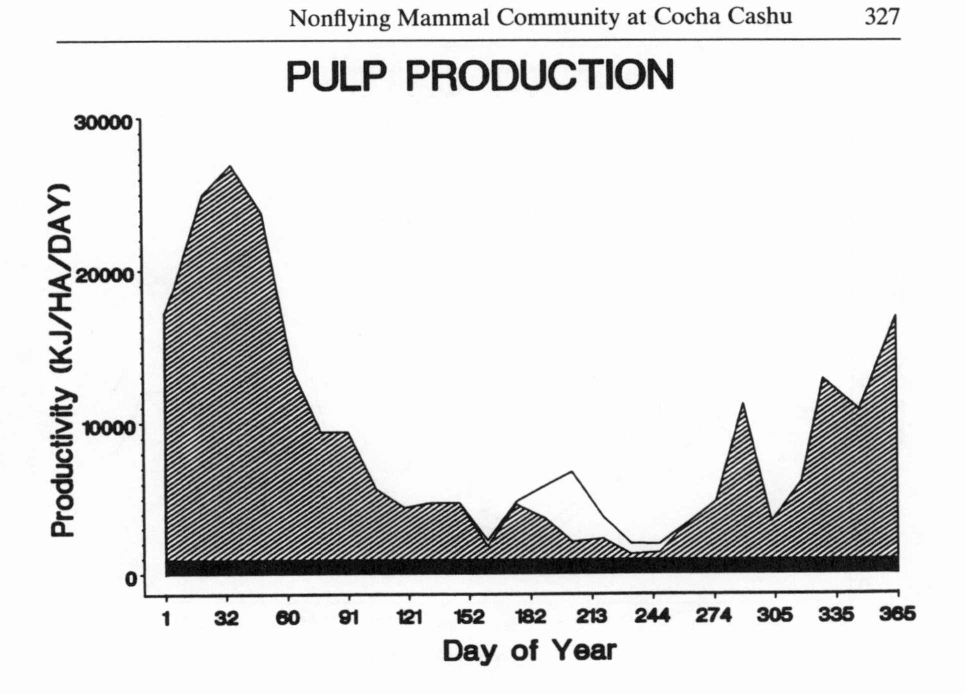
PULP PRODUCTION
30000
20000
10000
0
Productivity (KJ/HA/DAY)
1
32
60
91
121
152
182
213
244
274
305
335
365
Day of Year

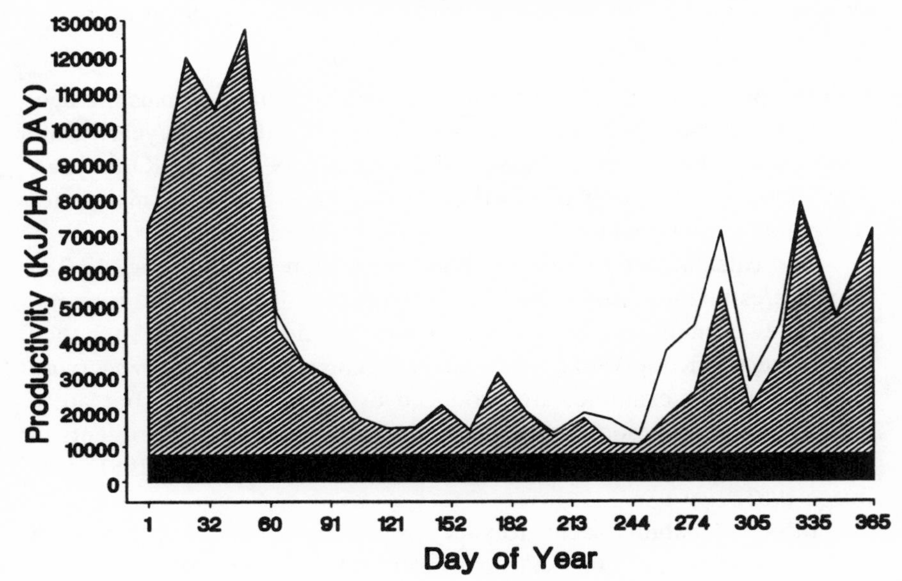
SEED PRODUCTION
130000
120000
110000
100000
90000
80000
70000
60000
50000
40000
30000
20000
10000
0
Productivity (KJ/HA/DAY)
1
32
60
91
121
152
182
213
244
274
305
335
365
Day of Year

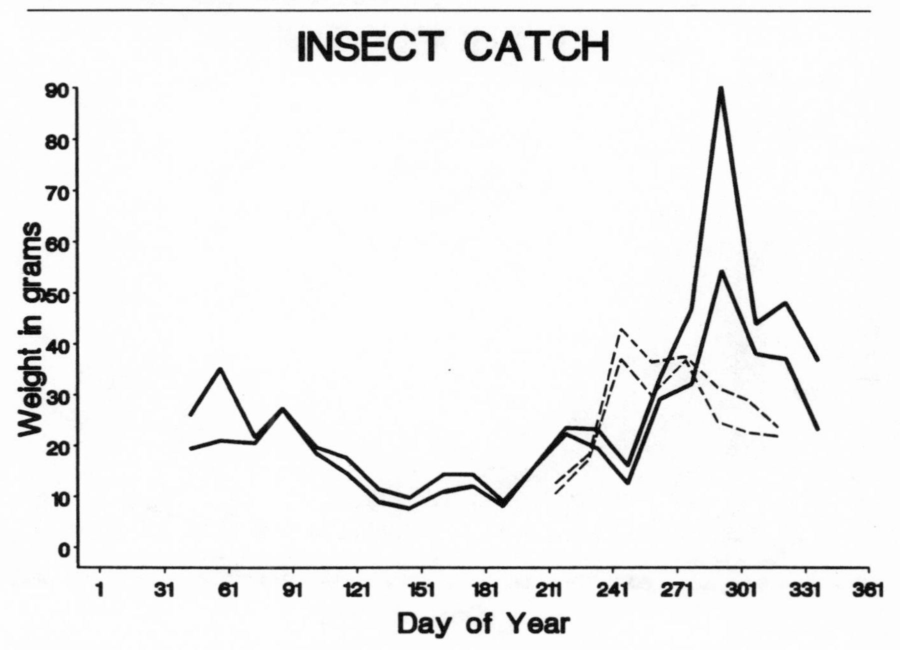

FIGURE 17.3. Seasonal variation in relative insect abundance at Cocha Cashu. Insects were collected at an ultraviolet light during 3–4 consecutive nights every two weeks. The *heavy solid line* is for 1981; the *lighter dashed line* is for 1980. The *upper line* for each year represents the maximum nightly catch per biweekly sampling period; the *lower line* is the second-highest catch per sampling period.

fruiting species was 8,678 KJ/ha/day. Altogether, fruit pulp plus the nectar of mammal-pollinated flowers provided 2.94×10^6 KJ/ha/year, while seeds of both fleshy and nonfleshy fruits provided 15.17×10^6 KJ/ha/year. The relatively low energy production in fruit pulp is due to the high water content and low fat concentration of pulp compared to seed.

The calculated pattern of energy production in seeds (fig. 17.2B) may overestimate seasonal differences in food availability to consumers. The seeds of many species remain uneaten for many months after they are produced (Kiltie 1981). The seeds of a common palm (*Astrocaryum murumuru*), for example, are produced in February–March but are a major alternative resource for some primates, peccaries, and rodents in June–July, when fruit production nears its annual low (Kiltie 1981; Terborgh 1983, 1986).

Insect availability seems to vary strongly by season, with a peak of abundance in the dry-rainy season transition (September–November, fig.

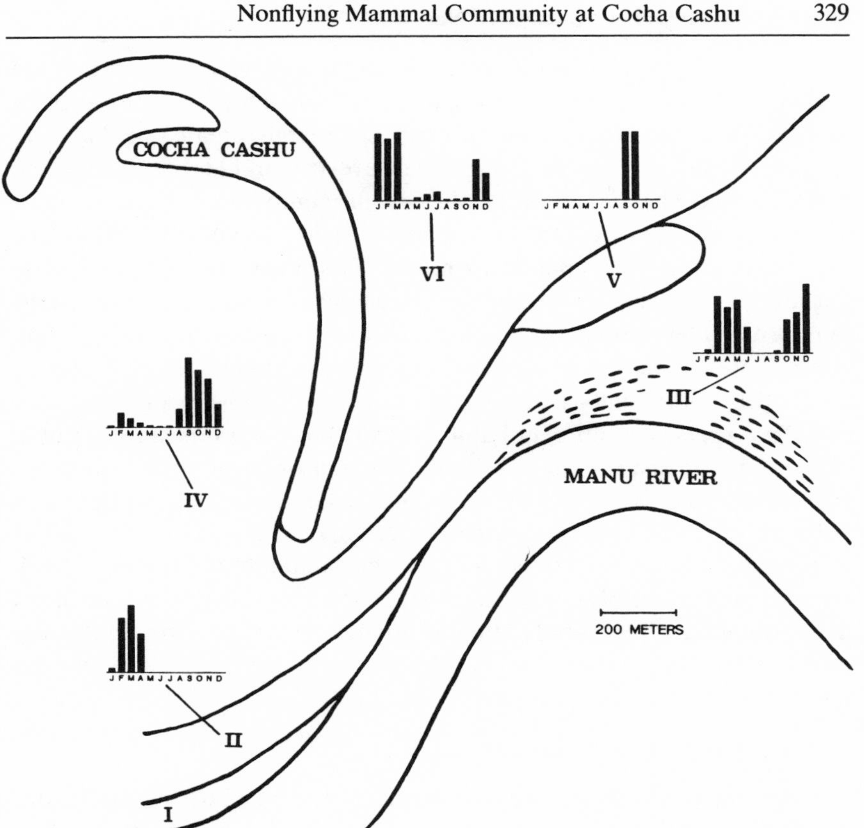

FIGURE 17.4. Spatial distribution of major plant associations at Cocha Cashu, with associated seasonal patterns of fruit production. The plant associations shown are: (I) *Tessaria-Ochroma-Gynerium* (beach); (II) *Cecropia-Ficus insipida* (levee); (III) *Allophylus-Combretum* (disturbed edges); (IV) *Guatteria-Brosimum-Inga* (seasonally flooded); (V) *Ficus trigona* (swamp); and (VI) *Dipteryx-Quararibea* (high ground). The bar graphs show the relative fruit availability by month in each habitat, estimated from data on phenology, crop size, and ripening rate of common plant species. The number of species represented in each habitat are: II = 14, III = 23, IV = 55, V = 3, VI = 66. Plant association I produces negligible amounts of fleshy fruits.

17.3). The total weight of insects caught is dominated by Orthoptera, Coleoptera, and Lepidoptera, all of which showed roughly coincident seasonal patterns of abundance (Janson, unpubl. data).

The study area at CC is subject to great abiotic disturbance. The Manu River meanders across its 10-km wide floodplain so rapidly that few parcels of land in the floodplain are older than a thousand years (Foster et al. 1986). Many low-lying remnants of former river channels are flooded seasonally, producing swamps hospitable to only a few specialized plant species. Finally, even in high-ground forest, treefalls create openings that disturb about 1–2% of the forest area every year (Terborgh, unpubl. data). A diversity of habitats in physical proximity results from this disturbance. In some places, a single transect of 700 m starting at the river's edge is sufficient to sample all six major plant associations (fig. 17.4; cf. Terborgh 1983). Even far from the river's edge, as many as three habitats can normally be sampled in a transect of 1,000 m in the floodplain: high-ground forest, swamp forest (in former river channels), and early successional forest in treefall gaps.

Plant species in different habitats at CC tend to produce ripe fruit at different times of year (fig. 17.4; see also Janson et al. 1981; Terborgh 1983). Although no one habitat produced large amounts of ripe fruit during more than five months per year, only for six weeks did all five major habitats fail to produce appreciable amounts of fruit (fig. 17.4; July–August). This period of fruit scarcity coincides with the flowering of the only common plants pollinated by nonflying mammals (fig. 17.2A; see Janson et al. 1981).

Effects of Floodplain Habitat

Habitat use by mammals may be restricted by nearly annual major flooding of habitats close to the river channel. In fact, most of the smaller terrestrial mammal species, including the dominant rodents (*Proechimys*), are limited largely to high-ground forest (table 17.1). Even in the high-ground forest, flooding may affect the species composition of mammals. Shallow standing water covering about one quarter of the high-ground forest may reduce the abundance of agoutis (*Dasyprocta*) and paca (*Agouti*) by limiting the number of permanently dry burrow sites. After the major flood of 1982, armadillos (*Dasypus novemcinctus*) were gone from the study area and did not recolonize it for at least two years (Emmons, unpubl. data). Surprisingly, the populations of other normally terrestrial species (such as *Proechimys* spp.) were not affected by this major flood.

The physical proximity of habitats enables mobile pulp-eating mammal species to choose a particular habitat when fruit production there is high relative to other habitats (Terborgh 1983). This ability to chose among habitats with generally distinct temporal patterns of fruit maturation (fig. 17.4) may allow high population densities to exist by reducing the impact of seasonal food shortages in any one habitat (Janson et al. 1981; Terborgh 1986). For instance, a variety of diurnal and nocturnal mammals turn to feeding on the nectar of a single common flower (*Combretum assimile*) in riverine and disturbed forest when fruit production in the unflooded forest reaches its nadir in June–July (Janson et al. 1981; Terborgh and Stern 1987).

The generally successional character of habitats at CC may permit high densities of certain species. Some primates species at CC clearly favor disturbed vegetation (*Saimiri*, *Callicebus*; Terborgh 1983). Rabbits (*Sylvilagus*) at CC frequent the river edges, swamps, and the small clearing surrounding the buildings; they are not found in mature terra firme forest.

Some mammal species are relative habitat specialists. The white-lipped peccary (*Tayassu pecari*) is nomadic and uses all habitats but seems to feed mostly in high-ground forest and in *Mauritia* palm swamps, in both of which it largely exploits palm seeds (Kiltie 1982). Many smaller rodent species appear to be restricted to high-ground forest, as are some smaller arboreal nonrodent species (table 17.1). Such species must be able to subsist on the resources available even during the period of lowest food production. Perhaps for this reason some of the small primate species have surprisingly low population densities, large home ranges, and specialize on a set of stable but unproductive plant species (Terborgh and Stern 1987).

Although annual pulp and seed production appear adequate to support the energy requirements of the CC mammal community, seasonal shortages of food do exist for many species (Janson 1984; Terborgh 1983, 1986). In particular, for fruit-pulp eaters, the energy production in pulp of seasonally fruiting species in July–August is 1,200 KJ/ha/day (fig. 17.2A). This level provides only 133 KJ/day per kg of mammal, a value far below that required by any known mammal (Nagy 1987). Not surprisingly, virtually all fruit-pulp eaters turn to alternative resources in the dry season: nectar, palm seeds, leaves, insects, and aseasonally fruiting species (Terborgh 1986). In July–August, nectar yields an average of 1,913 KJ/ha/day, whereas aseasonally fruiting species yield an average of 964 KJ/ha/day in fruit pulp, together more than twice the energy production in pulp of seasonally ripening species in the dry season (fig. 17.2A). A similar seasonal shortage of fruit has been reported for agouti and paca on Barro Colorado Island (BCI), Panama (Smythe et al. 1982).

In contrast to pulp-eaters, seed-eating mammals should not experience a period of acute food shortage if all seeds were edible. Even in the biweekly period of lowest seed production (early September), 5,238 KJ/ha/day is available from seeds of seasonally fruiting species. Since the expected daily consumption by seed-eating mammals is only 2,008 KJ/ha/day (table 17.1), the lowest seasonal seed production would still appear adequate to meet energy demands (even allowing for only 50% assimilation, as assumed by Smythe et al. 1982). In addition, aseasonally fruiting species yield an average of 7,715 KJ/ha/day (fig. 17.2B). And some seeds remain available for months after they are produced, especially if cached by rodents (Kiltie 1981). The high production of seeds in the rainy season may thus provide an additional buffer against low production in the dry season. Finally, predominantly seed-eating mammals often consume substantial amounts of fruit pulp, insects, or leaves.

The apparent oversupply of seeds relative to energy demands of seed-eating mammals contrasts with Smythe et al. (1982) at BCI, who concluded that seed availability was seasonally limiting to the biomass of agoutis (*Dasyprocta punctata*) and pacas (*Agouti paca*). The difference derives almost entirely from the much higher biomass of terrestrial seed-eaters on BCI (about 12 kg/ha just for agouti and paca, Smythe et al. 1982) versus CC (3.96 kg/ha, including all rodents and peccaries).

In spite of the apparent excess of seasonal seed production over seed consumption by mammals, it would be premature to say that seed-eating mammals at CC are not food-limited. In contrast to fruit pulp, which is designed to attract and be eaten by consumers (e.g., Janson 1983), seeds must escape consumption in large numbers. Because of high seedling mortality, the net reproductive output of individual plants would become vanishingly small if seed-eating mammals consistently ate as much as 62% of annual seed production, as pulp-eaters may do for fruit pulp. Given such strong selection, it is not surprising that most seeds are not easily available to consumers—many are small and hard to find, while others are inedible by mammals because of heavy physical or chemical protection. As is seen for mammalian forest folivores and forest leaves (Leigh and Smythe 1978), the relatively low biomass of mammalian granivores relative to total seed production is an expected result. Just the opposite pattern is expected and observed for consumers of easily obtained and palatable fruit pulp—a biomass capable of consuming most of pulp production. If we knew more about the availabilities and palatabilities of different seed types to mammalian consumers, fig. 17.2B could be adjusted to reflect more properly the energy actually available to seed-eating mammals.

Data on seasonal availability of insects at CC are insufficient to estimate the possibility of absolute seasonal shortages for insectivorous mammals. The results of night-light captures, however, do suggest the presence of marked seasonal fluctuations in the abundance of insects (fig. 17.3). Given such overall seasonality, it is interesting that the large-bodied insectivorous mammals at CC are not generalists, unlike most large pulp-eating species. The specialization of large insectivores is probably related to their use of nests of social insects (especially Hymenoptera), which are not likely to fluctuate in abundance seasonally as much as do nonsocial insects.

Comparison with Other Study Areas

The most striking aspect of the mammal community at CC is the presence of large mammals (and birds) that have been hunted to near extinction in more populated areas of Amazonia, including giant otters (*Pteronura brasiliensis*), jaguar (*Felis onca*), ocelot (*F. pardalis*), spider monkeys (*Ateles paniscus*), woolly monkeys (*Lagothrix lagothricha*), tapir (*Tapirus terrestris*), capybara (*Hydrochaeris hydrochaeris*), and white-lipped peccaries (*Tayassu pecari*). These large species include the most important predators, large-seed dispersers, and seed-eaters in Amazonia, so that their absence may be expected to lead to many additional changes in community composition.

Lack of hunting at CC has allowed high population densities of large primates to survive (table 17.1). Large primates are dominant competitors for ripe-fruit trees against smaller primates (Wright 1985; Terborgh and Janson 1986; Terborgh and Stern 1987). This competition may explain why population densities of small primate species at CC are actually less than those of many larger species, in contrast to the usual ecological pattern. In fact, in areas where larger primates are absent, smaller species can achieve much higher densities (e.g., *Callicebus*, Mason 1968; *Saguinus* sp., Freese et al. 1982).

Medium-sized terrestrial mammals are scarce at CC relative to BCI (Emmons 1984). This relative scarcity may be because such species are included in the diets of all felids at CC, whereas smaller or larger species are fed on by fewer cat species (fig. 17.1). On BCI, large predators are extinct, and their predation has not been replaced by human hunting, perhaps allowing medium-sized species to achieve high densities (Terborgh and Winter 1978; Eisenberg et al. 1979; Emmons 1984). Similarly, the notable scarcity of sloths at CC may be due to the frequent use of the area by large birds of prey, including harpy eagles (*Harpya harpija*) and

Guianan crested eagles (*Morphnus guianensis*), at least the former of which is known to include sloths in its diet (Rettig 1978). As noted above, however, medium-sized terrestrial mammal densities may also be reduced somewhat at CC because of the 20–30% loss in dry land area in the rainy season due to riverine flooding and poor drainage.

The relative scarcity of medium-sized mammals at CC may help to explain several additional characteristics of the mammal community at CC. First, small terrestrial mammals are more common at CC than on BCI. Small- and medium-sized terrestrial mammals overlap broadly in diet (both feed heavily on seeds), and evidence from other study areas suggests that larger species are competitively superior to smaller ones (Emmons 1984). Thus, if the density of medium-sized species is reduced by relatively heavy predation, the smaller species may have experienced density compensation (cf. MacArthur 1972). Second, the relative scarcity of medium-sized terrestrial mammals may contribute to the observed mismatch between seed production and consumption by mammals. The biomass of seed-eating mammals is dominated by medium and large terrestrial mammals. In addition to the relative scarcity of medium-sized mammals, one of the larger terrestrial seed-eating mammals, the collared peccary (*Tayassu tajacu*), is a favored prey of jaguar and may thus be maintained at lower densities than the habitat could support (Emmons 1987). Heavy predation on major seed-eating mammals, however, is only one possible explanation of the relatively low abundance of seed-eating mammals relative to seed production. Even if the biomass of seed-eaters were several times its present value at CC, seed-eaters would still only consume a small fraction of the total energy produced in seeds (most of which may not be available to consumers; see above).

At least seventy species of nonflying mammals are found at Cocha Cashu. Several factors favor the coexistence of such a high diversity of mammals: (1) a high annual production of fruits and seeds; (2) a variety of closely spaced habitats with differing temporal peaks of fruit production; and (3) an undisturbed abundance of large predators. Fruit and insect production in the high-ground forest vary more than tenfold seasonally, both reaching a minimum in the dry season (June–August). Such large variation might force some species to become locally extinct if alternative resources were unavailable. Energy from fruit pulp during the dry season, for example, is much less than the energy demands of predominantly frugivorous mammals, leading most frugivores to use alternative resources then (nectar, seeds, leaves, and insects). In contrast, the biomass of predominantly seed-eating mammals consumes less than 20% of the energy in seeds even when production is lowest. The apparent scarcity of seed-eating mammals relative to seed production is probably due to adaptations of seeds to avoid discovery and consumption. In addition, the densities of important seed-eating mammals may be reduced because of high predation pressure from felids. The lack of human hunting and relatively high densities of large predators may help to explain distinctive patterns of species abundances at Cocha Cashu.

REFERENCES

Adams, C. F. 1975. Nutritive values of American foods in common units. Agriculture Handbook 456. U.S. GPO, Washington, D.C.

Charles-Dominique, P., M. Atramentowics, M. Charles-Dominique, H. Gérard, A. Hladik, C. M. Hladik, and M. S. Prévost. 1981. Les mammifères frugivores arboricoles nocturnes d'une forêt guyanaise: inter-relations plantes-animaux. Rev. Ecol. (Terre et Vie) 35: 341–435.

Crespo, J. A. 1982. Ecología de la communidad de mammíferos del Parque nacional Iguazú, Misiones. Rev. Mus. Argentino de Ciencas Naturales "Bernardino Rivadavia." Ecol. Ser. 3: 45–161.

Eisenberg, J. F. and R. W. Thorington. 1973. A preliminary analysis of a Neotropical mammal fauna. Biotropica 5: 150–161.

Eisenberg, J. F., M. A. O'Connell, and P. V. August. 1979. Density, productivity, and distribution of mammals in two Venezuelan habitats. In J. F. Eisenberg (ed.), *Vertebrate Ecology in the Northern Neotropics*. Smithsonian Inst. Press, Washington, D.C., pp. 187–207.

Emmons, L. H. 1981. Ecology and resource partitioning among nine species of African rainforest squirrels. Ecol. Monogr. 50: 31–54.

———. 1982. Ecology of *Proechimys* (Rodentia, Echimyidae) in southeastern Peru. Trop. Ecol. 23: 280–290.

———. 1984. Geographic variation in densities and diversities of non-flying mammals in Amazonia. Biotropica 16: 210–222.

———. 1987. Comparative feeding ecology of felids in a Neotropical rainforest. Behav. Ecol. Sociobiol. 20: 271–283.

Emmons, L. H., A. Gautier-Hion, and G. Dubost. 1983. Community structure of the frugivorous-folivorous forest mammals of Gabon. J. Zool. Lond. 199: 209–222.

Foster, R. B., J. Arce, and T. S. Wachter. 1986. Dispersal and the sequential plant communities in Amazonian Peru floodplain. In A. Estrada and T. H. Fleming (eds.), *Frugivores and Seed Dispersal*. W. Junk, Dordrecht, pp. 357–370.

Freese, C. H., P. G. Heltne, R. N. Castro, and G. Whitesides. 1982. Patterns and determinants of monkey densities in Peru and Bolivia, with notes on distribution. Int. J. Primatol. 3: 53–90.

Glanz, W. E. 1982. The terrestrial mammal fauna of Barro Colorado Island: census and long-term changes. In E. G. Leigh, Jr., A. S. Rand, and D. M. Windsor (eds.), *The Ecology of a Tropical Forest: Seasonal Rhythms and Long-Term Changes*. Smithsonian Inst. Press, Washington, D.C., pp. 455–468.

———. (This volume). Neotropical mammal densities: how unusual is the community on Barro Colorado Island, Panama?

Goldizen, A. W. and J. W. Terborgh. 1986. Cooperative polyandry and helping behavior in saddle-backed tamarins (*Saguinus fuscicollis*). In J. G. Else and P. C. Lee (eds.), *Primate Ecology and Conservation*. Cambridge Univ. Press, Cambridge, pp. 191–198.

Guillotin, M. 1982. Rhythmes d'activité et regimes alimentaires de *Proechimys cuviert* et *Oryzomys capito velutinus* (Rodentia) en forêt guyanaise. Rev. Ecol. (Terre et Vie) 36: 337–371.

Hames, R. B. 1979. A comparison of the efficiency of the shotgun and the bow in Neotropical forest hunting. Human Ecol. 7: 219–252.

Janson, C. H. 1983. Adaptation of fruit morphology to dispersal agents in a Neotropical forest. Science 219: 187–189.

———. 1984. Female choice and mating system of the brown capuchin monkey *Cebus apella* (Primates: Cebidae). Z. Tierpsych. 65: 177–200.

———. 1985. Aggressive competition and individual food consumption in wild brown capuchin monkeys (*Cebus apella*). Behav. Ecol. Sociobiol. 18: 125–138.

———. 1986. The mating system as a determinant of social evolution in capuchin monkeys (*Cebus*). In J. G. Else and P. C. Lee (eds.), *Primate Ecology and Conservation*. Cambridge Univ. Press, Cambridge, pp. 169–179.

Janson, C. H., J. W. Terborgh, and L. H. Emmons. 1981. Non-flying mammals as pollinating agents in the Amazonian forest. Biotropica 13(suppl.): 1–6.

Janson, C. H. and J. W. Terborgh. 1986. Censusing primates in rainforest, with special reference to the primate community at Cocha Cashu Biological Station, Peru. In M. A. Ríos (ed.), *Reporte Manu*. Centro de Datos para la Conservación, Univ. Agraria, La Molina, Peru, pp. 1–48.

Janzen, D. H. 1974. Tropical blackwater rivers, animals, and mast fruiting by the Dipterocarpaceae. Biotropica 6: 69–103.

Kiltie, R. A. 1981. Distribution of palm fruits on a rain-forest floor—why white-lipped peccaries forage near objects. Biotropica 13: 141–145.

———. 1982. Bite force as a basis for niche differentiation between rain forest peccaries (*Tayassu tajacu* and *T. pecari*). Biotropica 14: 188–195.

Kiltie, R. A. and J. W. Terborgh. 1983. Observations on the behavior of rain forest peccaries in Peru: why do white-lipped peccaries form herds? Z. Tierpsych. 62: 241–255.

Leigh, E. G., Jr. and N. Smythe. 1978. Leaf production, leaf consumption, and the regulation of folivory on Barro Colorado Island. In G. G. Montgomery (ed.), *The Ecology of Arboreal Foliovores*. Smithsonian Inst. Press, Washington, D.C., pp. 33–50.

MacArthur, R. H. 1972. *Geographical Ecology*. Princeton Univ. Press, Princeton, N.J.

Mason, W. A. 1968. Use of space by callicebus groups. In P. C. Jay (ed.), *Primates: Studies in Adaptation and Variability*. Holt, Rinehart, and Winston, New York, pp. 200–216.

McFarland, M. J. 1986. Ecological determinants of fission-fusion sociality in *Ateles* and *Pan*. In J. G. Else and P. C. Lee (eds.), *Primate Ecology and Conservation*. Cambridge Univ. Press, Cambridge, pp. 181–190.

Meggers, B. J. 1971. *Amazonia: Man and Culture in a Counterfeit Paradise*. Aldine and Atherton, Chicago.

Milton, K. 1980. *The Foraging Strategy of Howling Monkeys*. Columbia Univ. Press, New York.

Nagy, K. A. 1987. Field metabolic rate and food requirement scaling in mammals and birds. Ecol. Monogr. 57: 111–128.

Odum, E. P., C. E. Connell, and L. B. Davenport. 1962. Population energy flow of three primary consumer components of old-field ecosystems. Ecol. 43: 88–96.

Patton, J. L., B. Berlin, and E. A. Berlin. 1982. Aboriginal perspectives of a mammal community in Amazonian Peru: knowledge and utilization patterns among the Aguaruna Jívaro. In M. A. Mares and H. H. Genoway (eds.), *Mammalian Biology in South America.* Pymatuning Symp. Ecol. 6: 111–128.

Pine, R. H. 1973. Mammals (exclusive of bats) of Belem, Para, Brazil. Acta Amazônica 3(2): 48–79.

Rettig, N. L. 1978. Breeding behavior of the harpy eagle (*Harpia harpyja*). Auk 95: 629–643.

Smythe, N., W. E. Glanz, and E. G. Leigh, Jr. 1982. Population regulation in some terrestrial frugivores. In E. G. Leigh, Jr., A. S. Rand, and D. M. Windsor (eds.), *The Ecology of a Tropical Forest: Seasonal Rhythms and Long-Term Changes*. Smithsonian Inst. Press, Washington, D.C., pp. 227–252.

Terborgh, J. 1983. *Five New World Primates: A Study in Comparative Ecology*. Princetion Univ. Press, Princeton, N.J.

———. 1986. Community aspects of frugivory in tropical forests. In A. Estrada and T. H. Fleming (eds.), *Frugivores and Seed Dispersal*. W. Junk, Dordrecht, pp. 371–384.

Terborgh, J. W., J. W. Fitzpatrick, and L. Emmons. 1984. An annotated checklist of bird and mammal species of Cocha Cashu Biological Station, Manu National Park, Peru. Fieldiana (Zool.), n.s., 21: 1–29.

Terborgh, J. W. and C. H. Janson. 1983. Ecology of primates in southeastern Peru. Nat. Geog. Soc. Res. Rep. 15: 655–662.

———. 1986. The socio-ecology of primate groups. Ann. Rev. Ecol. Syst. 17: 111–135.

Terborgh, J. W., C. H. Janson, and M. Brecht. 1986. Cocha Cashu: its vegetation, climate, and resources. In M. A. Ríos (ed.), *Reporte Manu*. Centro de Datos para la Conservación, Univ. Agraria, La Molina, Peru, chap. 1.

Terborgh, J. W. and M. Stern. 1987. The surreptitious life of the saddle-backed tamarin. Amer. Sci. 75: 260–269.

Terborgh, J. W. and B. Winter. 1978. Some causes of extinction. In M. E. Soulé and B. A. Wilcox (eds.), *Conservation Biology: An Evolutionary-Ecological Perspective*. Sinauer, Sunderland, Mass., pp. 119–133.

White, F. 1986. Census and preliminary observations on the ecology of the black-faced black spider monkey (*Ateles paniscus chamek*) in Manu National Park, Peru. Amer. J. Primatol. 11: 125–132.

Wilson, D. E. (This volume). Mammals of La Selva, Costa Rica.
Wright, P. C. 1985. The costs and benefits of nocturnality for *Aotus trivirgatus* (the night monkey). Ph.D. diss., CUNY.
———. 1986. Ecological correlates of monogamy in *Aotus* and *Callicebus*. In J. G. Else and P. C. Lee (eds.), *Primate Ecology and Conservation*. Cambridge Univ. Press, Cambridge, pp. 159–167.

18

Estimation of Mammalian Densities in Continuous Forest North of Manaus

JAY R. MALCOLM

Comparison of community characteristics across sites offers one way to identify processes that structure tropical forest communities. Unfortunately, comparisons among Neotropical mammalian communities are difficult at best. Few sites have been sampled, even for particular mammalian groups, and if one considers the whole class, only a few sites have been studied in detail (Fleming 1975; Eisenberg 1980). At the same time, attempts to characterize the mammalian community at one site involve many technical problems. Any given census method reveals only a fraction of the number of species present, and it provides reasonable estimates of population densities for even a smaller fraction. For some taxa, particularly those that frequent the forest canopy, current census methods are inadequate (Bourlière 1973; August and Fleming 1984). We know little about the biases inherent in particular methods. Should it be desirable to compare sites sampled using different methodologies (which, unfortunately, will often be the case) or to attempt to make statements about the whole mammalian community, the nature of these biases should be investigated.

During research on the Minimum Critical Size of Ecosystems (MCSE) project, I censused mammalian populations in continuous forest

Several field assistants cheerfully led me around and climbed trees with me. I would especially like to thank R. Cardoso, J. Lopes, L. Reis, M. Santos, and O. Souza. I would also like to thank R. Bierregaard, M. Carleton, L. Emmons, L. Joels, T. Lovejoy, J. Rankin, A Rylands, and B. Zimmerman for their assistance during the study. Funding was provided by World Wildlife Fund-U.S., the Instituto Nacional de Pesquisas da Amazônia, the Natural Sciences and Engineering Research Council of Canada, and a Tinker Summer Research Award.

using two common methods: terrestrial live-trapping and nocturnal strip-censuses. In addition, I developed a method to census the little-known arboreal small mammal fauna (Malcolm, in press). One purpose of this chapter is to present density estimates calculated from these census data and to compare density estimates among the methods. Did estimates agree in an absolute sense, or, if not, at least in a relative sense?

A second purpose is to compare the density estimates with those from other Neotropical forest sites. Of seven Amazonian sites censused by Emmons (1984), the MCSE reserves area exhibited the poorest and least abundant fauna, a result Emmons (1984) attributed to the poor soils of the area. In contrast, my subsequent trapping during a seven-month visit to the MCSE reserves revealed a richer and much more abundant fauna (Malcolm 1988). Abundances were close to mean abundances at Emmons's (1984) richest site (Cocha Cashu, Peru). Given an additional eighteen months of trapping in the MCSE reserves, it is now possible to present a more robust test of Emmons's (1984) result. More generally, Gentry and Emmons (1987) found that density and species richness of fertile understory plants varied predictably with rainfall and soil fertility across thirteen Neotropical forest sites. Of the sites sampled, the MCSE reserves exhibited by far the fewest species and individuals of understory shrubs in fruit or flower, as correlated with the poor soils, low rainfall, and relatively strong dry season in the area. Gentry and Emmons (1987) suggested that understory plant fertility may provide a relative index of overall ecosystem productivity and, as such, correlate with certain characteristics of animal populations. Given low primary productivity in the MCSE reserves, one might expect low mammalian standing biomass and therefore low densities (Eisenberg 1980). Are mammal densities in the MCSE reserves low compared to densities in other Neotropical forests?

Materials and Methods

The MCSE reserves area is in primary, upland (terra firme) forest on moderately rugged terrain approximately 80 km north of Manaus. The area is dissected by small creeks that form the headwaters of tributaries of three small rivers: the Cuieiras, the Preto da Eva, and the Urubú. All sites are thus far from large rivers and their associated riverine habitats (*várzea* and *igapó*). Habitats associated with low-lying terrain (such as palm swamps) are rare and small in area. Rainfall is relatively seasonal. Monthly rainfall in the Manaus area (seventy-year average) ranged 42–162 mm between June and November and 211–300 mm between December and May. The yearly average was 2,105 mm (Ribeiro and Adis, ms). Soils are nutrient-poor yellow latisols. General descriptions of the site are found in Lovejoy et al. (1986) and Lovejoy and Bierregaard (this volume), and of the flora in Rankin-de-Merona et al. (this volume). Hunting is recent and restricted to areas near roads (ZF-3 and BR-174) and farm headquarters. The fauna in most of the continuous forest of the area is probably almost pristine (see also Emmons 1984). General accounts of many of the area's mammal species can be found in Husson (1978).

Small Mammal Trapping

During the twenty-five-month period November 1983–November 1985, I set live-traps on the ground at eleven sites in continuous forest (10-ha reserves 1201, 1204, 1205, 1208, and 1210; 100-ha reserves 1301, 1302, and 2303; 1,000-ha reserve 3402; and two other sites). Nine of the eleven sites were at least 1 km from areas where forest had been clear-cut. Two sites (reserves 1301 and 2303) abutted clear-cut along one side and were contiguous with continuous forest in other directions. To minimize any edge effect, traplines were oriented perpendicular to the edge, with nearest traps 140 m from the edge. The greatest distance between trap sites was approximately 27 km. Sites were visited on average 2.7 times (range: 1–4), with an average interval of seven months between visits (range: four to twelve months).

The basic sampling unit was a trapline of fifteen trap stations at 20-m intervals. At each site, traplines were established along parallel trails. Configurations in 10- and 100-ha plots are described in Malcolm (1988). In the 1,000-ha plot (reserve 3402), I used the same configuration used in 100-ha plots, except that the distance between adjacent trails was 250 m and the distance between two traplines on a trail was 126 m. At the two sites outside reserves, I used two parallel traplines, with 200 between traplines at one site, and 300 m between traplines at the other.

Each trap station consisted of a Tomahawk trap (14 cm × 14 cm × 40 cm) and a Sherman trap (8 cm × 8 cm × 23 cm) placed 2–4 m apart and 4–6 m perpendicular to the trail. Traps were checked and rebaited with peanut butter and banana each morning for nine nights, and captures were identified, measured, and released (Malcolm 1988). Voucher specimens are deposited at the Instituto Nacional de Pesquisas da Amazônia. *Proechimys* species, were identified only to genus. I have recently learned to distinguish live-captures of the two species in the reserves. Preliminary results indicate that *P. guyannensis* is much more common than *P. cuvieri*.

I used two types of trap stations in lines of fifteen to trap arboreal mammals (Malcolm, in press). Average trap height was 14.7 m. In the first method, the two traps, the Sherman on top of the Tomahawk, were tied onto two poles, a tree was climbed, and the two poles were secured to the tree. In the second method, a frame with two pulleys was nailed to the tree. The two traps were ferried up and down the tree using the pulleys. Since success rates did not appear to differ between the methods (Malcolm, in press), I present combined data. When set, traps were baited with banana, peanut butter, and a cloth sack containing peanut butter and raisins. Since insects rarely chewed through the cloth sack, it provided a long-term bait source. Traps were rebaited only when they had a capture or were otherwise disarmed. During the twenty-five months, I trapped at three sites in continuous forest (reserves 1201 and 1205, and an area outside reserves), using 3 or 4 traplines per site. Four of the lines were sampled more than once, providing seventeen trapline samples (Malcolm, in press). Variation in trap effort among traplines due to vagaries of weather was small and, to simplify analyses, ignored.

The central problem in estimating densities of open populations by trapping is to estimate the area the traps sample. The area is defined by the area over which individuals are trappable; clearly, if an individual is trappable over a relatively large area, the sampling area of a trap is larger than if an individual is trappable over a relatively small area. Given that only traps are used, the only method to estimate the area over which individuals are trappable is to use the distribution of distances between recaptures of individuals. Most authors have used some index of recapture distances—for example, the average distance between successive captures (e.g., Fleming 1971) or the maximum distance between captures (e.g., O'Farrell et al. 1977). Unfortunately, apparently no theory suggests which, if any, of these indices is appropriate. I therefore developed a simple model of the area of trappability of an individual (and hence of the sampling area of a trap). The model is formulated for a trapline of t traps, but it can be easily extended to any trap configuration.

First, assume that an individual of a given species uses a circular area of radius r during some time period such that a trap set within the circle will catch the animal, whereas a trap set outside the circle will not catch the animal. A circle of radius r therefore defines the sampling area of a single trap. Further, assume that if several traps are set within the circle, the individual will be caught in at least the two most distant. Given this second assumption, the distribution of maximum distances between captures can be used to calculate r for the species. For example, given two traps less than $2r$ units apart and a circular sampling area of radius r centered at each trap, then the ratio of the number of individuals caught in both traps to the number of individuals caught in only one trap equals the ratio of the area common to both circles to the area common to only one circle. The density of the species is the number of individuals captured divided by the area bounded by the two circles. In appendix 18.1, I derive the distribution of maximum distances between recaptures on a trapline of t traps, for any r.

To estimate r for each species, I used a numerical method. For each individual recaptured during a nine-night trapping session, I calculated the maximum distance between captures on a trapline; r was the value (to the nearest meter) between 10 m and the greatest individual maximum distance between captures that minimized the summed differences between expected and observed distributions of maximum capture distances. For species with fewer than five recaptured individuals, I did not calculate r, or estimate densities. For *Marmosa parvidens*, *Monodelphis brevicaudata*, *Metachirus nudicaudatus*, and *Rhipidomys mastacalis*, I was able to obtain more than five recaptures by including recapture data from 10- and 100-ha forest fragments.

Expected and observed distributions are shown for four representative species in fig. 18.1. Although the model predicted the shape of observed distributions reasonably well, several of its limitations should be made clear. As with all methods that use distributions of recapture distances, the model assumes that capture itself does not influence the area over which an individual is trappable. Without the aid of field techniques other than trapping, this assumption cannot be tested. It seems likely, however, that frequent capture restricts movements; hence, more accurate density estimation might be possible by using only individuals recaptured several days after initial capture. The model's assumption of equal sized and shapes "areas of trappability" is obviously an approximation. Part of the variability among individuals will be attributable to age and sex; hence, r estimates for different age/sex classes might lead to more accurate density estimation. Unfortunately, in this study, most species were infrequently captured; sample sizes in the various age/sex/capture classes would thus become prohibitively small. Because of this deterministic nature, the model failed to predict long-distance movements. Expected distributions tended to be shifted to the left relative to observed distributions (fig. 18.1). As a result, my estimates are probably slight over

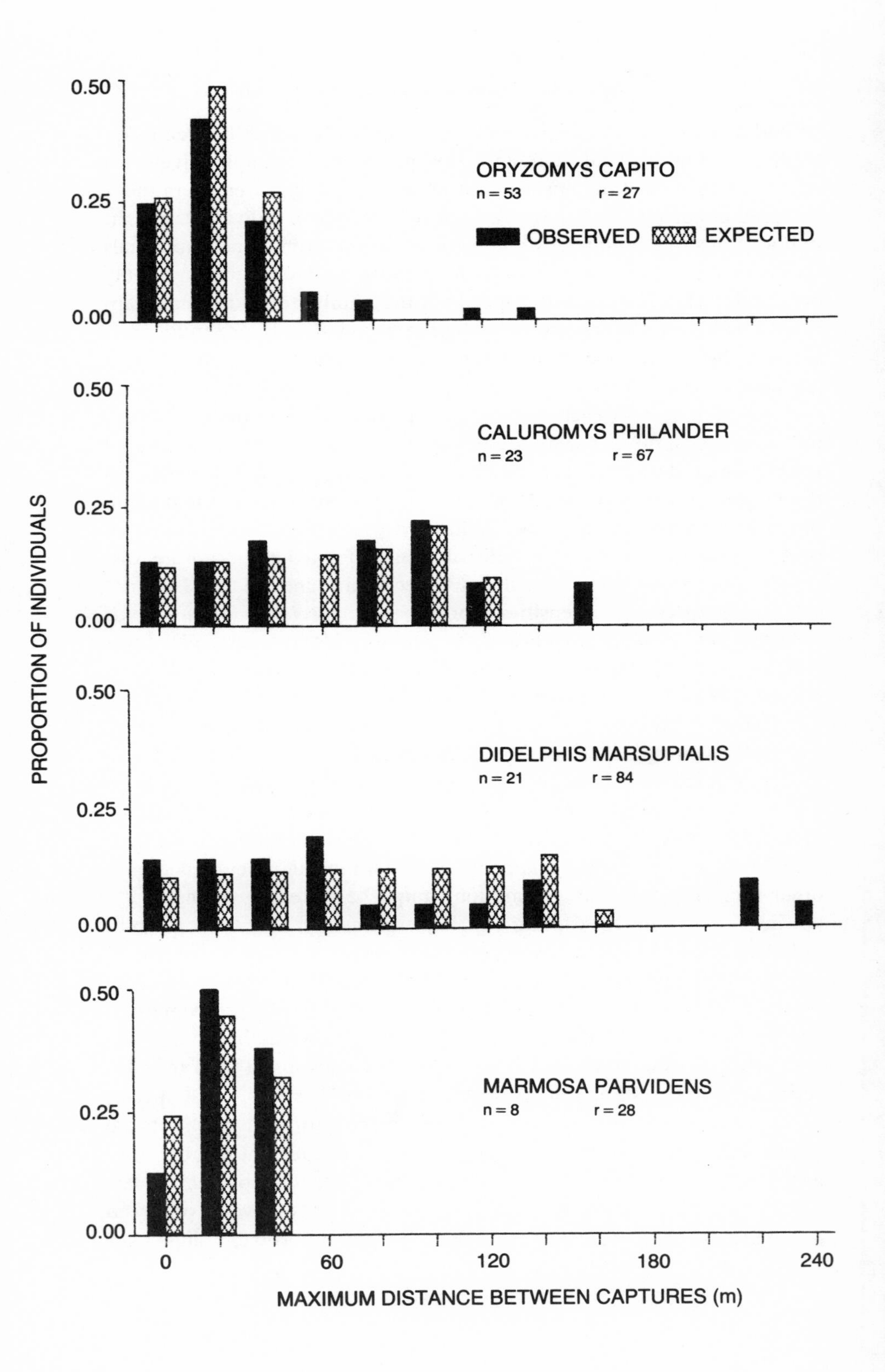
ORYZOMYS CAPITO
n = 53
r = 27
OBSERVED
EXPECTED
CALUROMYS PHILANDER
n = 23
r = 67
DIDELPHIS MARSUPIALIS
n = 21
r = 84
MARMOSA PARVIDENS
n = 8
r = 28
PROPORTION OF INDIVIDUALS
0.50
0.25
0.00
0
60
120
180
240
MAXIMUM DISTANCE BETWEEN CAPTURES (m)

estimates in comparison with other estimates from trapping. Since I was interested in whether MCSE populations had lower densities than other Neotropical populations, overestimation would lead to a more conservative test. More realistic models can easily be envisioned (for example, by incorporating more general shape parameters or by allowing the probability of capture to vary within the area of trappability); the model presented here has the advantage of being perhaps the simplest (and hence most tractable) of the many imaginable. Finally, the density estimates I present depend greatly on the calculated r values, since traps were arranged in lines. In compensation, however, traplines sample a greater area per trap than do grids or assessment lines. In light of these shortcomings, the density estimates presented must be regarded as approximate.

Given r, equation [3] of appendix 18.1 provided the sampling area of a trapline. Density was estimated as the number of individuals captured (whether recaptured or not) divided by the sampling area of the traplines. When r was greater than 50 m, trapline sampling areas in 10-ha plots overlapped along their sides, and when r was greater than 63 m or 75 m, trapline sampling areas in respectively 100-ha plots and the 1,000-ha plot overlapped at their ends. In these cases, density was calculated as the number of individuals captured divided by the area bounded by the trapline sampling areas. To simplify calculations, the sampling area of a trapline was approximated as a $2(t - 1)z$ by $2r$ rectangle, with half-circles of radius r at each end. For $z = 10$, the approximation is good.

Nocturnal Strip Censusing

Trails within continuous forest were walked at a slow pace (approximately 2 kph), between 7 P.M. and 10 P.M., and both sides of the trails were scanned with a headlamp. Routes were chosen to minimize trail overlap within and between nights and thus the probability of viewing the same individual more than once. Data recorded included species (if possible) and estimates of the distance from the observer to the mammal when the mammal was first sighted (sighting distance) and the perpendicular distance from the mammal to the trail (perpendicular distance). I present

←

FIGURE 18.1. Observed distributions of maximum distances between recaptures compared with "best fit" expected distributions. See text for calculation of expected distributions. Comparisons are shown for four representative species. n = number of individuals recaptured, r = best fit radius of trappability.

TABLE 18.1 Density estimates from terrestrial trapping (18,495 station nights), arboreal trapping (2,040 station nights), and nocturnal strip-censuses (302 km) in continuous forest of the MCSE reserves area

	Terrestrial trapping				Arboreal trapping			
	Individuals captured (*N*)	Individuals recaptured (*N*)	*r* (m)	Density (/ha)	Individuals captured (*N*)	Individuals recaptured (*N*)	*r* (m)	Density (/ha)
Marmosa cinerea	28	20	60	0.048	30	12	90	0.264
M. murina	4	2	—	—	1	0	—	—
M. parvidens	35	8[3]	28	0.144	0	—	—	—
Monodelphis brevicaudata	30	12[3]	27	0.129	0	—	—	—
Didelphis marsupialis	59	21	84	0.070	1	0	—	—
Metachirus nudicaudatus	52	7[3]	24	0.256	0	—	—	—
Caluromys philander	5	2	—	—	56	23	67	0.687
Oryzomys capito	94	53	27	0.404	0	—	—	—
O. macconnelli	94	45	28	0.387	0	—	—	—
O. paricola	7	1	—	—	3	1	—	—
O. bicolor	4	2	—	—	11	0	—	—
Rhipidomys mastacalis	4	0	—	—	17	11[3]	36	0.417
Proechimys spp.	235	74	26	1.055	0	—	—	—
Mesomys hispidus	0	—	—	—	7	0	—	—
Isothrix pagurus	2	0	—	—	1	0	—	—
Agouti paca	0	—	—	—	0	—	—	—
Mazama spp.	0	—	—	—	0	—	—	—
Dasypus spp.	0	—	—	—	0	—	—	—
Unidentified	0	—	—	—	0	—	—	—

TABLE 18.1 (*Continued*)

	Nocturnal strip-census			
	Mean sighting distance (m)	Number of individuals seen (/10 km)	Density[1] (/ha)	Robinson and Redford (1986)
Marmosa cinerea	4.0 (3)[2]	0.07	0.009	0.250 ± 0.17 (3)
M. murina	—	0	—	0.430 (1)
M. parvidens	1.9 (10)	0.36	0.095	—
Monodelphis brevicaudata	2.3 (4)	0.03	0.007	0.630 (1)
Didelphis marsupialis	6.4 (25)	0.53	0.041	0.553 ± 0.24 (8)
Metachirus nudicaudatus	6.6 (47)	0.96	0.073	0.083 ± 0.07 (2)
Caluromys philander	—	0	—	0.598 ± 0.94 (3)
Oryzomys capito	2.0 (33)	0.76	0.190	2.769 ± 1.16 (9)[4]
O. macconnelli	2.1 (14)	0.40	0.095	—
O. paricola	—	0	—	—
O. bicolor	2.0 (1)	0.03	0.008	—
Rhipidomys mastacalis	—	0	—	1.720 ± 0.56 (2)[4]
Proechimys spp.	3.7 (159)	3.44	0.465	3.345 ± 2.10 (6)
Mesomys hispidus	—	0	—	—
Isothrix pagurus	—	0	—	—
Agouti paca	12.6 (7)	0.17	0.007	0.275 ± 0.20 (8)
Mazama spp.	14.5 (4)	0.07	0.002	0.104 ± 0.13 (4)[5]
Dasypus spp.	9.1 (14)	0.33	0.018	0.219 ± 0.15 (8)[6]
Unidentified	8.0 (44)	0.89	0.056	

Sources: Densities from trapping were calculated using within-trapline recaptures (see text for details). Mean sighting distances are from sightings in all primary forest habitats, including forest fragments. Mean densities [± 2SEM (n)] calculated by Robinson and Redford (1986) from a variety of other Neotropical forest sites are also shown.

Notes :[1] King's estimate
[2] Sample size in parenthesis
[3] Includes recaptures in forest fragments
[4] Densities from Robinson and Redford (1986) are for the genus
[5] Densities from Robinson and Redford (1986) are for *M. qouazoubira*
[6] Densities from Robinson and Redford (1986) are for *D. novemcinctus*

data from the forty-three-month period September 1983–March 1987. We censused access trails and trails within the same nine reserves that were sampled by terrestrial trapping. Reserves were sampled an average 7.4 (range: 3–14) nights. In total, 302 km were walked on approximately 60 km of trail. On most nights, two people worked together. I, or at least one of two trained field assistants, took part in all surveys. I used King's method (Glanz 1982) to calculate densities. To calculate mean sighting distances, I used all sightings in primary forest habitat (continuous forest and forest fragments).

Results and Discussion

Comparison of Densities from Terrestrial and Arboreal Trapping

Only *Marmosa cinerea* was caught often enough on the ground and in the trees to allow comparison of densities from the two methods. Density estimates from terrestrial trapping were much lower than estimates from arboreal trapping (table 18.1), despite a lower *r* value from terrestrial trapping. *Caluromys philander* and *Rhipidomys mastacalis* frequented the ground even less, and density estimates from terrestrial trapping would even more seriously underestimate arboreal densities. More important, trapping at heights close to 2 m (the usual "arboreal" sample) failed to capture these arboreal species (Malcolm, in press). Evidently, canopy trapping or similar methods are required to census nocturnal, arboreal small mammals. Comparison of densities among sites based on terrestrial or near-terrestrial trapping will probably be meaningless, perhaps even misleading. For example, if strictly arboreal species center their activities in the zone where the majority of light is intercepted by vegetation, then as canopy openness increases, one might expect near-terrestrial trapping to provide better density estimates. In this case, a positive correlation as revealed by near-terrestrial trapping between arboreal species abundance or diversity and canopy openness would reflect changes in vertical use of space, and not changes in community structure. Although we have set few arboreal traps as yet in the MCSE reserves, it is interesting to note that populations of arboreal small mammals are among the most dense in the forest and that arboreal small mammal biomass may approach that of terrestrial populations. We have also caught more strictly arboreal species than terrestrial species. In forests with greater development of vertical strata, one might expect an even richer and more abundant canopy fauna (Eisenberg 1980; Eisenberg et al.

1979). August (1983) found no significant correlation between development of vertical strata and small mammal density at Masagural, Venezuela, perhaps because canopy trapping was not used.

Comparison between Strip Census and Trapline Densities

King's estimates were in all cases much lower than estimates derived from trapline data (table 18.1). Evidently, King's method does not provide realistic density estimates. Keller's method (Glanz 1982) is not much of an improvement, given that densities calculated from the two methods are very close (Glanz 1982). Not surprisingly, estimates of population densities of arboreal mammals were in particularly poor agreement. Density of *M. cinerea* from strip censuses was only 18% of that from terrestrial trapping (which in turn was only 18% of that from arboreal trapping). We never saw *C. philander* or *R. mastacalis*, even though they are perhaps the second and third most abundant nonflying mammals in the forest. We scanned the overstory at night infrequently, so it is not surprising that we rarely saw arboreal species. Although absolute agreement was poor, estimates did agree in a relative sense. Excluding the three arboreal species listed above, correlation between trapline density and strip-census density was high ($rs = 0.90$, $n = 7$, $p = 0.03$) and stronger than correlation between number of individuals/10 km and trapline density or between number of individuals/10km and trapline abundance (respectively, $rs = 0.64$, $p = 0.12$, and $rs = 0.72$, $p = 0.08$). It appears that some of the residual variation can be explained by a species' size and behavior. In spite of its small size, density estimates for *M. parvidens* were relatively close, perhaps because this species frequents small saplings of the undergrowth or because Sherman traps did not consistently catch them. Density of *M. brevicaudata* was seriously underestimated; however, this species appears to be at least partly diurnal (Charles-Dominique et al. 1981; pers. obs.). Leaving out these two species, and combining data for the two *Oryzomys* species, then as body size increased, agreement between estimates generally improved.

Comparison with Abundance and Diversity from Other Amazonian Sites

I have not yet undertaken a detailed analysis of temporal variation in small mammal abundances during the twenty-five months of trapping, in general, however, abundances of the three common terrestrial rodent taxa (*Proechimys* spp., *O. capito*, and *O. macconnelli*) declined during

TABLE 18.2 Individuals per 100 station nights (SN) of terrestrial trapping

	MCSE					Cocha Cashu	Tambopata
	Feb. 1982– Jul. 1982	Nov. 1983– Mar. 1984	Apr. 1984– Nov. 1985	Nov. 1983– Nov. 1985	All	Three years	1979
Opossums							
Didelphis	0.12	0.56	0.24	0.32	0.29	1.00	0
Metachirus	0.03	0.29	0.28	0.28	0.24	0.54	0.70
Rodents							
Proechimys	0.60	2.36	0.93	1.27	1.17	2.28	3.20
Oryzomys	0.06	2.54	0.61	1.07	0.91	2.78	0.70
Total SN	3,434	4,455	14,040	18,495	21,929	2,763	434

Sources: Results of February–July 1982 in the MCSE reserves and from Cocha Cashu and Tambopata are from Emmons (1984; table 5). Emmons (1984) used one Tomahawk per trap station; I used one Tomahawk and one Sherman per trap station.

TABLE 18.3 Nocturnal census distance and number of individuals seen/10 km of census

	Census distance (km)	Opossums	Small rodents (< 1 kg)	Large terrestrial mammals	Canopy mammals	Total
Limoncocha	est. 30	8.0	12.0	2.0	12.3	34.3
Cocha Cashu	116.4	4.8	24.6	2.7	7.2	39.3
Tambopata	27.3	2.6	9.5	2.9	2.2	17.2
Yanamono	15.0	2.7	6.7	0	10.0	19.3
Ecuador latosols	16.9	1.8	10.6	2.9	6.5	21.9
Mishana	18.0	2.2	9.4	1.7	2.2	15.5
MCSE						
(Emmons 1984)	65.0	0.6	1.7	2.0	0.6	4.9
(this study)	302.0	2.0	4.6	0.6	0	8.1[1]
(combined)	367.0	1.7	4.1	0.8	0.1	7.5

Source: Data from first six sites are from Emmons (1984: table 4). See Emmons (1984: table 4) for species included in the mammalian groups. Diurnal species are excluded.

Note: [1] Includes 0.9 unidentified mammals/10 km

the first seven months of study and remained more or less stable thereafter, whereas temporal variation in abundances of other taxa was small or nonexistent. Compared to Emmons's (1984) trap results from the MCSE reserves, I obtained higher capture rates, even during the eighteen-month period when densities were relatively low and stable (table 18.2). Similarly, we saw more mammals/10 km during the forty-three months of nocturnal censuses than when Emmons (1984) visited the reserves (table 18.3). As Emmons (1984) observed at Cocha Cashu, temporal variation in abundance of *Proechimys* was less dramatic than of *Oryzomys* species. This may be related to the success of *Proechimys* in Neotropical forests; in most communities, they appear to be the most abundant terrestrial small mammal (Emmons 1982). Neither *M. nudicaudatus* nor *D. marsupialis* showed great changes in abundance during the twenty-five months I trapped in the reserves, a result in marked contrast to changes in abundance of *Proechimys* and *Oryzomys* species. Surprisingly, these marsupials were apparently much less abundant when Emmons (1984) visited the reserves. At this point, it is difficult even to guess why these four taxa show contrasting patterns of variation in abundance. Several authors have suggested that fruit production limits some Neotropical mammal populations (Smythe 1986; Smythe et al. 1982; Foster 1982). If so, differences in population dynamics should be attributable to variation in resource utilization among species (such as the types of fruits eaten or the importance of fruit in the diet), to differences in the responses of populations to variation in resource abundance (such as population growth rates, or density dependent effects), and/or to differences in competitive abilities. Obviously, detailed information on diets and resource distributions are required to test this idea.

Emmons's (1984) general conclusion of low densities in the MCSE reserves relative to other Amazonian sites remained true, however. Combining data for my twenty-five months of trapping or for the thirty-one months when one or the other of us was trapping in the reserves, the four commonly caught genera were less abundant in the MCSE reserves than at Cocha Cashu, and three of the four genera were less abundant than at Tambopata (table 18.2). We also saw fewer mammals/10 km of night survey than Emmons saw at other Amazonian sites, a result that appeared consistent across a variety of taxa (table 18.3).

Altogether, 20 species of mammals have been trapped in the MCSE reserves, and an additional 31 have been seen by myself or other biologists (table 18.4). Emmons's (1984) estimate of 68 nonflying species in the MCSE reserves seems high but is probably reasonable for the region north of Manaus. Certainly, the MCSE reserves have fewer species than

TABLE 18.4 Native species of nonflying mammals trapped (t) or seen (s) in the MCSE reserves area, September 1983–March 1987

Marsupialia
Marmosa cinerea (t)
M. murina (t)
M. parvidens (t)
Monodelphis brevicaudata (t)
Philander opossum (t)
Didelphis marsupialis (t)
Metachirus nudicaudatus (t)
Caluromys philander (t)
C. lanatus (t)

Primata
Saguinus midas (s)
Alouatta seniculus (s)
Pithecia pithecia (s)
Chiropotes satanas (s)
Cebus apella (s)
Ateles paniscus (s)

Xenarthra
Choloepus didactylus (s)
Bradypus tridactylus (s)
Myrmecophaga tridactyla (s)
Tamandua tetradactyla (s)
Cyclopes didactylus (s)
Priodontes maximus (s)
Dasypus novemcinctus (s)
D. kappleri (s)

Rodentia
Sciurus gilvigularis (s)
Oryzomys capito (t)
O. macconnelli (t)
O. paricola (t)
O. bicolor (t)
Neacomys guianae (t)
Rhipidomys mastacalis (t)
Coendou prehensilis (s)
Dasyprocta leporina (s)
Myoprocta acouchy (s)
Agouti paca (s)
Proechimys guyannensis (t)
P. cuvieri (t)
Mesomys hispidus (t)
Isothrix pagurus (t)
Echimys sp. (t)

Carnivora
Speothos venaticus (s)
Nasua nasua (s)
Potos flavus (s)
Eira barbara (s)
Lutra longicaudis (s)
Felis concolor (s)
Panthera onca (s)

Perissodactyla
Tapirus terrestris (s)

Artiodactyla
Tayassu tajacu (s)
T. pecari (s)
Mazama americana (s)
M. gouazoubira (s)

Manu National Park, Peru, and it seems very likely that the region north of Manaus has fewer species as well. It should be stressed that the species list I present for the MCSE reserves is for one habitat type: upland forest. As close as 80 km away, at the Balbina dam site, *Philander opossum*, a species of arboreal echimyid (apparently *Makalata armata*), an as of yet unidentified species of arboreal cricetid, and *P. cuvieri* all appear to be abundant (Silva and Gribel, p.c.). These species are rare or have not yet been caught in the MCSE reserves. Any comparison of species diversity across sites must obviously consider habitat diversity.

Comparison with Densities from Other Neotropical Forest Sites

For 8 of the species I captured, Robinson and Redford (1986) provided mean densities (and standard deviations) calculated from a variety of other Neotropical sites. I compared densities in the MCSE reserves with their mean density and with densities ±2 SEM from the mean. Densities of 5 of 8 species were less than the mean minus 2 SEM (table 18.1). I found only one density estimate lower than the estimates I calculated for these 5 species: Guillotin (1982) estimated 0.30 *O. capito* per ha in a four-year-old secondary forest (Arbocel) in French Guiana. Density estimates of the other 3 species were above the mean. Mean densities of *C. philander* and *M. cinerea* are probably underestimates, since Robinson and Redford (1986) evidently included estimates from terrestrial or near-terrestrial trapping. Arboreal trapping of these 2 species in French Guiana gave estimates roughly twice those from the MCSE reserves (Charles-Dominique et al. 1981). Density of *M. nudicaudatus* in the MCSE reserves was close to the mean from two other sites.

Primate densities appear to be very low as well. Rylands and Keuroghlian (1988) used the repeat transect method to estimate primate densities in the MCSE reserves and found that densities were the lowest, or among the lowest, yet recorded (see also Peres, in press). As before, I compared their estimates with Robinson and Redford's (1986) mean. Of the 6 species, only 2 (*Saguinus midas* and *Pithecia pithecia*) had densities that fell above the mean minus 2 SEM. Even for these 2 species, density estimates from other sites cited by Rylands and Keuroghlian (1988) were all higher.

Densities of mammals in the MCSE reserves appear to be the lowest yet recorded from any Neotropical forest site. Whether the extremely low densities can be attributed solely to poor soils and a relatively strong dry season requires further testing. The ordination of Neotropical mammal communities according to soil and rainfall parameters seems a possibility well worth investigating.

REFERENCES

August, P. V. 1983. The role of habitat complexity and heterogeneity in structuring tropical mammal communities. Ecol. 64: 1495–1507.

August, P. V. and T. H. Fleming. 1984. Competition in Neotropical small mammals. Acta Zool. Fennica 172: 33–36.

Bourlière, F. 1973. The comparative ecology of rain forest mammals in Africa and tropical America: some introductory remarks. In B. J. Meggers, E. S. Ayensu, and W. D. Duckworth (eds.), *Tropical Forest Ecosystems in Africa and South America: A Comparative Review*. Smithsonian Inst. Press, Washington, D.C., pp. 279–292.

Charles-Dominique, P. M., M. Atramentowicz, M. Charles-Dominique, H. Gerard, A. Hladik, C. M. Hladik, and M. F. Prevost 1981. Les mammifères frugivores arboricoles nocturnes d'une forêt guyanaise: inter-relations plantes-animaux. Rev. Ecol. (Terre et Vie) 35: 341–435.

Eisenberg, J. F. 1980. The density and biomass of tropical mammals. In M. E. Soulé and B. A. Wilcox (eds.), *Conservation Biology: An Evolutionary-Ecological Perspective*. Sinauer, Sunderland, Mass., pp. 35–55.

Eisenberg, J. F., M. A. O'Connell, and P. V. August. 1979. Density, productivity, and distribution of mammals in two Venezuelan habitats. In J. F. Eisenberg (ed.), *Vertebrate Ecology in the Northern Neotropics*. Smithsonian Inst. Press, Washington, D.C., pp. 187–207.

Emmons, L. H. 1982. Ecology of *Proechimys* (Rodentia, Echimyidae) in southeastern Peru. Trop. Ecol. 23: 280–290.

———. 1984. Geographic variation in densities and diversities of non-flying mammals in Amazonia. Biotropica 16: 210–222.

Fleming, T. H. 1971. Population ecology of three species of Neotropical rodents. Misc. Publ. Mus. Zool., Univ. Mich. 143: 1–77.

———. 1975. The role of small mammals in tropical ecosystems. In F. B. Golley, K. Petrusewicz, and L. Ryskowski (eds.), *Small Mammals: Their Productivity and Population Dynamics*. Cambridge Univ. Press, Cambridge, pp. 269–298.

Foster, R. B. 1982. The seasonal rhythm of fruitfall on Barro Colorado Island. In E. G. Leigh, Jr., A. S. Rand, and D. M. Windsor (eds.), *The Ecology of a Tropical Forest: Seasonal Rhythms and Long-Term Changes*. Smithsonian Inst. Press, Washington, D.C., pp. 151–172.

Glanz, W. E. 1982. The terrestrial mammal fauna of Barro Colorado Island: censuses and long-term changes. In E. G. Leigh, Jr., A. S. Rand, and D. M. Windsor (eds.), *The Ecology of a Tropical Forest: Seasonal Rhythms and Long-Term Changes*. Smithsonian Inst. Press, Washington, D.C., pp. 455–468.

Gentry, A. H. and L. H. Emmons. 1987. Geographical variation in fertility, phenology, and composition of Neotropical forests. Biotropica 19: 216–227.

Guillotin, M. 1982. Place de *Proechimys cuvieri* (Rodentia, Echimyidae) dans les peuplements micromammaliens terrestres de la forêt guyanaise. Mammalia 46: 299–318.

Husson, A. M. 1978. The mammals of Suriname. E. J. Brill, Leiden.

Lovejoy, T. E., and R. O. Bierregaard, Jr. (This volume). Central Amazonian forests and the Minimum Critical Size of Ecosystems Project.

Lovejoy, T. E., R. O. Bierregaard, Jr., A. B. Rylands, J. R. Malcolm, C. E. Quintela, L. H. Harper, K. S. Brown, Jr., A. H. Powell, G. V. N. Powell, H. O. R. Schubart, and M. B. Hays. 1986. Edge and other effects of isolation of Amazon forest fragments. In M. E. Soulé (ed.), *Conservation Biology: The Science of Scarcity and Diversity*. Sinauer, Sunderland, Mass., pp. 257–285.

Malcolm, J. R. 1988. Small mammal abundances in isolated and non-isolated primary forest reserves near Manaus, Brazil. Acta Amazônica 18: 67–83.

———. In press. Comparative abundances of Neotropical small mammals by trap

height. J. Mammal.
O'Farrell, M. J., D. W. Kaufman, and D. W. Lundahl. 1977. Use of live-trapping with the assessment line method for density estimation. J. Mammal. 58: 575–582.
Peres, C. A. Primate community structure in western Brazilian Amazonia. Prim. Conserv.: *in press*.
Rankin-de-Merona, J. M., R. W. Hutchings H., and T. E. Lovejoy. (This volume). Tree mortality and recruitment over a five-year period in undisturbed upland rainforest of the Central Amazon.
Robinson, J. G. and K. H. Redford. 1986. Body size, diet, and population density of Neotropical forest mammals. Amer. Nat. 128: 665–680.
Rylands, A. B. and A. Keuroghlian. 1988. Primate populations in continuous forest and forest fragments in central Amazonia. Acta Amazônica 18: 291–307.
Smythe, N. 1986. Competition and resource partitioning in the guild of Neotropical terrestrial frugivorous mammals. Ann. Rev. Ecol. Syst. 17: 169–188.
Smythe, N., W. E. Glanz, and E. G. Leigh, Jr. 1982. Population regulation in some terrestrial frugivores. In E. G. Leigh, Jr., A. S. Rand, and D. M. Windsor (eds.), *The Ecology of a Tropical Forest: Seasonal Rhythms and Long-Term Changes*. Smithsonian Inst. Press, Washington, D.C., pp. 227–238.

APPENDIX 18.1. The area of overlap of two circles, each with radius r, as a function of half the distance between the centers of the two circles (x) is:

$$[1] \quad o(x) = 2[r^2 \cos^{-1}(x/r) - x\sqrt{(r^2 - x^2)}] \qquad 0 \leq x < r.$$

If $x \geqslant r$, let $o(x) = 0$. Given t traps on a trapline, with $2z$ units between adjacent traps and a circular sampling area of radius r centered at each trap, then the area in common to exactly j circles is:

$$[2] \quad a(j) = (t - (j - 1)) \cdot o((j - 1)z) - 2(t - j) \cdot o(jz) + (t - (j + 1)) \cdot o((j + 1)z)$$

for $j = 1$ to m, where m is such that $(m - 1)z \leqslant r \leqslant mz$, or $m = t$, whichever is less. The total sampling area (A) of the trapline is the area bounded by the t circles

$$[3] \quad A = \sum_{j=1}^{m} a(j)$$
$$= t \cdot o(0) - (t - 1) \cdot o(z)$$
$$= t\pi r^2 - 2(t - 1)[r^2 \cos^{-1}(z/r) - z\sqrt{(r^2 - z^2)}].$$

The proportion of individuals with maximum distance $2(j - 1)z$ between captures is therefore:

$$[4] \quad p(j) = \frac{a(j)}{A}.$$

19

Neotropical Mammal Communities

JOHN F. EISENBERG

The ecology of Neotropical mammals is still poorly understood. Hundreds of species have not been the subject of a single ecological investigation. Yet interest in these mammals by ecologists has swelled during the past thirty years, and we have accumulated much information on the ecology and behavior of many species. Given the pressing need to understand how tropical forest ecosystems function in order to implement long-range conservation strategies, it seems timely that we attempt a synthesis at this, albeit imperfect, stage of our understanding.

The analysis of mammalian communities seems to involve two primary groups of researchers: those interested in bat communities and those devoted to the study of nonflying mammals. This derives in part from the fact that special techniques must be employed to capture, mark, and recapture bats. Great technical problems must be overcome because bats are nocturnal and can move rapidly, not only horizontally but also vertically. In spite of these difficulties, significant advances have been made by Bonaccorso (1979), Bradbury and Vehrencamp (1976a,b, 1977a,b), Gardner (1977), Wilson (1979), and Fleming (1982), to mention only a few.

Students of the nonflying species of mammalian communities seem to fall into two subgroups: those interested in primates and those interested in small to medium-sized mammals that are readily live-trapped for mark and release. Thus, the species that have been studied have a certain bias, and this bias is reflected in the information available for synthesis and development of testable hypotheses. Bats in the lowland Neotropics probably account for better than 50% of the species found in a given area. Wilson's data are consistent with data deriving from a vast number of surveys (Wilson, this volume; Eisenberg 1989).

In this volume, we are concerned primarily with the lowland tropics, but steep elevational gradients are found within the geographically defined tropics, and the structure of mammalian communities can alter rapidly when one samples over an elevational gradient. Species of the order Chiroptera, for example, may account for over 50% of species richness in the lowland tropics of Venezuela. But bats are very sensitive to the ambient temperature, and even at 1,000 m elevation the number of bat species declines by 50%. Marsupials and rodents also decline with increasing altitude. At mid-elevations—approximately 2,500 m—however, a new set of rodent species adapted to the rigors of lower temperatures usually appears. Lowland study sites subjected to seasonal temperature declines deriving from cold winds sweeping downslope from the Andean cordillera may also lack certain temperature-sensitive mammals in their faunas (see Handley 1976; Eisenberg 1989).

The species composition of mammal communities can also vary with latitude. In Central America, for example, rodent genera deriving from relatives in North America interdigitate with genera that show their closest affinities with South American forms. And in Central America some of the more typical Neotropical genera are absent. This results from a combination of factors, but the Andes mountains and the major rivers of South America have barred and continue to block free dispersal between Central and South America (Eisenberg 1979, 1989). Among the 7 species of primates in Panama and Costa Rica, 5 have been recorded in sympatry. Thirteen sympatric species have been recorded by Terborgh (1983) and his workers in Amazonian Peru.

Although the species composition among Neotropical communities can vary, some general principles seem to hold for all areas. Leaving the Chiroptera aside for the moment, let us turn to the nonflying mammals. It is possible in the lowland Neotropics to locate tracts of savanna. If one samples from savannas to gallery forests, the following trends can be noted: Vegetational complexity, both vertically and horizontally, increases as one proceeds from savanna through brush land to semideciduous forest and finally to the evergreen gallery. The species richness of nonflying mammals also increases along the same gradient. Interestingly, August's results (1981, 1983) from Venezuela indicate that the number of guilds increases dramatically with increased vegetational cover and complexity and that in areas of high vegetational diversity, higher species richness levels are achieved not by adding additional guilds but rather by adding more species within a guild.

A second generalization can be made when forest tracts of different age and complexity are compared. As the moisture gradient changes from open savanna to brush woodland and thence through low stature, dry deciduous forest to multistratal, tropical evergreen forest, the composition of the mammalian fauna changes with respect to substrate use. Second-growth forests and areas subjected to extreme seasonal drought often show a lower arboreal biomass value than areas of mature forest growth with a less seasonal schedule of plant productivity (Eisenberg et al. 1979). If one examines standing crop biomass for the species of any tropical community, it is evident that the largest proportion of biomass is contributed by herbivores and the lowest by carnivores. In this characteristic the biomass pyramid derived from studies of tropical communities conforms to those pyramids established for temperate zone mammalian communities (Eisenberg 1980).

Robinson and Redford (1986b) tabulated density estimates for 121 species of South American mammals. They demonstrated that population densities of Neotropical mammal species are predictably related to their body mass and dietary specializations. As a general rule, population density declines with increasing body mass. In contrast, much of the scatter around the regression line could be accounted for by considering dietary specialization. Carnivores exist at lower densities than do herbivores, regardless of body size (see also Eisenberg et al. 1979).

Robinson and Redford (1986a) also calculated the intrinsic rate of population increase (r_m) for the same set of species. This rate declines with increasing body size. Significant scatter around the regression line can be explained by phylogenetic affinity rather than dietary specialization. They suggest that r_m is inextricably tied to life-history strategies that derive in part from phylogenetic affinity.

These results underscore a further need for studies of secondary productivity, since a species of small size may at any one time have a low standing crop biomass but over the annual cycle contribute significantly in terms of secondary productivity. Eisenberg et al. (1979) demonstrated that the cane rat *Zygodontomys* may exist at an average biomass of less than 20 kg/km^2 but exhibit a productivity of 181 kg/km^2 per year. On the other hand, in the same habitat, the howler monkey *Alouatta seniculus* may exist with an average standing crop biomass of 365 kg/km^2 but exhibit an annual productivity of 66 kg/km^2 per year.

Kinnaird and Eisenberg (1989) used the same data base from Ro-

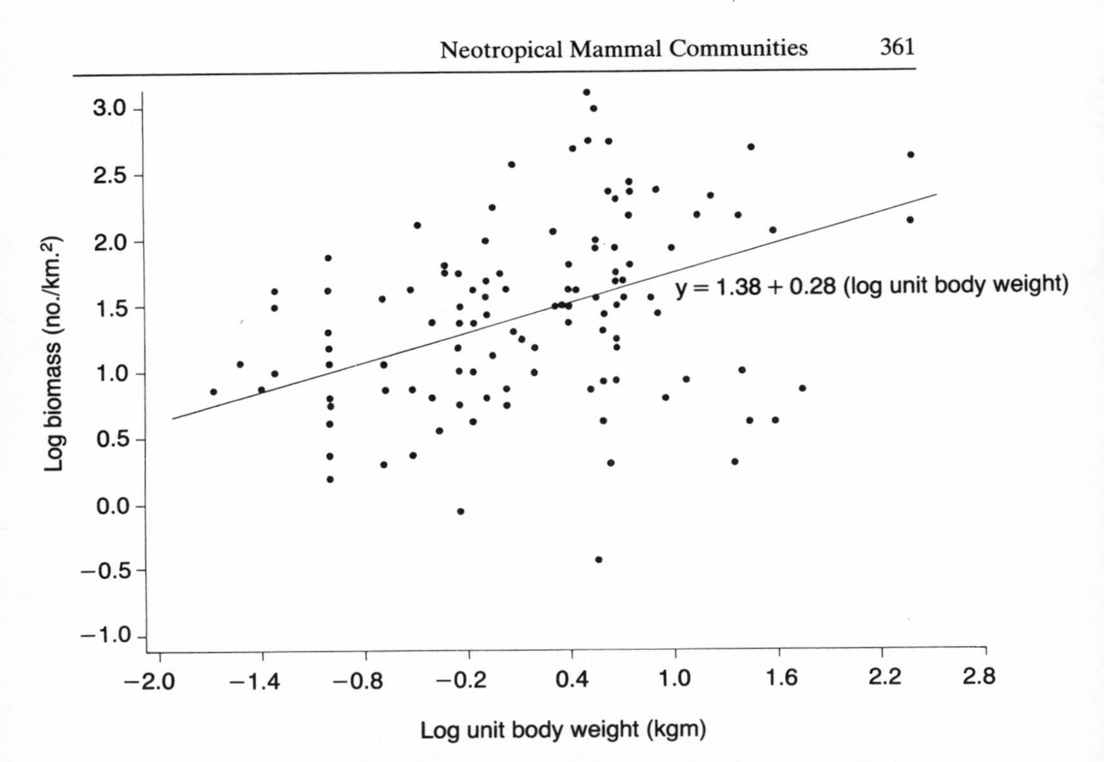

FIGURE 19.1. Regression analysis of Neotropical mammalian biomass against mean body weight for Neotropical mammals. Data base from Robinson and Redford (1986a). Note the high degree of scatter around the regression line. (From Kinnaird and Eisenberg, 1989.)

binson and Redford (1986a,b) for calculating the relation between standing crop biomass and body size. Plotting standing crop biomass against average body size produces a positive correlation. The larger taxa comprise the greater percentage of the standing crop biomass. Again, the regression exhibits considerable scatter. Dietary specialization explains much of the variability. The dietary specialization of a species is often tied to phylogenetic affinity, and both may well be an excellent predictor of a species' potential biomass within a community and thus its overall impact on community dynamics (figs. 19.1, 19.2).

I conclude, then, that a fundamental organization exists within tropical mammal communities regardless of the species composition. As determined for temperate zone ecosystems, primary consumers contribute the most significant proportion to standing crop biomass. In mature, multistratal tropical rainforests, however, the arboreal components of the community may contribute a far greater percentage to standing crop.

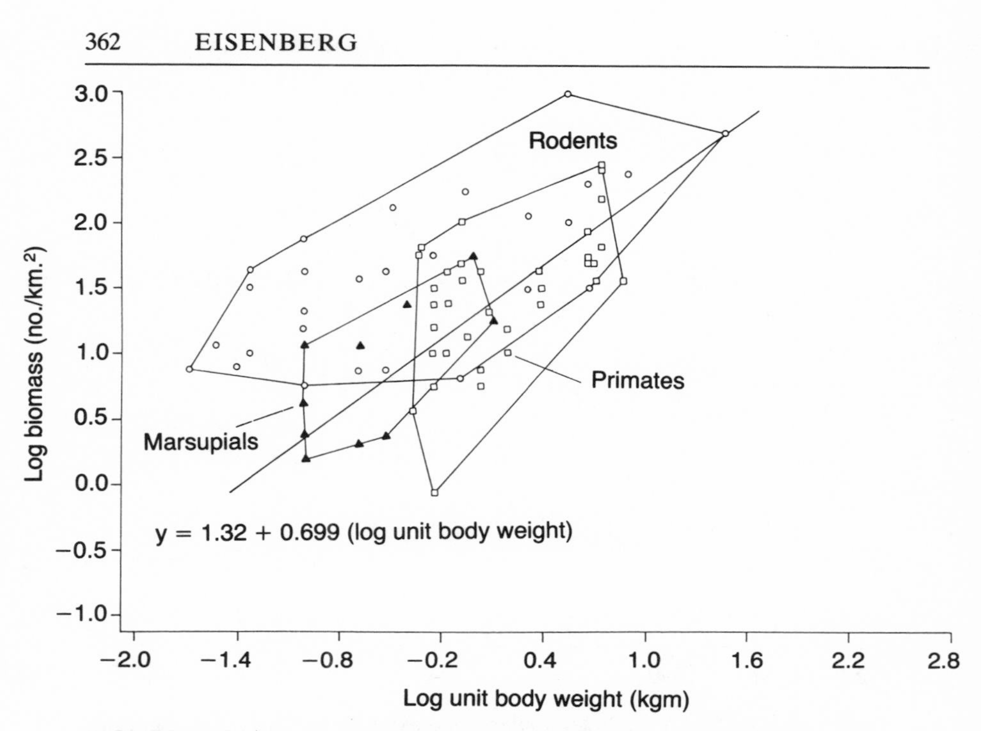

FIGURE 19.2. Plot of biomass against body weight for selected mammalian species derived from the data set in fig. 19.1. Here the species are plotted according to taxonomic affinity. Much of the scatter around the regression line can be accounted for by phylogenetic affinity or trophic adaptation. (From Kinnaird and Eisenberg, 1989.)

biomass than the terrestrial components. This second result stands in marked contrast to many of the temperate zone mammal communities studied to date (Eisenberg 1980).

A Comparison of the Four Study Sites

The four study sites that are the subject of this volume differ in many features. Barro Colorado Island (BCI) and La Selva both lie in Central America. The species included in these two areas obviously differ from those in the Amazon Basin—namely, Cocha Cashu and the research area north of Manaus. La Selva includes a steep elevational gradient up to 2,600 m, whereas the other study areas are at modest elevations. Because of the gradient, some premontane species mix with typical lowland forms. La Selva is the smallest area, at approximately 7.5 km² (Wilson, this volume). BCI is an artificial island isolated at the completion of the Panama Canal, when water from the Chagres River was partially impounded to create Lake Gatun. The insular characteristics of BCI, however, have perhaps been somewhat exaggerated since it is scarcely 300 m at its closest point from the mainland and such large species as deer, tapir, and peccary can easily swim to it (Glanz 1982, this volume).

The history of land use patterns for the areas also differs markedly. Approximately 30–40% of BCI was cleared for agricultural purposes before it was isolated. The gradual succession of the clear-cut area over the past seventy or so years has had a profound impact on the avifaunal and mammalian species composition. Forms adapted to second growth or more open habitats have been disfavored and have thus declined in numbers. These facts are evident from the early records established by Chapman and others when compared to present-day censusing efforts (Glanz 1982; Willis 1974). La Selva has also had its share of human disturbance. Parts of the area have only recently come under protection. The area 40 km north of Manaus has clearly been influenced by human activities, but of course the aim of the research effort at the Manaus site is to clarify the minimum critical size of ecosystems (MCSE), and perturbations are thus an active part of the study (Malcolm, this volume). Clearly, Cocha Cashu is the most nearly pristine of the four areas (Terborgh 1983).

Collecting efforts, reintroductions, and censusing activities have different histories in the four areas. Early on, BCI was subjected to intensive collecting, so that some long-term, baseline data exist. Attempts at reintroductions were also made on BCI to restore species that either had been extirpated through poaching or had not been isolated on the island at its formation. On BCI, Montgomery and Sunquist (1978) made an intensive effort to census the sloth populations. To my knowledge, no such comparable effort has been undertaken at the other study sites. In his work at Manaus, Malcolm (this volume) has attempted an elaborate sampling scheme for increasing our knowledge of small arboreal mammals. Though a comparable effort is underway at Cocha Cashu, work on arboreal species and their numbers is only beginning at La Selva and, aside from the sloths, has been attempted only sporadically on BCI.

Aside from BCI, the sites differ in choice of area and sampling procedure. At the Manaus site studied by Malcolm, 52 nonflying mammal species have been recorded; however, only 80 km away exist the very different *várzea* and *igapó* habitats. The nonflying mammal list of Manaus could jump to 68 species if these habitats were included. Malcolm (this volume) demonstrates that densities of even common species fluctuate and that short-term studies are probably inadequate for understanding mammalian communities. Censusing techniques need to be refined because of differences in observer biases and sampling schemes. In spite of all the cautions, it seems reasonable to assume that the densities for most nonflying mammals are low at the study sites described by Malcolm (this volume).

There are 50 nonflying and 63 flying species of mammals at La Selva. Determination of bat species' richness requires sophisticated techniques and long years of effort. Bat density estimates are even more demanding and as yet determined for only a few species in Panama. Just a very few species of bats (< 10) account for most of the individuals sampled at La Selva (Wilson, this volume). Sixty-seven nonflying species of mammals are found at Cocha Cashu, Peru. The bats of Cocha Cashu have not been adequately studied (Janson and Emmons, this volume).

The four areas have definite similarities. All are within the geographically defined tropics. As a result, average temperatures above 20°C prevail and annual differences in photoperiod are minimal. All four areas have a limited drought period that synchronizes the flowering and fruiting of the major tree species that support mammals. The drought imposes a seasonal flux in the availability of fruits, nuts, and seeds and a period of varying duration when resources for primary and secondary consumers are limited (Glanz 1982; Janson and Emmons, this volume). How comparable the soil fertility is from one area to another cannot be answered at present; however, the soils at BCI, Cocha Cashu, and La Selva appear to have pockets or areas of exceptional fertility. Primary productivity values, where they have been measured, appear to be comparable for lowland tropical areas with similar soil types (Leigh and Windsor 1982). Cocha Cashu is unique in that it is an alluvial plain that annually receives considerable sedimentary deposits, thus recharging the soil. The high water table in the primary study area at Cocha Cashu probably ameliorates the effects of its brief dry season. In contrast, the low density of fig trees (*Ficus*) at the Manaus sites and the relatively low densities of vertebrate secondary consumers suggests lowered primary productivity (Malcolm, this volume).

Factors Contributing to Variation in Species Richness

In spite of the cautions outlined in the previous section, I am delighted at the comparability of results. To highlight some influences on species richness, I shall discuss the *llanos* of Venezuela (Eisenberg et al. 1979). One Venezuelan site, Masaguaral, is at approximately 100 m elevation. It floods during the rainy season but experiences a pronounced drought for some three to four months when virtually no rain falls. Masaguaral is bisected by the Rio Guarico, and evergreen gallery forest predominates on the banks. Thus, as one proceeds west from the river, one passes through tropical evergreen forest, dry deciduous forest, and finally savanna woodland. One would anticipate in this somewhat harsh environment with reduced vegetational cover that species richness would be lower than that recorded in sites with a more even distribution of rainfall. We have recorded 75 species of mammals, of which 42 are bats, and 223 species of resident birds from Masaguaral. The area sampled is approximately 10 km^2. If we consider, however, that the gallery forest on the Rio Guarico is potentially continuous with the premontane forest on the north coast range of Venezuela—in short, a "ribbon of life"—then the colonization during favorable climatic regimes (such as long-term increases in rainfall) could yield a different figure. If the gallery forest of the Rio Guarico were continuous and wide enough, some 14 additional nonflying mammalian species could be added, thus raising the potential total to 47. The 14 additional species occur only slightly north of Masaguaral. My point here is only to illustrate that given comparability in evenness and quality of rainfall, soil fertility and drainage, then at low altitudes between 12° north and south latitudes, plant productivity will be comparable, as will be the structure of mammalian communities in terms of species richness and biomass.

Emmons (1984) has commented on the wide variation in the species richness of mammal communities within the Amazon Basin. Given equal rainfall patterns, variations in soil fertility could limit primary productivity and in turn limit the density and diversity of mammalian species. This hypothesis was reinforced by the studies of Gentry and Emmons (1987), where they concluded that differences in understory fertility were an important indicator of potential productivity within lowland tropical systems.

Conclusions and Recommendations

The number of resident bird species for the four study areas discussed in this volume seems to range from 259 to 286. Where measured, the number of bat species ranges from 60 to 65, and the number of nonflying mammals ranges from 40 to 67. The low figure of 40 for BCI does not include 11 extinctions through succession and human disturbance (Glanz, this volume). The nonflying mammalian genera at La Selva show some elements from North America, but with respect to the genera of marsupials, edentates, carnivores, perissodactylans, and artiodactylans, the faunas are very similar for all four areas. Cocha Cashu has a high species richness when primates are considered. This surely has to do in part with biogeographical factors. An assessment of differences in standing crop biomass among the four study areas must await further research. Clearly, the site at Cocha Cashu is extraordinarily rich. This might be attributable to not only its pristine condition but also to very favorable soil fertility and the site's location on an alluvial plain. Ultimately, the standing crop biomass of consumers rests squarely upon the primary productivity, and primary productivity is a function of temperature, photoperiod, availability of water, and soil fertility.

We have made an important first step in the comparison of Neotropical ecosystems, but much more must be done. The La Selva site has been well collected, but studies of mammalian community ecology are in their infancy. Malcolm's efforts at Manaus emphasize the need for long-term efforts and the need for more studies of the arboreal, nocturnal, small mammal community. The low densities of mammals at Manaus offers genuine challenges to plant ecologists to elucidate the environmental constraints on primary productivity. The publication by Emmons (1984) offers promising, testable hypotheses to account for the variations in mammalian densities and species richness found throughout the Amazon Basin. Finally, we should strive for a uniform strategy of sampling and monitoring at long-term study sites. Only then will we achieve the ability to make time comparisons.

The chapters in this volume point out some intriguing general patterns, as well as genuine gaps in our knowledge. I trust that an understanding of what we do not know will lead to gathering information for a more complete synthesis in perhaps twenty years. Given the present rate of deforestation, twenty years may be all the time we have left to work in some forests.

REFERENCES

August, P. V. 1981. Population and community ecology of small mammals in northern Venezuela. Ph.D. diss., Boston Univ.

———. 1983. The role of habitat complexity and heterogeneity in structuring tropical mammal communities. Ecol. 64: 1495–1507.

Bonaccorso, F. 1979. Foraging and reproductive ecology in a Panamaniam bat community. Bull. Fla. State Mus. (Biol. Sci.) 24(4): 359–408.

Bradbury, J. W. and S. L. Vehrencamp. 1976a. Social organization and foraging strategies in emballonurid bats. Behav. Ecol. Sociobiol. 1(1): 357–381.

———. 1976b. Social organization and foraging strategies in emballonurid bats. Behav. Ecol. Sociobiol. 1(2): 383–404.

———. 1977a. Social organization and foraging strategies in emballonurid bats. Behav. Ecol. Sociobiol. 2(3): 1–17.

———. 1977b. Social organization and foraging strategies in emballonurid bats. Behav. Ecol. Sociobiol. 2(4): 19–30.

Eisenberg, J. F. 1979. Habitat economy and society: some correlations and hypotheses for the Neotropical primates. In I. S. Bernstein and E. O. Smith (eds.), *Primate Ecology and Human Origins: Ecological Influences on Social Organization*. Garland Press, New York, pp. 215–262.

———. 1980. The density and biomass of tropical mammals. In M. E. Soulé and B. A. Wilcox (eds.), *Conservation Biology: An Evolutionary-Ecological Perspective*. Sinauer, Sunderland, Mass., pp. 35–55.

———. 1989. *Mammals of the Neotropics*. Vol. 1, The Northern Neotropics: Panama, Colombia, Venezuela, Guyana, Suriname, French Guiana. Univ. of Chicago Press, Chicago.

Eisenberg, J. F., M. A. O'Connell, and P. V. August. 1979. Density, productivity, and distribution of mammals in two Venezuelan habitats. In J. F. Eisenberg (ed.), *Vertebrate Ecology in the Northern Neotropics*. Smithsonian Inst. Press, Washington, D.C., pp. 187–207.

Emmons, L. H. 1984. Geographic variation in densities and diversities of nonflying mammals in Amazonia. Biotropica 16: 210–222.

Fleming, T. H. 1982. Foraging strategies of plant visiting bats. In T. Kunz (ed.), *Biology of Bats*. Plenum Press, New York, pp. 287–326.

Gardner, A. E. 1977. Feeding habits. In R. J. Baker, J. K. Jones, and D. C. Carter (eds.), *Biology of Bats of the New World Family Phyllostomidae*, pt. 2. Spec. Publ. Mus. Texas Tech. Univ. 13, pp. 293–350.

Gentry, A. H. and L. H. Emmons. 1987. Geographic variation in fertility, phenology and composition of the understory of Neotropical forests. Biotropica 19: 216–227.

Glanz, W. E. 1982. The terrestrial mammal fauna of Barro Colorado Island: censuses and long-term changes. In E. G. Leigh, Jr., A. S. Rand, and D. M. Windsor (eds.), *The Ecology of a Tropical Forest: Seasonal Rhythms and Long-Term Changes*. Smithsonian Inst. Press, Washington, D.C., pp. 455–468.

———. (This volume). Neotropical mammal densities: how unusual is the community on Barro Colorado Island, Panama?

Handley, C. O., Jr. 1976. Mammals of the Smithsonian Venezuelan Project. Brigham Young Univ. Sci. Bull., Biol. Ser. 20(5): 1–91.

Janson, C. H. and L. H. Emmons. (This volume). Ecological structure of the nonflying mammal community at Cocha Cashu Biological Station, Manu National Park, Peru.

Kinnaird, M. and J. F. Eisenberg. 1989. A consideration of body size, diet, and population biomass for Neotropical mammals. In K. H. Redford and J. F. Eisenberg (eds.), *Advances in Neotropical Mammalogy*. Sand Hill Crane Press, Gainesville, Fla., pp. 595–604.

Leigh, E. G., Jr. and D. M. Windsor. 1982. Forest production and regulation of primary consumers on Barro Colorado Island. In E. G. Leigh, Jr., A. S. Rand, and D. M. Windsor (eds.), *The Ecology of a Tropical Forest: Seasonal Rhythms and Long-Term Changes*. Smithsonian Inst. Press, Washington, D.C., pp. 111–122.

Malcolm, J. R. (This volume). Estimation of mammalian densities in continuous Forest north of Manaus.

Montgomery, G. G. and M. E. Sunquist. 1978. Habitat selection and use by two-toed and three-toed sloths. In G. G. Montgomery (ed.), *The Ecology of Arboreal Foliovores*. Smithsonian Inst. Press, Washington, D.C., pp. 329–359.

Robinson, J. G. and K. H. Redford. 1986a. Body size, diet and population density of Neotropical forest mammals. Amer. Nat. 128: 665–680.

———. 1986b. Intrinsic rate of natural increase in Neotropical forest mammals: relationship to phylogeny and diet. Oecologia 68: 516–520.

Terborgh, J. 1983. *Five New World Primates: A Study in Comparative Ecology*. Princeton Univ. Press, Princeton, N.J.

Willis, E. 1974. Populations and local extinctions of birds on Barro Colorado Island, Panama. Ecol. Monogr. 44: 153–169.

Wilson, D. E. 1979. Reproductive patterns. In R. J. Baker, J. K. Jones, and D. C. Carter (eds.), *Biology of Bats of the New World Family Phyllostomidae*, pt. 3. Spec. Publ. Mus. Texas Tech. Univ. 13, pp. 317–378.

———. (This volume). Mammals of La Selva, Costa Rica.

PART V

Reptiles and Amphibians

20

The Herpetofauna of La Selva, Costa Rica

CRAIG GUYER

Amphibians and reptiles are conspicuous components of the vertebrate community in the Neotropics. Herpetofauna populations often reach high densities, relative to birds and mammals, perhaps because ectothermy allows efficient conversion of food energy to biomass (Pough 1983). Such high population densities permit amphibians and reptiles to be important regulators of arthropod abundance (Schoener and Spiller 1987; Pacala and Roughgarden 1984) and provide a prey base for other tropical vertebrates (Greene 1988). Amphibians and reptiles thus represent an important link in the food web of Neotropical communities.

Research on amphibians and reptiles at La Selva, Costa Rica, has provided important insights into the structure and function of this vertebrate group within tropical wet forest ecosystems. In this chapter I explore the composition and dynamics of the La Selva herpetofauna relative to three questions: (1) Does the species composition at La Selva differ from other Neotropical sites? (2) Do habitat preferences of the major amphibian and reptilian groups differ at La Selva from those at other Neotropical sites? (3) What community-level patterns are indicated by detailed studies of population dynamics of amphibians and reptiles at La Selva?

Amphibians and reptiles at La Selva inhabit a forest environment characterized by seasonal rainfall with decreased amounts of rain from January through April (dry season) and in September (*veranillo*), as well as increased rain throughout the rest of the year (wet season). With a mean annual rainfall of about 4,000 mm, the La Selva environment has been characterized as transitional between lowland and premontane wet forest (Holdridge 1967). Descriptions can be found elsewhere of forest composition (Frankie et al. 1974) and dynamics (Hartshorn 1980; Denslow 1980; Lieberman et al. 1985). Amphibians and reptiles at La Selva are known to peak in abundance during the dry season. This peak correlates with patterns of increased air temperature, decreased humidity, decreased rainfall, and increased leaf-litter depth (Lieberman 1986).

Species Composition

Species lists are now available for several sites in the Neotropics. La Selva, through the activities of repeated sampling from Organization for Tropical Studies (OTS) courses and long-term monitoring over the past five years, has a known fauna of 48 species of amphibians (Donnelly 1990) and 86 species of reptiles (Guyer 1990), most of these belonging to four families (Hylidae, Leptodactylidae, Iguanidae, and Colubridae; table 20.1). Species lists can be used to determine faunal uniqueness by studying patterns of endemism or shared species among sites. The geographic region to which La Selva belongs has been analyzed with track analysis by Duellman (1966), Müller (1973), and Savage (1966, 1982). All concluded that the region including La Selva was part of an area of eastern Caribbean lowland wet forest extending from Tamaulipas, Mexico, to central Panama (Savage 1966). Although this conclusion is based on concordant distributions of individual species, the method lacks rigor because it is unknown how many concordant distributions are needed before a pattern is generated that differs from what might be expected to occur due to chance alone.

Another method for analyzing such distributional data has been advanced by Simberloff and Connor (1979). This method arranges data into an area by species matrix using 0's to represent absence and 1's to indicate presence of each species at all sites. Such a matrix of observed distributions can be used to compare, for example, the observed number of species shared between two sites to an expected number of shared species based on randomly arranged matrixes of similar form. By constraining the row and column totals to equal those of the original matrix, a considerable amount of biological reality can be retained by this randomization procedure. These constraints insure that wide-ranging species remain wide ranging in the randomized data (and narrowly distributed species remain so) and that each site maintains the species diversity that was observed.

Faunal lists of reptiles and amphibians are available from several Neotropical sites. I have chosen lists from eight reasonably well sampled

TABLE 20.1 Number of species within each family of amphibians and reptiles found at La Selva

Amphibia	Reptilia
Gymnophiona	Crocodylia
Caeciliidae 1	Crocodylidae 1
Caudata	Testudinata
Plethodontidae 3	Emydidae 2
Anura	Kinosternidae 2
Bufonidae 3	Sauria
Centrolenidae 6	Anguidae 3
Dendrobatidae 2	Gekkonidae 4
Hylidae 13	Iguanidae 13
Microhylidae 1	Scincidae 2
Leptodactylidae 16	Teiidae 2
Ranidae 3	Xantusiidae 1
	Serpentes
	Boidae 2
	Colubridae 46
	Micruridae 3
	Tropidophiidae 1
	Viperidae 4

sites (table 20.2). These data were used to construct a species-by-area matrix of Neotropical amphibians and reptiles from which the observed similarity between the herpetofaunas at La Selva and other Neotropical sites are compared relative to what might be expected based on chance alone. For the three major groups of amphibians and reptiles (frogs, lizards, and snakes) a consistent pattern involving greater than expected number of shared species occurs between La Selva and each of the following sites: Tortuguero, Rincón de Osa, and Las Cruces, all in Costa Rica, and Barro Colorado Island (BCI), Panama (table 20.3). Greater similarity was observed for salamanders and caecilians between La Selva, Tortuguero, and Las Cañas, Costa Rica, as well as for crocodilians and turtles between La Selva, Tortuguero, and BCI. These results indicate the integrity of the eastern Caribbean lowland herpetofauna as derived from previous track analysis (Savage 1966, 1982). The biological phenomena shaping distribution patterns of amphibian and reptilian species found at La Selva thus seem to act over a wide area in Caribbean wet forest of Central America, including mid-elevation sites (that is, Las Cruces).

For most amphibian and reptilian groups I found no differences between observed and expected numbers of species shared between La Selva and Monteverde or Las Cañas (table 20.3). The pattern of similarity of La Selva to these two sites thus cannot be distinguished from ran-

TABLE 20.2 Neotropical herpetofaunas compared with the fauna at La Selva

Site	Amphibians	Reptiles	Total	Life Zone	Source
Tortuguero, Costa Rica	33	59	92	Lowland wet	Lahanas, p.c.
Rincón de Osa, Costa Rica	48	70	118	Lowland wet	Savage and McDiarmid, p.c.
Las Cruces, Costa Rica	32	39	71	Midelevation wet	Scott et al. 1983
Barro Colorado Island, Panama	53	68	121	Lowland wet	Myers and Rand 1969
Monteverde, Costa Rica	30	35	65	Cloud forest	Hayes et al. 1988
Las Cañas, Costa Rica	22	57	79	Dry forest	Scott et al. 1983
Santa Cecilia, Ecuador	94	91	185	Wet forest	Duellman 1978
Iquitos, Peru	68	142	210	Lowland wet	Dixon and Soini 1975, 1977

TABLE 20.3 Results of comparisons of observed and expected number of species shared between La Selva and other Neotropical sites

	Tortuguero	Rincón de Osa	Barro Colorado Island	Las Cruces	Monteverde	Las Cañas	Santa Cecilia	Iquitos
Frogs	+	+	+	+	0	0	–	ND
Salamanders and Caecilians	+	0	0	0	0	+	0	ND
Lizards	+	+	+	+	0	0	–	–
Snakes	+	+	+	+	0	0	–	–
Crocodilians and Turtles	+	0	+	ND	ND	0	0	0

ND = No data
\+ = Significantly greater shared species than expected at random
– = Significantly fewer shared species than expected at random
0 = No difference of observed number from random expectation
Source: Tests were based on z-scores of single observed values vs. expected distributions generated via 500 randomizations

dom expectation. It is important to note that this analysis merely fails to reject the hypothesis that the observed number of shared species could be due to random effects. It does not rule out biological explanations of the distribution of La Selva amphibians and reptiles relative to cloud forest (Monteverde) and dry forest (Las Cañas) sites in Central America.

For the larger groups (frogs, lizards, and snakes) a consistent pattern of decreased similarity in amphibian and reptilian species composition was observed between La Selva and the Amazonian lowland faunas of Iquitos, Peru, and Santa Cecilia, Ecuador (table 20.3). This indicates that, despite presumed similarities in forest structure of these lowland wet forest sites, faunal interchange of amphibians and reptiles across the Panamanian portal restricts similarity based on random distributions of the species, a finding that corroborates Savage's (1966, 1982) contention that the fauna of the eastern Caribbean wet forest is distinctive and endemic to Central America. Additional samples from Colombia and Ecuador are needed to test Savage's contention that the eastern Caribbean herpetofauna of Central America is continuous with wet forest sites along the Pacific coast of northern South America.

Habitat Preferences

A common feature of written descriptions of Neotropical herpetofaunas is an attempt to place species into broad ecological categories. Often these categories comprise species of similar morphologies (e.g., Williams 1983; Vitt and Vangilder 1983; Miyamoto 1982). These descriptions are especially common for lizards, snakes, and leptodactylid frogs. A second categorization scheme, based on reproductive mode, is frequently used for amphibian species (Duellman and Trueb 1986). These two sets of categories provide a basis for describing patterns of habitat preference among assemblages of Neotropical amphibians and reptiles.

At La Selva, lizards, snakes, and frogs can be categorized into groups representing species found near water (aquatic), within the layer of fallen leaves on the ground (leaf litter), on the surface of the ground (terrestrial), and at various levels within the vegetation (arboreal). For each of the major amphibian and reptilian groups, most species at La Selva reside within the leaf litter or among the various levels of vegetation (table 20.4). Given the tremendous height and biomass of wet forest vegetation at La Selva, as well as the extent and seasonal pattern of leaf litter (Frankie et al. 1974), it is not surprising that most species occupy these sites.

TABLE 20.4 Observed and mean expected (± 95% confidence limits) number of species in habitat preference and reproductive mode categories for La Selva amphibians and reptiles

Lizards				Snakes			
Habitat preference category	Observed	Expected	Z-score	Habitat preference category	Observed	Expected	Z-score
Leaf litter	8	7.0 ± 0.15	0.57	Leaf Litter	22	22.6 ± 0.28	−0.21
Terrestrial	3	4.2 ± 0.13	−1.25	Terrestrial	9	8.5 ± 0.17	0.27
Trunk	5	4.9 ± 0.13	0.07	Semiarboreal	7	8.4 ± 0.18	−0.81
Crown	3	3.7 ± 0.12	−0.51	Arboreal	15	12.8 ± 0.23	0.92
Twig	2	1.2 ± 0.07	0.96	Aquatic	2	2.7 ± 0.13	−0.52
Aquatic	4	2.5 ± 0.10	1.27				

Leptodactylid frogs				Frogs			
Habitat preference category	Observed	Expected	Z-score	Reproductive mode category	Observed	Expected	Z-score
Leaf litter	4	3.2 ± 0.10	0.66	Water	16	17.8 ± 0.23	−1.09
Shrub	5	4.3 ± 0.12	0.52	Foam	2	3.3 ± 0.11	−1.06
Arboreal	4	3.0 ± 0.08	1.04	Leaf	10	7.0 ± 0.17	1.54
Aquatic	3	4.9 ± 0.12	−1.38	Terrestrial	16	13.6 ± 0.19	0.63

Note: Expected distributions based on 500 randomizations

Reproductive modes of amphibians at La Selva fall into four major categories: (1) species that lay eggs in water in which the tadpoles emerge, grow, and eventually metamorphose; (2) species that create floating foam nests into which eggs are deposited; (3) species that lay eggs on vegetation above a water source into which tadpoles drop upon hatching; and (4) species that lay eggs on land and that transport them to water upon hatching or that hatch directly into froglets. The presence of such diverse modes of reproduction within the herpetofauna at La Selva is undoubtedly related to the high annual rainfall and relatively short dry season, which allows for deposition of eggs on leaf surfaces out of water as well as on land (table 20.4).

From these data one might ask whether the relative frequencies of species exhibiting these habitat preferences and reproductive modes at La Selva differ from those at other Neotropical sites. I used a G-test of independence on counts of habitat and reproductive mode categories at La Selva versus counts at other Neotropical sites to determine if amphibians and reptiles at La Selva differed in habitat preference or reproductive modes. In general, no differences were observed between amphibians and reptiles at La Selva and those at other Neotropical sites (table 20.5). The two notable exceptions were in the direction of increased numbers of terrestrial and leaf breeders at La Selva relative to the assemblages at Las Cañas and for increased numbers of leaf-litter and arboreal leptodactylid frogs at La Selva relative to frogs at Las Cañas. Thus, despite differences in climate and forest structure among these Neotropical sites, the relative proportions of various amphibian and reptilian habitat and reproductive mode preferences cannot be distinguished from those at other sites, except for frogs at La Selva versus those of dry forest sites. This could be interpreted in at least two ways. It may be that the categories traditionally used to describe groups within Neotropical amphibian and reptilian assemblages are too crude to allow meaningful differentiation among sites. Alternatively, the factors responsible for determining the relative number of species within each habitat type at La Selva may occur over great geographic distances and diverse habitat types.

Another way to test whether the observed habitat preferences of La Selva amphibians and reptiles have unique characteristics is to compare these values with expected values based on randomly created assemblages. To perform this, I repeated the area by species randomization procedure and recorded the species "colonizing" La Selva during each randomization. From these lists, I generated expected counts within each

TABLE 20.5 G-test statistics for tests of independence for numbers of species in habitat preference and reproductive mode categories observed at La Selva versus other Neotropical sites

	Tortuguero	Rincón de Osa	Barro Colorado Island	Las Cruces	Monteverde	Los Cañas	Santa Cecilia	Iquitos
Frogs (reproductive modes)	1.24	2.21	0.61	1.36	8.61*	10.03*	1.61	—
Leptodactylids	2.93	1.32	0.65	1.39	2.85	15.74*	2.40	—
Lizards	1.65	1.21	1.53	1.18	9.48	9.32	8.89	5.23
Snakes	4.31	0.86	5.42	4.05	2.52	2.38	1.15	3.52

* = $P < 0.05$

habitat preference category based on published descriptions of the ecology of each species. Thus, each species was assumed to remain unchanged by "colonization." These random assemblages were used to generate an expected distribution of habitat preferences. The observed values were compared to these expected distributions with z-scores (Zar 1974) to determine if the single observed value differed significantly from random expectation.

The above analysis yielded no significant difference from random expectation for any habitat preference category in any group of amphibians and reptiles at La Selva (table 20.4). Again, these findings may indicate that the categories are too crude or that all Neotropical sites tend to have the same relative proportions within these habitat preference categories. The latter would indicate that factors shaping habitat preference are widespread throughout the Neotropics.

Community-Level Patterns

Much of the community-level research on amphibians and reptiles at La Selva has centered on leaf-litter organisms (Scott 1976; Lieberman 1986). This represents a habitat preferred by many species of lizards, snakes, and leptodactylid frogs. There is a strong seasonal component to this habitat at La Selva because leaves tend to accumulate during the dry season, when many trees lose their leaves, and tend to decompose during the wet season, when leaf drop is reduced and mechanical and biological degradation increases (Frankie et al. 1974; Lieberman 1986). This results in a seasonal cycle of abundance of arthropods within the litter, with abundance peaking late in the dry season (Lieberman and Dock 1982).

In spite of the clear seasonal aspect to the leaf-litter habitat, population dynamics of species of leaf-litter amphibians and reptiles from La Selva and surrounding areas have been variously described as stable and constant to cyclic (e.g., *Norops humilis*—Fitch 1973; Lieberman 1986). Additionally, a variety of factors have been proposed as being responsible for regulating leaf-litter amphibian and reptilian abundance in the Neotropics. Principle among these are competition (Wright 1979), predation (Andrews 1979; Talbot 1979; Wright et al. 1984), food (Guyer 1988), and nest or rearing sites (Andrews 1982; Donnelly 1987).

A problem with interpreting these data is a lack of consistency with respect to area sampled and timing of samples. In addition, serial samples of sufficient size to describe cycles of abundance are known from only a very few species at La Selva. One exceptional case is *Norops humilis*, a

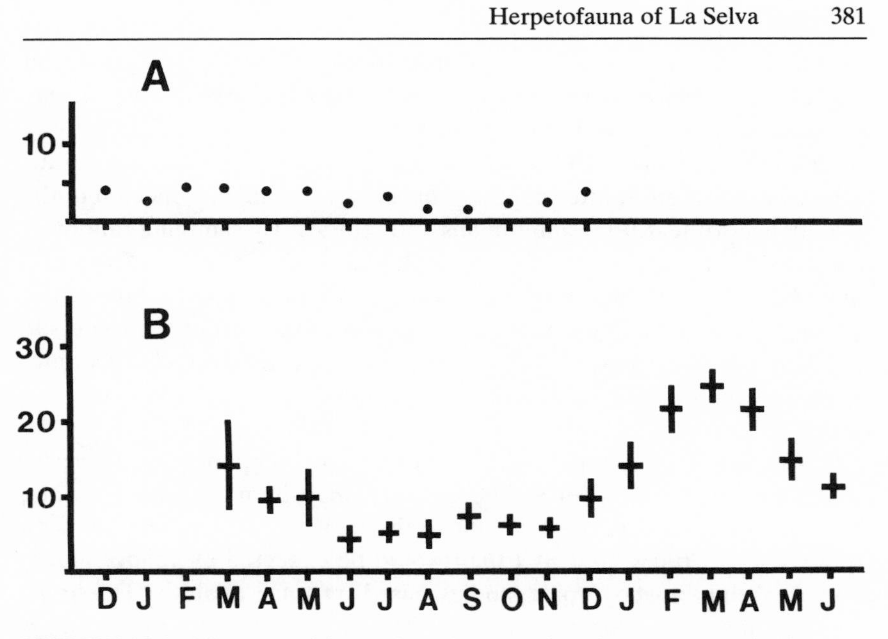

FIGURE 20.1. Mean monthly number of *Norops humilis* captured in (*top*) randomly selected litter plots (after Lieberman 1986) and (*bottom*) weekly samples of a repeatedly censused plot (after Guyer 1988).

leaf-litter anole. It can serve as an example of how spacing and timing of samples affect interpretations of stability and predictability of population density. *Norops humilis* has been studied at La Selva as part of a survey of the structure and function of the leaf-litter amphibian and reptilian community (Lieberman 1986). These data represent 8 m × 8 m litter-plot samples taken over thirteen continuous months at randomly selected sites in primary forest and a fallow cacao plantation. Data are also available from a detailed population study in a fallow cacao plantation (Guyer 1988). These data represent repeated weekly sampling of small study plots over seventeen continuous months.

The pattern generated by the two sampling regimes differ dramatically (fig. 20.1). Monthly samples at random sites show a weak but significant trend toward increased abundance during the dry season (Lieberman 1986). Weekly samples from single study plots also show significantly increased abundance during the dry season (Guyer 1988). These data indicate, however, that litter amphibian and reptilian populations may display predictable cycles of abundance tied to cycles of leaf drop. I have hypothesized (Guyer 1988) that the appropriate scale for studying amphibian and reptilian population or community dynamics in primary forest at La Selva is the area under single forest trees. Because tree species display differences in timing and intensity of leaf drop (Frankie et al. 1974), the litter community should reflect these differences by resembling a mosaic of sites with different litter depths. This in turn should be reflected in the abundances of leaf-litter amphibians and reptiles. By sampling randomly chosen sites throughout the forest, highly predictable patterns of seasonal abundance may be smoothed or obscured. Mark-recapture data representing repeated samples of amphibians and reptiles from the same site are sorely needed to expand knowledge of community structure of amphibians and reptiles at La Selva.

Amphibians and reptiles form an important component of the vertebrate community at La Selva. The species composition of amphibians and reptiles at La Selva is more similar to the faunas at Tortuguero, Rincón de Osa, Las Cruces, and BCI than would be expected by chance alone, corroborating biogeographic studies based on track analysis (Duellman 1966; Müller 1973; Savage 1966, 1982). The fauna at La Selva differs significantly in species composition from lowland wet forest sites in Amazonian South America. This may reflect the separation between Central and South America until the Miocene (Savage 1982). No difference from random expectation was found for the number of shared species between La Selva and Monteverde or Las Cañas. Any biological factor describing how species composition of the amphibians and reptiles of La Selva came about must include eastern lowland wet forest of Central America.

When grouped into ecological categories based on habitat preference or reproductive mode, relative frequencies of species within each category, in general, were found not to differ between La Selva and other Neotropical sites. The exception to this trend was increased numbers of leaf-litter and arboreal leptodactylid frogs and increased diversity of reproductive modes at La Selva relative to dry forest sites. The observed frequencies of habitat and reproductive modes at La Selva did not differ from random expectation. These findings indicate either that the traditionally described habitat categories are too crude or that the factors regulating community structure at this level are widespread throughout the Neotropics. Recent population studies of leaf-litter amphibians and reptiles at La Selva indicate that some species exhibit predictable cycles of abundance and that the appropriate scale for studying population and community structure within this habitat is the area under single forest trees. Detailed population studies based on mark-recapture data are needed to expand knowledge of community ecology of amphibians and reptiles at La Selva.

REFERENCES

Andrews. R. M. 1979. Evolution of life histories: a comparison of *Anolis* lizards from matched island and mainland habitats. Breviora 454: 1–51.

———. 1982. Spatial variation in egg mortality of the lizard *Anolis limifrons*. Herpetologica 38: 165–171.

Andrews, R. M., A. S. Rand, and S. Guerrero. 1982. Seasonal and spatial variation in the annual cycle of a tropical lizard. In G. J. Rhodin and K. Miyata (eds.), *Advances in Herpetology and Evolutionary Biology: Essays in Honor of E. E. Williams*. Harvard Univ. Press, Cambridge, Mass., pp. 441–454.

Denslow, J. S. 1980. Gap partitioning among tropical rainforest trees. Biotropica 12(suppl.): 47–55.

Dixon, J. R. and P. Soini. 1975. The reptiles of the upper Amazon Basin, Iquitos region, Peru. I. Lizards and amphisbaenians. Contrib. Biol. Geol. Milwaukee Pub. Mus. 4: 1–58.

———. 1977. The reptiles of the upper Amazon basin, Iquitos region, Peru. II. Crocodiles, turtles, and snakes. Contrib. Biol. Geol. Milwaukee Pub. Mus. 12: 1–91.

Donnelly, M. A. 1987. Territoriality in the strawberry poison-dart frog *Dendrobates pumilio*. Ph.D. diss., Univ. of Miami, Coral Gables.

———. 1991. The biology of the La Selva amphibians. In L. A. McDade, K. S. Bawa, H. A. Hespenheide, and G. S. Hartshorn (eds.), *La Selva: Ecology and Natural History of a Neotropical Rainforest*. Univ. of Chicago Press, Chicago.

Duellman, W. E. 1966. The Central American herpetofauna: an ecological perspective. Copeia 1966: 700–719.

———. 1978. The biology of an equatorial herpetofauna in Amazonian Ecuador. Misc. Publ. Mus. Nat. Hist. Univ. Kans. 65: 1–352.

Duellman, W. E. and L. Trueb. 1986. *Biology of Amphibians*. McGraw-Hill, New York.

Fitch, H. S. 1973. Population structure and survivorship in some Costa Rican lizards. Occas. Pap. Mus. Nat. Hist. Univ. Kans. 18: 1–41.

Frankie, G. W., H. G. Baker, and P. A. Opler. 1974. Comparative phenological studies of trees in tropical wet and dry forest in the lowlands of Costa Rica. J. Ecol. 62: 881–919.

Guyer, C. 1988. Demographic effects of food supplementation in a tropical mainland anole, *Norops humilis*. Ecol. 69: 350–361.

———. 1990. The reptile fauna of La Selva: comparisons with other Neotropical sites. In L. A. McDade, K. S. Bawa, H. A. Hespenheide, and G. S. Hartshorn (eds.), *La Selva: Ecology and Natural History of a Neotropical Rainforest*. Sinauer, Sunderland, Mass.

Hartshorn, G. S. 1980. Neotropical forest dynamics. Biotropica 12(suppl.): 23–30.

Hayes, M. P., J. A. Pounds, and W. Timmerman. In press. An annotated field list and guide to the amphibian and reptile fauna for the Monteverde region of the Cordillera de Tilaran, Costa Rica.

Holdridge, L. 1967. *Life Zone Ecology*. San José, Costa Rica.

Lieberman, D., M. Lieberman, R. Peralta, and G. S. Hartshorn. 1985. Mortality patterns and stand turnover rates in a tropical forest in Costa Rica. J. Ecol. 73: 915–924.

Lieberman, S. S. 1986. Ecology of the leaf litter herpetofauna of a Neotropical rainforest: La Selva, Costa Rica. Acta Zool. Mex., n.s., 15: 1–71.

Lieberman, S. S. and C. F. Dock. 1982. Analysis of the leaf litter arthropod fauna of a lowland tropical evergreen forest sites (La Selva, Costa Rica). Rev. Biol. Trop. 30: 27–34.

Miyamoto, M. M. 1982. Vertical habitat use by *Eleutherodactylus* frogs (Leptodactylidae) at two Costa Rican localities. Biotropica 14: 141–144.

Müller, P. 1973. The dispersal centres of terrestrial vertebrates in the Neotropical realm. Biogeographica 2: 1–244.

Myers, C. W. and A. S. Rand. 1969. Checklist of amphibians and reptiles of Barro Colorado Island, Panama, with comments of faunal change and sampling. Smithsonian Contrib. Zool. 10: 1–11.

Pacala, S. and J. Roughgarden. 1984. Control of arthropod abundance by *Anolis* lizards on St. Eustatius (Neth. Antilles). Oecologia 64: 160–162.

Pough, F. H. 1983. Amphibians and reptiles as low-energy systems. In W. P. Aspey and S. I. Lustick (eds.), *Behavioral Energetics: The Cost of Survival in Vertebrates*. Ohio State Univ. Press, Columbus, pp. 141–188.

Savage, J. M. 1966. The origins and history of the Central American herpetofauna. Copeia 1966: 719–766.

———. 1982. The enigma of the Central American herpetofauna: dispersal or vicariance. Ann. Mo. Bot. Gard. 69: 464–547.

Schoener, T. W. and D. A. Spiller. 1987. Effects of lizards on spider populations: manipulative reconstruction of a natural experiment. Science 236: 949–952.

Scott, N. J. 1976. The abundance and diversity of the herpetofauna of tropical forest litter. Biotropica 8: 41–58.

———. 1982. The herpetofauna of forest litter plots from Cameroon, Africa. In N. J. Scott (ed.), *Herpetological Communities*. Wildlife Res. Rep. 13, U.S. Fish and Wildlife Serv., Washington, D.C., pp. 145–150.

Scott, N. J., J. M. Savage, and D. C. Robinson. 1983. Checklist of amphibians and reptiles. In D. H. Janzen (ed.), *Costa Rican Natural History*. Univ. of Chicago Press, Chicago, pp. 367–374.

Simberloff, D. S. and E. F. Connor. 1979. Q-mode and R-mode analyses of biogeographic distributions: null hypotheses based on random colonization. In G. P. Patil and M. Rosenzweig (eds.), *Contemporary Quantitative Ecology and Related Econometrics*. International Co-operative Publishing House, Fairfield, Md., pp. 123–138.

Talbot, J. J. 1979. Time budget, niche overlap, inter- and intraspecific aggression in *Anolis humilis* and *Anolis limifrons* from Costa Rica. Copeia 1979: 472–481.

Toft, C. A. 1980. Feeding ecology of thirteen syntopic species of anurans in a seasonal tropical environment. Oecologia 45: 131–141.

Vitt, L. J. and L. D. Vangilder. 1983. Ecology of a snake community in northeastern Brazil. Amphibia-Reptilia 4: 273–296.

Williams, E. E. 1983. Ecomorphs, faunas, island size, and diverse end points in island radiations of *Anolis*. In R. B. Huey, E. R. Pianka, and T. W. Schoener (eds.), *Lizard Ecology: Studies of a Model Organism*. Harvard Univ. Press, Cambridge, Mass., pp. 326–370.

Wright, S. J. 1979. Competition between insectivorous lizards and birds in central Panama. Amer. Zool. 19: 1145–1156.

Wright, S. J., R. Kimsey, and C. J. Campbell. 1984. Mortality rates of insular *Anolis* lizards: a systematic effect of island size? Amer. Nat. 123: 134–142.

Zar, J. H. 1974. *Biostatistical Analysis*. Prentice-Hall, Englewood Cliffs, N.J.

21

The Herpetofauna of Barro Colorado Island, Panama: An Ecological Summary

A. STANLEY RAND AND CHARLES W. MYERS

Barro Colorado Island (BCI), in central Panama, is a partially submerged hill that was isolated from the adjacent mainland by the formation of Gatun Lake in 1914. Owing its existence to the construction of the Panama Canal, BCI comprises 1,500 ha and has a maximum elevation of 137 m above Gatun Lake, or 164 m above sea level. It is the most intensively studied site of the world's tropical forest ecosystems (e.g., Croat 1978; Leigh, Rand, and Windsor 1982).

Early herpetological studies on BCI emphasized faunal surveys. Eighty-three of the 106 species of amphibians and reptiles recorded from BCI had been collected by 1930. Another 12 species were taken by 1940, and a further 8 were added by 1967 (Myers and Rand 1969:8). Three additional species have been identified on BCI since Myers and Rand (1969) published their checklist.

Over the past few decades, most herpetological research done on BCI has focused on only part of the fauna. Heatwole and Sexton (1966) and Toft (1980, 1981) looked at a variety of litter inhabitants, but most studies have concentrated on one or another of the common species of lizards and frogs, and we have little information about less frequently encountered members of other groups.

Three species have been particularly intensively studied on BCI: *Anolis limifrons*, a small insectivorous lizard of the forest understory, has

For unpublished observations on the herpetofauna of BCI and adjacent areas, we thank Fernando Crastz, Bonifacio De León, Robert L. Dressler, Roberto Ibáñez, Fidel and César Jaramillo, James R. Karr, and Neal G. Smith. We also acknowledge our indebtedness to numerous authors whose papers were consulted but not explicitly cited herein.

received attention from R. M. Andrews, S. Guerrero, A. S. Rand, and O. J. Sexton; *Iguana iguana*, a large herbivorous lizard of the forest canopy and forest edge, has been studied by B. C. Bock, G. B. Burghardt, B. A. Dugan, H. W. Greene, A. S. Rand, and O. J. Sexton; and *Physalaemus pustulosus*, a forest-floor frog that breeds in small temporary pools, has been studied by B. D. Brattstrom, A. S. Rand, M. J. Ryan, O. J. Sexton, and M. D. Tuttle. In nearby Gamboa, communication in three small tree frogs (*Hyla ebraccata, H. microcephala,* and *H. phlebodes*) that breed in temporary marshes and ponds and at the edges of permanent bodies of water, have received extensive study by P. M. Narins, J. J. Schwartz, T. L. Taigen, and K. D. Wells.

In this chapter we try to summarize what is known in a broad sense about the ecology and natural history of amphibians and reptiles on BCI and adjacent parts of the mainland. The many question marks in tables 21.1 (amphibians) and 21.2 (reptiles) will perhaps reinforce Myers's (1972: 205) assertion that "most biologists would be genuinely shocked at how little we really know about the biota" of this long-studied region. It seems safe to assume that reasonably complete life-history and ecological studies will never be made of some of the particularly rare species—for example, the little snake *Tantilla albiceps*, still known to science only from the holotype specimen collected on BCI in 1925! Nonetheless, many species are amenable to study in ways limited only by the imagination and energy of the investigator, and we hope that this summary will encourage further work.

Biogeography

Although the Isthmus of Panama was established perhaps as recently as 3–5 million years ago, this narrow land bridge has come to support a herpetofauna of about 400 species (Duellman and Myers 1980), this number being somewhere in the range of 3–4% of the world's living species of amphibians and reptiles. Where did this rich fauna come from?

Savage (1982:table 4) recognized four historical assemblages in the Central American herpetofauna. About 25% of the genera recorded herein from the BCI region are assigned by Savage to an Old Northern Unit, about 32% to a Middle American Unit, and about 43% to a South American Unit. This does not mean that one-fourth of the genera migrated all the way from North America to central Panama (and farther) once the isthmus was in place—Savage's Old Northern genera had a secondary center of evolution in Central America. As also emphasized by

Vanzolini and Heyer (1985), few *species* of amphibians and reptiles dispersed over the Panamanian land bridge all the way from North to South America or vice versa (in part because of the obstacle of a richly endemic Central American fauna already in place). There was, however, considerable dispersal between Central America and northern South America. Thus, the Panamanian lowland herpetofauna, including that of BCI, is a mixture of species typical of adjacent parts of Central America and South America. (The same holds true for the highland herpetofaunas of Panama, except that there is a higher frequency of endemic species in the uplands.)

The lowland Central American herpetofauna can be roughly divided between two ecological assemblages (Duellman 1966): (1) that associated with more mesic forests, primarily on the Atlantic coast (but also including the Golfo Dulce region of southwestern Costa Rica), and (2) that associated with more xeric forests and open formation, primarily on the Pacific coast. These assemblages (comprising species originating from both Central and South America) are largely separated by the cordilleras that carry the continental divide through Central America, but BCI is situated in a structural depression across the center of Panama, and some representatives of the Atlantic and Pacific assemblages meet there. Authors sometimes seem to assume that these two assemblages simply intersect in central Panama, on their way to or from correspondingly wet or dry areas on the opposite coasts of northern South America. This is true in a historic sense only, for the present-day "Panamanian X" is a broken letter (Myers 1972): the dry Pacific-side assemblage is largely absent in the eastern half of Panama, which separates a number of Panamanian species from conspecific populations in arid northern Colombia. Eastern Panama must once have provided a relatively dry Pleistocene corridor between Pacific-western Panama and extreme northern South America. Although eastern Panama is again an area of humid forest, it nonetheless forms a geographic gap even in the ranges of some taxa representative of the wet Atlantic assemblage (e.g., Myers, Daly, and Martínez 1984:19).

Therefore, BCI and the adjacent lowlands support a complex herpetofauna that has been derived at different times and in different climates from varied ecogeographic sources. But it bears mentioning that the BCI area does *not* seem to be a region in which different assemblages of related species are in secondary contact. Such zones are best characterized by hybridization events, which have not been detected in the local herpetofauna. The BCI faunal mosaic has few pairs of closely related species.

Recent history. The damming of the Río Chagres, which created Gatun Lake and turned BCI into an island, drowned the extensive floodplain habitats that provided marshes, breeding ponds for frogs, and sandbanks for turtles, iguanas, and crocodilians. Many of these species now use human-made and maintained clearings and ponds around the margins of Gatun Lake. In dry years, the floodplains probably also provided moist refuges that are no longer available. The lake has essentially eliminated recolonization of BCI by species that have become extinct there. This has probably most affected such species as *Dendrobates minutus* (now in the genus *Minyobates*) and *Rana warschewitschii*, for which BCI has only marginal habitats. These are two of the five species that Myers and Rand (1969) considered to have become extinct on BCI since the 1920s and 1930s.

The other three extirpated species (*Bufo granulosus, Anolis auratus,* and *Gonatodes albogularis*) were probably victims of succession. All are common in more open habitats and occur on the adjacent mainland. Since the early days of the canal, the forest has gradually matured, with reduction of clearings and young second-growth on BCI and surrounding areas. Of these species, the two lizards have been reestablished by humans in the laboratory clearing on BCI.

Present Fauna

Excluding the 3 extirpated anurans, 103 species are currently listed for BCI (tables 21.1, 21.2): 1 caecilian, 2 salamanders, 30 frogs, 2 crocodilians, 5 turtles, 1 amphisbaenian, 22 lizards, and 40 snakes. This adds 5 species to the 1969 checklist: the 2 reintroduced lizards mentioned above; a small, as yet unnamed, anole of the *fuscoauratus* group that was overlooked earlier because of its close resemblance to *A. limifrons*; a small frog (*Hyla microcephala*) that has successfully invaded marshes along the margin of the island; and a coral snake mimic (*Erythrolamprus bizona*), of which a single individual was found in the forest. The last species is a conspicuous diurnal forager unlikely to have been overlooked (even at low density) for over half a century, and it almost certainly represents a new colonist, as we predicted in 1969. Although we know of no extinctions since 1969, no attempts have been made to search for the less common species. Two of the 10 species scored as rare (a category including possibly extinct species) in the 1969 checklist have been collected since then (*Phrynohyas venulosa* and *Rhinoclemmys funerea*).

TABLE 21.1 Ecological summary of 52 species of amphibians from Barro Colorado Island and adjacent areas[1]

Taxon	BCI[2]	Apr[3]	Abun[4]	Habitat[5]	Diel[6]	Food[7]	Season[8]	Rep. Mode[9]	Pond[10]
GYMNOPHIONA (1 species)									
Oscaecilia ochrocephala	X	con	I	Fos	?	?	?	?	–
CAUDATA (2)									
Oedipina complex	X	con	I	Fos?&Gnd	N	I	?	TeDD	–
O. parvipes	X	con	I	Fos?&Gnd	N	I	?	TeDD	–
ANURA (49)									
Bufonidae (4)									
Bufo granulosus	–	con	I	Gnd	N	I	pw	Let	temp
B. haematiticus	–	con	U	Gnd	N	I-A	ew	Set	–
B. marinus	X	con	U	Gnd	N	I,V	pd	Let	perm
B. typhonius alatus	X	var	C	Gnd	D	I-A	ew	Set	–
Centrolenidae (5)									
Centrolenella granulosa	–	con	C	Bsh	N	I	pw	AeSt	–
C. colymbiphyllum	–	con	U	Bsh	N	I	pw	AeSt	–
C. fleischmanni	X	con	C	Bsh	N	I	pw	AeSt	–
C. prosoblepon	X	con	R	Bsh	N	I	pw	AeSt	–
C. spinosa	X	con	U	Bsh	N	I	pw	AeSt	–
Dendrobatidae (4)									
Dendrobates auratus	X	var	U	Gnd&Tr	D	I-A	pw	TeAt	temp
Colostethus flotator	X	con	C	Gnd	D	I-Gen	pb	TeSt	–
C. inguinalis	–	con	U	AqMrg	D	I-Gen	pb	TeSt	–
C. talamancae	–	con	U	Gnd	D	I	pb	TeA(&S?)t	–
Hylidae (17)									
Agalychnis calcarifer	X	con	R	Tr	N	I	?	AeA(&L?)t	temp
A. callidryas	X	var	C	Tr	N	I	pw	AeLt	t&p
A. spurrelli	X	var	R	Tr	N	I	pw?	AeA&Lt	temp
Hyla boans	–	con	U	Tr	N	I	pd	SetCB	–
H. crepitans	–	con	U	Bsh	N	I	pw	Let	temp

H. ebraccata	–	var	C	Bsh	N	I	pw	AeLt	t&p
H. phlebodes	X	con	U	Bsh	N	I	pw	Let	t&p
H. microcephala	X*	con	C	Bsh	N	I	pb	Let	t&p
H. rosenbergi	–	con	U	Bsh&Tr	N	I	pb	SetCB	t&p
H. rufitela	X	con	U	Bsh?	N	I	pw	Let	temp
Ololygon boulengeri	X	con	C	Bsh	N	I	pb	Let	perm
O. rostrata	–	con	U	Bsh	N	I	pw	Let	perm
O. rubra	–	con	C	Bsh	N	I	ew	Let	temp
O. staufferi	–	con	C	Bsh	N	I	pw	Let	temp
Phrynohyas venulosa	X	var	I(C)	Tr	N	I	ew	Let	temp
Smilisca phaeota	X	var	R(I)	Bsh&Tr	N	I	pw	Let	temp
S. sila	X	var	U	Bsh&Tr	N	I	pd	Set	–
Leptodactylidae (16)									
Eleutherodactylus biporcatus	X	con	R	Gnd	ND	I	?	TeDD	–
E. bransfordii	–	var	U	Gnd	ND	I-NA	?	TeDD	–
E. bufoniformis	X	con	U	AqMrg	N	I-NA	?	TeDD	–
E. cerasinus	X	var	I	Gnd&Bsh	ND?	I-NA	?	TeDD	–
E. crassidigitus	X	var	I	Gnd&Bsh	ND?	I-NA	pw	TeDD	–
E. diastema	X	con	C	Bsh	N	I	pw	TeDD	–
E. fitzingeri	X	var	C	Gnd&Bsh	N	I-NA	pw	TeDD	–
E. gaigeae	X	con	R	Gnd	N	I	?	TeDD	–
E. ridens	X	var	I	Gnd&Bsh	N	I	?	TeDD	–
E. taeniatus	X	var	C	Gnd&Bsh	N	I-NA	?	TeDD	–
E. vocator	–	con	C	Gnd & Bsh	N	I-Gen	pw	TeDD	–
Leptodactylus insularum	X	con	R	Gnd	N	I	pb	LfeLt	t&p
L. labialis	–	con	C	Gnd	ND	I	pw	LfeLt	temp
L. poecilochilus	–	var	I	Gnd	N	I	ew	LfeLt	temp?
L. pentadactylus	X	con	C	Gnd	N	I,V	pw	LfeLt	temp
Physalaemus pustulosus	X	var	C	Gnd	ND	I-A	pb	LfeLt	temp
Microhylidae (1)									
Chiasmocleis panamensis	–	con	U	Fos	N	I	ew	Let	temp

TABLE 21.1 (*Continued*)

Taxon	BCI[2]	Apr[3]	Abun[4]	Habitat[5]	Diel[6]	Food[7]	Season[8]	Rep. Mode[9]	Pond[10]
Ranidae (2)									
Rana warschewitschii	–	con	U	AqMrg	D	I	e?w	Set	–
R. vaillanti	X	con	I	AqMrg	ND	I	p?w	Let	perm

Notes: [1]Ecological data include observations from the BCI area, with some extrapolations from other regions (especially for rare species)

[2]BCI: X = species known on BCI (* indicates a species added since Myers and Rand, 1969);—= additional species known from the adjacent lowlands, including the BCNM (peninsulas adjacent to BCI) and Parque Nacional Soberania (including the Pipeline Road area, Summit Gardens, Madden Forest, and the areas between them)

[3]Apr (appearance): con = constant, meaning little variation; var = noticeably variable or polymorphic color patterns not explained by sex or age

[4]Abun (abundance): C = common—one can find many individuals; U = usual—can find by looking in the appropriate habitat and season; I = infrequent—not predictable; R = rarely seen, some species possibly extinct on BCI. Added symbol in ( ) gives status in areas adjacent to BCI where different. Many species rare in the central lowlands are common in adjacent highlands or in other lowland parts of Panama.

[5]Habitat (nonbreeding habitat if different): AqMrg = aquatic margin, riparian; Bsh = bush and/or forest understory; Fos = fossorial; Gnd = ground and litter, terrestrial; Tr = tree, canopy

[6]Diel, time of activity: D = diurnal; N = nocturnal; ND = variably active day or night

[7]Food: I = insects and/or other invertebrates; V = small vertebrates such as frogs. Following categories based on Toft (1981); I-A = mainly ants; I-NA = non-ants; I-Gen = generalist, both ants and non-ants.

[8]Season, meaning seasonality of reproduction: p = prolonged breeding season; e = explosive breeding; w = wet season; d = dry season; b = both wet and dry

[9]Rep. Mode, type of reproduction: A = arboreal; L = lentic; S = stream; T = terrestrial; e = eggs; fe = eggs in foam nest; t = tadpoles; CB = constructed basin; DD = direct development. *Note*: At (arboreal tadpoles) refers to small pools in treeholes or in cavities on logs.

[10]Pond, for pond-breeding anurans only: temp = temporary pond; perm = permanent; t&p = both

TABLE 21.2 Ecological summary of 82 species of reptiles from Barro Colorado Island and adjacent areas[1]

Taxon	BCI[2]	Apr[3]	Abun[4]	Habitat[5]	Diel[6]	Food[7]	Season[8]	Rep. Mode[9]	Mimic[10]
CROCODYLIA (2 species)									
Caiman crocodilus	X	con	C	Aq	ND	FPT	e –dw	O	
Crocodylus acutus	X	con	C	Aq	ND	M&?	d –ww	O	
TESTUDINES (5)									
Chelydra acutirostris	X	con	R	Aq	?	O	?	O	
Rhinoclemmys annulata	X	con	I	Gnd	D	H	?	O	
R. funerea	X	con	R	Aq	D	O	? –w–	O	
Kinosternon leucostomum	X	con	I	Aq	N	O	P	O	
Trachemys scripta	X	con	C	Aq	D	O	P –dw	O	
AMPHISBAENIA (1)									
Amphisbaenidae (1)									
Amphisbaena fuliginosa	X	con	R	Fos	?	I	?	O	
SAURIA (27)									
Gekkonidae (5)									
Gonatodes albogularis	X*	con	U	TrTr	D	I	p –ww	O	
Hemidactylus frenatus	–	con	C	Edif	N	I	p	O	
Lepidoblepharis sanctaemartae	X	con	I	Gnd	D	I	p?	O	
Sphaerodactylus lineolatus	X	con	U	TrTr	D	I	p	O	
Thecadactylus rapicauda	X	con	U	TrTr	N	I	p?	O	
Iguanidae (15)									
Anolis auratus	X*	con	C	Bsh	D	I	p www	O	
A. biporcatus	X	con	I	TrLm	D	I	p?	O	
A. capito	X	var	R	TrTr	D	I	p?	O	
A. frenatus	X	con	U	TrTr	D	I	p www	O	
A. limifrons	X	var	C	Bsh	D	I	p www	O	
A. lionotus	X	con	R	AqMrg	D	I	p?	O	
A. pentaprion	X	con	I	TrLm	D	I	p? ––w	O	
A. poecilopus	–	con	U	AqMrg	D	I	p www	O	
A. tropidogaster	–	con	U	Bsh	D	I	p www	O	

TABLE 21.2 (*Continued*)

Taxon	BCI[2]	Apr[3]	Abun[4]	Habitat[5]	Diel[6]	Food[7]	Season[8]	Rep. Mode[9]	Mimic[10]
A. vittigerus	X	var	I	TrLm	D	I	p?	O	
A. sp. (*fuscoauratus* group)	X*	var	I	Bsh	D	I	p?	O	
Basiliscus basiliscus	X	con	C	AqMrg	D	IPH	p bbb	O	
Corytophanes cristatus	X	con	I	Bsh&TrTr	D	I	p?	O	
Iquana iquana	X	con	C	TrLm	D	H	e wdw	O	
Polychrus gutturosus	X	con	I	TrLm	D	IH	?	O	
Scincidae (1)									
Mabuya unimarginata	X	con	U	Gnd	D	I	?	V	
Teiidae (5)									
Ameiva ameiva	–	con	C	Gnd	D	I	p --w	O	
A. festiva	X	con	C	Gnd	D	I	p --w	O	
A. leptophrys	X	con	C	Gnd	D	I	p --w	O	
Gymnophthalmus speciosus	–	con	U	Gnd&Fos	D	I	p wdw	O	
Leposoma southi	X	con	R	Gnd	D	I	p www	O	
Xantusiidae (1)									
Lepidophyma flavimaculatum	X	con	I	Gnd	N?	I	e? --w	V	
SERPENTES (47 species)									
Anomalipididae (2)									
Anomalepis mexicanus	X	con	I	Fos	?	I	?	O?	
Liotyphlops albirostris	X	con	I	Fos	ND	I	?	O?	
Boidae (3)									
Boa constrictor	X	con	U	Gnd&Tr	ND	MBL	e --w	V	
Corallus annulatus	X	con	R	TrLm	N	MB	?	V	
Epicrates cenchria	X	con	I	TrLm&Gnd	N	MBF	?	V	
Colubridae (37)									
Amastridium veliferum	X	con	I	Gnd	D	?	?	O?	
Chironius carinatus	X	con	I	Gnd&Bsh	D	F	?	O	
C. grandisquamis	X	con	I	Gnd&Bsh	D	FU	?	O	
Clelia clelia	–	con	I	Gnd	ND	SLM	?	O	

Coniophanes fissidens	X	con	U	Gnd	D	FLSUI	?		O	
Dendrophidion percarinatum	X	con	U	Gnd	D	LF	?		O	
Dipsas variegata	–	con	I	Bsh(&Tr?)	N	G	?		O	
Drymarchon corais	X	con	I	Gnd	D	LSFMB	?		O	
Enulius flavitorques	X	con	I	Gnd	N?	?	?		O	
E. sclateri	X	con	I	Gnd	N?	?	?		O	
Erythrolamprus bizona	X*	con	R	Gnd	D	S	?		O	coral
Geophis hoffmanni	–	con	R	Gnd	N	I	?		O	
Imantodes cenchoa	X	con	U	Bsh&TrLm	N	LF	?		O	
I. gemmistratus	X	con	R	Gnd&Bsh	N	L	?		O	
I. inornatus	–	con	R	Bsh&TrLm	N	L	?		O	
Lampropeltis triangulum	–	con	R	Gnd	D	(LSMB)?	?		O	coral
Leptodeira annulata	X	con	I(U)	Gnd&Bsh	N	F	?		O	
L. septentrionalis	X	con	I	Bsh&TrLm	N	F	?		O	
Leptophis ahaetulla	X	con	I	Bsh&TrLm	D	F	?	–ww	O	
Liophis epinephelus	X	con	U	Gnd	D	F	?	–w–	O	
Mastigodryas melanolomus	X	con	I	Gnd	D	LFM	?		O	
Ninia maculata	X	con	I	Gnd	D	?	?		O	
Oxybelis aeneus	X	con	U	Bsh&TrLm	D	L	?		O	
O. fulgidus	–	con	I	Bsh&TrLm	D	LB	?		O	
Oxyrhopus petola	X	con	I	Gnd	N	LM	?		O	coral
Pliocercus euryzonus	X	con	I	Gnd	ND	FPU	?		O	coral
Pseudoboa neuwiedii	X	con	I	Gnd	N	L	?		O	
Pseustes poecilonotus	X	var	U	Gnd	D	BM	?		O	
Rhadinaea decorata	X	con	I	Gnd	D	FLU	?		O	
R. fulviceps	X	con	I	Gnd	ND	?	?		O	
Siphlophis cervinus	X	con	I	Gnd&Bsh	N	L	?		O	
Spilotes pullatus	X	var	U	Gnd&TrLm	D	BM	?		O	
Stenorrhina degenhardtii	X	con	I	Fos&Gnd	D	I	?		O	
Tantilla albiceps	X	?	R	Gnd (Fos?)	?	?	?		O?	
T. armillata	X	con	U	Gnd	D	I	?		O	
Trimetopon barbouri	X	con	I	Gnd	?	?	?		O	
Xenodon rabdocephalus	X	con	I	Gnd	D	F	?		O	viper

TABLE 21.2 (*Continued*)

Taxon	BCI[2]	Apr[3]	Abun[4]	Habitat[5]	Diel[6]	Food[7]	Season[8]	Rep. Mode[9]	Mimic[10]
Crotalidae (3)									
Bothrops atrox asper	X	con	R(U)	Gnd	ND	MLF	e --w	V	
B. schlegelii	X	var	R	Bsh&TrLm	N	FLMB	?	V	
Lachesis muta	–	con	I	Gnd	N	M	?	O	
Elapidae (2)									
Micrurus mipartitus	X	con	I	Gnd	ND	LS	?	O	coral
M. nigrocinctus	X	con	U	Gnd	ND	LSC	?	O	coral

Notes: [1]Ecological data include observations from the BCI area and from other regions (particularly for snakes, whose natural history is especially poorly known)

[2]BCI: X = species known on BCI (* indicates species added since Myers and Rand, 1969); – = additional species known from the adjacent lowlands, including the BCNM (peninsulas adjacent to BCI) and Parque Nacional Soberania (including the Pipeline Road area, Summit Gardens, Madden Forest, and the areas between them)

[3]Apr (appearance): con = constant, meaning little variation; var = noticeably variable or polymorphic color patterns not explained by sex or age

[4]Abun (abundance): C = common—one can find many individuals; U = usual—can find if you look in the appropriate habitat and season; I = infrequent—not predictable; R = rarely seen, some species possibly extinct on BCI. Added symbol in ( ) gives status in adjacent area where different. Many species rare in the central lowlands are common in adjacent highlands or in other lowland parts of Panama.

[5]Habitat (nonbreeding, nonsleeping habitat if different): Aq = aquatic; AqMrg = aquatic margin, riparian; Bsh = bush and/or forest understory; Edif = edificarian (exotic introduction); Fos = fossorial; Gnd = ground and litter, terrestrial; Tr = tree, canopy; TrLm = tree limb, canopy; TrTr = tree trunk

[6]Diel, time of activity: D = diurnal; N = nocturnal; ND = variably active day or night; H = heliotherm

[7]Food: B = birds; C = caecilians; F = frogs; G = gastropods; H = herbivorous; I = insects and/or other invertebrates; L = lizards; M = mammals; O = omnivorous; P = fish; S = snakes; T = turtles; U = salamanders. Range of food taken by many snakes is probably much greater than shown.

[8]Season, meaning seasonality of reproduction: p = prolonged breeding season; e = explosive breeding; w = wet season; d = dry season; b = both wet and dry. Following above, seasons of mating, oviposition, and emergence (or birth) of young are given sequentally where known (– = no data).

[9]Rep. Mode, type of reproduction: O = oviparous; V = viviparous

[10]Mimic: coral = coral-snake pattern; viper = resembles *Bothrops atrox*

Adding the species known from the adjacent mainland (the adjacent peninsulas that are part of the Barro Colorado Nature Monument, and the Parque Nacional Soberania, which includes Pipeline Road, Summit Garden, and Madden Forest) to the BCI list brings the total to 134 and gives a more balanced picture of the fauna of the area because of the wider range of habitats represented. Doing so adds 6 genera (5 of snakes, 1 of frogs) and 31 species (19 frogs, 7 snakes, and 5 lizards).

Although our knowledge of the adjacent region is much less complete than that of BCI, the species listed in tables 21.1 and 21.2 nonetheless provide a reasonably complete census of the lowland amphibian and reptile faunas in the *middle* of the central part of the Isthmus of Panama. Distributional details are complex, however, and it is important to remember that additional species are found on the northern (Atlantic) and southern (Pacific) sides of the Canal Zone, while still more are found in the low highlands to either side of the canal.

Notes on the taxonomy. In the 1969 checklist, we parenthetically included synonymous names and misidentifications that have been used in the extensive literature on BCI. This guide is still useful in determining what species were involved in earlier studies, but there have been additional changes since 1969. Most involve only some modification in spelling or choice of a different species name in the same genus (e.g., *Rana palmipes* to *R. vaillanti*), and such replacements should cause no trouble in the case of small genera. Generic and other changes that may be less obvious to the nonspecialist follow; the current proper name is given first, and the name given in 1969 follows in parentheses:

Frogs: *Colostethus flotator* (*nubicola*)*; *Eleutherodactylus cerasinus* (*cruentus*); *E. crassidigitus* (*longirostris*); *E. ridens* (*molinoi*); *E. taeniatus* (*ockendeni*); *Ololygon* (*Hyla*, part); *Physalaemus* (*Engystomops*)
Turtles: *Rhinoclemmys* (*Geoemyda*); *Trachemys* (*Pseudemys*)
Snakes: *Liophis epinephelus* (*Leimadophis epinephalus*); *Mastigodryas* (*Dryadophis*)

We have avoided subspecies names in tables 21.1 and 21.2 even more than in the 1969 checklist, mainly because patterns of geographic variation do not readily conform to the notion of regional subspecies in this

* *Colostethus flotator* was formerly in the synonymy of *C. nubicola*, but Ibáñez has demonstrated that they are sibling species that are microsympatric at some localities. BCI is the type locality of *flotator*, so the identification of that population is assured.

part of the world. We have retained such names only in two cases—one where there is unpublished reason to suspect that the subspecies name may become the proper species name (*Bufo typhonius alatus* under study by M. S. Hoogmoed, Rijksmuseum van Naturerlijke Historie, Leiden) and another where the name is being widely used as a species name with inadequate published documentation (*Bothrops atrox asper*).

Guyer and Savage (1986) recently proposed a major reclassification of the large lizard genus *Anolis*. Under their arrangement, the reader should be aware that all *Anolis* in table 21.2 become *Norops*, with the exception of *A. frenatus*, which becomes *Dactyloa*. We have not adopted this arrangement for two reasons: (1) there has been no time for other specialists to respond to the Guyer and Savage paper in terms of either their interpretation or the adequacy of the data (it is their interpretation, not the data, that is new); (2) even though an ecologist or other nonsystematist might know the species of "*Anolis*" in a given study area, there is no way to decide to which of the resurrected genera all the species belong—the characters are mainly osteological, and no comprehensive list of species was attempted.

Populations

Habitat. Local amphibians and reptiles occupy all available habitats: aquatic, riparian, subterranean, terrestrial, bush and understory, trunks of large trees, and canopy (see Leigh, Rand, and Windsor 1982 for descriptions of the physical and biotic settings of BCI).

Abundance. Species can be assigned to rough abundance categories. COMMON species are those that are conspicuous in the right habitat. For example, on BCI, *Basiliscus* at the boat dock and along edges of the coves, *Smilisca sila* calling at night during the dry season along stream beds, and *Trachemys scripta* basking on emergent logs in the lake. By day, one is seldom out of earshot of a calling *Colostethus flotator* along any forested stream. At any time of year at least a few *Physalaemus* are calling at ponds in the laboratory clearing. Near Gamboa, breeding choruses of small *Hyla* contain hundreds of males of several species; choruses including ten to fifteen species are predictable at the beginning of the rains. USUAL species are those of which at least one example can almost always be found if the appropriate habitat is searched. INFREQUENT species are those that have repeatedly turned up but cannot predictably be found. RARE species are those that are seldom encountered. Some "rare" species might be extinct on BCI, but others might actually be abundant.

Fossorial amphibians and reptiles, for example, are rarely seen, and we have no knowledge of their real densities.

It is seldom clear why some species are more common than others, and the problem is particularly acute in the case of snakes generally. One sees snakes infrequently on BCI—an average of one a day being a generous estimate (exceptional days of several sightings stand out in one's memory)—whereas they seem more common nearby on Pipeline Road. Although it is tempting to speculate that the absence of big carnivores on BCI has allowed small generalist carnivores, such as coatis, that eat snakes to increase in abundance and thus depress snake populations, the difference might simply be related to habitat diversity—particularly the greater abundance of ecological "edge" along Pipeline Road. In any case, one has to work hard to find snakes in *any* lowland humid forest habitat. Snakes are seen so infrequently that biologists are generally astounded to learn that there are many more species of snakes than lizards in a given lowland forest; only part of the reason can be attributed to differences in trophic level (see Myers and Rand 1969:9 for further discussion).

Quantitative estimates for a few common species of frogs and lizards are available. *Physalaemus pustulosus*: Ryan, Tuttle, and Taft (1981) reported 425 males calling at one time in a pond about 10 m × 20 m. *Iguana iguana*: Rand (1968) reported approximately 200 females gathering to nest in a 10 m × 10 m clearing on Slothia, an islet just off BCI. *Anolis limifrons*: At a high population density, Andrews and Rand (1982) estimated 150 individuals (including 86 adults) resident in a 25 m × 35 m plot.

There have been two systematic samplings of litter fauna: In eight 8 m × 8 m quadrats on BCI, in which the litter was searched, Heatwole and Sexton (1966) found a total of 15 species, with a mean of 12.3 individuals of reptiles and amphibians per quadrat. More than two-thirds of the individuals were *Anolis limifrons*; other species were represented by at most 1 or 2 individuals per quadrat. Toft (1980) reported a mean of 8.1 frogs per 36 m^2 census plot during the dry season on Pipeline Road.

Andrew (p.c.) recorded all reptiles and amphibians seen during censuses of *Anolis limifrons* in permanent quadrats on BCI and Pipeline Road, during each December, 1983–86. She also censused one quadrat at La Selva for two years. Both species and individuals are more common on Pipeline Road than on BCI, whereas the results for La Selva and Pipeline Road were very similar.

Population fluctuations. Major annual population fluctuations have been documented in three species on BCI and observed more casually in a number more. These fluctuations seem of as great or greater magnitude than those in populations of temperate reptiles and amphibians. The best studied species is *Anolis limifrons* (Andrews and Rand 1982, unpubl. data). The population followed since 1971 on a small study area has shown peaks in 1971 and 1976–78 and lows in 1975 and 1983, with a pattern of gradual declines and rapid increases. The greatest change between years was a sevenfold increase between 1975 and 1976. There seems to be general synchrony among populations on different parts of BCI but not in adjacent small islands and mainland areas (Andrews, in prep.).

The number of *Iguana iguana* nesting on Slothia has decreased markedly since they were first studied in the late 1960s. This is a local reduction; populations are still high only 3 km away on the other side of BCI. The change may be due to the appearance of nearby alternative nesting sites or to degradation of nearby juvenile habitat as vegetation has changed around the laboratory clearing. Since the number nesting on Slothia did not increase after our intensive study stopped in 1984, the decline was probably not due to the study itself.

For the forest tree frog *Hyla rufitela*, Rand, Ryan, and Troyer (1983) documented a dramatic increase in the number and distribution of places from which the species was heard calling. Since 1983 numbers have again declined, and the new calling sites, at least those around the laboratory clearing, have been abandoned.

Seasonality

The strong alternation of wet and dry seasons on BCI is reflected in the behavior of the reptiles and amphibians and is particularly conspicuous in the distribution and activity of the litter fauna and in the timing of reproduction.

As the dry season progresses, leaf litter accumulates, the understory dries out, and moist litter becomes increasingly restricted to bottoms of ravines. By season's end, concentrations of terrestrial frogs can be found in the stream bottoms (Toft 1980). When the rains come, these animals disperse. Toft (1980) found also that frogs were more abundant in the litter during the dry season even on upland plots in her study.

Some species breed opportunistically throughout the year, but even these are influenced by the seasons. *Physalaemus pustulosus*, for example, breeds whenever it finds a small body of standing water. During the dry season this is often a pool left in an intermittent stream; during the wet season it is likely to be a puddle atop a fallen log.

Other species have very restricted and synchronous breeding seasons. *Iguana* females, for example, lay just one clutch per year, and all females lay in a six-week period in the early dry season. An even more extreme example is *Phrynohyas venulosa*, a large forest-canopy tree frog that breeds only during the first heavy rains of the wet season and whose breeding seems as explosive as that of spadefoot toads in the deserts of the American Southwest. *Bufo typhonius* is another explosive breeder on BCI, although its chorusing seems not to be correlated with rainfall (Wells 1979).

Most species are intermediate and have a more prolonged breeding season. Breeding of forest species seems to occur primarily during the wet season, whereas breeding of the lake and lake-edge species generally occurs during the dry season, presumably when water levels are low and potential sites are exposed and less likely to flood. We know for some species (e.g., all *Anolis*, the geckos, and *Trachemys scripta*) that females lay more than one clutch per season, but for most we do not know the frequency of laying. As one would expect from the seasonal nature of breeding activity, populations that have been studied (*Anolis limifrons, Iguana iguana, Bufo typhonius*) show a seasonal peak of recruitment. Breeding in *Iguana* is timed so that young emerge when food is most abundant. This may also be true of *Bothrops atrox* and *Boa constrictor*, whose young are seen predominantly at the beginning of the rainy season. Conditions for incubation are probably as important for reptiles as are conditions for the growth and survival of tadpoles for anurans (Heyer 1976). Predator satiation may promote synchronous breeding or recruitment in some species (Greene et al. 1978; Wells 1979).

The rain itself seems to trigger breeding activity in some species (*Phrynohyas*), whereas in others (*Iguana*), different cues must be involved and day length may be implicated.

Diel Cycles

The herpetofauna includes diurnal species, nocturnal species, and others for which there is no clearcut pattern of activity. The last category seems to include both ambush hunters (e.g., *Boa constrictor*) and active foragers (e.g., *Micrurus nigrocinctus*), but virtually nothing is known about what triggers activity in species found abroad both day and night.

One might expect there to be seasonal and/or ontogenetic differences in the diel cycles of some such species.

Jaeger, Hailman, and Jaeger (1976) documented bimodal peaks of diel activity for the diurnal frog *Colostethus flotator* (under the name *C. nubicola*) on BCI. Jaeger and Hailman (1981) reported on the phototactic behavior of five additional nocturnal and diurnal frogs on the island and suggested that photic cues may aid in niche partitioning among species.

Life-History Strategies

A tabulation of the anurans by Duellman's (1989) classification of reproductive modes shows that several modes are represented on BCI, though a few of the most spectacular, such as pouch-brooding, are missing. Most species have some mechanism to prevent eggs from being submerged in water—laying eggs in a foam nest, as a surface film (sometimes in a protected basin constructed by the frog [e.g., see Kluge 1981]), on a leaf overhanging water, or in leaf litter—thus avoiding both aquatic predators and anoxic conditions. In the largest genus, *Eleutherodactylus*, development is direct, with the tadpole stage passed within a terrrestrial egg. Most frogs that breed in ponds do so in some sort of temporary pond, though some use both temporary and permanent water and a few breed primarily in permanent water. The use of temporary ponds probably reduces tadpole mortality from aquatic predators, particularly fish (Heyer 1976).

We have essentially no information on clutch size and number in the local populations of anurans.

Complex frog choruses—competition for signaling space. Because of differences in microhabitat and in timing calls, most male frogs call near others only of their own and perhaps one to three other species. At certain times and places, however, particularly in temporary ponds at the beginning of the rainy season, choruses of ten to fifteen species are encountered. Frogs probably compete for signal space, which has led to the evolution of adaptations in frequency, time, and spatial domains to reduce interference between heterospecific calls (Schwartz and Wells 1985).

Reptiles. The reptiles are predominately egg-layers, with several live-bearers among the snakes and lizards (table 21.2). The clutch size in many local lizards (anoles, geckos, and microteids) ranges from one to two eggs, and up to about seventy eggs in *Iguana*. Other species have intermediate-sized clutches. Central Panamanian populations of *Lepidophyma* seem to be composed only of females reproducing parthenogenetically (Telford and Campbell 1979). We know little about reproductive behavior and clutch or litter size in the local snakes, nor is there much direct information about the local behavior of the two crocodilians. Moll and Legler (1971:table 91) and Mittermeier (1971) summarized information on the local turtles.

Food and Foraging

Niche/guild. Most amphibians and reptiles appear to be generalist predators that feed on a range of prey species and types. *Iguana* is unusual in being a folivore, and several snakes, such as the snail-eating *Dipsas*, appear to have specialized diets. More commonly, species differ in the size of food taken and where and how it is sought. Toft (1980) argued that forest-floor frogs can be arranged along a continuum from those that forage actively on very small prey to those that sit-and-wait and ambush larger prey. She recognized three groups along this continuum: ant specialists, generalists, and non-ant specialists. In the nine species in her sample from Pipeline Road (Toft 1981), she found one ant specialist, two generalists, and six non-ant specialists. Similar differences among BCI lizards include extreme sit-and-wait predators, such as *Corytophanes*, and active foragers, such as *Ameiva*.

Competition for food. Competition for food between species, particularly among the insectivores, is probably diffuse and likely involves elements of the fauna other than reptiles and amphibians, such as birds (Wright 1979) and arachnids. Though species may occasionally be limited by food, as *Anolis limifrons* sometimes seems to be (Andrews, in prep.), intraspecific competition for food is probably rare. Heyer (1976) thought that interspecific competition is probably unimportant among the tadpoles that he studied on BCI. On Pipeline Road, however, Jaslow (1982) found that stream tadpoles (*Hyla boans* and *Smilisca sila*) competed in stream enclosures at densities encountered in the wild.

Predation and Antipredatory Devices

Not surprisingly, predation, particularly on egg (Andrews, in prep.) and larval stages (Heyer 1976) is a significant source of mortality. The importance of predation for adult BCI reptiles and amphibians is evidenced by the variety of antipredatory mechanisms exhibited. These include crypsis, polymorphism, aposematic coloration and mimicry, startle and threat displays, toxic and/or noxious skin secretions, venom, and adaptations in vocalizations and calling behavior of frogs.

Cryptic coloration and polymorphism. Most species are colored and patterned so that they somewhat match the background against which they normally rest. Beyond this, even when one discounts sexual dimorphism and ontogenetic changes, many of the common amphibians and reptiles show great individual variability in appearance. Some have distinct morphs, whereas others show more idiosyncratic individual differences, such as spot and blotch patterns, often asymmetrically arranged. Some vary individually in color, whereas others show rapid color change. Even the flash colors that are displayed briefly by a fleeing animal may be relevant.

Some of this variability may have evolved to facilitate intraspecific individual recognition. Some is probably aspect diversity (Rand 1967), evolved as the result of apostatic selection generated by visually hunting predators. Curiously, some polymorphic forms (*Bufo typhonius, Dendrobates auratus, Phrynohyas venulosa*) have toxic skin secretions that are noxious to many (not all) predators. Either the aspect diversity idea does not apply in such cases or some visually hunting predators can cope with the chemical defenses.

Warning coloration. Conspicuous coloration in toxic or noxious species is seen only in the black and bright metallic green *Dendrobates auratus,* the red and black, or red, yellow, and black coral snakes (*Micrurus* and their less toxic mimics). Possibly the black *Bufo marinus* tadpoles, which contain noxious bufodienolides (Daly, Myers, and Whittaker 1987), should be included here. In shallow water, these tadpoles form aggregations in which their black bodies are highly conspicuous.

Mimicry. The case for mimicry among snakes is best made by (1) evidence that some predators innately avoid coral snake patterns, and (2) striking examples of concordant geographic pattern variation among unrelated species (for review, see Greene and McDiarmid 1981). The two true coral snakes on BCI are quite different in banding pattern (red-yellow-black-yellow in *Micrucrus nigrocinctus*; red-black-red in the local population of *M. mipartitus*, which is geographically variable in Panama).

The bicolored *mipartitus* pattern of black and red is shared by local populations of the slightly venomous *Pliocercus euryzonus* and *Oxyrhopus petola*; all three are infrequently seen nocturnal species (*Pliocercus* has also been found by day in other areas). The tricolored *nigrocinctus* pattern is imperfectly matched by the mildly venomous *Erythrolamprus bizona* (red-black-white [or pale yellow]-black-red); these are largely diurnal snakes, although the *Micrurus* has also been found active by night. The nonvenomous, diurnal *Lampropeltis triangulum* is patterned like the *Erythrolamprus*, except that the interspaces between the double black bands usually seem to be red in lowland populations of this rare snake.

Several other snakes, though not banded, are conspicuously red and black (*Pseudoboa neuwiedii*, young *Clelia clelia*, and *Siphlophis cervinus*), although we do not suggest that these are mimics (the identical pattern in juvenile *Clelia* and the related *Pseudoboa* is probably a primitive one in the *Clelia-Pseudoboa* series). *Siphlophis* is uniquely patterned, with black and yellow bands and a broad red or orange dorsal stripe.

In contrast to the coral snake mimics, *Xenodon rabdocephalus* is strikingly similar in color and pattern to *Bothrops atrox*. There is no evidence of any sort of mimicry among the frogs or lizards.

Threat or startle displays. Several species of diurnal arboreal snakes have similar displays when frightened; they expand the anterior portion of the neck laterally and/or open the mouth exposing a contrasting color. These include *Leptophis ahaetulla*, *Oxybelis aeneus*, *Pseustes poecilonotus*, and *Spilotes pullatus*. Three related terrestrial species display dorsoventral hoods (tribe Xenodontini: *Erythrolamprus bizona*, *Liophis epinephelus*, *Xenodon rabdocephalus*). A number of snakes, both terrestrial and arboreal, vibrate their tails against the substrate as an obvious threat.

Many lizards and snakes, as well as crocodilians and turtles, threaten with open mouths and various other defensive postures. The crested lizard *Corytophanes* has a particularly impressive display of turning partly sideways to a predator and increasing its apparent size by laterally compressing its body, elevating its crest, expanding a throat fan, and threatening to bite.

Antipredator characteristics of frog vocalizations. Some calling frogs are exceedingly difficult to locate, whether because they call from concealment or because of characteristics of the call itself, and it is supposed that this is in part an adaptive response for avoiding predators that hunt by sound. Such predators at BCI include a giant toad (Jaeger 1976), philander opossums (Tuttle, Taft, and Ryan 1981), and bats (Tuttle and Ryan 1981). Tuttle and Ryan (1982) suggested that, in *Smilisca sila*, calls of varying complexity, synchronization of calls, and choice of call sites in response to ambient light and high noise level may all be related to avoiding bat predation. Lynch and Myers (1983:556–57) similarly suggested that call variability, a burst of calling at dusk, and infrequency of nocturnal calls serve as ways to avoid predators in *Eleutherodactylus crassidigitus* and *E. fitzingeri*.

Parasitism

All the reptiles and amphibians of BCI are presumed to be affected by, or to at least harbor, both ecto- and especially endoparasites. Some endoparasites, such as liver nematodes in *Anolis limifrons*, seem invariably fatal, whereas others, such as malarial organisms in the same species, have much more limited effects on their host (Rand et al. 1983). Some parasites, like the anole malarias, seem quite host specific (Guerrero et al. 1977); others, such as arboviruses and the bacteria *Salmonella* and *Arizona*, can infect a wide variety of vertebrates (Craighead, Shelokov, and Peralta 1962; Kourany, Myers, and Schneider 1970; Kourany and Telford 1981). Ectoparasites are conspicuous even to the casual observer. One regularly sees ticks on the shells of terrestrial turtles and on iguanas; orange mites can be conspicuous on anoles and other lizards and snakes. Blood-sucking diptera (at least mosquitoes and sandflies) feed on both lizards and frogs. Indeed, a surprising number of photographs of calling male frogs, show, on close inspection, mosquitoes in the process of feeding. We have little idea of the impact of parasites and/or haematophagous insects on either individuals or populations of amphibians and reptiles.

Impact on the Forest Ecosystem

Amphibians and reptiles occur throughout the area, and many species are often conspicuous and abundant. Unlike birds and mammals, the herpetofauna is not important in seed dispersal, pollination, or, except locally, in herbivory. At least in years of peak population density, lizards and frogs of the forest understory are probably important insectivores and important prey for carnivorous birds and bats. Lizards and frogs are always important prey for snakes, many of which eat little or nothing else.

Some snakes are nest-predators and may be a major source of mortality among birds. Snakes, in turn, are important prey for some forest hawks. During its peak years, *Anolis limifrons* is surely the most abundant vertebrate on BCI, although because of its small size it does not contribute the greatest biomass.

Reptiles and amphibians are potential reservoirs for diseases that may affect other organisms. The importance of this—indeed, the impact of disease in general in tropical forest—is almost completely unknown. Amphibians and reptiles are such a pervasive, albeit little-understood, influence in this tropical ecosystem that we find it impossible to visualize the chain reactions that would occur if the herpetofauna were suddenly to disappear.

REFERENCES

Andrews, R. M. and A. S. Rand. 1982. Seasonal breeding and long-term population fluctuations in the lizard *Anolis limifrons*. In E. G. Leigh, Jr., A. S. Rand, and D. M. Windsor (eds.), *The Ecology of a Tropical Forest: Seasonal Rhythms and Long-Term Changes*. Smithsonian Inst. Press, Washington, D.C., pp. 405–411.

Craighead, J. E., A. Shelokov, and P. H. Peralta. 1962. The lizard: a possible host for eastern equine encephalitis virus in Panama. Amer. J. Hyg. 76: 82–87.

Croat, T. B. 1978. Flora of Barro Colorado Island. Stanford Univ. Press, Stanford.

Daly, J. W., C. W. Myers, and N. Whittaker. 1987. Further classification of skin alkaloids from Neotropical poison frogs (Dendrobatidae), with a general survey of toxic/noxious substances in the Amphibia. Toxicon 25: 1023–1095.

Duellman, W. E. 1966. The Central American herpetofauna: an ecological perspective. Copeia 1966: 700–719.

———. 1989. Tropical herpetofaunal communities: patterns of community structure in Neotropical rainforests. In M. L. Harmelin-Vivien and F. Bourlière (eds.), *Vertebrates in Complex Tropical Systems*. Springer-Verlag, New York, pp. 61–88.

Duellman, W. E. and C. W. Myers. 1980. The Panamanian herpetofauna: historical biogeography and patterns of distribution. In W. G. D'Arcy (ed.), *Botany and Natural History: A Symposium Signalling the Completion of the "Flora of Panama."* Univ. of Panama, April 14–17, 1980, Abstracts and Program, pp. 46–48.

Etheridge, R. E. 1960. The relationships of the anoles (Reptilia: Sauria: Iguanidae): an interpretation based on skeletal morphology. Ph.D. diss., Univ. of Michigan.

Greene, H. W., G. M. Burghardt, B. A. Dugan, and A. S. Rand. 1978. Predation and the defensive behavior of green iguanas (Reptilia, Lacertilia, Iguanidae). J. Herpetol. 12: 169–176.

Greene, H. W. and R. W. McDiarmid. 1981. Coral snake mimicry: does it occur? Science 213: 1207–1212.

Guerrero, S., C. Rodriguez, and S. C. Ayala. 1977. Prevalencia de hemoparásitos en lagartijas de la isla de Barro Colorado, Panamá. Biotropica 9: 118–123.

Guyer, C. and J. M. Savage. 1986. Cladistic relationships among anoles (Sauria: Iguanidae). Syst. Zool. 35: 509–531.

Heatwole, H. and O. J. Sexton. 1966. Herpetofaunal comparisons between two climatic zones in Panama. Amer. Midland Nat. 75: 45–60.

Heyer, W. R. 1976. Studies in larval amphibian habitat partitioning. Smithsonian Contrib. Zool. 242.

Ibáñez D., R. 1982. El estado específico de dos ranas del género *Colostethus* (Anura: Dendrobatidae). Lic. Biol. thesis, Univ. of Panama.

Jaeger, R. G. 1976. A possible prey-call window in anuran auditory perception. Copeia 1976: 833–834.

Jaeger, R. G. and J. P. Hailman. 1981. Activity of Neotropical frogs in relation to ambient light. Biotropica 13: 59–65.

Jaeger, R. G., J. P. Hailman, and L. S. Jaeger. 1976. Bimodal diel activity of a Panamanian frog, *Colostethus nubicola*, in relation to light. Herpetologica 32: 77–81.

Jaslow, A. P. 1982. Factors affecting the distribution and abundance of tadpoles in a lowland tropical stream. Ph.D. diss., Univ. of Michigan.

Kluge, A. G. 1981. The life history, social organization, and parental behavior of *Hyla rosenbergi* Boulenger, a nest-building gladiator frog. Misc. Publ. Mus. Zool., Univ. Mich. 160.

Kourany, M., C. W. Myers, and C. R. Schneider. 1970. Panamanian amphibians and reptiles as carriers of *Salmonella*. Amer. J. Trop. Med. Hyg. 19: 632–638.

Kourany, M. and S. R. Telford. 1981. Lizards in the ecology of salmonellosis in Panama. Appl. Envir. Microbiol. 45: 1248–1253.

Leigh, E. G., Jr., A. S. Rand, and D. M. Windsor (eds). 1982. *The Ecology of a Tropical Forest: Seasonal Rhythms and Long-Term Changes*. Smithsonian Inst. Press, Washington, D.C.

Lynch, J. D. and C. W. Myers. 1983. Frogs of the *fitzingeri* group of *Eleutherodactylus* in eastern Panama and Chocoan South America (Leptodactylidae). Bull. Amer. Mus. Nat. Hist. 175: 481–572.

Mittermeier, R. A. 1971. Notes on the behavior and ecology of *Rhinoclemys annulata* Gray. Herpetologica 27: 485–488.

Moll, E. O. and J. M. Legler. 1971. The life history of a Neotropical slider turtle, *Pseudemys scripta* (Schoepff), in Panama. Los Angeles Co. Mus. Nat. Hist., Sci. Bull. 11.

Myers, C. W. 1972. The status of herpetology in Panama. In M. L. Jones (ed.), The Panamic biota: some observations prior to a sea-level canal. Bull. Biol. Soc. Washington 2: 199–209.

Myers, C. W., J. W. Daly, and V. Martínez. 1984. An arboreal poison frog (*Dendrobates*) from western Panama. Amer. Mus. Nov. 2783.

Myers, C. W. and A. S. Rand. 1969. Checklist of amphibians and reptiles of Barro Colorado Island, Panama, with comments on faunal change and sampling. Smithsonian Contrib. Zool. 10.

Rand, A. S. 1967. Predator-prey interactions and the evolution of aspect diversity. Atas do Simposio sobre Biota Amazônica (Zool.) 5: 73–83.

Rand, A. S., S. Guerrero, and R. M. Andrews. 1983. The ecological effects of malaria on populations of the lizard *Anolis limifrons* on Barro Colorado Island, Panama. In A. G. J. Rodin and K. Miyata (eds.), *Advances in Herpetology and Evolutionary Biology*. Mus. Comp. Zool., Cambridge, Mass., pp. 455–471.

Rand, A. S., M. J. Ryan, and K. E. Troyer. 1983. A population explosion in a tropical tree frog: *Hyla rufitela* on Barro Colorado Island, Panama. Biotropica 15: 72–73.

Ryan, M. J., M. D. Tuttle, and L. K. Taft. 1981. The costs and benefits of frog chorusing behavior. Behav. Ecol. Sociobiol. 8: 273–278.

Savage, J. M. 1982. The enigma of the Central American herpetofauna: dispersals or vicariance? Ann. Mo. Bot. Gard. 69: 464–547.

Savage, J. M. and J. Villa. 1986. Introduction to the herpetofauna of Costa Rica. Contrib. Herpet. Soc. Stud. Amphib. Rept. 3.

Schwartz, J. J. and K. D. Wells. 1985. Intra- and interspecific vocal behavior of the Neotropical treefrog *Hyla microcephala*. Copeia 1985: 27–38.

Telford, S. R., Jr. and H. W. Campbell. 1979. Ecological observations on an all female population of the lizard *Lepidophyma flavimaculatum* (Xantusiidae) in Panama. Copeia 1979: 379–381.

Toft, C. A. 1980. Seasonal variation in populations of Panamanian litter frogs and their prey: a comparison of wetter and drier sites. Oecologia 47: 34–38.

———. 1981. Feeding ecology of Panamanian litter anurans: patterns in diet and foraging mode. J. Herp. 15: 139–144.

Tuttle, M. D. and M. J. Ryan. 1981. Bat predation and the evolution of frog vocalizations in the Neotropics. Science 214: 677–678.

———. 1982. The role of synchronized calling, ambient light, and ambient noise, in anti-bat-predator behavior of a treefrog. Behav. Ecol. Sociobiol. 11: 125–131.

Tuttle, M. D., L. K. Taft, and M. J. Ryan. 1981. Acoustical location of calling frogs by philander opossums. Biotropica 13: 233–234.

Vanzolini, P. E. and W. R. Heyer. 1985. The American herpetofauna and the interchange. In F. G. Stehli and S. D. Webb (eds.), *The Great American Biotic Interchange*. Plenum Press, New York, pp. 475–487.

Wells, K. D. 1979. Reproductive behavior and male mating success in a Neotropical toad, *Bufo typhonius*. Biotropica 11: 301–307.

Williams, E. E. 1976. South American anoles: the species groups. Pap. Avul. Zool., São Paulo 29: 259–268.

Wright, S. J. 1979. Competition between insectivorous lizards and birds in central Panama. Amer. Zool. 19: 1145–1156.

22

A Preliminary Overview of the Herpetofauna of Cocha Cashu, Manu National Park, Peru

LILY B. RODRIGUEZ AND JOHN E. CADLE

To date there has been no detailed systematic survey of the herpetofauna of Manu National Park. Collections from the park are small; they are limited to a few sampling sites within the park boundaries, allowing only preliminary comparisons with other tropical lowland forest sites. We present here an overview of the herpetofauna of Cocha Cashu and nearby sites within the park.

Early work on the herpetofauna of Cocha Cashu focused on black caimans (*Melanosuchus niger*; Otte 1978), and the caimans remain the only species of the herpetofauna whose ecology and behavior has been studied in any detail (most recently by Herron 1985). A concerted effort to catalog the diversity of amphibians and reptiles at Cocha Cashu began with Cadle's work in 1983–84 and has continued since 1985 at a more intensive level through Rodriguez's work on anurans. Due to restrictions on collecting, the presence of some species, particularly reptiles, is so far documented only through photographs or site records, which in some

We thank Alwyn Gentry for the opportunity to present our preliminary observations on the magnificent herpetofauna of Cocha Cashu. LBR is especially indebted to J. Terborgh for encouraging her work at Cocha Cashu and for providing partial support through a World Wildlife Fund (WWF) grant. Our work in Peru has been possible through the help and cooperation of the Ministerio de Agricultura, Dirección General Forestal y de Fauna (Ing. Marco Romero Pastor and Biólogas Mariza Falero S. and Rosario Acero). We have also received numerous courtesies from Dra. Nellie Carillo de Espinoza of the Museo de Historia Natural "Javier Prado" in Lima. LBR received support from the Smithsonian Institution for work at the U.S. National Museum and from the WWF to attend the symposium. JEC was supported by NSF grant BSR-84-00166 during manuscript preparation. Helpful comments on an early version of this paper were provided by D. Davidson, M. S. Foster, C. Toft, B. Zimmerman, and especially W. R. Heyer.

cases are inadequate for positive species identification. Studies of the natural history and ecology of the component species are still embryonic.

Since we regard this overview of the Cocha Cashu herpetofauna as preliminary, we do not attempt to consider in detail habitat or other resource use by components of the herpetofauna. These subjects are under investigation for the anurans by Rodriguez. Here we characterize the herpetofauna at Cocha Cashu in terms of its diversity and compare the fauna with that of other lowland forest sites in South America. We also discuss aspects of anuran reproductive biology and spatial distribution at Manu.

Species Composition of the Manu Herpetofauna

We have assembled from the vicinity of Cocha Cashu all records of amphibians and reptiles that we consider reliable (table 22.1). Most species are recorded from Cocha Cashu itself, a field station on the floodplain of the Rio Manu that has been the center of most biological work at Manu (see Terborgh 1985). The vicinity of Cocha Cashu includes a variety of lowland forest habitats. Additional records are from the upland forest (higher, somewhat hilly terrain that is not seasonally flooded) at Pakitza, a guard post 10 km south of Cocha Cashu, and from an area of upland forest on the west bank of the Rio Manu opposite Cocha Cashu. The herpetofauna as presently known comprises 133 species, of which 79 are amphibians and 54 are reptiles (table 22.1).* Because of systematic problems, or the lack of voucher specimens, several of these species are not yet precisely identified. Most of these problem species belong to the large genera *Hyla* and *Eleutherodactylus*.

Not all components of the herpetofauna at Cocha Cashu are equally well known. Rodriguez's work on anuran ecology has resulted in a better knowledge of the frog fauna than of any of the other groups. Snakes and lizards remain poorly known, and numerous additions to their species lists can be expected. Judging from the diversity seen in other well-sampled sites in the upper Amazon Basin (Iquitos, Peru, and Santa Cecilia, Ecuador), our snake list identifies at most 60% of the species eventually to be expected at a site such as Cocha Cashu. Our lizard list identifies perhaps only 50% of lizard species. Two species of turtles, *Phrynops gibbus* and *Chelus fimbriatus*, are known from localities near Manu and

* Recent collections at Cocha Cashu (table 22.1) increase the species total to 146 (82 amphibians, 64 reptiles). These additional species are not included in this discussion. The herpetofauna of Pakitza is being intensively studied as part of the Smithsonian BIOLAT project. Many additional species now known from there are not incorporated into this discussion.

TABLE 22.1 Species composition of the herpetofauna of Cocha Cashu, the upland forest across the Rio Manu from it, and those from Pakitza noted during Cadle's 1984 fieldwork there

Species	HAB[1]	VER[2]	ACT[3]	RM[4]
Plethodontidae				
Bolitoglossa altamazonica	PF			
Caeciliidae				
Oscaecilia bassleri	PF			
Bufonidae				
Bufo guttatus	PF	T	N	1
B. marinus	PF,SM,L	T	N	1
B. typhonius[5]	U,SM	T	D	1
Dendrobatidae				
Colostethus marchesianus	PF,FS,U	T	D	6
Colostethus sp.[6]	U	T	D	6
* *Dendrobates ventrimaculatus*[5]				
Epipedobates femoralis	PF,U	T	D	6
E. pictus	PF,U,SM	T	D	6
E. trivittatus	PF	T	D	6
E. sp. nov.	U,SM	T	D	6
Hylidae				
Agalychnis craspedopus	PF	A	N	4
Hemiphractus scutatus	PF	T	N	10
Hyla bifurca	PF	A	N	4
H. boans	SM,FS	A	N	3
H. calcarata	FS	A	N	1
H. fasciata	PF,FS	A	N	1
H. geographica	FS	A	N	1
H. granosa	FS	A	N	1
H. lanciformis[6]	U,C	A,T	N	1
H. leali	L,PF,FS	A	N	1
H. leucophyllata	PF,FS,L,FP	A	N	4
H. minuta[5]	U,FS	A	N	1
H. parviceps	PF,FS	A	ND	1
H. punctata	L	B	N	1
H. rhodopepla	FS,U	A	N	4
H. "*riveroi*"	FS	A	N	1
H. sarayacuensis	PF,FS,FP	A	N	4
H. triangulum	L	B	N	4
H. sp. (*geographica* group)	PF	A	N	1
H. koechlini	PF,FP	A	N	1
H. sp. 2	PF,FS	A	N	1
Scarthyla ostinodactyla	PF,FS	B	D	1
Ololygon cruentomma	PF,FP	A	N	1
O. epacrorhina	PF	A	ND	1
O. garbei	L,FS	B,A	N	1
O. rubra	PF	A	N	1
Osteocephalus leprieurii	PF	A	N	1

O. taurinus	PF	A	N	1
Phrynohyas coriacea	FS	A	N	1
**P. resinifictrix*				
P. venulosa	PF	A	N	1
Phyllomedusa atelopoides	PF	T	N	4
P. palliata	FS	A	N	4
**P. tomopterna*				
P. vaillanti	PF,FS	A	N	4
P. sp. (*bolivianus* group)	PF	A	N	4
Sphaenorhynchus dorisae	PF,L	B	N	1
S. lacteus	PF,L	B	N	1
Leptodactylidae				
Adenomera andreae	PF	T	ND	7
A. hylaedactyla	PF,SM	T	ND	7
Ceratophrys cornuta	PF,FS	T	N	1
Ceratophryinae sp.	PF	T		1
Edalorhina perezi	PF,FP	T	D	5
Eleutherodactylus "altamazonicus"	PF,U	B	ND	8
E. "carvalhoi"	PF,U	B	N	8
E. fenestratus	PF,U,SM	B	N	8
E. mendax	PF	B	N	8
E. ockendeni	PF	B	ND	8
E. peruvianus	PF,U	B	ND	8
E. toftae	PF	B	ND	8
E. ventrimarmoratus	PF			8
E. sp. 1 (fitzingeri group)	PF	B	ND	8
E. sp. 2 (fitzingeri group)	PF	B	N	8
E. sp. 3 (green)	PF	B	N	8
Ischnocnema quixensis	PF	T	N	8
Leptodactylus bolivianus	C	T	ND	5
L. knudseni[6]	U	T	N	5
L. mystaceus	PF,FP	T	N	5
L. pentadactylus	PF,L	T	N	5
L. podocipinus	PF,FS	T	N	5
L. rhodomystax	U	T	N	5
**L. rhodonotus*				
L. wagneri	PF,FP,L	T	ND	5
Lithodytes lineatus	U	T	N	5
**Phyllonastes myrmecoides*				
Physalaemus petersi	PF,FS,FP,L	T	N	5
Microphylidae				
Chiasmocleis ventrimaculata	PF,FP	T	ND	1
Ctenophryne geayi	PF	T	N	1
Elachistocleis sp.	PF	T	N	1
Hamptophryne boliviana	PF,FS	T	ND	1
Gekkonidae				
Gonatodes hasemani	C	A	D	
G. humeralis	PF	A	D	
**Pseudogonatodes* sp.[5]				
Thecadactylus rapicaudus	PF	A	N	

TABLE 22.1 (*Continued*)

Species	HAB[1]	VER[2]	ACT[3]	RM[4]
Iguanidae				
Anolis chrysolepis	PF	TB	D	
A. fuscoauratus	PF	B	D	
**A. ortoni*				
A. punctatus	PF	A	D	
Enyalioides cf. *laticeps*	PF	A	D	
Ophryoessoides sp.	PF	T	D	
Plica plica	PF	A	D	
P. umbra	PF	A	D	
Teiidae				
***Alopoglossus buckleyi*				
Ameiva ameiva	C	T	D	
Kentropyx altamazonicus	PF	T	D	
***Kentropyx pelviceps*				
**Neusticurus ecpleopus*[5]				
***Prionodactylus argulus*				
***P. eigenmanni*				
**P. manicatus*				
* *P.* sp.				
Tupinambis teguixin	D(L)	T	D	
Bachia sp.	PF	F	D	
Anguidae				
Diploglossus fasciatus	PF	T	D	
Scincidae				
Mabuya bistriata	PF,C	T	D	
Aniliidae				
* *Anilius scytale*				
Boidae				
Corallus caninus	PF	A	N	
C. enydris	PF	A	N	
Epicrates cenchria	PF	T,A	ND	
Colubridae				
Atractus sp. (like *major*)	PF	T		
Chironius exoletus[6]	PF	T	D	
C. scurrulus	PF	T	D	
Clelia clelia	PF	T	ND	
Dipsas catesbyi	PF	A	N	
Drepanoides anomalus	PF	T	N	
Drymarchon corais	PF	T	D	
Helicops angulatus[6]	R	Aq	N	
H. polylepis	R	Aq		
Imantodes cenchoa	PF	A	N	
Leptodeira annulata	PF	A	N	
Leptophis ahaetulla	PF	A	D	
Liophis cobella[6]	PF	T	D	
L. reginae[6]	PF	T	D	
Oxybelis argenteus	PF	A	D	

Oxyrhopus formosus[6]	PF	T	D
O. petola[6]	PF	T	D
Pseudoboa coronata	PF	T	ND
Rhadinaea brevirostris	PF	T	D
* *Rhinobothryum lentiginosum*			
Siphlophis cervinus	PF	T	N
Tripanurgos compressus	PF	A	N
Xenodon severus	PF	T	D
Xenopholis scalaris	PF	T	N
Viperidae			
Bothrops atrox	PF,C	T	ND
Lachesis muta	PF	T	N
Elapidae			
Micrurus lemniscatus	PF	T	N
M. surinamensis[6]	PF	Aq	
Kinosternidae			
Kinosternon scorpioides	PF,FS,FP	Aq	
Pelomedusidae			
Podocnemis unifilis	L	Aq	D
Phrynops nasutus	PF,FP	Aq	D
Chelidae			
Platemys platycephalus	FS,FP,FS	Aq	D
Testudinidae			
Geochelone denticulata	PF	T	D
Crocodylidae			
Melanosuchus niger	L,R	Aq	ND
Caiman crocodylus	L,R	Aq	ND

Notes: [1]HAB = General Habitat: PF, primary alluvial forest; SM, stream margin; FP, forest pools; L, lakes; FS, forest swamp; U, upland forest; C, clearing within forest; R, river

[2]VER = Vertical Distribution: T, terrestrial; A, arboreal; B, low vegetation; Aq, aquatic; F, fossorial

[3]ACT = Activity Pattern: N, nocturnal; D, diurnal

[4]RM = Reproductive Mode, given for anurans only, following Crump (1974): 1, eggs and larvae in body of water; 3, eggs and larvae in constructed basin; 4, eggs on vegetation above water; 5, eggs in a foam nest; 6, eggs terrestrial, larvae carried to water; 7, eggs and tadpoles in a terrestrial foam nest; 8, direct development, terrestrial; 10, direct development, eggs attached to dorsum of female

[5]Species presently known from the forest directly across the Rio Manu from Cocha Cashu, but not at Cocha Cashu itself

[6]Species presently known from the Pakitza guard post, but not from Cocha Cashu or the forest opposite

* = Specimens collected too late for inclusion in the discussion or computations in this chapter; noted for completeness only

** = Specimens from Cocha Cashu located in the Museum of Comparative Zoology (Harvard University) after final revision of this chapter; they are included for completeness but are not represented in the discussion or computations here

TABLE 22.2 Herpetofaunal species richness at five sites

Taxa	Tambopata	Panguana	Iquitos	Belém	Manu
Caecilians	—	1	3	3	1
Salamanders	—	—	1	1	1
Anurans	70	63	54	37	78
Turtles	5	—	9	3	5
Amphisbaenians	1	—	2	—	—
Lizards	22	—	38	24	16
Snakes	39	—	88	47	31
Crocodilians	4	—	4	1	2
Total Amphibians	70	64	58	41	79
Total Reptiles	71	—	141	75	54
Total Species	141	64	199	120	133

Sources: Tambopata, Peru (McDiarmid and Cocroft, unpubl. data); Panguana, Peru (Schluter 1984); Iquitos, Peru (Dixon and Soini 1975 and unpubl. data); Belém, Brazil (Crump 1971; Duellman 1978); Manu (Cocha Cashu and Pakitza), Peru (this chapter)
Note: For comparisons to La Selva, BCI, Manaus, and Santa Cecilia, see table 24.1

are to be expected within the park. It should be noted, however, that all of these expectations are based on observations of a few well-studied sites, and as yet we have little knowledge of site-to-site variation in herpetofaunal diversity, even within such a small area as southern Peru. For example, despite reasonably intensive sampling, the snake diversity recorded for Tambopata, Peru (also in Departamento Madre de Dios), is low relative to that of other upper Amazonian sites sampled (see table 22.2 and Duellman, this volume, table 24.1); whether the Tambopata observations should be considered predictive for Manu, which is geographically close, is unclear from present information.

Examination of the species list for Cocha Cashu reveals that the composition of the herpetofauna is typical of that of other sites in the western Amazon Basin. That is, most species present at Manu are widely distributed in the Amazon Basin or inhabit forests of the western part. This general pattern suggests that as with mammalian assemblages (Emmons 1984), and in contrast to patterns shown by insects and plants (Erwin 1983; Gentry 1982), there is little wholesale turnover in species composition of the herpetofauna among sites within Amazonia. The herpetofauna of Cocha Cashu does include some noteworthy records of several species, however: (1) documented occurrences of species for which records are sparse or rare (*Bolitoglossa altamazonica, Diploglossus fasciatus*); (2) significant range extensions or marginal records (*Agalychnis craspedopus*, *Phrynops nasutus*, *Kinosternon scorpioides*, *Oscaecilia bassleri*, *Gona-*

todes hasemani, *Sphaenorhynchus dorisae*); and (3) newly discovered or recently described species (*Epipedobates* sp. nov., *Phyllomedusa atelopoides*, *Phyllomedusa bolivianus*-group, *Eleutherodactylus* spp.). Most of these records are undoubtedly more a reflection of our poor knowledge of Amazonian herpetofaunas in general than any uniqueness of Cocha Cashu, since some of the same species have been discovered at nearby sites in Peru (e.g., reserves at Tambopata and Cuzco Amazonico, both southeast of Manu in Madre de Dios). We suspect that few of our records will be found to be true endemics.

Species Diversity and Comparisons to Other Sites

Because the task of inventorying the herpetofauna of Manu is still in its infancy, generalizations about the species richness of the fauna must be made cautiously. This is especially true in view of the very different sampling regimes for the Neotropical sites that have been studied. Nevertheless, comparisons of the species richness of several Neotropical sites (table 22.2; see also Duellman, this volume, table 24.1) reveal that the herpetofauna of Manu is comparatively rich. With the exceptions of snakes and lizards, which we consider too poorly sampled at present for conclusions to be drawn, amphibian and reptile diversity at Manu is known to be greater than that at any other site except Tambopata, Peru, and Santa Cecilia, Ecuador. We consider this observation significant in view of the much more intensive sampling at Tambopata and Santa Cecilia; it suggests that at least the amphibian fauna of Cocha Cashu, when more completely known, may have a higher species richness than any of these other sites. Comparable projections for reptiles are not possible, given our current data for Manu.

Comparisons of the herpetofauna of Cocha Cashu with those of other Peruvian and Brazilian sites in terms of Faunal Resemblance Factors (Duellman 1966) are shown in table 22.3; Duellman (this volume, table 24.2) reports these comparisons with La Selva, Barro Colorado Island (BCI), Manaus, and Santa Cecilia. As expected, Cocha Cashu is most similar to Tambopata, which is also the closest site geographically (approximately 200 km to the southeast). Other Amazonian localities are more similar to Cocha Cashu than are either of the Central American localities. Interestingly, those sites in the western Amazon Basin close to the base of the Andes are more similar to Manu than are sites further removed from the Andean front. Panguana and Santa Cecilia are thus compositionally more similar to Manu than are Iquitos, Manaus, or

TABLE 22.3 Faunal Resemblance Factors between the Manu herpetofauna and those of four other sites. Comparisons to La Selva, BCI, Manaus, and Santa Cecilia are found in Duellman (chap. 24, table 24.2). The number of species shared between Manu and other sites is followed in parentheses by the Faunal Resemblance Factor (Duellman 1965).

Site	Total Herpetofauna *N* (FRF)	Anurans* *N* (FRF)	Reptiles *N* (FRF)
Panguana, Peru	—	43 (.59)	—
Iquitos, Peru	74 (.43)	26 (.37)	46 (.47)
Tambopata, Peru	88 (.71)	52 (.69)	36 (.58)
Belém, Brazil	—	18 (.31)	10 (.16)

Source: FRF from Duellman (1965); see also note to table 22.2
Notes: * Inclusion of the caecilian and salamander known from Cocha Cashu does not change the community coefficients; both species are shared with Iquitos, and the salamander is also shared with Belém

Belém. This pattern reflects the large number of amphibian and reptile species whose distributions apparently encompass the upper part of the Amazon Basin but do not extend into the basin's central region (Duellman 1978). Endemism is apparently higher in the western Amazon Basin for both amphibians and reptiles (Lynch 1979; Dixon 1979). (There are problems with the interpretation of the data base for many of these distribution patterns [see Heyer 1988], but for lack of alternative data, our inferences are presently based on apparent patterns.)

Factors controlling the species richness of herpetofaunal assemblages are poorly understood. Undoubtedly, the interplay of historical and current ecological variables is involved, although even defining these variables may prove difficult. In comparing the species richness of Santa Cecilia with that of Iquitos, BCI, and Belém, Duellman (1978) noted that Santa Cecilia had a proportionally greater representation of frogs in its fauna than any of the other sites. He attributed the anuran diversity to the more equable annual moisture regime at Santa Cecilia, which, in contrast to the other sites, has no distinct dry season. However, as noted above (see table 22.2), the presently known species diversity of frogs at Cocha Cashu is nearly the same as that of Santa Cecilia, even though there is a distinct and prolonged dry season at Manu (April through November or December, depending on the year; see Terborgh 1985). These observations suggest that the determinants of anuran diversity at these sites involves more than simply moisture equability.

Other influences on anuran diversity are difficult to evaluate because quantitative studies at various sites are lacking. Historical factors in particular cannot be evaluated until much more is known about speciation patterns in various parts of the Neotropics. More amenable to quantitative studies presently are factors related to habitat complexity and productivity. Inger (1980) suggested that differences in productivity could account for differences in the richness of frogs and lizards between forests of Central America and Southeast Asia. Both diversity and density of mammals are lower at Manaus than at Manu, as documented by Emmons (1984). The differences between the two sites appeared to be related to more stressful environmental conditions for mammals at Manaus, mediated through poor soils. But a strong association between productivity and density such as the one seen in mammals is probably not to be expected in anuran populations, since all frogs are secondary consumers and most of the mammals Emmons studied were primary consumers. Nevertheless, if such an association exists for the frogs, then there should be a correlation among sites between diversity and population density (see Emmons 1984). The necessary data could be gathered for several sites, though it would require correction for long-term variation in densities.

Spatial and Temporal Distribution of the Herpetofauna

As noted, Cocha Cashu is the only site within Manu for which extensive observations of the herpetofauna are available. The site offers a variety of microhabitats that vary in the extent to which they are used by the herpetofauna. (For a brief description of habitats at Cocha Cashu, see Terborgh 1985.) Most of our comments concerning microhabitat use pertain to frogs, since these are the best-known component of the herpetofauna. Frogs were collected by means of transect sampling along the 50 km of trails at Cocha Cashu and through the upland forest on the opposite bank. During a total of 115 days that spanned both wet and dry seasons, 60 species, or 85% of the frog fauna, were collected (fig. 22.1). This is similar to the species accumulation curve obtained by Duellman (1978) for the entire herpetofauna at Santa Cecilia (75% of species in 100 man-days of collecting).

Habitat distributions and species richness with respect to habitat are

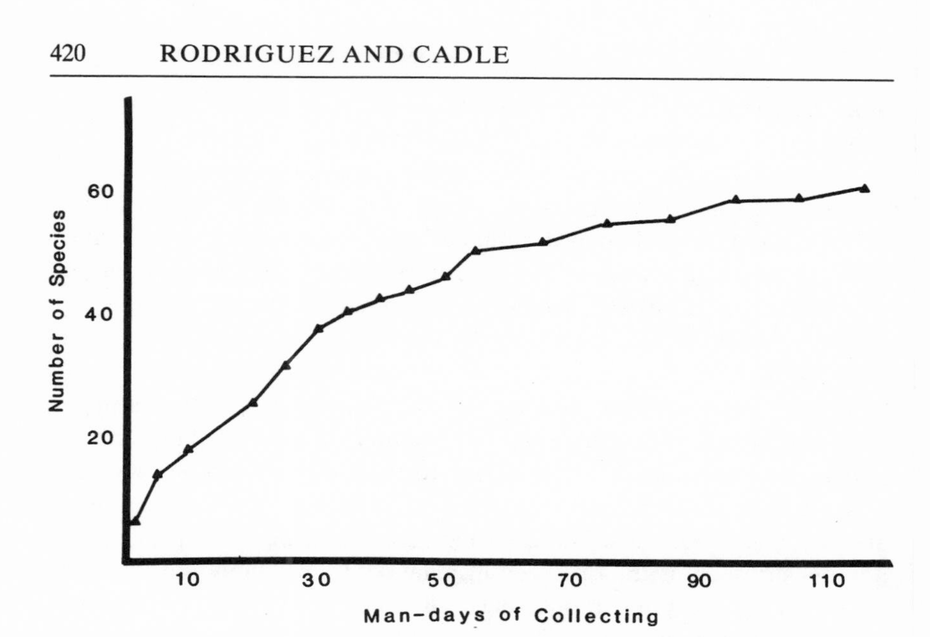

FIGURE 22.1. Species accumulation curve relative to man-days of collecting for anurans at Cocha Cashu.

difficult to quantify with our current data, since sampling has not been randomized with respect to habitats. Approximately 80% of the frog species have been found in primary alluvial forest, but this is also the most extensive and most thoroughly sampled habitat at Cocha Cashu. Some species have thus far been encountered only at Pakitza and on the west bank opposite Cocha Cashu. These well-drained upland forests may support species not found in the alluvial forests at Cocha Cashu. Examples of species thus far found only in these sites are *Leptodactylus knudseni*, *Bufo typhonius*, *Hyla lanciformis*, *Epipedobates* sp. nov., and *Colostethus* sp. If such apparent microspatial differences in species composition reflect true patchy distributions, then the beta diversity of anurans at Manu may be greater than presently realized. Rodriguez is currently conducting randomized frog sampling with respect to habitat to explore this possibility. Other differences among microhabitats were observed for similar macrohabitats. Cocha Totora, approximately 1.5 km from Cocha Cashu, and another small pond only 250 m from Cocha Cashu, were both covered with floating vegetation (*Pistia*) and were used by the two species of *Sphaenorhynchus*. At Cocha Cashu, in contrast, with only emergent vegetation, only *S. lacteus* was observed.

The marked seasonality of rainfall clearly influences the activity of some amphibians and reptiles at Manu. The increase in breeding activity of anurans is certainly the most obvious effect of the onset of rains. However, other species are also affected, though these effects are not documented to the extent that they are for the frogs. Several species of turtles, including *Phrynops nasutus*, *Platemys platycephala*, and *Kinosternon scorpioides*, have been observed exclusively or most frequently during the rainy season. These species are generally observed only during the rainy season in flooded forest pools and streams. In January 1984, many hatchling *P. nasutus*, as well as adult *P. platycephala*, were observed soon after heavy rains filled forest pools and streams. Habitat associations and activity patterns of these species during the dry season are unknown.

Much more work remains to be done on the temporal pattern of activity of the Manu herpetofauna. As noted above, the Manu herpetofauna is similar in diversity and composition to that of Santa Cecilia, even though Santa Cecilia is far less seasonal than Cocha Cashu. This suggests that the annual reproductive and other behavioral cycles documented by Duellman (1978) for Santa Cecilia will differ substantially at Manu. Currently there are insufficient data from Cocha Cashu to document this difference. However, such data will have an important bearing on interpreting the evolution of life history parameters, such as fecundity and timing of reproduction, in these tropical species. There has been an emphasis in the literature (e.g., Duellman 1978) on comparing aseasonal humid tropical sites with more seasonal temperate sites, in attempts to understand the evolution of particular reproductive modes and activity patterns. We suggest that a more realistic comparison might be between aseasonal humid tropical forests, such as Santa Cecilia, and seasonal humid ones, such as Cocha Cashu. This would go a long way toward eliminating the vast historical differences of the communities being compared and would allow more meaningful comparison of assemblages with comparable diversities.

Anuran Reproduction

Using Crump's (1974) classification of anuran reproductive modes, eight modes can be distinguished at Cocha Cashu (see tables 22.1 and 22.4). Not surprisingly, nearly half of the frog species use the most common mode of reproduction for anurans: egg deposition and larval development in water (Duellman and Trueb 1986). The remaining species each use one of the other seven modes. Our observations of reproductive behavior indicate that swamps and forest pools are used much more

TABLE 22.4 Frequency distribution of modes of reproduction for Anurans at three sites

Mode of reproduction*	Santa Cecilia N (%)	Manaus N (%)	Manu N (%)
1	35 (41)	16 (41)	33 (42)
2	2 (2)	2 (5)	0
3	1 (1)	1 (3)	1 (1)
4	13 (15)	5 (13)	11 (14)
5	9 (8)	6 (15)	10 (13)
6	6 (7)	3 (8)	6 (8)
7	1 (1)	1 (3)	2 (3)
8	17 (20)	3 (8)	14 (18)
9	1 (1)	1 (3)	0
10	1 (1)	0	1 (1)

Source: Data for Santa Cecilia, Crump (1974); for Manaus; B. Zimmerman (p.c.)
Note: * See table 22.1 for explanation of modes

heavily than streams and lakes by species with aquatic tadpoles (36 versus 13 of the species for which we have data).

The percentages of species using each mode of reproduction are similar among Santa Cecilia, Manaus, and Cocha Cashu (table 22.4). The one notable exception to this general pattern is the lower percentage of terrestrial breeders (Mode 8; mostly species of *Eleutherodactylus*) at Manaus, where there are proportionally somewhat more species using foam nests in which tadpoles are aquatic (Mode 5). Although one might suggest that the difference in proportion of terrestrial breeders between Santa Cecilia and Manaus is a simple function of rainfall (more than 4 m/yr for Santa Cecilia and only somewhat more than 2 m/yr for Manaus; Duellman 1978; Emmons 1984), this explanation fails to account for the same difference between Manaus and Cocha Cashu, which have very similar rainfall regimes. Terrestrial reproduction such as seen in *Eleutherodactylus* requires a continually moist substratum for successful completion of egg development. It is possible that soil conditions at Manaus (white sand) are not conducive to such a resproductive mode. Unlike at Manaus, the alluvial soils at Cocha Cashu are poorly drained and support a dense understory (Emmons 1984), which would contribute to greater water retention by soils. Soils at Cocha Cashu may thus retain sufficient moisture for terrestrial reproduction, whereas those at Manaus do not.

We emphasize that this explanation for the distribution of Mode 8 reproduction does not incorporate the phylogenetic aspects of reproductive mode. Mode 8 reproduction is characteristic of eleutherodactyline frogs, which are most speciose in the western Amazon Basin and the northern Andes. Lynch (1979) documented the high species density and endemicity of species with Mode 8 reproduction in these areas. Thus, it is quite possible that the differences observed between Cocha Cashu and Manaus are related primarily to historical differences in the evolution of the frog fauna.

Obviously, the relative importance of ecological and historical factors in the distribution of reproductive modes could be better grasped through more comparative work between localities that have contrasting ecological conditions and are within one morphoclimatic domain. Such studies might reveal much about the extent to which frog distribution and reproductive mode selection are influenced by environmental factors (see Heyer 1988).

Discussion and Future Work

The high herpetofaunal diversity for Cocha Cashu is especially notable given the much lower sampling intensity there compared to other upper Amazonian sites. This suggests that, as with bird populations (Terborgh et al. 1984), the site may have one of the highest species densities for amphibians and reptiles anywhere in the world. The causes of such high species diversity are unclear. Because most amphibians and reptiles at Manu are widely distributed either throughout the Amazon Basin or along the western Andean front, the high diversity cannot be explained as a function of high beta diversity. Moreover, a direct relation to rainfall pattern or intensity is not apparent, since such sites as Santa Cecilia receive approximately twice as much rainfall as Cocha Cashu (distributed more evenly through the year) but have comparable herpetofaunal diversities. Microhabitat differences mediated through soil types could in part explain variations between sites for certain portions of the herpetofauna, as discussed previously for terrestrially breeding frogs. We suggest, however, that any explanations accounting for differences in diversity among sites will likely have a large historical component, and phylogenetic studies of particular groups are urgently needed before these hypotheses can be evaluated. Until such studies become available, explanations for intersite variability in species composition will remain incomplete. In addition, uniform sampling procedures must be developed to quantify differences among sites more accurately, both on a micro-scale (as between Cocha Cashu and Pakitza) and on a macro-scale.

The inventory of the herpetofauna of Manu Park is far from complete. Because studies of community dynamics, species interactions, and biogeographic similarities are dependent on detailed systematic inventories, we advocate the continued survey of Cocha Cashu and other sites within Manu. Only through documenting species presence by collecting specimens can taxonomic assignments be verified and used to form the basis for further ecological and evolutionary investigations. Many species recorded from the park are poorly known taxonomically and represent significant additions to knowledge of distribution patterns (see above). The development of a reference collection of the herpetofauna should be a focus of further efforts to survey the Manu fauna, since such a collection will be a primary resource for the investigation of many other questions.

A number of the species of Manu that are uncommon in parts of their ranges are represented by large populations at Cocha Cashu and other areas within the park. Examples are *Podocnemis unifilis*, *Melanosuchus niger*, *Caiman crocodylus*, *Corallus enydris*, *Geochelone denticulata*, and *Platemys platycephala*. These populations would make good subjects for ecological studies (Herron 1985, Otte 1978). Because of the wealth of ecological data accumulating for other species at Cocha Cashu (mostly through the efforts of J. Terborgh and associates), opportunities exist for the detailed investigation of species interactions. An example is recent observation of jaguar predation on *Geochelone denticulata* and *Podocnemis unifilis* (Emmons 1988). Manu's protection from hunting pressures and other exploitations make it ideal for such research.

REFERENCES

Crump, M. 1971. Quantitative analysis of the ecological distribution of a tropical herpetofauna. Occas. Pap. Mus. Nat. Hist. Univ. Kans. 3: 1–62.

———. 1974. Reproductive strategies in a tropical anuran community. Misc. Publ. Mus. Nat. Hist. Univ. Kans. 61: 1–68.

Dixon, J. R. 1979. Origin and distribution of reptiles in lowland tropical rainforests of South America. In W. E. Duellman (ed.), *The South American Herpetofauna: Its Origin, Evolution, and Dispersal*. Monogr. Mus. Nat. Hist. Univ. Kans. 7, pp. 217–240.

Duellman, W. E. 1965. A biogeographic account of the herpetofauna of Michoacan, Mexico. Univ. Kans. Publ. Mus. Nat. Hist. 15: 627–709.

———. 1978. The biology of an equatorial herpetofauna in Amazonian Ecuador. Misc. Publ. Mus. Nat. Hist. Univ. Kans. 65: 1–352.

———. (This volume). Herpetofaunas in Neotropical rainforest: comparative composition, history, and resource use.

Duellman, W. E. and L. Trueb. 1986. *Biology of Amphibians*. McGraw-Hill, New York.

Emmons, L. H. 1984. Geographic variation in densities and diversities of nonflying mammals in Amazonia. Biotropica 16: 210–222.

———. 1988. Jaguar predation on chelonians. J. Herp. 23: 311–314.

Erwin, T. L. 1983. Beetles and other insects of tropical forest canopies at Manaus, Brazil, sampled by insecticidal fogging. In S. L. Sutton, T. C. Whitmore, and A. C. Chadwick (eds.), *Tropical Rain Forest: Ecology and Management*. Brit. Ecol. Soc. Spec. Publ. 2, pp. 59–75.

Gentry, A. H. 1982. Neotropical floristic diversity: phytogeographical connections between Central and South America, Pleistocene climatic fluctuation, or an accident of the Andean orogeny? Ann. Mo. Bot. Gard. 69: 557–593.

Herron, J. C. 1985. Population status, spatial relations, growth, and injuries in black and spectacled caimans in Cocha Cashu. Undergraduate thesis, Princeton Univ.

Heyer, W. R. 1988. On frog distribution patterns east of the Andes. In W. R. Heyer and P. E. Vanzolini (eds.), *Proceedings of a Workshop on Neotropical Distribution Patterns Held 12–16 January 1987*. Acad. Bras. de Ciencias, Rio de Janeiro, pp. 245–273.

Inger, R. F. 1980. Densities of floor-dwelling frogs and lizards in lowland forests of Southeast Asia and Central America. Amer. Nat. 115: 761–770.

Lynch, J. D. 1979. The amphibians of the lowland tropical forests. In W. E. Duellman (ed.), *The South American Herpetofauna: Its Origin, Evolution, and Dispersal*. Monogr. Mus. Nat. Hist. Univ. Kans. 7, pp. 189–215.

Myers, C. W. and A. S. Rand. 1969. Checklist of amphibians and reptiles of Barro Colorado Island, Panama, with comments on faunal change and sampling. Smithsonian Contrib. Zool. 10: 1–11.

Otte, K. C. 1978. *Untersuchungen zu biologie des mohrenkaiman (*Melanosuchus niger *Spix 1825) aus dem Nationalpark Manu (Peru)*. Priv. publ., Munich and Lima.

Scott, N. J., J. M. Savage, and D. C. Robinson. 1983. Checklist of reptiles and amphibians. In D. H. Janzen (ed.), *Costa Rican Natural History*. Univ. of Chicago Press, Chicago, pp. 367–374.

Terborgh, J. 1983. *Five New World Primates: A Study in Comparative Ecology*. Princeton Univ. Press, Princeton, N.J.

Terborgh, J., J. W. Fitzpatrick, and L. H. Emmons. 1984. An annotated checklist of bird and mammal species of Cocha Cashu Biological Station, Manu National Park, Peru. Fieldiana (Zool.), n.s., 21: 1–29.

Zimmerman, B. L. and R. O. Bierregaard, Jr. 1986. Relevance of the equilibrium theory of island biogeography and species-area relations to conservation with a case from Amazonia. J. Biogeogr. 13: 133–143.

23

Frogs, Snakes, and Lizards of the INPA-WWF Reserves near Manaus, Brazil

BARBARA L. ZIMMERMAN AND MIGUEL T. RODRIGUES

Reptiles and amphibians of the Instituto Nacional do Pesquisas da Amazonia-World Wildlife Fund (INPA-WWF) primary forest reserves were surveyed during many months in 1980 and from 1983 through 1987 with the ultimate aim of understanding minimum critical population sizes or area and habitat requirements for self-sustaining populations (Zimmerman and Bierregaard 1986).

As Amazonian rainforest is increasingly devastated, it becomes imperative to select and protect areas that will preserve representative primary forest ecosystems. Such land can be chosen to maximize the number of species and interactions protected when something is understood about the diversity and ecology of intact forest. In this chapter, we report the results of the species survey and describe relative abundances and such basic ecological parameters as food and habitat preference (including breeding site for frogs) for each species. We also present general data on relative abundances. Because the methods and resulting data are distinct for frogs, snakes, and lizards, we treat each group independently.

Methods and Study Site

Frogs were surveyed between February and July 1980, in March 1983, in July 1984, and, except for four months, between November 1984–July 1987. Two or three people walked an average of 4.5 km of trail

This research was made possible by World Wildlife Fund–U.S., Dr. P. E. Vanzolini, and World Wildlife Fund–Canada. Invaluable field help was provided by Manuel Rego, Ocirio Perreira, and Laercio dos Reis.

between 1,900 and 2,230 hours per survey and recorded all acoustic and visual observations of frogs. When a frog was seen, species, time of observation, SVL (snout-vent length), vertical position, microhabitat, and approximate location were noted. In addition to the above data when frogs were heard the approximate number of callers was noted. Examples of vocalization by each species were recorded on tape.

The study was performed almost entirely in continuous forest reserves and was concentrated in reserves 1301, 1401, 1501, 3402, 1207, and 3304 (before isolation) (see Lovejoy and Bierregaard, this volume). Within these reserves, approximately 65 unique kms of trail were surveyed, of which 40 were walked regularly. Except in reserves 1301 and 3304, where trails were separated by just 200 m, most trails walked were separated by at least 500 m. Reserve 1301 has about a 500-m section of edge along one border. Short sections of edge also border second growth in areas 1401 and 3402.

In this paper, *upland* refers to all terra firme forest, excluding stream valleys; *streamside* refers to sharply cut valleys with streams, usually bordered by narrow floodplain. Frogs bred in two main types of ground-level aquatic habitat: stream valleys containing streams and landlocked, upland pools. Streams were divided into two classes: small (less than 2 m wide) and large (more than 2 m wide). Upland pools fell into five roughly delimited classes: large (greater than 5 m × 5 m area), temporary pools that flood annually; large, temporary pools that flood only in the wettest years; small (up to approximately 4 m × 4 m area), temporary pools that flood annually; large, permanent ponds; and small, permanent or almost permanent pools. Associated with the trail system were sixteen isolated upland pools of various types, two streams more than 3 m wide, and at least 30 trail crossings of small streams between 1 and 2 m wide (table 23.2). Data are also presented from two pools in the area of reserve 1207. After deforestation in the area in 1983, one pool was in a 10-ha forest fragment and the other was approximately 200 m from clear-cut forest. All other pools and all trail sections were located in continuous primary forest.

Data on (precise) abundance, geographic distribution, and seasonality are not considered here. Vocalizations by species are presented and discussed elsewhere (Zimmerman 1983; Zimmerman and Hödl 1983; Zimmerman and Bogart 1984, 1988). We examine species richness, sampling success, and breeding habitat and comment generally on relative abundances.

Snakes were surveyed in the INPA-WWF area between January and July 1980, March 1983 and July 1984, November 1984 and July 1985, and January and July 1987. Two survey methods were used. BLZ walked forest trails during the day and at night, often with one, and occasionally with two, other people. Species, time, and location of all snakes observed were recorded (except in 1980 when no records were kept). Snakes were collected only if their taxonomy was uncertain or if their species was not yet represented in the voucher collection for the area. In addition, between May and August 1983 and May and July 1984, several hundred hectares of primary forest in the reserve area were clear-cut. We bought snakes found by men doing the clearing.

The walking survey was performed principally in the area of reserves 3304, 1301, 1401, 3402, and 1501. (Some kilometers were walked in 1202, 2303, and 1302, but sampling effort was so poor in these areas compared to the others that they were not considered in the test for significance among reserve abundance differences.) Primarily continuous primary forest was surveyed, although all principal sampling areas except 1501 were adjacent to cattle pasture or second growth along relatively short sections of their borders. A few kilometers of survey in reserve 3304 during and immediately after its isolation by clear-cutting are included in analyses. Reserve areas 1301 and 1401 are separated by only 2 km at their closest points, but all other areas are separated by more than 6 km. Snakes were bought from men clear-cutting forest adjacent to areas 3304, 2303, and 1301. Voucher specimens of each species collected are in the collections of the Museu de Zoologia da Universidade de São Paulo (MZUSP) in São Paulo, Brazil. Doctor P. E. Vanzolini of the MZUSP identified many of the specimens.

Lizards were surveyed by MTR, who walked trails for 35 days in 1985, 1986, 1987. The study site was the same as that described for snakes.

Results and Discussion: Frogs

Forty-two frog species that occur principally in primary forest were found in the INPA-WWF reserves area (table 23.1). Table 23.1 does not include *Bufo marinus*, *B. granulosus*, *Hyla* sp. (*minuta*-like), *H. marmorata*, *H. multifasciata*, *Ololygon garbei*, or *O. rubra*, species that are associated with cleared habitat but are sometimes found a few hundred meters into primary forest that borders disturbed areas. Although included on the list, *Phyllomedusa vaillanti* was not found in the INPA-

TABLE 23.1 Species of frogs and associated natural history data from the INPA-WWF Reserves near Manaus

Family and Species	Gen Hab[1]	Act[2]	Food[3]	Br Hab[4]	Vert P[5]	Repro[6]	RA[7]	Call Freq[8]
Bufonidae								
Atelopus pulcher	SV,L	D		Str*	0	2*	C	CSM
Bufo dapsilis	PF,L	ND		SP	0	1	UC	SSM
Dendrophryniscus minutus	SV,L	D	A	SP*	0	1*		
Centrolenidae								
Centrolenella oyampiensis	SV,Al	N	I	Str	<1.5	18	C	CTY
Dendrobatidae								
Colostethus marchesianus	PF,L	D	I,A,T	P,SP	0	14	Ub	CSM,SSM
Colostethus n. sp.	PF,L	D	I,A,T	Ter	0	14*	Ub	CSM,SSM
Epipedobates femoralis	UF,L	D	I,A,T	P	0	14	C	CTY
Hylidae								
Hyla boans	PF,Ah	N	I	Str	0–3	3	C	CSM
H. geographica	SV,Al	N	I	SP,Str	0–1.5	1,2	C	STY,CTY
H. granosa	SV,Ah	N	I	SP,Str*	0–1	1*,2*	C	CTY
H. minuta	UF,A	N	I	T		1	C	CSM
H. sp. (*microcephala* gr.)	SV*,A	N		SP	1–2	1*	UC	SSM
H. *brevifrons*(-like)	UF*,A	N		T	1.5–2.5	1*	UC	SSM
Ololygon cruentomma	PF,A	N	I	T	1–3	1	UC	SSM
Osteocephalus sp.	Sv*,A	N		Str*SP*		1		
O. buckleyi	PF,Ah	ND		TrHo	>2	4	Ub	CSM
O. taurinus	PF,Ah	N	I	T,P,SP	0–3	1	C	STY
Phrynohyas resinifictrix	PF,Ah	N		TrHo	>5	4	C	CTY
P. coriacea	A							
Phyllomedusa bicolor	PF,Ah	N	I	T,P,SP	>2	18	C	STY
P. tarsius	UF,A	N	I	T,P+	1.5–2.5	18	C	CTY,STY
P. tomopterna	UF,A	N	I	T,P+	>2.5	18	C	CTY,STY
P. vaillanti	PF	N	I	SP,S	1–2.5	18		STY

TABLE 23.1 (*Continued*)

Family and Species	Gen Hab[1]	Act[2]	Food[3]	Br Hab[4]	Vert P[5]	Repro[6]	RA[7]	Call Freq[8]
Leptodactylidae								
Adenomera andreae	PF,L	ND	I,A	Ter	0	22	Ub	CTY,STY
Ceratophrys cornuta	PF,L	ND*	ATMF	T	0	1	R	R
Eleutherodactylus fenestratus	PF,L,Al	ND	I*	Ter	1	17	Ub	CTY,STY
E. sp. 1	PF,Al	N	I*	Ter*	1–2	17*20*	C	CSM
E. sp. 2	PF	N	I*			17*20*		
Leptodactylus knudseni	PF,L,sub	N	ATMF	T,P	0	21*	UC	SSM
L. mystaceus	UF,L	N		T,Ter	0	22	C	SSM
L. pentadactylus	PF,L,sub	N	ATMF	S,Subterr	0,−0	8*22*	UC	SSM
L. rhodomystax	PF,L	N	I	P,T,SP	0	8/21	C	SSM
L. riveroi	SV,L,sub*	N		SP	−0	8*9*	UC	SSM
L. stenodema	UF,L,SL	N	I	Ter*	0	22*	C	STY
L. wagneri	SV,L	N	I	SP	0	8	C	STY
Lithodytes lineatus	PF,L,sub*	N	I		−0*			
Microhylidae								
Chiasmocleis shudikarensis	PF,L,sub	N	A*	T	0	1	UC	SSM
C. sp. (*hudsoni* gr.)	PF,L,sub	N	A*	T	−0	1	C	SSM
Ctenophryne geayi	PF,L,sub*							
Synapturanus mirandiriberoi	PF,sub	ND	A*T*	Subterr	−0	15	Ub	SSM
S. salseri	PF,sub	ND	A*T*	Subterr	—	15	Ub	SSM
Pipidae								
Pipa arrabali	aquatic	ND	AGPF	P,SP,S*	−0	11	UC	STY

Notes:

Gen Hab = General Habitat of Occurrence

PF = Primary forest; all forest including stream valleys

UF = Upland forest; all forest excluding stream valleys

SV = Stream valleys
L = Litter layer
Sub = Subterranean or sub leaf litter
A = Arboreal, relative height unknown
Al = Arboreal, low heights
Ah = Arboreal, high heights
Act = Diel Activity
N = Nocturnal
D = Diurnal
ND = Nocturnal but at least some type of active behavior occurs during the day
Food = Stomach content data from Duellman (1978)
I = Insects
A = Primarily ants
T = Termites
ATMF = Anything that moves and fits
ATPF = Anything palatable
Br Hab = Breeding habitat
S = In streams
T = Temporary (less than 4 months with water), landlocked upland pools
P = Permanent or semipermanent (at least 8 months with water), landlocked upland pools
SP = Pools beside streams that are usually temporary
TH = Treeholes
Ter = Terrestrial, leaf litter
Subterr = Subterranean
+ = Preferred
Vert P = Vertical position of calling males expressed as meters from ground
0 = Sub leaf litter or subterranean
Repro = Reproductive mode abbreviations from Duellman (1985), data from p. obs.; Hero, p.c.; Hodl, p.c. Juncá, p.c.; Crump 1974; Duellman 1978; Heyer and Bellin 1973; and Pyburn 1975
1 = Aquatic eggs and feeding tadpoles in lentic water
2 = Aquatic eggs and tadpoles in lotic water
3 = Aquatic eggs and early larval stages in natural or constructed basins, subsequent to flooding, feeding tadpoles in ponds or streams

4 = Aquatic eggs and feeding tadpoles in water in tree holes or aerial plants
8 = Eggs in foam nest on pond; feeding tadpoles in pond
9 = Eggs in foam nest in pool; feeding tadpoles in stream
11 = Eggs imbedded in dorsum of aquatic female; eggs hatch into froglets
14 = Eggs on ground; eggs hatch into feeding tadpoles that are carried to water by adult
18 = Eggs arboreal; eggs hatch into feeding tadpoles that drop into ponds or streams
22 = Eggs in foam nest in burrow; tadpoles complete development in nest
RA** = Relative abundance of calling males:
R = Rare; individuals or groups heard to call on very few nights at intervals of more than one year and at only a subset of the seemingly available breeding habitat
UC = Uncommon; heard at some of the available breeding sites suitable for that species but does not necessarily appear at each site every year
C = Common; calling individuals occur at all the available breeding habitat suitable for that species and call for many months
Ub = Ubiquitous; widespread, terrestrially breeding species that call over many months
Call Freq** = How frequently a local population or individuals of a species call throughout a year:
Rare = One or two nights a year or less at only a subset of the available breeding habitat
SSM = Sporadically during a specific period lasting months
CSM = Continuously during a specific period lasting months
STY = Sporadically throughout the year
CTY = Continuously throughout the year (although call frequency may vary considerably among months)
* = probable but uncertain
blank = nothing known about the feature
** Relative abundance of calling males and annual calling frequency of a species are not behaviors that can be classified into rigid categories because such variation and combinations of vocalization traits exist among the species. "RA" and "Call Freq" in this table are meant to give the reader a general idea of the chance of seeing and/or hearing a species.

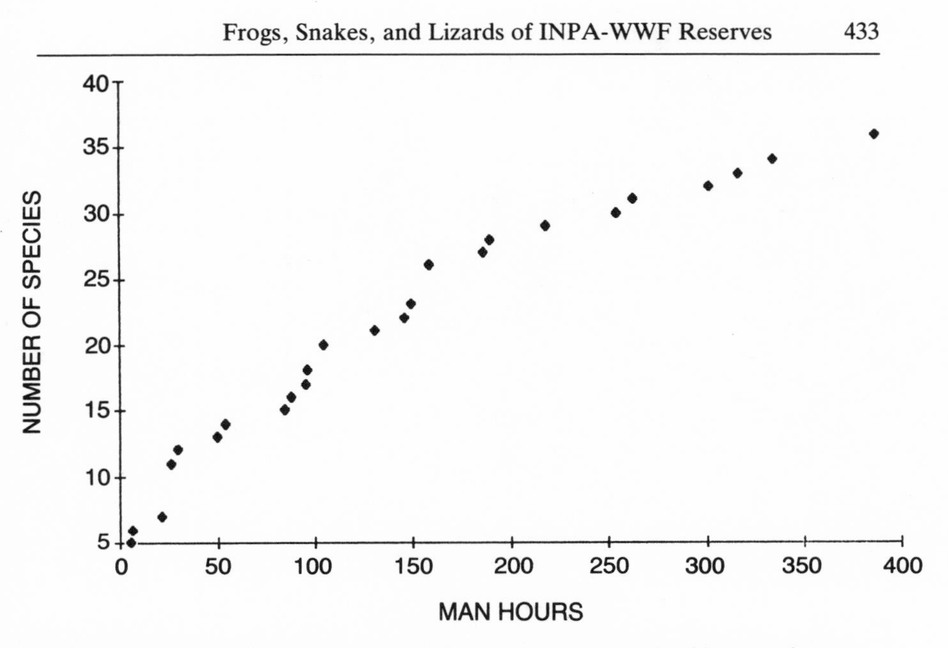

FIGURE 23.1. Cumulative number of frog species encountered by one observer versus man-hours of walking in the INPA-WWF reserves near Manaus.

WWF reserves. We include it in table 23.1 because it is a forest species, breeds at a site within 8 km of reserve 3304, and is common in Reserve Ducke, 50 kms south of the INPA-WWF reserve system (Hero and Magnusson, p.c.). First records for central Amazonia are *Pipa arrabali*, *Centrolenella oyampiensis*, *Bufo dapsilis*, *Ololygon cruentomma*, *Phrynohyas coriacea*, *Leptodactylus stenodema*, *Chiasmocleis shudikarensis*, and *Ctenophryne geayi*. A new species of *Chiasmocleis* belonging to the *hudsoni* group was also found (Heyer, p.c.).

A curve of number of species versus man-hours of sampling (fig. 23.1) reaches an asymptote at 45.6 species (equation fitted: number of species = 45.6 × man-hours / [129.7 + man-hours]).

Characteristic combinations of species bred at aquatic sites on the ground. The two main breeding habitats, streams (and associated pools) and upland pools, had few species in common (table 23.2). Several other species have terrestrial development or other reproductive features that allow them to breed throughout the forest without being obviously associated with any subhabitat. These species are *Adenomera andreae*, *Eleutherodactylus fenestratus*, *E.* sp. 1, *Leptodactylus stenodema*, *Osteocephalus buckleyi* (Zimmerman and Bogart 1988), *Phrynohyas re-*

TABLE 23.2 Species of frogs that bred at different aquatic sites, 1984–1987, in the INPA-WWF Reserves near Manaus

Habitat type	1501	1401	3402	1207 and area
Upland, large, temporary, floods annually	C5** *Oloygon cruentomma* *Hyla minuta* *Phyllomedusa tarsius* *P. tomopterna* *Chiasmocleis shudikarensis* *Chiasmocleis n.* sp. Unident. *microhylid* (t) CAMPO GRANDE* *Phyllomedusa tomopterna* *P. tarsius* *Hyla minuta* *Leptodactylus* sp. (t)			LAGINHO DO SORVETE*** *Ololygon cruentomma* *Hula minuta* *Hyla brevifrons* (-like) *Hyla* sp. (like *minuta*) *Phyllomedusa tarsius* *Phyllomedusa tomopterna* *Osteocephalus taurinus* *Leptodactylus mystaceus* *Chiasmocleis shudikarensis* *Chiasmocleis n.* sp. *Leptodactylus* sp. + (t) *Ceratophrys cornuta* (t) *Leptodactylus rhodomystax* (t)

Upland, large, temporary, does not flood annually		CERATOPHRYS POOL** *Hyla minuta* *Ololygon cruentomma* *Hyla brevifrons* (-like) *Phyllomedusa tarsius* *P. tomopterna* *Ceratophrys cornuta* *Leptodactylus mystaceus* *L. rhodomystax* *Chiasmocleis shudikarensis* *Chiasmocleis n.* sp. ESPINOZOAL* *Phyllomedusa tarsius* *P. tomopterna* *P. bicolor* *Hyla brevifrons* (-like) *Ololygon cruentomma* *Hyla minuta* *Leptodactylus riveroi* *L. mystaceus* *Leptodactylus* sp.
Upland, permanent large	CAMPO GRANDE ZF-3* *Hyla minuta* *H. rubra* *Phyllomedusa tomopterna* *P. tarsius* *Leptodactylus* sp.	LAGINHO 1207** *Phyllomedusa tomopterna* *Ololygon cruentomma* *Hyla garbei* *H. geographica*

TABLE 23.2. (*Continued*)

Habitat type	1501	1401	3402	1207 and area
Upland, small temporary, floods annually	PB18xGG* *Leptodactylus* sp.	P1.0*** *Phyllomedusa tarsius* *P. tomopterna* *P. bicolor* *Osteocephalus taurinus* *Leptodactylus rhodomystax* *Chiasmocleis shudikarensis* *Chiasmocleis n.* sp. *Colostethus* sp. *C. marchesianus* *Phyllobates femoralis* *Pipa arrabali* *Leptodactylus* sp. P1.5*** *Phyllomedusa tarsius* *P. tomopterna* *P. bicolor* *Osteocephalus taurinus* *Leptodactylus rhodomystax* *Chiasmocleis shudikarensis* *Chiasmocleis n.* sp. *Colostethus marchesianus* *Colostethus* sp. *Phyllobates femoralis* *Pipa arrabali*	POCOS CP-F* *Hyla minuta* *Leptodactylus mystaceus* *Chiasmocleis shudikarensis* *Chiasmocleis* sp. n.	

Upland, small, semipermanent or permanent	PBJ10** *Phyllomedusa bicolor* *P. tarsius* *Osteocephalus taurinus* *Pipa arrabali*		PB4** *Phyllomedusa tarsius* *P. tomopterna* *Leptodactylus* sp. *Colostethus marchesianus* *Colostethus* sp. *Phyllobates femoralis* (t)	
			PB1-2* *Phyllomedusa tomopterna* *P. tarsius*	
	PB18xGG-2* *Pipa arrabali* *Osteocephalus taurinus*		PB1 ARVORES* *Phyllomedusa tarsius* *P. tomopterna* *Phyllobates femoralis* (t)	
Stream, large		GAVIAOZINHO*** *Leptodactylus rhodomystax* *L. wagneri* *Osteocephalus taurinus* *Centrolenella oyampiensis* *Hyla granosa* *H. geographica* *Hyla* sp. (*microcephala* group) *Phyllomedusa bicolor* *Dendrophryniscus minutus* *Atelopus pulcher*		CP-F*** *Hyla boans* *H. granosa* *H. geographica* *Phyllomedusa bicolor* *Leptodactylus wagneri* *Atelopus pulcher* *Centrolenella oyampiensis* *Osteocephalus* sp.

TABLE 23.2. (*Continued*)

Habitat type	1501	1401	3402	1207 and area
Stream, small	various***	various***	various***	
	Centrolenella oyampiensis	*Centrolenella oyampiensis*	*Centrolenella oyampiensis*	
	Bufo dapsilis	*Bufo dapsilis*	*Bufo dapsilis*	
	Atelopus pulcher	*Atelopus pulcher*	*Atelopus pulcher*	
	Colostethus, 1 or 2 spp. (t)	*Colostethus*, 1 or 2 spp. (t)	*Colostethus*, 1 or 2 spp. (t)	
	Hyla granosa	*Hyla granosa*	*Hyla granosa*	
	H. geographica	*H. geographica*	*H. geographica*	
	Osteocephalus taurinus	*Osteocephalus taurinus*	*Osteocephalus taurinus*	
	Phyllomedusa bicolor	*Phyllomedusa bicolor*	*Phyllomedusa bicolor*	
	Leptodactylus rhodomystax	*Leptodactylus rhodomystax*	*Leptodactylus rhodomystax*	
	L. wagneri	*L. wagneri*	*L. wagneri*	
		L. riveroi		

Notes:
* Site surveyed on 5 or fewer nights
** Site surveyed on more than 5 but fewer than 20 nights
*** Site surveyed on more than 20 nights
+ Leptodactylus sp. refers to either *L. pentadactylus* or *L. knudseni*
(t) Tadpoles found but no calling males observed

sinifictrix, *Colostethus* n. sp. (F. Juncá, p.c.), *Synapturanus mirandiriberoi*, and *S. salseri*. *Leptodactylus mystaceus* breeds at localized upland sites that may or may not flood (including sites of certain temporary pools, table 23.2), and many of these sites are characterized by an abundance of a species of sedge. We know little or nothing in our area about breeding by *Osteocephalus* sp. (Zimmerman and Bogart 1988), *P. coriacea*, *Eleutherodactylus* sp. 2, *Lithodytes lineatus*, or *C. geayi*.

While walking trails at night through upland forest we most commonly saw *L. pentadactylus/knudseni* (at this time, we were unable to distinguish these two species by morphology alone), *Osteocephalus taurinus*, and *B. dapsilis*, whereas we most frequently heard *Osteocephalus buckleyi*, *E. fenestratus*, *E.* sp. 1, *P. resinifictrix*, and *L. stenodema*. At upland landlocked pool sites, the species encountered commonly throughout the year were *Phyllomedusa tarsius* and *P. tomopterna*. *L. pentadactylus/knudseni* were also frequently seen in stream valleys. *Hyla granosa*, *C. oyampiensis*, and *Leptodactylus wagneri*, as well as the species acoustically common in upland forest (except *L. stenodema*), were heard most frequently near streams.

The inventory of frogs is probably nearly complete because we have sampled long enough to encounter very rare and/or infrequent breeders (e.g., *Ceratophrys cornuta* and *Hyla* sp. *microcephala*-group). The curve for sampling success (fig. 23.1), based on six months of sampling in 1980 and occurrence of only 36 species (including 3 open habitat species that invade short distances into reserves), predicted that eventually we would find 47 species. This is very close to the number we found after extensive sampling (48 species, likewise including the disturbed habitat species).

Compared to the Tapajos National Park, a relatively thoroughly sampled site in the lower Amazon (Crombie, p.c.; Zimmerman, unpubl. data), INPA-WWF has 3 fewer species of dendrobatids. Compared to the Río Llullapichis study in the upper Amazon (Toft and Duellman 1979), INPA-WWF has 4 fewer species of dendrobatids and 7 fewer *Eleutherodactylus* species. Compared to Santa Cecillia (Duellman 1978), the richest frog fauna known for Amazon forest (only species found in primary forest considered), INPA-WWF lacks 3 dendrobatids, 13 *Eleutherodactylus*, and 6 *Hyla*. Richnesses of other genera are approximately equivalent at all 4 sites, except for the Río Llullapichis site, which has fewer *Leptodactylus*. The low number of *Hyla* in the INPA-WWF area relative to Santa Cecilia may in part reflect sampling error, because the small *Hyla* can be extremely difficult to find (for example, we know from a call that there is another small species of frog present in the INPA-WWF reserves, but we have never been able to catch it). Also, the sites may differ in availability of such subhabitats as forest "swamps." The low number of dendrobatids in the Manaus area is certainly real, because they do not obviously require specialized subhabitats in forest and are easy frogs to find. The most pronounced and enigmatic difference among the four sites is the lack of *Eleutherodactylus* in the central and lower Amazon compared to the upper Amazon (see Lynch 1980).

Attention to vocalization proved invaluable for extending the range and precision of the species survey. Some frogs in the INPA-WWF area are virtually never seen but commonly heard (e.g., *Synapturanus salseri*, *S. mirandiriberoi*, *Chiasmacleis* n. sp., and *Eleutherodactylus* sp. 1). Our list of species commonly seen differed greatly from the list of species commonly heard. Once we knew the calls, we could survey a site for breeding species quickly and accurately. When both visual and acoustic observation are considered, the species that seem genuinely to have relatively low populations in the INPA-WWF area are *Osteocephalus* sp., *P. vaillanti*, *Ceratophrys cornuta*, *L. lineatus*, and *C. geayi*.

Results of the survey of distribution of breeding (that is, calling) frogs in primary forest (table 23.2) demonstrate a clear division in the community: those that normally breed exclusively at upland sites, and those that normally breed in stream valley near streams. Only 6 species (besides terrestrially or tree-hole breeding ones) will breed at either upland or streamside pools. A feature of these ponds that seems important to the frogs is connection, or possible connection, with streams. Heyer et al. (1975), Hero and Magnusson of INPA in Manaus (unpubl. data), and others suggest that, as in the temperate zone, fish predators of tadpoles are a major influence on breeding site selection by frog species in the central Amazon. They hypothesize that the main evolutionary stimulus for species to breed exclusively in upland sites was to escape fish predation on their eggs and larvae.

Breeding habitat division may be even more finely tuned. Some species seem to prefer permanent or almost permanent pools, and others breed much more often in temporary pools. For example, *P. tarsius* and *P. tomopterna* occur more frequently at permanent or long-lasting pools, whereas *Hyla brevifrons*-like and *C. shudikarensis* are invariably found calling at temporary pools that generally do not last longer than three months (table 23.2). Pool size also seems to affect the species composition of the assemblage that breeds there. The effects of size and permanence of a pool probably exert their main influence by affecting predators of eggs and tadpoles. Essentially permanent landlocked upland pools are full of larval predators, including predatory tadpoles (notably large lepto-dactylids and *C. cornuta*), *P. arrabali* (Gascon, p.c.), dragonfly larvae, spiders, snakes, and turtles. There may be fewer such predators, or a period when predators are largely absent, in ephemerally flooding pools.

Aichinger (1987), however, found no clear delimitation between stream and isolated pond breeding habitat among 29 pool-breeding, primary forest species of frogs observed calling in Panguana, Peru, in the upper Amazon. Only 1 species bred exclusively at the stream; 18 species called at temporary and permanent ponds but not at the stream; and 10 species were heard at both stream and pools. There seems to be some sorting of species by breeding habitat at Panguana, but it is neither very similar to nor as pronounced as that observed in the INPA-WWF area. Some species differ strikingly in actual breeding habitat preference between Panguana and INPA-WWF. *Dendrophryniscus minutus*, *Hyla granosa*, and *L. wagneri*, for example, were never found calling at the stream in Panguana, whereas calling males of these species occurred extensively and exclusively at streams in the INPA-WWF area (though it must be noted that there are taxonomic problems with *L. wagneri* and *D. minutus*, so that these may not be the same species as those in the INPA-WWF area). In general, compared to our results, Aichinger (1987) found many fewer species associated exclusively with streams and a greater proportion that bred at both upland and streamside pools. In view of prevalent hypotheses that predators of tadpoles and eggs are the main force selecting for breeding site preference of Amazonian forest frogs (Heyer et al. 1975; Duellman 1978; Hero and Magnusson, p.c.), as is known in temperate zone anurans, it would be interesting to find out if distribution and abundances of predators of tadpoles in the Manaus area and Panguana could be correlated to distribution and abundances of calling male frogs in the respective areas. Are breeding site preferences of species fixed or flexible? If they are flexible in some species, what determines the reproductive strategy a species pursues in a certain area?

Results and Discussion: Snakes

As of February 1988; 62 species of snakes have been identified in the INPA-WWF area (table 23.3). The two specimens of *Micrurus lemniscatus* are from the outskirts of Manaus. We include it in the list because *M. lemniscatus* has an extensive Amazonian range and occurs in primary forest as well as in other habitats. Therefore it is almost certainly present in the INPA-WWF reserves. *Bothrops bilineatus* also was never collected in this study but because examples of this species collected from the Manaus area are in the Museu de Zoologia de Universidade de São Paulo and it is a widespread species but difficult to find, we have included it on our list.

One hundred twenty-six individuals of 28 species were seen along 1,661 km of nocturnal and diurnal trails walked by BLZ, usually accompanied by a second person (table 23.3). Sixty-five kms were unique trails, and some trails were walked many more times than others. Ninety-eight snakes were seen at night, 28 during the day. The clear-cut survey also produced 28 species: 20 species were shared with the BLZ survey, 8 species found by clear-cut workers were not encountered by BLZ walking, and BLZ saw 8 species not brought in by clear-cut workers (table 23.5). The species accumulation curve for the BLZ survey (fig. 23.2) includes an observation of an unidentified species seen near the beginning of the survey, but four observations of unidentified species are not included among abundances (table 23.5).

During the day a snake was encountered on average once per 35 km. Species most commonly encountered during the day were *Dendrophidion dendrophis* (6), *Erythrolamprus aesculapii* (4), *Bothrops atrox* (3), and *Liophis poecilogyrus* (3). At night, snakes were seen on average once per 7 km. Species most commonly encountered at night were *Leptodeira annulata* (27), *Oxybelis argenteus* (15), *Imantodes cenchoa* (7), *Dipsas catesbyi* (6), *Epicrates cenchria* (5), *B. atrox* (4), *Lachesis muta* (4), *Micrurus averyi* (4), and *Tripanurgos compressus* (4).

Comparisons of species compositions among different areas may be confounded by confused or incomplete taxonomic information. More than one species may be included under a single binomial. For example, there appear to be two species under *Atractus major* that cannot be distinguished using the literature. Particularly confused genera are *Atractus*, *Liophis*, and *Apostolepis*. Until these genera are revised, species lists of Amazonian forest snakes may be slightly inaccurate.

Some 88 species of snakes are known from the upper Amazon near Iquitos, Peru (Dixon and Soini 1977), and 76 species were taken from a large area near Belém in the lower Amazon (Cunha and Nascimento 1978). Duellman (1978) found 52 species in the Santa Cecillia area of Ecuador, but a lack of fossorial species in that survey indicates that the Santa Cecilia snake fauna may not be completely known. We also do not consider our list of 62 complete and believe that additional species of the less conspicuous fossorial and arboreal genera will be found. Specifically, the INPA-WWF area so far has no records of the apparently widespread

TABLE 23.3 Species of snakes and associated natural history data in the INPA-WWF Reserves near Manaus

Species	Habitat	Activity	Food
Anilidae			
Anilius scytale	F	ND	E,Amp+
Boidae			
Boa constrictor	G,T	N	M,B
Corallus caninus	T	N	M,B
Corallus enydris	B,T	N	M,B
Epicrates cenchria	G,T	N	M,B
Colubridae			
Apostolepis pymi	F	ND	S,C,Amp
Apostolepis sp.	F		
A. latifrons	F	ND	
A. major	F	ND	E+
A. snethlagae	F		
A. torquatus	F		
Atractus sp.	F		
Clelia cloelia	G	ND	S*,L,M
Chironius carinatus	G,T	D	M,B,F*
C. cinnamomeus	G,T	D	F+
C. fuscus	G,T	D	M,B,F*
C. scurrulus	G,T	D	F*
Dendrophidion dendrophis	G,T	D	F*,M,I
Dipsas catesbyi	B,T	N	G
D. indica	B,T	N	G
Drymarchon corais	G,B,T	D	F,L,S,M
Drymoluber dichrous	G,T	D	L*,F
Erythrolamprus aesculapii	G	D	F,S*,L
Helicops angulatus	A	ND	P*,F,L,t
H. hagmanni	A	ND	P*,F,t
Hydrops triangularis	A,AM		P,E
Imantodes cenchoa	B,T	N	F,L
Leptophis ahaetulla	B,T	D	F,L
Leptodeira annulata	G,B,T	N	F*,L,e
Liophis breviceps	G	D	F
L. cobella	G	ND	F
L. miliaris	G	ND	F
L. poecilogyrus	G	D	F
L. typhlus	G	D	F,A
Mastigodryas boddaerti	G,T	D	F,M,L,I
Oxybelis aeneus	T	D	F,L
O. argenteus	B,T	D	L,F
O. fulgidus	G,B,T	D	L,B
Oxyrhopus formosus	G	ND	L+
O. petola	G	ND	M+L+
O. trigeminus	G	N	M+
Philodryas viridissimus	T	D	F*,L,M,B
Pseustes poecilonotus	G,T	D	M,B

TABLE 23.3 (*Continued*)

Species	Habitat	Activity	Food
P. sulphureus	G,T	D	M,B
Rhadinea brevirostris	G	D	F,L*
Rhinobothryum lentiginosum	B,T	N	F*
Siphlophis cervinus	G,T	N	B,L+
Spilotes pullatus	G,T	D	M,B,L,F
Tantilla melanocephala	G,F	D	I
Tripanurgos compressus	B,T	N	L
Xenodon rhabdocephalus	G	ND	F
X. severus	G	ND	F
Elapidae			
Micrurus averyi	F,G	ND	
M. hemiprichii	F,G	N	Amp+
M. lemniscatus	G,F		S,Amp,C
M. spixii	F,G		S,Amp+
M. surinamensis	A,AM		P
Leptotyphlopidae			
Leptotyphlops septemstriatus	F		T,A
L. tenellus	F		T,A
Viperidae			
Bothrops atrox	G,B	ND	F,L,M
B. bilineatus	T		F,L,M*
Lachesis muta	G	N	M

Notes (abbreviations are taken or adapted from Duellman 1988):
Habitat:
F = fossorial
G = ground
B = bush (<1.5m)
T = trees
A = aquatic
AM = aquatic margin
Activity:
N = nocturnal
D = diurnal
ND = nocturnal and diurnal
Food (stomach content data from Cunha and Nascimento 1978; Duellman 1978; and pers. obs.):
E = earthworms
G = gastropods
F = frogs
L = lizards
S = snakes
P = fish
A = ants
T = termites
M = mammals
B = birds

C = caecilians
Amp = Amphisbaenids
I = insects
t = tadpoles
e = frog eggs
* = item most often found in stomachs
+ = one or few observations

TABLE 23.4 Survey of snakes in continuous forest of the INPA-WWF Reserves near Manaus

Reserve sampled	Dates sampled (*N*)	Snakes seen (*N*)	Observed during day	Observed at night	Kms walked day	Kms walked night
1202	1	1	0	1	0.0	2
1301	90	29	9	20	350.8	195.7
1302	6	4	4	0	26.0	23
1401	81	42	9	33	294.0	230.2
1501	19	11	3	8	35.1	49.1
2303	2	1	0	1	6.0	8.5
3304	46	23	7	16	184.0	94.9
3402	22	15	0	15	91.8	79.8
TOTAL	267	126	32	94	987.7	683.2

genera *Drepanoides*, *Pseudoboa*, *Pseudoeryx*, *Thamnodynastes*, *Xenopholis*, and *Typhlops*, which are present at both lower and upper Amazonian sites. Some absences are likely due to sampling error. The area sampled in the INPA-WWF study was less intensively sampled and quantitatively and qualitatively restricted compared to the other study areas, which spanned many more square kilometers (except Santa Cecilia) and included some cleared, second-growth, and riverine habitats as well as primary forest (aquatic habitats were relatively limited in this study, for example). Certain species that occur in primary forest may be easier to find in other habitats; there may also be exclusively open or disturbed habitat species.

The species accumulation curve for the walking survey (fig. 23.2) indicates that only slightly more than half the species eventually recorded in the INPA-WWF reserves were seen. Evidently, one would have to walk an enormous number of kilometers in order to encounter individuals representative of all the snake species.

Most species at lquitos, Santa Cecilia, and Belém (including more than 50% in each area shared with INPA-WWF) were found in other habitats as well as primary forest. The results from Belém (Cunha and

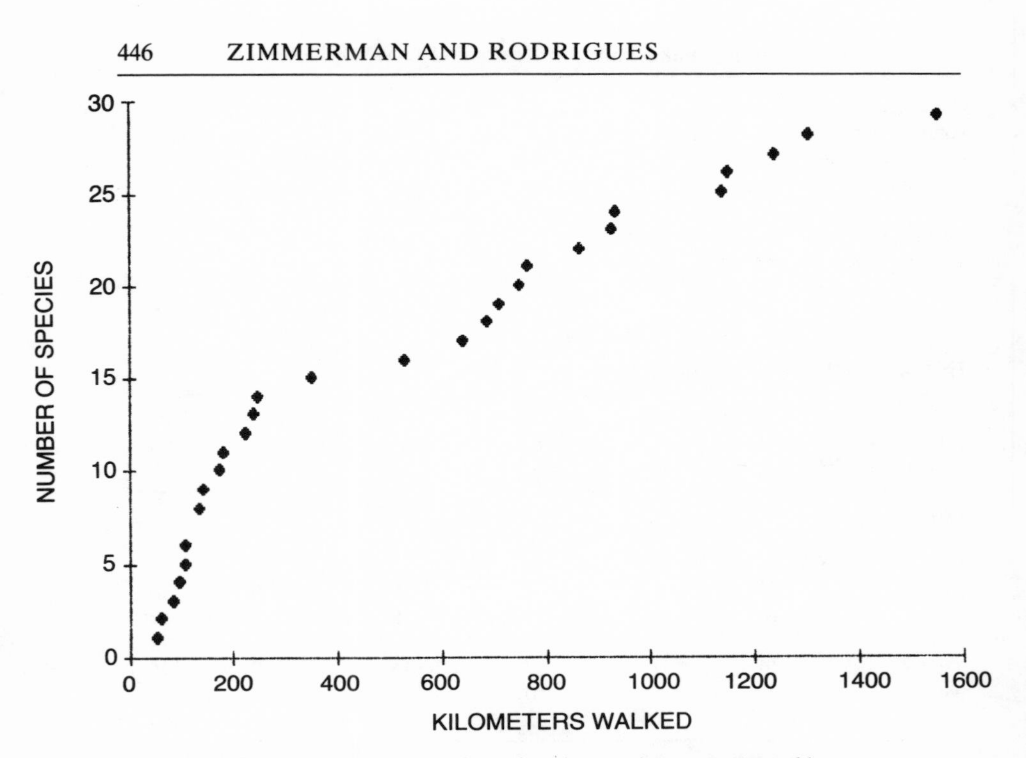

FIGURE 23.2. Cumulative number of snake species encountered by one observer versus number of kilometers walked in the INPA-WWF reserves near Manaus.

Nascimento 1978) are particularly striking. The survey encompassed about 45,000 km², and more than 90% of this area (where 8,000 specimens were collected) was drastically altered by humans. Only 5–10% of the area remained as primary forest, and most of the rest consisted of second growth and clearings. Apparently, even extensive deforestation that is subsequently replaced in part by second growth does not greatly alter snake species composition. At Belém, only 8 species, including 5 shared with INPA-WWF, were found exclusively in primary forest remnants, whereas 68 other species, including 44 shared with INPA-WWF area, ranged over primary forest, secondary forest of various ages, agricultural land, and other habitats. Species in the INPA-WWF reserves that were found virtually only in primary forest in any study are *Rhinobothryum lentiginosum*, *Siphlophis cervinus*, *T. compressus*, *Bothrops bilineatus*, and *L. muta*.

Any comparison of snake abundance will be difficult because of the extreme biases inherent in sampling. There seem to be no likely places or

TABLE 23.5 Abundances of INPA-WWF species of snakes found in primary forest of three Amazonian sites

Species	INPA-WWF BLZ	INPA-WF clear-cut	INPA-WWF totals	Santa Cecilia	Iquitos
Anilius scytale	1	4	5	6	17
Boa constrictor	0	2	2	0	7
Corallus caninus	0	1	1	6	2
C. enydris	2	1	3	0	10
Epicrates cenchria	5	2	7	3	1
Apostolepis pymi	0	2	2	—	—
Apostolepis sp.	0	1	1	—	—
Atractus latifrons	0	0	0	—	1
A. major	3	1	4	22	—
A. snethlagae	0	0	0	—	—
A. torquatus	1	0	1	—	—
Atractus sp.	0	0	0	—	—
Clelia clelia	2	1	3	—	4
Chironius carinatus	2	0	2	9	—
C. cinnamomeus	1	1	2	—	—
C. fuscus	0	0	0	2	—
C. scurrulus	0	0	1	5	15
Dendrophidion dendrophis	6	6	12	2	4
Dipsas catesbyi	6	0	6	12	19
D. indica	1	1	2	7	2
Drymarchon corais	0	0	0	—	2
Drymoluber dichrous	0	0	0	3	19
Erythrolamprus aesculapii	4	5	9	4	9
Helicops angulatus	0	0	0	7	—
H. hagmanni	1	0	1	—	—
Hydrops triangularis	1	0	1	—	—
Imantodes cenchoa	7	15	22	20	—
Leptodeira annulata	28	1	29	7	—
Leptophis ahaetulla	0	0	0	15	10
Liophis breviceps	0	0	0	—	4
L. cobella	0	0	0	2	—
L. miliaris	0	0	0	—	—
L. poecilogyrus	3	3	6	—	—
L. typhlus	0	0	0	—	5
Mastigodryas boddaerti	0	0	0	—	—
Oxybelis aeneus	0	0	0	—	—
O. argenteus	15	27	42	6	—
O. fulgidus	0	1	1	—	6
Oxyrhopus formosus	2	0	2	4	—
O. petola	0	0	1	5	13
O. trigeminus	2	0	2	—	0
Philodryas viridissimus	2	3	5	—	2
Pseustes poecilonotus	0	1	1	—	15
P. sulphureus	0	1	1	1	2

TABLE 23.5 (*Continued*)

Species	INPA-WWF BLZ	INPA-WF clear-cut	INPA-WWF totals	Santa Cecilia	Iquitos
Rhadinea brevirostris	1	4	5	4	10
Rhinobothryum lentiginosum	2	5	7	—	0
Siphlophis cervinus	0	0	0	1	1
Spilotes pullatus	0	0	0	—	—
Tantilla melanocephala	0	1	1	5	—
Tripanurgos compressus	4	3	7	1	7
Xenodon rhabdocephalus	0	0	0	—	—
X. severus	0	0	0	5	0
Micrurus averyi	5	2	7	—	—
M. hemiprichii	1	0	1	—	6
M. lemniscatus	0	0	0	2	0
M. spixii	0	0	0	7	27
M. surinamensis	0	0	0	1	5
Leptotyphlops septemstriatus	0	0	0	—	—
L. tenellus	0	0	0	—	—
Bothrops atrox	7	6	13	2	—
B. bilineatus	0	0	0	2	11
Lachesis muta	4	4	8	2	3

Sources: Data for INPA-WWF Reserves from BLZ walking and from clear-cut crews; for Santa Cecilia (forest only), Duellman (1978); for Iquitos, Dixon and Soini (1978)
Notes: — = Species not present or number of individuals caught in forest habitat unavailable
Abundances were significantly rank correlated in both BLZ and clearcut surveys (Spearman's test, $r = 0.583$, $p < 0.01$)

subhabitats (other than forest level) to look for most species, snakes make no sounds, many are cryptically, colored, many live out of our range in trees or underground, and many are nocturnal. Not only do the abundances of tropical snake remain unknown, but the possibility of comparing relative abundances among sites is slim owing to the few individuals collected and observed (table 23.5) in all but the most labor-intensive and lengthy studies (Cunha and Nascimento's 1978 survey). The data on abundance from this study and two other Amazonian sites (only primary forest records used) are presented in table 23.5 because they could provide a useful standard for future surveys at other sites. It is possible, for example, that sites could be found where snake abundances are significantly higher than those in table 23.5 and could therefore be considered important areas for conservation.

TABLE 23.6 Species of lizards and associated natural history data from the INPA-WWF Reserves near Manaus

Family and Species	Gen Hab	St Hab	Act	Food
GEKKONIDAE				
Coleodactylus amazonicus	TF,SV	LL	DS	I
Gonatodes humeralis	TF*,SV	Ar	DS	I
Thecadactylus rapicauda	TF	Ar	ND	I
IGUANIDAE				
Iguana iguana	SV	Ar	DS*H	H
Anolis chrysolepis	TF*,SV	LL,Ar*	DS	I
A. fuscoauratus	TF*,SV	LL,Ar*	DS	I
A. ortonii	TF*,SV	LL,Ar*	DS*,H	I
A. philopunctatus	TF	Ar	DS	I
Plica plica	TF	Ar	DS*H	I,A
P. umbra	TF	Ar	DS*H	I,A
Uracentron azureum	TF	Ar	DS*H	I,A
Uranoscodon superciliosum	SV	Am	DS	I
TEIIDAE				
Alopoglossus carinicaudatus	SV	LL	DS	I
Ameiva ameiva	TF,SV	LL,LB	DH	I,E
Arthrosaura reticulata	SV	LL	DS	I
Bachia cophias	SV	F	DS	I
B. panophia	SV	F	DS	I
Kentropyx calcarata	TF,SV	LL,LB	DH	I
Leposoma guianense	TF*,SV	LL	DS	I
L. pericarinatum	SV	LL	DS	I
Neusticurus bicarinatus	SV	LL	DS	I,A,t
Tretioscincus agilis	TF	LB	DH	I
Tupinambis nigropunctatus	TF	LL	DH	I,M,B,F,S,L
SCINCIDAE				
Mabuya bistriata	TF,SV	LB	DH	I

Notes:
Gen Hab = General Habitat:
TF = Terra firme upland forest
SV = Stream valley
* = Preferred habitat
ST Hab–Structural Habitat:
LL = Leaf litter
Ar = Arboreal
Am = Semiaquatic
F = Fossorial
LB = Fallen logs and branches
* = Preferred position
Act = Diel Activity:
DS = Diurnal shade
DH = Diurnal and heliophilic

DS*H = Diurnal and spends most time in shade but occasionally thermoregulates under sun
ND = Nocturnal and diurnal
Food:
I = Insects including spiders and other small arthropods
H = Herbivorous
A = Ants
M,B,F,S,L = Small mammals, birds, frogs, snakes, and lizards
t = Tadpoles
E = Worms

One snake per 7 km of night survey compared to one snake per 35 km of day survey gives the potentially misleading impression that the snake community is a great deal more active at night. Almost 15% of the snakes seen at night were *Oxybelis argenteus*, a strictly diurnal snake that is highly cryptic by day but not so at night, when it is easily found sleeping on small saplings. In addition, observation of the most common species seen during the walking survey, *Leptodeira annulata*, was undoubtedly correlated to habit. *L. annulata* is a dedicated frog eater and is usually found at ponds where frogs breed. The trails walked most frequently in this study led to frog breeding sites, and therefore sampling of *L. annulata* was not random, as we assume it was with other (nonaquatic) species. Finally, we know that half the INPA-WWF species are active during the day (table 23.3) and that more diurnal species are cryptically colored than are nocturnal species, which tend to have striking color patterns.

Perhaps it is informative to note that abundances recorded in the two INPA-WWF subsurveys (BLZ and CC in table 23.5) were correlated. That is, workers cutting down primary forest found more or less the same species in about the same proportions as one or two people walking through the forest. This provides some evidence that results from these two types of survey can be meaningfully compared.

Results and Discussion: Lizards

The lizard inventory is practically complete, though a few rare or inconspicuous species probably have not yet been found. These include *Polychrus marmoratus*, *Iphisia elegans*, and possibly additional species of *Bachia* and *Neusticurus*. The present list of 23 species (table 23.6) is similar to those for other well-surveyed Amazonian localities of roughly equivalent size (table 23.7). The small differences in species richness among Amazonian sites (table 23.7) can probably be attributed mostly to differences in sampling effort. One new species of *Anolis* was found during this survey (Rodrigues 1987).

TABLE 23.7 Species richness of lizards at thirteen sites in Amazonian forest

Locality	Species (*N*)	Observation Time	Source
Belém, Para, Brazil	21	some months	Crump 1971
Manaus, Amazonas, Brazil	23+	30 days	M. Rodrigues, unpubl. data
Rio Tapajos, Para, Brazil	24+	35 days	M. Rodrigues, unpubl. data
Rio Trombetas, Para, Brazil	17+	15 days	M. Rodrigues, unpubl. data
Lago Agrio, Napo, Ecuador	20	months	Duellman 1978
Santa Cecilia, Napo, Ecuador	25	many months	Duellman 1978
Centro Union, Iquitos, Peru	27	many months	Dixon and Soini 1975
Mishana, Iquitos, Peru	24	many months	Dixon and Soini 1975
Moropon, Iquitos, Peru	28	many months	Dixon and Soini 1975
Río Yuvineto, Loreto, Peru	21++	35 days	M. Rodrigues, unpubl. data
Río Zumun, Loreto, Peru	23++	30 days	M. Rodrigues, unpubl. data
French Guiana	28	many months	Hoogmoed and Lescure 1975
Suriname	32*	many months	Hoogmoed 1973

Notes:
+ Voucher specimens at Museu de Zoologia da Universidade de São Paulo
++ Voucher specimens at Musée d'Histoire Naturelle de Paris
* This figure becomes 20–24 species if an area of Surinamian forest only the size of the INPA-WWF area is considered

Species most frequently seen in INPA-WWF forest were *Gonatodes humeralis*, *Anolis chrysolepis*, *A. fuscoauratus*, *Plica umbra*, *P. plica*, *Kentropyx calcarata*, and *Mabuya bistriata*. Species most often seen at other Amazonian localities differ considerably, so that abundances of at least some species seems to vary over large geographical areas. *Polychrus marmoratus*, for example, was rare at most Amazonian localities but abundant at Río Zumun, Peru (Rodrigues, pers. obs.), *Arthrosaura reticulata* was common at the Rio Trombetas site but uncommon in the INPA-WWF reserves (Rodrigues, pers. obs.), *P. umbra* was relatively common in INPA-WWF forest but not so at Santa Cecilia (Duellman 1978), and so on.

Conclusions

Forty-two species of frogs that regularly occur throughout primary forest were found in the INPA-WWF reserves near Manaus. Most show a strict preference for breeding habitat that is either connected or has the possibility of connection to a stream or upland sites never connected with streams. Duration of water in pools and pool size also seem to influence composition of an assemblage of species that breed at a site. Comparison of the data with results from a similar study in the upper Amazon (Panguana) shows that breeding site preferences of some species may not be fixed over a large geographic area.

Sixty-two species of snake were identified from primary forest in the INPA-WWF reserves near Manaus. We believe additional species will be found. A snake was seen by one or two people on average every 35 km in the day and every 7 km at night. However, because of the special circumstances under which two common nocturnal species are encountered and the greater degree of cryptic coloration in diurnal snakes, the snake community is probably not much more active at night than during the day. Abundance data from INPA-WWF reserves and from primary forest at two other extensively surveyed upper Amazonian sites (lquitos and Santa Cecilia) show a similar pattern of very low numbers of individuals for most species. Relative abundances and species found by one observer (often accompanied by a second person) resembled abundances and species caught by several people during a clear-cut of primary forest in the same area. Including day and night surveys, the most commonly encountered species were: *Oxybelis argenteus*, *Leptodeira annulata*, *Imantodes cenchoa*, *Bothrops atrox*, *Dendrophidion dendrophis*, *Erythrolamprus aesculapii*, *Lachesis muta*, *Tripanurgos compressus*, *Micrurus averyi*, *Rhinobothryum lentiginosum*, and *Epicrates cenchria*.

Twenty-three species of lizards were identified from primary forest in the INPA-WWF reserves near Manaus. Richness of lizard species in the INPA-WWF reserves is similar to that found at other Amazonian sites.

REFERENCES

Aichinger, M. 1987. Annual activity pattern in a seasonal Neotropical environment. Oecologia 71: 583–592.

Crump, M. L. 1971. Quantitative analysis of the ecological distribution of a tropical herpetofauna. Occas. Pap. Mus. Nat. Hist. Univ. Kans. 3: 1–62.

———. 1974. Reproductive strategies in a tropical anuran community. Misc. Publ. Mus. Nat. Hist. Univ. Kans. 61: 1–68.

Cunha, O. R. and F. P. Nascimento. 1978. *Ofidios da Amazonia*. Publ. Avulsas Mus. Paraense Emilio Goeldi 31.

Dixon, J. R. and P. Soini. 1975. The reptiles of the upper Amazon Basin, Iquitos region, Peru I. Lizards and amphisbaenians. Contrib. Biol. Geol. Milwaukee Pub. Mus. 4: 1–58.

———. 1977. The reptiles of the upper Amazon Basin, Iquitos region, Peru. II. Crocodilians, turtles, and snakes. Contrib. Biol. Geol. Milwaukee Pub. Mus. 12: 1–91.

Duellman, W. E. 1978. The biology of an equatorial herpetofauna in Amazonian Ecuador. Misc. Pub. Mus. Nat. Hist. Univ. Kans. 65: 1–352.

———. 1989. Tropical herpetofaunal communities: patterns of community structure in Neotropical rainforests. In M. L. Harmelin-Vivien and F. Bourlière (eds.), *Vertebrates in Complex Tropical Systems*. Springer-Verlag, New York, pp. 61–88.

Heyer, W. R. and M. S. Bellin. 1973. Ecological notes on five sympatric Leptodactylus (Amphibia: Leptodactylidae) from Ecuador. Herpetologica 29: 66–72.

Heyer, W. R., R. W. McDiarmid, D. L. Weigmann. 1975. Tadpoles, predation, and pond habits in the tropics. Biotropica 7: 100–111.

Hoogmoed, M. S. 1973. Notes on the herpetofauna of Surinam: the lizards and amphisbaenians of Surinam. Biogeographica 4. W. Junk, The Hague, 1–9: 1–419.

Hoogmoed, M. S. J. Lescure. 1975. An annotated checklist of the lizards of French Guiana. Zool. Meded. 49: 141–171.

Lovejoy, T. E., and R. O. Bierregaard, Jr. (This volume). Central Amazonian Forests and the Minimum Critical Size of Ecosystems Project.

Lynch, J. D. 1980. A taxonomic and distributional synopsis of the Amazonian frogs of the genus *Eleutherodactylus*. Am. Mus. Nov. 2629: 1–24.

Pyburn, W. F. 1975. A new species of microhylid frog of the genus *Synapturanus* from southeastern Colombia. Herpetologica 31: 439–443.

Rodrigues, M. T. 1987. A new anole of the *punctatus* group from central Amazonia (Sauria, Iguanidae). Pap. Avul. Zool., São Paulo, 36: 333–336.

Toft, C. A. and W. E. Duellman. 1979. Anurans of the lower Río Llullapichis, Amazonian Peru. Herpetologica 35: 71–77.

Zimmerman, B. L. 1983. A comparison of structural features of calls of open and forest habitat frog species in the central Amazon. Herpetologica 39: 235–246.

Zimmerman, B. L. and R. O. Bierregaard, Jr. 1986. Relevance of the equilibrium theory of island biogeography and species-area relations to conservation with a case from Amazonia. J. Biogeogr. 13: 133–143.

Zimmerman, B. L. and J. P. Bogart. 1984. Vocalizations of primary forest frog species in the central Amazon. Acta Amazônica 14: 473–519.

———. 1988. Ecology and calls of four central Amazonian forest frog species. J. Herpetol. 22: 97–108.

Zimmerman, B. L. and W. Hodl. 1983. Distinction of *Phrynohyas resinifictrix* (Goeldi, 1902) based on acoustical and behavioural parameters. Zool. Anz. (Jenna) 5/6: 341–351.

24

Herpetofaunas in Neotropical Rainforests: Comparative Composition, History, and Resource Use

WILLIAM E. DUELLMAN

Information on the assemblages of amphibians and reptiles occuring at four sites in Neotropical rainforests has been presented in this volume by Guyer, Rand and Myers, Rodriguez and Cadle, and Zimmerman and Rodrigues. In this chapter I compare the herpetofaunas of these four sites. I also include data from Santa Cecilia, Ecuador, a site with the largest known herpetofauna in the Neotropics (Duellman 1978). Comparisons include species composition, historical faunal assemblages, resource use (microhabitat, diel activity, food, and, in the case of anurans, reproductive modes). The data used in this synthesis are tabulated in appendix 24.1.

Throughout this chapter I refer to the sites by abbreviated names: "La Selva" for the Estación Biológica La Selva, Costa Rica; "BCI" for Barro Colorado Island and the adjacent mainland in Panama; "Manaus" for the Minimum Critical Size of Ecosystems Study Area north of Manaus, Brazil; "Manu" for Cocha Cashu in the Parque Nacional Manu, Peru; and "Santa Cecilia" for Santa Cecilia, Ecuador (fig. 24.1). Santa Cecilia is situated at elevation 340 m on the Río Aguarico in the upper

I am grateful to Alwyn H. Gentry for the opportunity to participate in a very stimulating symposium and to contribute to this volume. My synthesis was possible because of the data collected and analyzed by John E. Cadle, Craig Guyer, Charles W. Myers, A. Stanley Rand, Lily B. Rodriguez, and Barbara L. Zimmerman, all of whom sent me advance copies of their manuscripts. My personal fieldwork at La Selva, BCI, and Santa Cecilia was supported at various times by grants from the National Science Foundation and the Museum of Natural History at the University of Kansas. I have never visited Manaus or Manu, but Barbara L. Zimmerman and Lily B. Rodriguez shared with me their experiences at those sites. Finally, I thank Linda Trueb for aid with the graphics and for her critical comments on my prose.

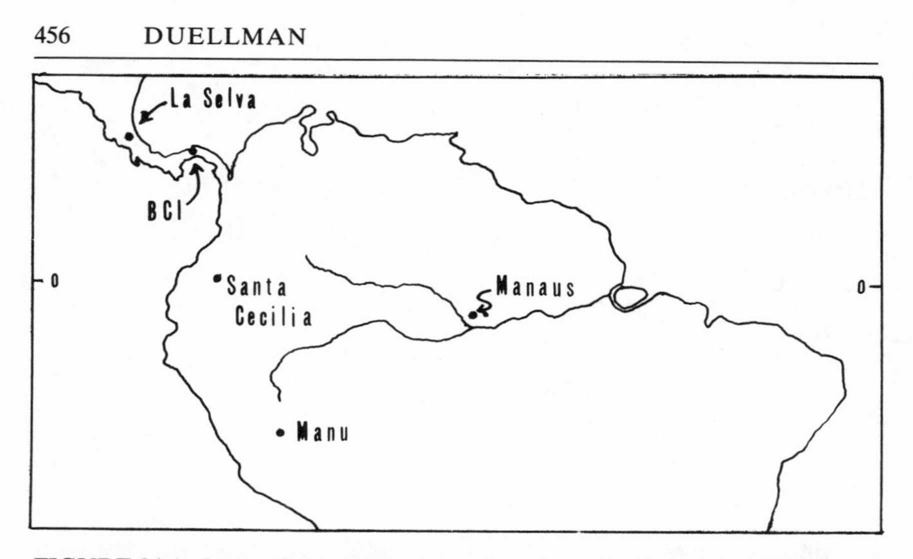

FIGURE 24.1. Lower Central America and northern South America showing the location of the five sites under comparison.

Amazon Basin (00°03′N, 76°59′W), Provincia de Napo, Ecuador; at the time of my studies the area was covered with primary rainforest growing on relatively nutrient-rich soils, and the annual rainfall was about 4,300 mm. The other sites are described elsewhere in this volume.

Of the five sites, the assemblages at La Selva and BCI are the best known; those at Santa Cecilia and Manaus follow. The herpetofauna, especially the lizards and snakes, at Manu is incompletely known; certainly such additional species as *Polychrus marmoratus*, *Dracaena guianensis*, *Iphisa elegans*, *Boa constrictor*, *Chironius fuscus*, *Dipsas indica*, *Drymoluber dichrous*, and *Imantodes lentiferus* should occur there.

Species Composition and Richness

The species composition at each site is based on the data provided in the individual faunal lists presented in this volume, plus my data (1978) for Santa Cecilia; the data for Santa Cecilia have been updated taxonomically (for example, the original material of *Gonatodes concinnatus* was found to consist of *G. concinnatus* and *G. humeralis*, and *Hyla favosa* is considered a synonym of *H. leucophyllata*). Species occurrence at each site is tabulated in appendix 24.1.

A synthesis of the data on species composition reveals that 131–185 (mean 143) species of amphibians and reptiles are known from the five

TABLE 24.1 Species richness at five sites

Taxon	La Selva		BCI		Manaus		Manu		Santa Cecilia	
	N	(%, D)*	*N*	(%, D)	*N*	(%, D)	*N*	(%, D)	*N*	(%, D)
Gymnophiona	1	(0.7 −0.6)	1	(0.7 − 0.5)	2	(1.5 +0.2)	1	(0.8 −0.5)	5	(2.7 +1.4)
Caudata	3	(2.3 +1.2)	2	(1.5 +0.4)	0	(0.0 −1.1)	1	(0.8 −0.3)	2	(1.1 ±0.0)
Anura	44	(32.7 −8.3)	49	(36.8 −4.2)	42	(31.6 −9.4)	75	(57.2 +16.2)	86	(46.5 +5.5)
Amphibia	48	(35.7 −7.7)	52	(39.1 −4.3)	44	(33.1 −10.20)	77	(58.8 +15.4)	93	(50.3 +6.9)
Testudines	4	(3.0 −0.4)	5	(3.7 +1.1)	2	(1.5 +1.1)	5	(3.8 +1.2)	6	(3.2 +0.6)
Crocodylia	1	(0.7 −0.6)	2	(1.5 +0.2)	2	(1.5 +0.2)	2	(1.5 +0.3)	2	(1.1 −0.2)
Sauria	25	(18.6 +1.6)	26	(19.8 +2.8)**	24	(18.0 +1.0)	16	(12.2 −4.8)	30	(16.2 −0.8)
Amphisbaenia	0	(0.0 −0.4)	1	(0.7 +0.4)	1	(0.8 +0.4)	0	(0.0 −0.4)	1	(0.6 +0.2)
Serpentes	56	(42.0 +7.1)	47	(35.2 +0.3)	60	(45.1 +10.2)	31	(23.7 −11.2)	53	(28.6 −6.3)
Reptilia	86	(64.3 +7.7)	81	(60.9 +4.3)	89	(66.9 +10.3)	54	(41.2 −15.4)	92	(49.7 −6.9)
TOTALS	134		133		133		131		185	

Notes: * Numbers in parentheses are percentages of total herpetofauna at site and the deviation from mean percentage from all sites
** Excluding the introduced *Hemidactylus frenatus*

sites. The average species assemblage in these lowland rainforests consists of 2 caecilians, 2 salamanders, and 59 anurans for a total of 63 species of amphibians, and 4 turtles, 2 crocodilians, 24 lizards, 1 amphisbaenian, and 49 snakes for a total of 80 species of reptiles. The contribution to the herpetofauna by caecilians, salamanders, turtles, crocodilians, and amphisbaenians at each site is less than 9% (table 24.1). The percentage contribution of anurans to the entire assemblage is much higher at the two sites in the upper Amazon Basin (Manu and Santa Cecilia) than at the others, although the high percentage (57.2) of anurans at Manu in part reflects the incomplete inventories of snake and lizard faunas. The fewest species (42) of anurans and their lowest percent of the fauna (33.1) is at Manaus. Only forest-dwelling frogs are compared here. Additional species occur in riparian and aquatic habitats. See Hödl [1977] for anurans living in the floating meadows. The relative contribution of lizards to the assemblages is nearly the same (16.2–19.8%) at four sites but lower at Manu (12.2%); the largest assemblage of lizards (30) is at Santa Cecilia. Snakes make up a good percentage of the fauna at La Selva (56 species, 42%) and Manaus (60, 45.1%); at Santa Cecilia, 53 species of snakes compose 28.6% of the herpetofauna.

Several algorithms have been proposed for making faunal comparisons (Cheetham and Hazel 1969). Jaccard (1928) proposed that a coefficient of community (CC) could be calculated by the formula

$$CC = C/(N_1 + N_2) - C,$$

where C is the number of species in common in two areas, N_1 is the number of species in the first area, and N_2 is the number of species in the second area. This is the same as Whittaker's (1970) coefficient of community,

$$CC = S_{ab}/(S_a + S_b) - S_{ab},$$

where S_{ab} is the species in common to areas *a* and *b*, S_a is the number of species in area *a*, and S_b is the number of species in area *b*. Simpson's coefficient (SC) was proposed by Simpson (1960) as the formula

$$C/N_1,$$

where C is the number of species in common between two areas and N_1 is the number of species in the smallest of the two faunas. Duellman (1965) suggested a modification of Pirlot's (1956) formula

$$FRF = 2C/(N_1 + N_2),$$

where FRF is Faunal Resemblance Factor, C is the number of species in common to two areas, N_1 is the number of species in the first area, and N_2 is the number of species in the second area. Peters (1968) also used the term *Faunal Resemblance Factor*. However, because the algorithm is equally applicable to animals and plants, I suggest using the term *Coefficient of Biogeographic Resemblance* (CBR).

As pointed out by Duellman (1965) and Dixon (1979), Simpson's formula does not account for the number of species in the larger fauna. The results of analyses using the CBR of Pirlot (1956) and Duellman (1965) and the CC of Jaccard (1928) and Whittaker (1970) are nearly identical, but the values for the CC are always lower than those for the CBR. The Coefficient of Biogeographic Resemblance has been used herein because it seems to be a more robust algorithm than either the Coefficient of Community or Simpson's formula and because it has been used for numerous herpetofaunal comparisons in the Neotropics (e.g., Dixon 1979; Duellman 1978; Hoogmoed 1979).

Numerical comparison of the herpetofaunal assemblages between La Selva and BCI reveals a moderately high CBR of 0.55, slightly greater than the average CBR among the Amazonian sites (0.42–0.57, mean = 0.48). The species composition at BCI includes Barro Colorado Island and the adjacent mainland [see app. 24.1]; the numbers are higher than those given by Duellman [1989], which included only those species occurring on the island). In contrast, the CBRs between the Central American and Amazonian sites are much lower (0.08–0.14, mean = 0.12) (table 24.2). When amphibians and reptiles are compared independently, the CBRs of amphibians generally are lower than those of reptiles (table 24.2). Although the CBRs are more nearly alike for lizards and snakes, the values for snakes are consistently higher than those for lizards, except for the comparison of Manaus and Manu (table 24.2).

The CBR values indicate that fewer amphibians are as widely distributed as reptiles among the sites. Only 8 species are known from all five sites—1 amphibian (*Leptodactylus pentadactylus*) and 7 reptiles (*Thecadactylus rapicauda*, *Clelia clelia*, *Imantodes cenchoa*, *Leptodeira annulata*, *Leptophis ahuetulla*, *Lachesis muta*, and *Caiman crocodilus*). Four additional species—1 amphibian (*Hyla boans*) and 3 reptiles (*Ameiva ameiva*, *Epicrates cenchria*, and *Siphlophis cervinus*)—occur at all sites except La Selva. Forty-five species (16 amphibians and 29 reptiles) occur at all three Amazonian sites, whereas 73 species (25 amphibians and 48 reptiles) occur at both Central American sites. The discrepancy between CBR values for amphibians and reptiles is caused mainly by the pro-

TABLE 24.2 Comparison of the herpetofaunas at five sites

	La Selva	BCI	Manaus	Manu	Santa Cecilia
		Amphibians and Reptiles			
La Selva	**134**	*0.55*	*0.13*	*0.08*	*0.08*
BCI	74	**133**	*0.14*	*0.14*	*0.13*
Manaus	17	19	**133**	*0.42*	*0.46*
Manu	11	19	56	**131**	*0.57*
Santa Cecilia	13	20	73	91	**185**
		Amphibians			
La Selva	**48**	*0.50*	*0.03*	*0.04*	*0.03*
BCI	25	**52**	*0.05*	*0.09*	*0.07*
Manaus	1	2	**44**	*0.40*	*0.38*
Manu	2	6	24	**77**	*0.60*
Santa Cecilia	2	5	26	48	**93**
		Anurans			
La Selva	**44**	*0.54*	*0.02*	*0.03*	*0.03*
BCI	25	**49**	*0.04*	*0.10*	*0.07*
Manaus	1	2	**42**	*0.41*	*0.38*
Manu	2	6	24	**75**	*0.58*
Santa Cecilia	2	5	24	47	**86**
		Reptiles			
La Selva	**86**	*0.57*	*0.18*	*0.13*	*0.12*
BCI	48	**81**	*0.20*	*0.19*	*0.17*
Manaus	16	17	**89**	*0.48*	*0.52*
Manu	9	13	34	**54**	*0.59*
Santa Cecilia	11	15	47	43	**92**
		Lizards and Amphisbaenians			
La Selva	**25**	*0.46*	*0.08*	*0.05*	*0.04*
BCI	12	**27**	*0.12*	*0.10*	*0.10*
Manaus	2	3	**25**	*0.49*	*0.39*
Manu	1	2	10	**16**	*0.43*
Santa Cecilia	1	3	11	10	**31**
		Snakes			
La Selva	**56**	*0.62*	*0.21*	*0.16*	*0.17*
BCI	32	**47**	*0.26*	*0.26*	*0.20*
Manaus	12	14	**60**	*0.46*	*0.57*
Manu	7	10	21	**31**	*0.69*
Santa Cecilia	9	10	32	29	**53**

Notes: **N** = species at each site
N = species in common between two sites
N = Coefficients of Biogeographic Resemblance

TABLE 24.3 Numbers of species of different taxonomic groups of the herpetofauna occurring at one or more sites

	Number of sites										
	One		*Two*		*Three*		*Four*		*Five*		
Group	*N*	(%)	*N*	(%)	*N*	(%)	*N*	(%)	*N*	(%)	*Total*
Gymnophiona	4	(57.1)	3	(42.9)	—		—		—		7
Caudata	8	(100.0)	—		—		—		—		8
Anura	111	(58.1)	46	(24.2)	31	(16.2)	2	(1.0)	1	(0.5)	191
Sauria	63	(74.1)	14	(16.5)	7	(8.2)	1	(1.1)	1	(1.1)	86
Amphisbaenia	—		—		1	(100.0)	—		—		1
Serpentes	69	(51.1)	37	(27.4)	19	(14.1)	5	(3.7)	5	(3.7)	135
Testudines	8	(57.2)	3	(21.4)	3	(21.4)	—		—		14
Crocodylia	2	50.0)	1	(25.0)	—		—		1	(25.0)	4
Total	265	(59.4)	104	(23.3)	61	(13.7)	8	(1.8)	8	(1.8)	446

TABLE 24.4 Distances between sites and Coefficients of Biogeographic Resemblance

		Coefficients of Biogeographic Resemblance		
Sites	Distance (km)	*Amphibians*	*Reptiles*	*Amphibians and Reptiles*
La Selva–BCI	540	0.50	0.57	0.55
BCI–Santa Cecilia	1,130	0.07	0.17	0.13
Santa Cecilia–Manu	1,520	0.60	0.59	0.57
Manaus–Manu	1,600	0.40	0.48	0.42
La Selva–Santa Cecilia	1,670	0.03	0.12	0.08
Manaus–Santa Cecilia	1,860	0.38	0.52	0.46
BCI–Manaus	2,550	0.05	0.20	0.14
BCI–Manu	2,710	0.09	0.19	0.14
La Selva–Manaus	3,090	0.03	0.18	0.13
La Selva–Manu	3,250	0.04	0.13	0.08

portionally higher percentage of snakes that have broad distributions (table 24.3).

Because of the comparably higher CBR values comparing the two Central American sites and the values comparing the two geographically closest Amazonian sites, CBRs were correlated with geographical distances among sites (table 24.4). Distances were approximated in a straight line and entirely overland. No positive correlation exists between CBRs and distances between sites. For example, BCI is closer to Santa Cecilia than Santa Cecilia is to Manu, yet the CBR (amphibians and reptiles) for the first pair is 0.13, whereas the CBR (amphibians and reptiles) for the second is 0.57.

Faunistic comparisons of the five sites thus show that the three Amazonian sites are more alike than any one of them is like the two Central American sites. Further, the herpetofauna of BCI has more in common with Manaus, especially with respect to reptiles, than to any other Amazonian site.

Historical Faunal Assemblages

Most biogeographic studies consider historical factors in the composition of biotas, but many works on specific communities ignore this important aspect of the biota. The histories of the biotas of the two Central American sites differ greatly from those at the South American sites because the two regions were separated by a seaway during most of the Cenozoic (see Smith 1985 and Donnelly 1985 for recent reviews of geological evidence).

In a lengthy discourse on the herpetofauna of Central America, Savage (1982) defined generalized tracks corresponding to four historical units of genera. Some genera were considered to be composites of two historical units. Working at the specific level, it has been possible to assign the species of amphibians and reptiles at the five sites to one of three units on the basis of phylogenetic relations of species of *Bufo* by Blair (1972), *Hyla* by Duellman (1970), and *Eleutherodactylus* by Lynch (1986). Genera of colubrid snakes were assigned on the basis of the subfamilial classification used by Dowling and Duellman (1978) and the biogeographic analysis by Cadle (1985). None of the species under consideration belongs to Savage's Young Northern Unit. The historical unit of each of the species is noted in appendix 24.1; the units are defined as follows:

Old Northern Unit. These taxa are derived from originally subtropical or warm-temperate groups that were widely distributed in the early Tertiary but were forced southward and fragmented into several disjunct components as a result of cooling, aridity, and orogeny in the late Cenozoic. With the probable exception of *Drymarchon corais* (Vanzolini and Heyer 1985), the Recent species of the Old Northern Unit in Mesoamerica and especially South America evolved in situ from northern stocks. Savage (1982) emphasized that the Central American component of this unit has been disjunct from other components throughout most of the later Tertiary and Quaternary and has evolved in situ in Central America.

Mesoamerican Unit. These taxa are derived from a generalized tropical American fauna isolated in tropical North and Central America during most of the Cenozoic. This unit developed in situ north of the Panamanian Portal and was restricted by orogenies and climatic changes in the late Cenozoic of Middle America. Savage (1982) argued that this unit differentiated because of a major vicariance event, the inundation of a putative late Cretaceous–Paleocene land bridge between Central and South America.

South American Unit. These taxa derive from a generalized tropical American fauna that evolved in situ in isolation in South America during most of the Cenozoic. Some of these taxa (e.g., xenodontine snakes) may have reached South America via Central America in the late Cretaceous (Savage 1982), but equally compelling evidence suggests that such taxa are members of the Gondwanan biota (Laurent 1979; Cadle 1984).

Some components of the Old Northern and Mesoamerican units may have dispersed via island-hopping from Central America to South American before the closure of the Panamanian Portal in the late Pliocene. Likewise, a few members of the South American Unit may have entered Central America by the same means during that time. However, the major interchange occurred after the Panamanian Portal closed in the late Pliocene.

The historical units defined here and their vicariance and dispersal is consistent with the geological evidence (Smith 1985; Donnelly 1985) and the paleontological evidence (Estes and Báez 1985). Furthermore, the proposed historical biogeography of the taxa under consideration here is not inconsistent with evidence from the herpetofauna as a whole (Duellman 1979; Vanzolini and Heyer 1985) or vertebrates in general (Bonaparte 1984).

Although most species at the five sites are members of the South American Unit, major discrepancies exist among different taxonomic groups (table 24.5). For example, all of the salamanders, half of the turtles, and about one-fourth of the snakes belong to the Old Northern Unit, whereas most of frogs (87.4%) and lizards (67.4%) belong to the South American Unit. Among the anurans, only members of the genus *Rana* belong to the Old Northern Unit.

TABLE 24.5 Numbers of taxa that occur in five Neotropical forests and their association with three historical units of the herpetofauna

Group	Old Northern N	Old Northern (%)	Mesoamerican N	Mesoamerican (%)	South American N	South American (%)	Total
Gymnophiona	—		1	(14.3)	6	(85.7)	7
Caudata	8	(100.0)	—		—		8
Anura	4	(2.1)	20	(10.5)	167	(87.4)	191
Amphibians	12	(5.8)	21	(10.2)	173	(84.0)	206
Sauria	5	(6.0)	22	(25.8)	59	(68.2)	86
Amphisbaenia	—		—		1	(100.0)	1
Serpentes	35	(25.9)	35	(25.9)	65	(48.2)	135
Testudines	7	(50.0)	—		7	(50.0)	14
Crocodylia	—		1	(25.0)	3	(75.0)	4
Reptiles	47	(19.6)	58	(24.3)	134	(56.1)	239
Total	59	(13.2)	79	(17.8)	307	(69.0)	445

TABLE 24.6 Numbers of species (amphibians/reptiles/total) that belong to each of the historical units of the herpetofauna and occur at each of five Neotropical forests

Unit	La Selva	BCI	Manaus	Manu	Santa Cecilia
Old Northern	6/29/35	4/21/25	0/17/17	0/7/7	3/14/17
Mesoamerican	17/39/56	11/30/41	0/10/10	1/6/7	0/12/12
South American	25/18/43	37/30/67	44/62/106	76/41/117	90/66/156
Total	48/86/134	52/81/133	44/89/133	77/54/131	93/92/185

Comparisons of the historical compositions of the herpetofaunal assemblages at each site (table 24.6) show a progressive decrease in the numbers of Old Northern and Mesoamerican taxa from La Selva to BCI and then to the South American sites, contrasted with a progressive decrease in South American taxa from the South American sites to BCI and then to La Selva.

The major Old Northern contributors to the herpetofaunas of the South American sites are colubrid snakes of the genera *Chironius*, *Dendrophidion*, *Drymarchon*, *Drymobius*, *Drymoluber*, *Leptophis*, *Mastigodryas*, *Oxybelis*, *Pseustes*, *Rhinobothryum*, and *Tantilla*. The major Mesoamerican contributors to these herpetofaunas are iguanid lizards of the *Norops* group of the genus *Anolis* and colubrid snakes of the genera *Dipsas*, *Imantodes*, and *Leptodeira*. Old Northern and Mesoamerican groups of amphibians contribute only minimally to the South American herpetofaunas.

Many taxonomic groups that contribute significantly to the herpetofaunas at the Central American sites are members of the Mesoamerican Unit that either do not occur at the South American sites or make only minor contributions there. These include frogs of the genera *Agalychnis* and *Smilisca*, lizards of the genus *Basiliscus*, and snakes of the genera *Coniophanes*, *Ninia*, *Rhadinaea*, *Sibon*, and *Trimetopon*.

Major South American groups of anurans that comprise significant parts of the herpetofaunas at the Central American sites include *Bufo*, *Centrolenella*, *Colostethus*, *Dendrobates*, *Hyla*, *Leptodactylus*, and *Ololygon*. Of the 18 species of *Eleutherodactylus* at the Central American sites, 10 are members of the South American clade of the genus. Reptiles of South American origin are comparatively fewer than amphibians of South American origin at these sites; of the 17 species of lizards, 7 are gekkonids and 6 are teiids. Only 17 species of South American snakes occur at the Central American sites.

Thus, although Old Northern, Mesoamerican, and South American taxa are present at all sites, anurans account for most South American taxa at the Middle American sites, and snakes account for most Old Northern taxa at the South American sites. Snakes and iguanid lizards of the *Norops* group of the genus *Anolis* comprise the majority of Mesoamerican taxa at the South American sites.

Resource Use

Schoener (1974) reviewed resource partitioning in ecological communities and identified three major ways in which similar species partition resources—habitat, time, and diet—and emphasized that the methods of resource partitioning differ notably between terrestrial ectotherms and endotherms. Pough (1980) suggested that these differences reflect in part the great difference in modal body size between ectothermic and endothermic tetrapods; he postulated that, because of basic differences in physiology and life history, endothermic vertebrates are likely to respond differently than ectotherms to habitat dimensions and that temporal partitioning may be much more important in ectotherms.

As emphasized by Duellman (1989), most species in herpetofaunal communities in Neotropical rainforests can be categorized as terrestrial or arboreal, and the species composition in each category differs greatly between night and day. Thus, it is necessary to analyze major habitat use with respect to diel activity.

Herpetofaunal communities in Neotropical rainforests are about equally divided between diurnal and nocturnal species; furthermore, about half of the species are terrestrial (including fossorial, leaf litter, and aquatic margin) and arboreal (bushes, tree trunks, and tree limbs). Although most are specific with respect to habitat (arboreal versus terrestrial) and diel activity, some diurnal species are active on the ground and in trees and bushes; included in this category are such snakes as *Oxybelis fulgidus*, *Pseustes poecilonotus*, *P. sulphureus*, and *Spilotes pullatus*. Several snakes (*Clelia clelia*, *Liophis cobella*, *Oxyrhopus formosus*, *petola*, *Pliocercus euryzonus*, *Pseudoboa coronata*, *Xenodon rhabdocephalus*, and *X. severus*) are active on the ground by day and night. Two snakes (*Bothrops atrox* and *Epicrates cenchria*) are active on the ground and in bushes and trees by night and day. A few terrestrial frogs are active on the ground by day and night; these include *Bufo haematiticus*, *Adenomera andreae* and *A. hylaedactyla*, *Leptodactylus fragilis*, *L. wagneri*, *Lithodytes lineatus*, *Vanzolinius discodactylus*, *Rana palmipes*, *R. warschewitschii*, and various species of *Eleutherodactylus* (e.g., *E. biporcatus, bransfordi, bufoniformis, gaigei, sulcatus, taeniatus*, and *vocator*). Other species of *Eleutherodactylus* (e.g., *E. altamazonicus*, *cerasinus*, *conspicillatus*, *crassidigitus*, *croceoinguinis*, *fenestratus*, *fitzingeri*, *lanthanites*, *ockendeni*, and *ridens*) are active on the ground by day and on bushes at night. In the following analysis of habitat use with respect to diel activity by anurans, lizards, and snakes, all species were scored in two or more categories. Those species that are active on the ground by day but sleep in bushes at night were scored only in the diurnal-terrestrial category.

Anurans are distinctive in having no diurnal-arboreal species. About equal numbers of anurans are active on the forest floor by day and night (table 24.7). On average, 30% of anurans (mostly dendrobatids but some bufonids and leptodactylids) are active on the ground by day, and 32% (mostly leptodactylids and microhylids but some bufonids and ranids) are active there by night. With the exception of Manaus (47.6%), most anurans (55.8–63.6%) are nocturnal and arboreal. This mainly reflects the large numbers of hylids and arboreal *Eleutherodactylus* at each site, as well as the presence of centrolenids at the Central American sites. Manaus has a paucity of arboreal frogs—16 hylids and 3 *Eleutherodactylus*—as contrasted to 35 hylids and 13 *Eleutherodactylus* at Manu and 38 hylids and 12 *Eleutherodactylus* at Santa Cecilia.

TABLE 24.7 Species in diel/major habitat categories at each of five sites*

	Diurnal		Nocturnal	
Site	*Terrestrial* (%)	*Arboreal* (%)	*Terrestrial* (%)	*Arboreal* (%)
		Anurans		
La Selva	25.0	0.0	36.3	63.6
BCI	38.5	0.0	34.6	55.8
Manaus	38.1	0.0	38.1	47.6
Manu	19.5	0.0	26.0	63.6
Santa Cecilia	26.7	0.0	25.6	62.7
		Lizards		
La Selva	45.8	50.0	4.2	4.2
BCI	38.5	53.9	3.8	3.8
Manaus	58.3	41.7	0.0	4.2
Manu	50.0	42.8	0.0	7.2
Santa Cecilia	63.3	40.0	0.0	3.3
		Snakes		
La Selva	67.9	23.2	21.4	16.1
BCI	74.5	23.4	25.5	19.1
Manaus	61.6	28.3	33.3	21.7
Manu	54.8	12.9	41.9	29.0
Santa Cecilia	60.4	15.1	32.1	26.4
		Total		
La Selva	48.4	20.2	23.4	30.6
BCI	52.5	20.8	23.0	31.1
Manaus	53.2	21.4	29.0	27.4
Manu	33.6	8.2	27.0	47.5
Santa Cecilia	37.9	11.8	23.1	40.8

Note: * Some species are in more than one category

Only 2 lizards are nocturnal; the arboreal *Thecadactylus rapicauda* occurs at all sites, and the terrestrial *Lepidophyma flavimaculatum* occurs at the two Central American sites. The proportion of diurnal-terrestrial lizards is greater at the South American than Central American sites. The high incidence of diurnal-terrestrial lizards at the South American sites derives from the species richness of small teiid lizards there as opposed to the Central American sites. The highest proportion (63.3%) of diurnal-terrestrial lizards is at Santa Cecilia; I have suggested (1987) that the presence of continuously humid leaf litter there may account for the abundance of small teiids. The differences in the numbers of diurnal-terrestrial species influences the proportionally higher percentage of diurnal-arboreal lizards at the Central American sites. The actual numbers of diurnal-arboreal lizards, however, do not differ markedly—14 at

BCI, 12 at La Selva and Santa Cecilia, and 10 at Manaus. The notation of only 6 species at Manu probably is an artifact of sampling.

Exclusive of Manu, where snake sampling has not equaled that at other sites, about two-thirds of snakes are diurnal-terrestrial, and about one-fourth are diurnal-arboreal, except that at Santa Cecilia only 15.1% fall in the second category. The proportions of both nocturnal-terrestrial and nocturnal-arboreal snakes is slightly higher at the South American sites than at Central American sites. At Manu and Santa Cecilia, the proportion of diurnal-arboreal snakes is less than that of nocturnal-arboreal species, whereas the reverse is true at the other sites.

Among colubrid snakes, the colubrines of the Old Northern Unit make up the majority of diurnal arboreal snakes at all sites, as well as a significant proportion of the diurnal terrestrial snakes. Yet, only one Old Northern colubrine, *Rhinobothyrum lentiginosum* at Manaus, is nocturnal. Most xenodontine colubrids of the South American Unit are terrestrial; only *Philodryas viridissimus* at Manaus is arboreal and diurnal. Most noctural arboreal snakes (*Dipsas*, *Imantodes*, *Leptodeira*, and *Imantodes*) are members of the Mesoamerican Unit.

With few exceptions, most lizards and anurans feed on small invertebrates, chiefly insects (appendix 24.1). Among lizards, exceptions are *Polychrus gutturosus* and the three species of *Basiliscus*, which feed on insects as juveniles and plant material as adults, the herbivorous *Iguana iguana*, and the omnivorous *Tupinambis teguixin*. The large teiid *Dracaena guianensis* is a feeding specialist on gastropods. Several species of lizards feed on small lizards in addition to arthropods; these include *Anolis chrysolepis*, *Ameiva ameiva*, *A. festiva*, *Kentropyx pelviceps*, *Leposoma parietale*, and *Neusticurus parietale. Enyalioides cofanorum*, *A. ameiva*, and *Bachia trisanale* include earthworms in their diets, and *Neusticurus bicarinatus* eats tadpoles in addition to insects.

Among anurans, exceptions are *Ceratophrys cornuta*, *Hemidactylus proboscideus* and *scutatus*, *Hyla calcarata*, *Leptodactylus knudseni*, and *L. pentadactylus*, all of which eat frogs in addition to arthropods. The aquatic *Pipa pipa* feeds on fish as well as aquatic insects, and *Eleutherodactylus bransfordi* and *L. pentadactylus* include gastropods in their diets.

However, most anurans and lizards feed on arthropods, of which insects are the most common prey. Prey size seems to be related to the size of the lizard or anuran, and most species feed on a variety of prey (Duellman 1978). Four lizards—*Plica plica*, *P. umbra*, *Uracentron azureum*, and *Neusticurus bicarinatus*—feed principally on ants but in-

TABLE 24.8 Snakes feeding on different prey at each of five sites*

Prey	La Selva (%)	BCI (%)	Manaus (%)	Manu (%)	Santa Cecilia (%)
Athropods	3.6	6.4	5.0	0.0	3.8
Earthworms	0.0	0.0	5.0	3.2	7.5
Mollusks	5.4	2.1	3.3	3.2	5.7
Fishes	1.8	2.1	6.6	6.4	3.8
Caecilians	1.8	2.1	5.0	3.2	3.8
Salamanders	3.6	6.4	0.0	0.0	0.0
Anurans	51.8	29.8	46.7	35.4	37.7
Lizards	41.1	44.7	33.3	42.2	34.0
Amphisbaenians	0.0	0.0	8.3	3.2	5.7
Snakes	16.1	14.9	10.0	9.7	5.7
Birds	14.3	27.7	21.7	16.1	17.0
Mammals	16.1	25.5	28.3	29.0	30.2

Note: * Some species are in more than one category

clude other kinds of insects. In contrast, numerous anurans (all families except centrolenids, ranids, and pipids) feed predominantly (6 species) or exclusively (19 species) on ants. Of those that feed predominantly on ants, 2 species of *Colostethus* and 2 of *Synapturanus* eat a high proportion of termites. *Physalaemus petersi* feeds only on termites.

In contrast to anurans and lizards, most snakes are dietary specialists, and the vast majority feed on vertebrates. Individual species are usually not restricted to a given species of prey, however, but tend to eat various species within one or few basic types of prey (e.g., anurans or lizards). Moreover, sympatric species of snakes that eat the same basic type of prey tend to concentrate on different species of prey (Arnold 1972). However, a few snakes take a variety of prey; *Spilotes pullatus*, for example, eats frogs, lizards, snakes, birds, and mammals.

The limited data on diets were combined from all sites to provide the dietary analysis (table 24.8). Anurans and lizards are the most commonly eaten prey at all five sites; more than half (52.8–64.3%) of the snakes at each site include anurans and/or lizards in their diets. However, the numbers (and proportions) of species of snakes feeding on anurans, lizards, or anurans and lizards varies among sites (table 24.9). Birds and mammals make up proportionately less of the diets of snakes at all five sites. Two families of snakes (Boidae and Viperidae) specialize on homeothermic prey. Of the 6 boids, 1 (*Eunectes murinus*) feeds exclusively on mammals and another (*Corallus annulatus*) eats only birds. Also, of the 6 viperids for which diets are known, all feed on mammals and 1 also takes birds. In

TABLE 24.9 Species of snakes feeding on anurans and lizards at each of five sites

Prey	La Selva N (%)	BCI N (%)	Manaus N (%)	Manu N (%)	Santa Cecilia N (%)
Anurans	28 (50.0)	18 (38.3)	27 (45.0)	11 (35.5)	19 (35.8)
Anurans but not lizards	16 (28.6)	8 (17.0)	14 (23.3)	5 (16.1)	10 (18.9)
Anurans and lizards	12 (24.0)	10 (21.3)	13 (21.6)	6 (19.4)	9 (17.0)
Lizards but not anurans	11 (19.6)	12 (25.5)	7 (11.7)	8 (25.9)	9 (17.0)
Lizards	23 (41.1)	22 (46.8)	20 (33.3)	14 (45.3)	18 (34.0)

contrast, of the 106 species of colubrid snakes known from the five sites, only 15 (14.2%) include mammals and only 10 (9.4%) include birds in their diets. The dipsadine snake genera *Dipsas* and *Sibon* feed only on arboreal gastropods. Several kinds of snakes (various colubrids and most species of *Micrurus*) eat other snakes.

Anurans and lizards are the most common items in the diets at all five sites; these are followed by mammals and birds. A few discrepancies are evident between Central American and South American sites. For example, no snakes at the Central American sites are known to eat fishes, which are eaten by *Micrurus surinamensis*, *Hydrops triangulatus*, and species of *Helicops* in South America. Likewise, at the Central American sites, no snakes are known to eat earthworms, whereas in South America *Atractus* seems to specialize on earthworms, and no snakes in Central America are known to eat amphisbaenians, whereas in South America amphisbaenians are eaten by *Anilius* and some species of *Apostolepis* and *Micrurus*. In contrast, species of *Rhadinaea* and *Pliocercus* in Central America feed on salamenders, but no snakes at the South American sites are known to eat salamanders. The reasons for some of these differences are correlated with the absence of the prey items at certain sites; for example, amphisbaenians are absent at La Selva and salamanders are absent at Manaus.

The figures for the percentage of snakes that feed on caecilians, amphisbaenians, and snakes are somewhat misleading. Several species of snakes feed on combinations of these prey. Thus, cylindrical-bodied vertebrates are eaten by 16.1% of the species of snakes at La Selva, 14.9% at BCI, 15.0% at Manaus, 9.7% at Manu, and 9.4% at Santa Cecilia.

Toft (1980a, 1981) pointed out the existence of different feeding guilds of anurans inhabiting the leaf litter at BCI and at a site in Peru; she noted that some species specialize on ants, some species commonly include ants in their diets, and other species seldom eat ants. I likewise identified (1978) feeding guilds of amphibians and reptiles at Santa Cecilia and compared (1989) those findings with data from other sites in Neotropical rainforests. With the exception of Toft's (1980b) work on forest-floor anurans, information on seasonal changes in diets is lacking. By examining the available data on diets with respect to habitat use and diel activity, it is possible to define several feeding guilds among the amphibians and reptiles at each site.

As noted by Toft (1981), anurans that are active on the forest floor by day may be grouped into active-foraging ant specialists and sit-and-wait strategists that feed mostly on larger prey. The number of species of diurnal frogs inhabiting the forest floor varies from 12 at La Selva to 23 at Santa Cecilia (fig. 24.2). At each site about 25% of anurans are actively foraging ant specialists (only 20% at Manu). Dendrobatids comprise the actively foraging ant specialists. A second major guild consists of frogs that are active on the forest floor at night; the numbers of such species at each site approximate the diurnal species, but the proportion of ant specialists is slightly less (fig. 24.2). An important difference between the diurnal and nocturnal guilds is that the ant specialists (mostly microhylids) in the nocturnal guild are not active foragers. The third major guild of anurans is the large nocturnal, arboreal assemblage, most, if not all, of which seem to be sit-and-wait strategists: 20 at Manaus, 27 at BCI, 28 at La Selva, 45 at Manu, and 54 at Santa Cecilia.

The two nocturnal lizards are active foragers—*Thecadactylus rapicauda* on tree trunks and limbs at all five sites, and *Lepidophyma flavimaculatum* in the leaf litter at the Central American sites. Both are insectivorous generalists. With the exception of the three arboreal species of *Basiliscus* at the Central American sites and the arboreal *Iguana iguana* at the Central American sites and Manaus, all of which are herbivourous as adults (juvenile *Basiliscus* feed on insects), and the omnivorous *Tupinambis teguixin* at the South American sites, most lizards are insectivorous generalists (see above).

Approximately half of the lizards at each site are arboreal. With the exception of the shade-inhabiting gekkonids, most species are heliophilic iguanids (*Anolis* have limited heliophilic tolerances.) The iguanids generally are sit-and-wait strategists, whereas the gekkonids are active foragers. Among the lizards active on the forest floor by day, the gekkonids

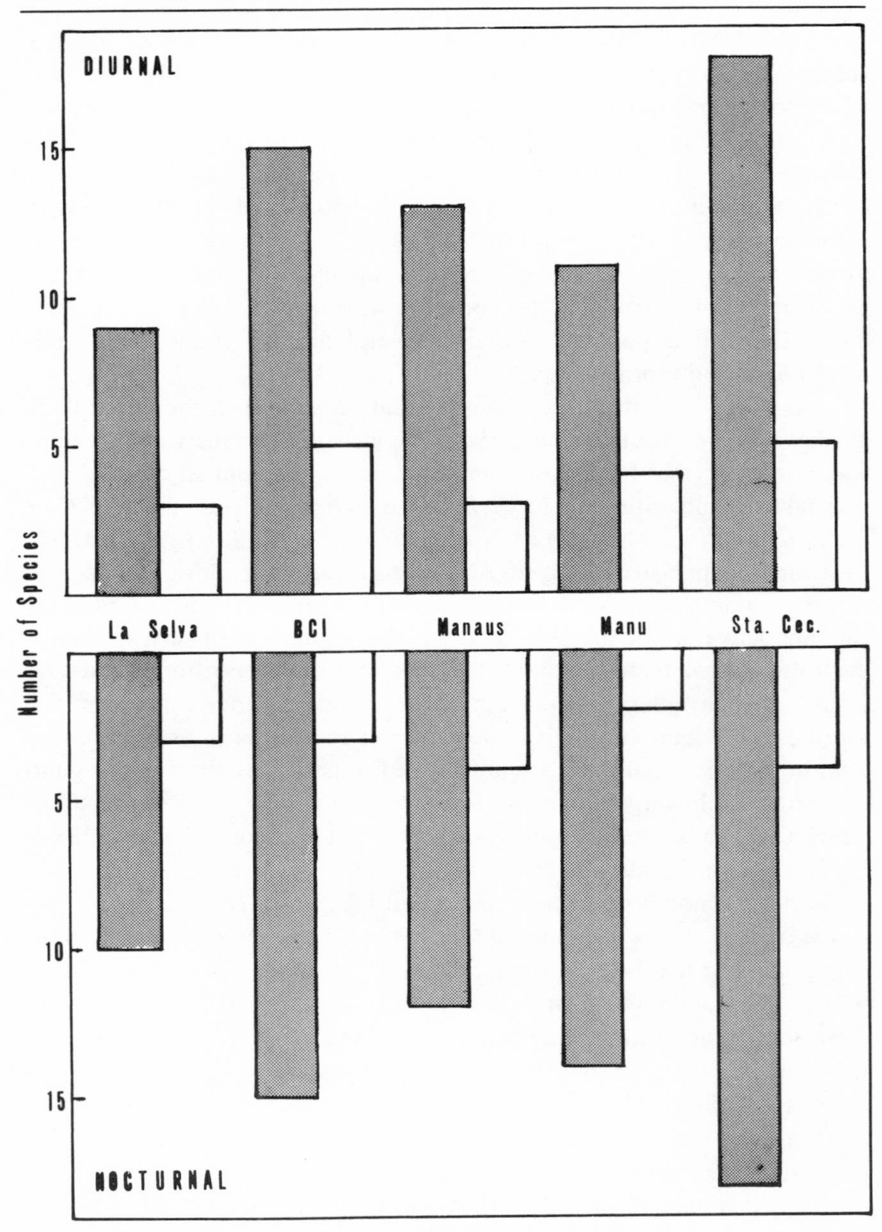

FIGURE 24.2. Diversity of terrestrial anurans at five sites. *Shaded bars* are insectivorous generalists; *open bars* are ant specialists.

TABLE 24.10 Numbers of species of anuran-eating snakes and species of anurans at each of five sites*

	Diurnal-terrestrial		Nocturnal-terrestrial		Nocturnal-arboreal	
Site	*Snakes*	*Frogs*	*Snakes*	*Frogs*	*Snakes*	*Frogs*
La Selva	20	12	4	13	5	28
BCI	12	20	4	18	5	27
Manaus	17	16	6	16	4	20
Manu	7	17	3	20	3	48
Santa Cecilia	12	23	3	22	5	54

Note: * Some species are in more than one category

and many microteiids remain in the shade. The iguanids, scincids, macroteids, and some microteids are heliophilic. With the exception of *Mabuya*, all are active foragers.

Three guilds of anuran-eating snakes are recognized: (1) diurnal and terrestrial, (2) nocturnal and terrestrial, and (3) nocturnal and arboreal. Most anuran-eating snakes at each site are diurnal and terrestrial, whereas few anurans at each site are diurnal and terrestrial (table 24.10). This apparent discrepancy between the diversity of predators and their prey results from the fact that many diurnal terrestrial snakes seek out nocturnal frogs that sleep by day. At each site relatively few species (3–5) of arboreal nocturnal snakes feed on the large number (20–54) of species of nocturnal arboreal frogs.

The guilds of lizard-eating snakes are categorized in the same way as anuran-eating snakes, except that there is a diurnal-arboreal guild. As expected, because most lizards are diurnal, most lizard-eating snakes are also diurnal (table 24.11). However, several nocturnal species of snakes at each site feed on diurnal species of lizards. The nocturnal terrestrial snakes seek sleeping diurnal terrestrial lizards, and the nocturnal arboreal snakes feed on sleeping species of *Anolis*.

There are five mammal-eating guilds of snakes: (1) diurnal and terrestrial, (2) nocturnal and terrestrial, (3) diurnal and arboreal, (4) nocturnal and arboreal, and (5) aquatic (*Eunectes murinus*). At each site, the number of nocturnal, terrestrial, mammal-eating snakes (4–7) approximates the number of diurnal, terrestrial mammal-eating snakes (3–8), except that 10 species are present at Manaus. The numbers of nocturnal, arboreal, mammal-eating snakes (2–6) are slightly less than the numbers of diurnal, arboreal, mammal-eating snakes (4–10) at each site, except there are 6 species in both categories at Santa Cecilia and only 1 diurnal,

TABLE 24.11 Numbers of species of lizard-eating snakes and species of lizards at each of five sites*

Site	Diurnal-terrestrial		Diurnal-arboreal		Nocturnal-terrestrial		Nocturnal-arboreal	
	Snakes	*Lizards*	*Snakes*	*Lizards*	*Snakes*	*Lizards*	*Snakes*	*Lizards*
La Selva	18	11	8	11	5	1	4	1
BCI	15	10	7	5	6	1	5	1
Manaus	12	16	8	10	4	0	4	1
Manu	9	8	2	7	7	0	3	1
Santa Cecilia	11	21	3	12	7	0	4	1

Note: * Some species are in more than one category

arboreal, mammal-eating snake at Manu. The tabulations in appendix 24.1 may provide an incorrect impression of the times of activity and feeding on mammals. Several species of mammal-eating snakes (e.g., *Boa constrictor*, *Epicrates cenchria*, *Bothrops asper*, and *B. atrox*) are active by day and night, and, with the exception of *B. asper*, all forage on the ground and in trees. Although they forage for diverse prey by day and night, the mammals in their diets are probably encountered at night. Obviously some species (e.g., *Bothrops bilineatus*, *Corallus caninus*, and *C. enydris*) are only active in trees at night and feed on nocturnal, arboreal mammals; *C. enydris* feeds primarily on bats.

Most bird-eating snakes are diurnal and arboreal; 2 (Manu) to 9 (Manaus) such species are present at each site. Also the 3 species of *Corallus* are nocturnal and feed on birds.

Other feeding guilds include the nocturnal, arboreal species of *Dipsas* and *Sibon* that feed on arboreal gastropods, nocturnal and diurnal snakes that feed on other snakes and other elongate prey (caecilians and amphisbaenians), and snakes (principally *Atractus*) that feed on earthworms.

Arnold (1972) demonstrated that the numbers of sympatric species of anurans, lizards, and snakes increase at lower latitudes and that positive correlations exist between the numbers of snakes that feed on anurans and lizards and the number of species of those types of prey. It is possible to examine the data on anurans, lizards, and their snake predators, but data are insufficient for mammals and birds and lacking for insects; consequently, the analysis involves only two groups of prey (anurans and lizards) and one group of predators (snakes).

No significant correlations exist between the numbers of snake predators on anurans and lizards (table 24.9) and the numbers of species of anurans and lizards at each site (table 24.1). The largest number of snakes feeding on anurans (26) occurs at Manaus, which has the smallest number of anurans (42). Likewise, at La Selva, where there are 44 anurans, 25 snakes feed on anurans. In contrast, 19 snakes feed on anurans at Santa Cecilia, where there are 86 anurans. Only 15 snakes feed on 50 anurans at BCI. Twenty-four species of lizards occur at La Selva and at Manaus, where there are 20 and 19 species of lizard-eating snakes, respectively. At BCI there are 27 lizards and 16 lizard-eating snakes, and at Santa Cecilia there are 30 lizards and 18 lizard-eating snakes. The snake and lizard faunas at Manu are insufficiently known to be of use in making comparisons.

The relative numbers of snake predators and their anuran and saurian prey at these five Neotropical sites do not support Arnold's (1972) broad latitudinal correlations between numbers of snakes and their prey. In fact, the greatest numbers of frog-eating snakes occur at sites with the fewest species of anurans. The same is true for lizards and the number of lizard-eating snakes. These data do not contradict Arnold's latitudinal gradients but show that there is considerable variation among low latitude sites. Perhaps at these sites significant correlations might be obtained between the abundances of snakes and their predators. However, Zimmerman and Rodrigues (this volume) reported low levels of snake abundance at Manaus, which has the greatest number of species of snakes feeding on the fewest number of species of frogs. Obviously, many more data are required before it will be possible to determine if the numbers (and/or abundance) of snakes are correlated with the numbers (and/or abundance) of species of lizards and anurans.

Arnold's (1972) generalization about sympatric species of snakes specializing on different species of prey within prey categories also may not hold true for many snakes in Neotropical rainforests. Among 10 frog-eating snakes at Santa Cecilia, for example, I reported (1978) only 1 species of frog being eaten by 4 species, 2 by 1 species, and 4 and 2 species. Moreover, 2 species of snakes (*Chironius carinatus* and *Leptodeira annulata*) had eaten 6 species of frogs each. Likewise, Vitt (1983) reported an anuran-eating guild of snakes in the Caatinga of Brazil in which 5 species of snakes had eaten 10 species of frogs. One snake (*Liophis poeciligyrus*) ate 7 species of frogs, and another (*Liophis lineatus*) ate 6 species of frogs; 2 others ate 3 species, and 1 ate 4.

Another spatial resource used by anurans is the site of egg deposition; this is closely associated with the reproductive modes of species. Reproductive mode may be defined as a combination of ovipositional and developmental factors, including oviposition site, ovum and clutch characteristics, rate and duration of development, stage and size of hatchling, and type of parental care, if any (Duellman and Trueb 1986). Many terrestrial modes (e.g., eggs and/or larvae exposed to air) are restricted to, or most common in, environments characterized by continuously high atmospheric moisture. This was noted in the fidelity of so-called forest modes to rainforests and cloud forests and the lack of fidelity of nonforest modes to those environments (Duellman 1988).

Of the 29 reproductive modes identified by Duellman and Trueb (1986), 16 are known among the anurans at the five sites under consideration. Reproductive modes are less diverse at the Central American sites (8 at La Selva and 9 at BCI) than at the South American sites (10 at Manu and 14 at both Manaus and Santa Cecilia). The comparatively fewer modes at the Central American sites results from the absence there of frogs having terrestrial eggs and aquatic eggs with no parental transport (*Synapturanus*), terrestrial foam nests with eggs having direct development (*Adenomera*), eggs in water-filled cavities in trees (*Osteocephalus buckleyi* and *Phrynohyas resinifictrix*), or maternal brooding of eggs (*Hemiphractus* and *Pipa*). The absence of these modes at the Central American sites is not because suitable reproductive sites are lacking but rather that frogs with these modes are restricted to South America (a species of *Hemiphractus* occurs in the highlands of Panama, and a species of *Pipa* occurs in eastern Darién, Panama).

The generalized reproductive mode (eggs and feeding tadpoles in lentic water) is usually by far the most common mode, but at La Selva there are two more species depositing terrestrial eggs that hatch as froglets than there are those having the generalized mode (table 24.12). Terrestrial eggs that hatch as froglets are characteristic of the species of *Eleutherodactylus* (also *Ischnocnema quixensis* at Manu and Santa Cecilia). All species that deposit eggs in a foam nest are members of the leptodactyline genera *Adenomera*, *Edalorhina*, *Leptodactylus*, and *Vanzolinius*. All species having terrestrial eggs that hatch into tadpoles transported to water by a parent are dendrobatids, and all of the species having arboreal eggs, the hatchlings of which develop in streams, are members of the genus *Centrolenella*. Arboreal eggs that hatch into larvae developing in ponds are characteristic of hylids of the genera *Agalychnis* and *Phyllomedusa*, as well as some species of *Hyla*.

Although many species of tadpoles could use the same ponds synchronously, different species of frogs breed at different times and in different kinds of ponds. Thus, only some of the lentic type tadpoles occupy the same pond concurrently. Studies on tadpoles at BCI (Heyer 1976) and at Santa Cecilia (Duellman 1978) showed differential spatial use of the water column and densities of aquatic vegetation. As many as 10 species of tadpoles were found in a single pond at the same time at Santa Cecilia (Duellman 1978), and Berger (1984) found 17 species of tadpoles existing synchronously in a temporary pond at Limoncocha near Santa Cecilia.

Discussion

Two points need to the addressed: (1) resource use and the structuring of communities of amphibians and reptiles, and (2) species richness at

TABLE 24.12 Reproductive modes of anurans at each of five sites

Mode	La Selva N	BCI N	Manaus N	Manu N	Santa Cecilia N
Eggs aquatic:					
Eggs and feeding tadpoles in lentic water	12	14	15	32	34
Eggs and feeding tadpoles in lotic water	3	5	1	1	1
Eggs and feeding tadpoles in natural or constructed basins	0	2	1	1	1
Eggs and feeding tadpoles in water in tree holes	0	0	2	0	2
Eggs in foam nest on water; tadpoles in lentic water	2	5	6	10	9
Eggs on dorsum of aquatic female; tadpoles in lentic water	0	0	1	0	0
Eggs on dorsum of aquatic female; hatch as froglets	0	0	0	0	1
Eggs terrestrial:					
Eggs on ground; feeding tadpoles washed into ponds	0	0	2	0	0
Eggs on ground; feeding tadpoles carried to lentic water	1	1	1	5	4
Eggs on ground; feeding tadpoles carried to lotic water	1	3	2	1	2
Eggs on ground; hatch as froglets	14	11	3	13	17
Eggs in foam nest in burrow; washed into water	0	0	2	0	1
Eggs in foam nest; hatch as froglets	0	0	1	2	1
Eggs on dorsum of female; hatch as froglets	0	0	0	1	1
Eggs arboreal:					
Tadpoles in lentic water	4	4	4	9	10
Tadpoles in lotic water	6	5	1	0	3

different sites. Ideally, problems of abundance and biomass should be considered, but little data exist for those parameters.

Empirical work on birds and resulting theories by MacArthur (1969, 1970, 1971) dominated community ecology for many years. MacArthur hypothesized that interspecific competition was the central influence on community structure. These ideas greatly influenced numerous studies of bird communities and pervaded herpetological studies, such as those on desert lizards by Pianka (summarized in 1986) and on herpetological communities in Thailand (Inger and Colwell 1977). However, Lowe-McConnell's (1975) work on tropical fish communities and Connell's (1975) experiments on tide pool communities offered alternate hypotheses, principally that predation is important in structuring communities. I emphasized (1978) that interspecific competition was not a primary factor in the organization of the herpetological community at Santa Cecilia. Pearson (1982) deemphasized the significance of interspecific competition in a comparative study of rainforest bird communities. In fact, competition models generally have received strong criticism (Connor and Simberloff 1986). Holt (1984) suggested that competition and predation are not alternate but complementary mechanisms of structuring communities.

Predation and climatic fluctuations (such as an unusually long dry season) seem to be the most important regulators of amphibian and reptile populations in Neotropical rainforests. Theoretically, populations will expand until they exceed the carrying capacity of the environment. At that time, limited resources cause interspecific (and intraspecific) competition. There are few such documented cases for Neotropical amphibians and reptiles. Rand et al. (1983) noted that the treefrog *Hyla rufitela* greatly increased in abundance and dispersed over BCI following a short dry season, but following several years of normal dry seasons the species was restricted to its former range (Rand and Myers, this volume). Likewise, the population of *Anolis limifrons* on BCI decreases after long dry seasons (Andrews and Rand 1982).

The tadpoles of many anurans develop in temporary ponds. As I noted (1978) for Santa Cecilia, where there is no pronounced dry season, a few days without rain would cause some shallow ponds to dry up and entire cohorts of tadpoles to be lost. Although sound empirical evidence is limited for species in tropical rainforests, predation on anuran eggs and larvae, and on young frogs and lizards, must be a major cause of mortality. Hero and Magnusson (1987) noted the importance of fish predation on tadpoles at Manaus and suggested that differential predation on tadpoles influences the choice of breeding sites in some anurans.

Competition for food may be diffuse at certain times of the year. During dry times, the abundance and diversity of insects diminish. However, Toft's (1980a) work on diurnal terrestrial frogs showed the lowest similarities in diets among feeding guilds in the dry season, when food was less abundant. She suggested that at the time of limited food supply the frogs specialized more and thereby avoided competition.

Much empirical evidence on populations within communities needs to be gathered and analyzed before broad generalities about causal factors of community structure can be postulated. As emphasized by Pough (1980), the modest energy requirements of amphibians and reptiles allow them to exploit adaptive zones unavailable to endothermic vertebrates. Thus, as I have pointed out (1989), we do not expect amphibians and reptiles to behave like birds, and we should not expect interactions within their communities to be like those of birds.

There are distinct differences in species composition and richness among the five sites, as well as differences in the historical components principally between the two Central American sites and the three South American sites. Moreover, the differences in the histories of the Mesomerican and South American biotas are explainable by geological evidence primarily concerned with the history of land connections between Middle and South America. But what about the great differences in species richness?

Paleoclimatological and geological evidence have dispelled the long-held idea that the lowland tropical rainforests have persisted unchanged for millennia. Haffer (1969) and Vuilleumier (1971) proposed a model of late Cenozoic and Quaternary climatic and ecological fluctuations correlated with Pleistocene glacial and interglacial periods. According to this model, during glacial times much of the Neotropical region now covered by rainforest supported savannas, and rainforest persisted in isolated areas that received rainfall in sufficient quantities for the forest to survive. In contrast, during interglacial (pluvial) periods, the forests expanded and coalesced, and the savannas became restricted to areas of low rainfall or soils having limited nutrients.

Haffer (1969, 1974) postulated Pleistocene forest refugia that served not only as areas of refuge during glacial periods but were also centers of differentiation because of the isolation of populations in different refugia. Further paleoclimatic and geological evidence in support of the model were provided by Haffer (1979) and contributors to a volume edited by Prance (1982). The forest refugia model was applied to reptiles by Vanzolini and Williams (1970) and Dixon (1979) and to amphibians by Lynch (1979) and myself (1982).

The refuge theory, however, has been criticized by several investigators (see Connor 1986 for a review). Although palynological data demonstrate an impressive sequence of vegetational changes in the northern Andes (Van der Hammen et al. 1973; Van Geel and Van der Hammen (1973), palynological data from the lowland American tropics are scarce and conflicting. Holocene pollen profiles from Rondônia, Brazil (Absy and Van der Hammen 1976), can be interpreted as illustrating alternating savanna and forest vegetation (Van der Hammen 1982). Absy (1982) and Irion (1982) interpreted pollen profiles from central Amazonia as showing fluctuations in water level, not climate, a conclusion supported by mineralogical and geochemical evidence from four lakes formed during the Holocene in central Amazonian Brazil (Irion 1982). However, Van der Hammen (1982) noted that these sites are associated with major river floodplains that may always have supported gallery forest. After the formulation of the refuge theory, new geological evidence for Tertiary and Quaternary fluvial perturbations (Salo et al. 1986; Räsänen et al. 1987) and Holocene flooding (Campbell et al 1985; Colinvaux et al. 1985) in the upper Amazon Basin has provided alternative (or complementary; Campbell and Frailey 1984) explanations for high species diversity in the upper Amazon Basin.

How have these changes influenced distributions and speciation? Only three of the forest refuges proposed by Haffer (1969, 1974) are relevant to the five sites considered here. La Selva lies within the Caribbean Costa Rican Refuge, Santa Cecilia is within the Napo Refuge, and Manu is at the edge of the Ucayali Refuge. These refugia also correspond to centers of origin and dispersal postulated by Müller (1973). Manaus and BCI are not closely associated with any refuges, and these two sites have comparably low species richness. La Selva is much richer than BCI, whereas Manaus has the most depauperate fauna of the Amazonian sites. Likewise, the two sites in the upper Amazon Basin (Manu and Santa Cecilia) are in regions of postulated fluvial perturbations, whereas Manaus is in a region of hypothesized Holocene flooding. Major changes in river courses could be responsible for disjunctions in distributions with subsequent speciation resulting in the high species diversity observed at the sites in the upper Amazon Basin. On the other hand, Holocene flooding may have resulted in the elimination of some taxa in central Amazonia.

Present climatic conditions also presumably affect the species richness, especially that of amphibians. La Selva and Manu have rather short dry seasons and have high numbers of amphibian species, and Santa Cecilia has the most equable climate and the largest number of amphibian species. In contrast, BCI and Manaus have less equable climates and lower species diversity, especially of amphibians. Other factors also may influence the species diversity. The Recent history of BCI (formation of Gatun Lake and the island) has certainly influenced the biota of the island; several species occur on the adjacent mainland but not on the island (Rand and Myers, this volume). Relatively nutrient-poor soils at Manaus seem to have a negative effect on overall productivity, which is reflected in lower species diversity. Manaus has the largest snake fauna of the five sites, however, even though the abundance of snakes is very low (Zimmerman and Rodrigues, this volume).

All of the sites under comparison, except Santa Cecilia, are protected, so there is reasonably good assurance that those herpetofaunal assemblages will persist. This is important not only for conservation but also for continuing studies of these organisms. However, human exploitation of the rainforest at Santa Cecilia has resulted in the destruction of the forest at the site of highest known herpetofaunal diversity in the world. For conservation of a large herpetofaunal assemblage and for the potential of crucial ecological studies on amphibians and reptiles, it is imperative that a large reserve be established close to the equator and near the base of the Amazonian slopes of the Andes. Only in this area is there constant day length and evenly distributed rainfall throughout the year.

REFERENCES

Absy, M. L. 1982. Quaternary palynological studies in the Amazon Basin. In G. T. Prance (ed.), *Biological Diversification in the Tropics*. Columbia Univ. Press, New York, pp. 67–73.

Absy, M. L. and T. van der Hammen. 1976. Some palaeoecological data from Rondônia, southern part of the Amazon Basin. Acta Amazônica 3: 293–299.

Andrews, R. M. and A. S. Rand. 1982. Seasonal breeding and long-term population fluctuations in the lizard *Anolis limifrons*. In E. G. Leigh, Jr., A. S. Rand, and D. M. Windsor (eds.), *The Ecology of a Tropical Forest: Seasonal Rhythms and Long-Term Changes*. Smithsonian Inst. Press, Washington, D.C., pp. 405–411.

Arnold. S. J. 1972. Species densities of predators and their prey. Amer. Nat. 106: 220–236.

Berger, T. J. 1984. Community ecology of pond-dwelling anuran larvae. Ph.D. diss., Univ. of Kansas.
Blair, W. F. 1972. *Bufo* of North and Central America. In W. F. Blair (ed.), *Evolution in the Genus Bufo*. Univ. of Texas Press, Austin, pp. 93–101.
Bonaparte, J. 1984. El intercambio faunístico de vertebrados continentales entre America del Sur y del Norte a fines del Cretacico. Mem. III Congr. Latinoamer. Paleo., pp. 438–450.
Cadle, J. E. 1984. Molecular systematics of Neotropical xenodontine snakes. III. Overview of xendontine phylogeny and the history of New World snakes. Copeia 1984: 641–652.
———. 1985. The Neotropical colubrid snake fauna (Serpentes: Colubridae): lineage components and biogeography. Syst. Zool. 34: 1–20.
Campbell, K. E., Jr. and D. Frailey. 1984. Holocene flooding and species diversity in southwestern Amazonia. Quatern. Res. 21: 369–375.
Campbell, K. E., Jr., D. Frailey, and J. Arellano L. 1985. The geology of the Río Beni: further evidence for Holocene flooding in Amazonia. Contr. Sci. Nat. Hist. Mus. Los Angeles Co. 364: 1–18.
Cheetham, A. J. and J. E. Hazel. 1969. Binary (presence-absence) similarity coefficients. J. Paleo. 43: 1130–1136.
Colinvaux, P. A., M. C. Miller, K. Liu, M. Steinitz-Kannan, and I. Frost. 1985. Discovery of permanent Amazon lakes and hydraulic disturbance in the upper Amazon Basin. Nature 313: 42–45.
Connell, J. H. 1975. Some mechanisms producing structure in natural communities: a model and evidence from field experiments. In M. L. Cody and J. M. Diamond (eds.), *Ecology and Evolution of Communities*. Harvard Univ. Press, Cambridge, Mass., pp. 460–490.
Connor, E. F. 1986. The role of Pleistocene forest refugia in the evolution and biogeography of tropical biotas. Trends Ecol. Evol. 1: 165–168.
Connor, E. F. and D. Simberloff. 1986. Competition, scientific method, and null models in ecology. Amer. Sci. 74: 155–162.
Dixon, J. R. 1979. Origin and distribution of reptiles in lowland tropical rainforests of South America. In W. E. Duellman (ed.), *The South American Herpetofauna: Its Origin, Evolution, and Dispersal*. Monogr. Mus. Nat. Hist. Univ. Kans. 7, pp. 217–240.
Donnelly, T. W. 1985. Mesozoic and Cenozoic plate evolution of the Caribbean region. In F. G. Stehli and S. D. Webb (eds.), *The Great American Biotic Interchange*. Plenum Press, New York, pp. 89–121.
Dowling, H. G. and W. E. Duellman. 1978. *Systematic Herpetology*. HISS, New York.
Duellman, W. E. 1965. A biogeographic account of the herpetofauna of Michoacán, México. Univ. Kans. Publ. Mus. Nat. Hist. 15: 627–709.
———. 1970. *The Hylid Frogs of Middle America*. Monog. Mus. Nat. Hist. Univ. Kans. 1.
———. 1978. The biology of an equatorial herpetofauna in Amazonian Ecuador. Misc. Publ. Mus. Nat. Hist. Univ. Kans. 65: 1–352.
———. 1979. The South American herpetofauna: a panoramic view. In W. E.

Duellman (ed.), *The South American Herpetofauna: Its Origin, Evolution, and Dispersal. Monogr.* Mus. Nat. Hist. Univ. Kans. pp. 1–28.

———. 1982. Quaternary climatic-ecological fluctuations in the lowland tropics: frogs and forests. In G. T. Prance (ed.), *Biological Diversification in the Tropics*. Columbia Univ. Press, New York, pp. 389–402.

———. 1987. Lizards in an Amazonian rain forest community: resource utilization and abundance. Nat. Geog. Res. 2: 489–500.

———. 1988. Patterns of species diversity in Neotropical anurans. Ann. Mo. Bot. Gard. 75: 79–104.

———. 1989. Tropical herpetofaunal communities: patterns of community structure in Neotropical rainforests. In M. L. Harmelin-Vivien and F. Bourlière (eds.), *Vertebrates in Complex Tropical Systems*. Springer-Verlag, New York, pp. 61–88.

Duellman, W. E. and L. Trueb. 1986. *Biology of Amphibians*. McGraw-Hill, New York.

Estes, R. and A. Báez. 1985. Herpetofaunas of North and South America during the Late Cretaceous and Cenozoic: evidence for interchange? In F. G. Stehli and S. D. Webb (eds.), *The Great American Biotic Interchange*. Plenum Press, New York, pp. 139–197.

Haffer, J. 1969. Speciation in Amazonian forest birds. Science 165: 131–137.

———. 1974. Avian speciation in tropical South America. Publ. Nuttall Ornithol. Club. 14: 1–390.

———. 1979. Quaternary biogeography of tropical lowland South America. In W. E. Duellman (ed.), *The South American Herpetofauna: Its Origin, Evolution, and Dispersal*. Monogr. Mus. Nat. His. Univ. Kans. 7, pp. 107–140.

Heyer, W. R. 1976. Studies in larval amphibian habitat partitioning. Smithsonian Contrib. Zool. 242: 1–27.

Hero, J.-M. and W. E. Magnusson. 1987. Predation, palatability, and the distribution of tadpoles. Program and Abstracts, Joint Ann. Meet. Soc. Study Amphib. Rept. and Herpetol. League, Veracruz, Mexico.

Hödl, W. 1977. Call differences and calling site segregation in anuran species from central Amazonian floating meadows. Oecologia 28: 351–363.

Holt, R. D. 1984. Spatial heterogeneity, indirect interactions, and the coexistence of prey species. Amer. Nat. 124: 377–406.

Hoogmoed, M. S. 1979. The herpetofauna of the Guianan region. In W. E. Duellman (ed.), *The South American Herpetofauna: Its Origin, Evolution, and Dispersal*. Monogr. Mus. Nat. Hist. Univ. Kans. 7, pp. 241–279.

Inger, R. F. and R. K. Colwell. 1977. Organization of contiguous communities of amphibians and reptiles in Thailand. Ecol. Monogr. 47: 229–253.

Irion, G. 1982. Minerological and geochemical contribution to climatic history in central Amazonia during Quaternary time. Trop. Ecol. 23: 76–85.

Jaccard, P. 1928. Die statistische-floristische Methode als Grundlage der Pflanzensoziologie. Hanb. Biol. Abietsm. Abderhalden 1928. 11: 165–202.

Laurent, R. F. 1979. Herpetofaunal relationships between Africa and South America. In W. E. Duellman (ed.), *The South American Herpetofauna: Its Origin, Evolution, and Dispersal*. Monogr. Mus. Nat. Hist. Univ. Kans. 7, pp. 55–71.

Lowe-McConnell, R. H. 1975. *Fish Communities in Tropical Freshwaters: Their Distribution, Ecology, and Evolution*. Longman, New York.

Lynch, J. D. 1979. The amphibians of the lowland tropical forests. In W. E. Duellman (ed.), *The South American Herpetofauna: Its Origin, Evolution, and Dispersal*. Monogr. Mus. Nat. Hist. Univ. Kans., pp. 189–215.

———. 1986. The definition of the Middle American clade of *Eleutherodactylus* based on jaw musculature (Amphibia: Leptodactylidae). Herpetologica 42: 248–258.

MacArthur, R. H. 1969. Species packing and what competition minimizes. Proc. Nat. Acad. Sci. 64: 1369–1371.

———. 1970. Species packing and competitive equilibrium for many species. Theoret. Pop. Biol. 1: 1–11.

———. 1971. Patterns in terrestrial bird communities. In D. S. Farner et al. (eds.). *Avian Biology*. Academic Press, New York, 1: 189–221.

Müller, P. 1973. The dispersal centres of terrestrial vertebrates in the Neotropical realm. Biogeographica, 2. Junk, The Hague.

Pearson, D. 1982. Historical factors and bird species richness. In G. T. Prance (ed.), *Biological Diversification in the Tropics*. Columbia Univ. Press, New York, pp. 441–452.

Peters, J. A. 1968. A computer program for calculating degree of biogeographical resemblance between areas. Syst. Zool. 13: 64–69.

Pianka, E. R. 1986. *Ecology and Natural History of Desert Lizards*. Princeton Univ. Press, Princeton, N.J.

Pirlot, P. L. 1956. Les formes européene du genre *Hipparion*. Inst. Geol. Barcelona, Mem. y Commun. 14: 1–150.

Pough, F. H. 1980. The advantages of ectothermy for tetrapods. Amer. Nat. 115: 92–112.

Prance, G. T. (ed.). 1982. *Biological Diversification in the Tropics*. Columbia Univ. Press, New York.

Rand, A. S. and C. W. Myers. (This volume). The herpetofauna of Barro Island, Panama: an ecological summary.

Rand, A. S., M. J. Ryan, and K. E. Troyer. 1983. A population explosion in a tropical tree frog: *Hyla rufitela* on Barro Colorado Island. Biotropica 15: 72–73.

Räsänen, M. E., J. S. Salo, and R. J. Kalliola. 1987. Fluvial perturbance in the western Amazon Basin: regulation by long-term sub-Andean tectonics. Science 238: 1398–1400.

Salo, J., R. Kalliola, I. Häkkinen, Y. Mäkinen, P. Niemelä, M. Pahukka, and P. D. Coley. 1986. River dynamics and the diversity of Amazon lowland forest. Nature 322: 254–258.

Savage, J. M. 1982. The enigma of the Central American herpetofauna: dispersals or vicariance? Ann. Mo. Bot. Gard. 69: 464–547.

Schoener, T. W. 1974. Resource partitioning in ecological communities. Science 185: 27–39.

Simpson, G. G. 1960. Notes on the measurement of faunal resemblance. Amer. J. Sci. 258: 300–311.

Smith, D. L. 1985. Caribbean plate relative motions. In F. G. Stehli and S. D.

Webb (eds.), *The Great American Biotic Interchange*. Plenum Press, New York, pp. 17–48.

Toft, C. A. 1980a. Feeding ecology of thirteen syntopic species of anurans in a seasonal tropical environment. Oecologia 45: 131–141.

———. 1980b. Seasonal variation in populations of Panamanian litter frogs and their prey comparison of wetter and dryer sites. Oecologia 47: 34–38.

———. 1981. Feeding ecology of Panamanian litter anurans: patterns in diet and foraging mode. J. Herpetol. 15: 139–144.

Van der Hammen, T. 1982. Paleoecology of tropical South America. In G. T. Prance (ed.), *Biological Diversification in the Tropics*. Columbia Univ. Press, New York, pp. 60–66.

Van der Hammen, T., J. H. Werner, and H. van Dommelen. 1973. Palynological record of the upheaval of the northern Andes: a study of the Pliocene and Lower Quaternary of the Colombian Eastern Cordillera and the early evolution of its biota. Rev. Palaeobot. Palynol. 16: 1–122.

Van Geel, B. and T. van der Hammen. 1973. Upper Quaternary vegetation and climatic sequence of the Fúquene area. Palaeogeog. Palaeoclim. Palaeoecol. 14: 9–92.

Vanzolini, P. E. and W. R. Heyer. 1985. The American herpetofauna and the interchange. In F. G. Stehli and S. D. Webb (eds.), *The Great American Biotic Interchange*. Plenum Press, New York, pp. 475–487.

Vanzolini, P. E. and E. E. Williams. 1970. South American anoles: the geographic differentiation and evolution of the *Anolis chrysolepis* species group (Sauria: Iguanidae). Arq. Zool. São Paulo, 19: 1–124.

Vitt, L. J. 1983. Ecology of an anuran-eating guild of terrestrial tropical snakes. Herpetologica 39; 52–56.

Vuillemier, B. S. 1971. Pleistocene changes in the fauna and flora of South America. Science 173: 771–780.

Whittaker, R. H. 1970. *Communities and Ecosystems*. Macmillan, London.

Zimmerman, B. and M. T. Rodrigues. (This volume). Frogs, snakes, and lizards of the INPA-WWF reserves near Manaus, Brazil.

APPENDIX 24.1. Species of Amphibians and Reptiles Occurring at Five Sites in Neotropical Rainforest

Abbreviations: Sites: BCI = Barro Colorado Island, Panama; LS = La Selva, Costa Rica; Manu = Manu National Park, Peru; MCSE = Minimum Critical Size of Ecosystems Area, Manaus, Brazil; SC = Santa Cecilia, Ecuador. Origin: MA = Mesoamerican; ON = Old Northern; SA = South American. Habitat: A = aquatic; B = bushes; F = fossorial; G = ground; L = leaf litter; R = riparian (aquatic margin); T = trees. Diel activity: B = diurnal and nocturnal; D = diurnal; N = nocturnal. Food: A = ants; Ae = anuran eggs; Al = anuran larvae; Am = amphisbaenids; B = birds; C = caecilians; E = earthworms; F = frogs; G = gastropods; H = herbivorous; I = insects and other small arthropods; I(A) = insects (mostly ants); L = lizards; M = mammals; O = omnivorous; P = fishes; S = snakes; T = turtles; Te = termites; U = salamanders. In columns under sites, + = present, − = presumably absent.

Taxon	Origin	Habitat	Diel	Food	Site				
					LS	BCI	MCSE	Manu	SC
GYMNOPHIONA									
Caeciliidae									
Caecilia disossea	SA	F	?	?	−	−	−	−	+
C. tentaculata	SA	F	?	E	−	−	+	−	+
Gymnopis multiplicata	MA	F	?	?	+	−	−	−	−
Microcaecilia albiceps	SA	F	?	?	−	−	−	−	+
Oscaecilia bassleri	SA	F	?	?	−	−	−	+	+
O. ochrocephala	SA	F	?	I	−	+	−	−	−
Siphonops annulatus	SA	F	?	?	−	−	+	−	+
CAUDATA									
Plethodontidae									
Bolitoglossa altamazonica	ON	B,L	N	I	−	−	−	+	−
B. colonnea	ON	B,G,L	N	I	+	−	−	−	−
B. equatoriana	ON	B	N	I	−	−	−	−	+
B. peruviana	ON	B,L	N	I	−	−	−	−	+
O. complex	ON	F,L	?	I	−	+	−	−	−
O. parvipes	ON	F,L	?	I	−	+	−	−	−
O. pseudouniformis	ON	G,L	?	I	+	−	−	−	−
O. uniformis	ON	G,L	?	I	+	−	−	−	−
ANURA									
Bufonidae									
Atelopus pulcher	SA	L	D	?	−	−	+	−	−
Bufo (typhonius) alatus	SA	G,L	D	A	−	+	−	−	−
B. coniferus	MA	G,L	D	I	+	−	−	−	−

B. dapsilis	SA	L	B	?	–	–	+	–	–
B. (guttatus) glaberrimus	SA	G	N	I	–	–	–	+	+
B. granulosus	SA	G,L	N	I	–	+	–	–	–
B. haematiticus	SA	G,L	B	A	+	+	–	–	–
B. marinus	SA	G,L	N	I	+	+	–	+	+
B. typhonius	SA	G,L	D	A	–	–	–	+	+
Dendrophryniscus minutus	SA	L	D	A	–	–	+	–	+
Centrolenidae									
Centrolenella albomaculata	SA	B	N	I	+	–	–	–	–
C. colymbiphyllum	SA	B	N	I	+	+	–	–	–
C. fleischmanni	SA	B	N	I	–	+	–	–	–
C. granulosa	SA	B	N	I	+	+	–	–	–
C. midas	SA	B	N	I	–	–	–	–	+
C. munozorum	SA	B	N	I	–	–	–	–	+
C. oyampiensis	SA	B	N	I	–	–	+	–	–
C. prosoblepon	SA	B	N	I	+	+	–	–	–
C. pulverata	SA	B	N	I	+	–	–	–	–
C. resplendens	SA	B	N	?	–	–	–	–	+
C. spinosa	SA	B	N	I	+	+	–	–	–
C. valerioi	SA	B	N	I	+	–	–	–	–
Dendrobatidae									
Colostethus flotator	SA	G,L	D	I	–	+	–	–	–
C. inguinalis	SA	R	D	I	–	+	–	–	–
C. marchesianus	SA	L	D	I(A,Te)	–	–	+	+	+
C. sauli	SA	R	D	I	–	–	–	–	+
C. talamancae	SA	G,L	D	I	–	+	–	–	–
C. sp.	SA	L	D	I(A,Te)	–	–	+	+	–
Dendrobates auratus	SA	G,L	D	A	–	+	–	–	–

Taxon	Origin	Habitat	Diel	Food	Site				
					LS	BCI	MCSE	Manu	SC
D. pumilio	SA	G,L	D	A	+	−	−	−	−
D. quinquevittatus	SA	L	D	A	−	−	−	−	+
Epipedobates femoralis	SA	L	D	I	−	−	+	+	+
E. sp. nov.	SA	L	D	?	−	−	−	+	−
E. parvulus	SA	L	D	I	−	−	−	−	+
E. pictus	SA	L	D	I(A)	−	−	−	+	+
E. trivittatus	SA	L	D	A	−	−	−	+	−
Phyllobates lugubris	SA	G,L	D	I	+	−	−	−	−
Hylidae									
Agalychnis calcarifer	MA	B,T	N	I	+	+	−	−	−
A. callidryas	MA	B,T	N	I	+	+	−	−	−
A. craspedopus	MA	T	N	I	−	−	−	+	−
A. saltator	MA	B,T	N	I	+	−	−	−	−
A. spurrelli	MA	T	N	I	−	+	−	−	−
Hemiphractus proboscideus	SA	B	N	F,I	−	−	−	−	+
H. scutatus	SA	L	N	F,I	−	−	−	+	−
Hyla alboguttata	SA	B	N	I	−	−	−	−	+
H. bifurca	SA	B	N	I	−	−	−	+	+
H. boans	SA	B,T	N	I	−	+	+	+	+
H. bokermanni	SA	B	N	I	−	−	−	−	+
H. brevifrons	SA	B	N	I	−	−	−	−	+
H. calcarata	SA	B,T	N	F,I	−	−	−	+	+
H. crepitans	SA	B	N	I	−	+	−	−	−
H. ebraccata	SA	B	N	I	+	+	−	−	−
H. fasciata	SA	B,T	N	I	−	−	−	+	+

H. geographica	SA	B,T	N	I	−	−	+	+	+
H. granosa	SA	B,T	N	I	−	−	+	+	+
H. lanciformis	SA	B,G	N	I	−	−	−	+	+
H. leali	SA	B	N	I	−	−	−	+	+
H. leucophyllata	SA	B,T	N	I	−	−	−	+	+
H. loquax	SA	B	N	I	+	−	−	−	−
H. marmorata	SA	B,T	N	I	−	−	−	−	+
H. microcephala	SA	B	N	I	−	+	−	−	−
H. minuta	SA	B	N	I	−	−	+	+	+
H. parviceps	SA	B	N	I	−	−	−	+	+
H. phlebodes	SA	B	N	I	+	+	−	−	−
H. punctata	SA	B	N	I	−	−	−	+	+
H. rhodopepla	SA	B	N	I	−	−	−	+	+
H. riveroi	SA	B	N	I	−	−	−	+	+
H. rosenbergi	SA	B,T	N	I	−	+	−	−	−
H. rossalleni	SA	B	N	?	−	−	−	−	+
H. rufitela	SA	B	N	I	+	+	−	−	−
H. sarayacuensis	SA	B	N	I	−	−	−	+	+
H. triangulum	SA	B	N	I	−	−	−	+	+
H. (*microcephala* gr.)	SA	B	N	I	−	−	+	−	−
H. (*parviceps* gr.)	SA	B	N	I	−	−	+	+	−
H. (*geographica* gr.)	SA	B	N	?	−	−	−	+	−
H. (*pauiniensis*-like)	SA	B	N	I	−	−	−	+	−
Nyctimantis rugiceps	SA	T	N	I	−	−	−	−	+
Ololygon boulengeri	SA	B,T	N	I	+	+	−	−	−
O. cruentomma	SA	B	N	I	−	−	+	+	+
O. elaeochroa	SA	B	N	I	+	−	−	−	−
O. epacrorhina	SA	B	N	I	−	−	−	+	−
O. funerea	SA	G,B,T	N	I	−	−	−	−	+
O. garbei	SA	B	N	I	−	−	−	+	+

Taxon	Origin	Habitat	Diel	Food	Site				
					LS	BCI	MCSE	Manu	SC
O. rostrata	SA	B	N	I	–	+	–	–	–
O. rubra	SA	G,B,T	N	I	–	+	–	+	+
O. staufferi	SA	B	N	I	–	+	–	–	–
Osteocephalus buckleyi	SA	B,T	B	I	–	–	+	–	+
O. leprieurii	SA	T	N	I	–	–	–	+	+
O. taurinus	SA	T	N	I	–	–	+	+	+
O. sp.	SA	T	N	?	–	–	+	–	–
Phrynohyas coriacea	SA	B,T	N	I	–	–	+	+	+
P. resinifictrix	SA	T	N	I	–	–	+	–	–
P. venulosa	SA	T	N	I	–	+	–	+	+
Phyllomedusa atelopoides	SA	G	N	I	–	–	–	+	–
P. bicolor	SA	T	N	I	–	–	+	–	–
P. palliata	SA	B,T	N	I	–	–	–	+	+
P. tarsius	SA	T	N	I	–	–	+	–	+
P. tomopterna	SA	B,T	N	I	–	–	+	–	+
P. vaillanti	SA	B,T	N	I	–	–	+	+	+
P. (*boliviana*-like)	SA	T	N	?	–	–	–	+	–
Scarthyla ostinodactyla	SA	B	N	I	–	–	–	+	–
Smilisca baudinii	MA	B	B	I	+	–	–	–	–
S. phaeota	MA	B,T	N	I	+	+	–	–	–
S. sila	MA	B,T	N	I	–	+	–	–	–
S. sordida	MA	B	N	I	+	–	–	–	–
Sphaenorhynchus carneus	SA	B	N	A	–	–	–	–	+
S. dorisae	SA	B	N	A	–	–	–	+	–
S. lacteus	SA	B	N	A	–	–	–	+	+

Letodactylidae									
Adenomera andreae	SA	L	B	I	−	−	+	+	+
A. hylaedactyla	SA	L	B	I	−	−	−	+	−
Ceratophrys cornuta	SA	L	B	F,I	−	−	+	+	+
Edalorhina perezi	SA	L	D	I	−	−	−	+	+
Eleutherodactylus acuminatus	SA	B,T	N	A	−	−	−	−	+
E. altae	SA	B	N	I	+	−	−	−	−
E. altamazonicus	SA	B,L	B	I	−	−	−	+	+
E. biporcatus	MA	G,L	B	I(A)	+	+	−	−	−
E. bransfordi	MA	G,L	B	I,G	+	+	−	−	−
E. bufoniformis	MA	R	B	I	−	+	−	−	−
E. carvalhoi	SA	B	N	I	−	−	−	+	−
E. caryophyllaceus	SA	B	N	I	+	−	−	−	−
E. cerasinus	SA	B,G,L	B	I	+	+	−	−	−
E. conspicillatus	SA	B,G	B	I	−	−	−	−	+
E. crassidigitatus	MA	B,G,L	B	I	+	+	−	−	−
E. croceoinguinis	SA	B,G	B	I	−	−	−	−	+
E. cruentus	SA	B	N	I	+	−	−	−	−
E. diadematus	SA	B,T	N	I	−	−	−	−	+
E. diastema	SA	B,T	N	I	+	+	−	−	−
E. fenestratus	SA	B,L	B	I	−	−	+	+	−
E. fitzingeri	MA	B,G,L	B	I	+	+	−	−	−
E. gaigei	SA	G,L	B	I	−	+	−	−	−
E. lacrimosus	SA	B,T	N	I	−	−	−	−	+
E. lanthanites	SA	B,G,T	B	I	−	−	−	−	+
E. martiae	SA	B	N	I	−	−	−	−	+
E. mendax	SA	B	N	I	−	−	−	+	−
E. mimus	MA	G,L	N	I	+	−	−	−	−
E. nigrovittatus	SA	L	D	I	−	−	−	−	+
E. noblei	MA	G,L	N	I	+	−	−	−	−
E. ockendeni	SA	B,L	B	I	−	−	−	+	+

Taxon	Origin	Habitat	Diel	Food	Site				
					LS	BCI	MCSE	Manu	SC
E. paululus	SA	B	N	I	–	–	–	–	+
E. pseudoacuminatus	SA	B,G,T	N	I	–	–	–	–	+
E. quaquaversus	SA	B	N	I	–	–	–	–	+
E. ridens	SA	B,G,L	B	I	+	+	–	–	–
E. rugulosus	MA	R	N	I	+	–	–	–	–
E. peruvianus	SA	B	B	I	–	–	–	+	–
E. sulcatus	SA	L	B	I	–	–	–	–	+
E. talamancae	MA	B,G	N	I	+	–	–	–	–
E. taeniatus	SA	G,L	B	I	–	+	–	–	–
E. toftae	SA	B	B	I	–	–	–	+	–
E. variabilis	SA	B	N	I	–	–	–	–	+
E. ventrimarmoratus	SA	B	N	I	–	–	–	+	–
E. vocator	SA	G,L	B	I	–	+	–	–	–
E. sp. 1	SA	B	B	I	–	–	–	+	–
E. sp. 2	SA	B	N	I	–	–	–	+	–
E. sp. 3	SA	B	N	?	–	–	–	+	–
E. sp. 4	SA	B	N	I	–	–	+	–	–
E. sp. 5	SA	B	N	I	–	–	+	–	–
Ischnocnema quixensis	SA	G	N	I	–	–	–	+	+
Leptodactylus (labialis) fragilis	SA	G,L	B	I	–	+	–	–	–
L. insularum	SA	G,L	N	I	–	+	–	–	–
L. kundseni	SA	F,L	N	F,I	–	–	+	+	–
L. melanonotus	SA	G,L,R	N	I	+	+	–	–	–
L. mystaceus	SA	G,L	N	I	–	–	+	+	+
L. pentadactylus	SA	F,G,R	N	F,G,I	+	+	+	+	+
L. podicipinus	SA	G	N	I	–	–	–	+	–
L. rhodomystax	SA	G,L	N	I	–	–	+	+	+

L. riveroi	SA	F,L	N	?	–	–	+	–	–
L. stenodema	SA	G,L	N	I	–	–	+	–	+
L. wagneri	SA	G,L	B	I	–	–	+	+	+
Lithodytes lineatus	SA	G,L	B	I	–	–	+	+	+
Physalaemus petersi	SA	L	B	Te	–	–	–	+	+
P. pustulosus	SA	G,L	B	A	–	+	–	–	–
Vanzolinius discodactylus	SA	R	B	I	–	–	–	–	+
Microhylidae									
Chiasmocleis anatipes	SA	F,G,L	N	A	–	–	+	–	+
C. bassleri	SA	L	N	A	–	–	–	–	+
C. shudikarensis	SA	F,L	N	A	–	–	+	–	–
C. ventrimaculata	SA	L	B	A	–	–	–	+	+
Ctenophryne geayi	SA	F,L	N	I	–	–	+	+	+
Elachistocleis ovalis	SA	F,L	N	I	–	+	–	+	–
Gastrophryne pictiventris	MA	G,L	?	A	+	–	–	–	–
Hamptophryne boliviana	SA	L	N	A	–	–	–	+	+
Synapturanus mirandaribeiroi	SA	Γ	B	A,Te	–	–	+	–	–
S. salseri	SA	F	B	A,Te	–	–	+	–	–
Syncope antenori	SA	L	N	I	–	–	–	–	+
Pipidae									
Pipa arrabali	SA	A	B	I	–	–	+	–	–
P. pipa	SA	A	B	P	–	–	–	–	+
Ranidae									
Rana palmipes	N	R	B	I	–	–	–	–	+
R. taylori	ON	R	N	I	+	–	–	–	–
R. vaillanti	ON	R	N	I	+	+	–	–	–
R. warschewitschii	ON	G,L,R	B	I	+	+	–	–	–

Taxon	Origin	Habitat	Diel	Food	Site				
					LS	BCI	MCSE	Manu	SC
SAURIA									
Anguidae									
Celestus sp.	ON	T	D	I	+	–	–	–	–
Diploglossus bilobatus	SA	G,L	D	I	+	–	–	–	–
D. fasciatus	SA	G	D	?	–	–	–	+	–
D. monotropis	SA	G,L	D	I	+	–	–	–	–
Gekkonidae									
Coleodactylus amazonicus	SA	L	D	I	–	–	+	–	–
Gonatodes alboguttatus	SA	T	D	I	–	+	–	–	–
G. concinnatus	SA	T	D	I	–	–	–	–	+
G. hasemanni	SA	T	D	I	–	–	–	+	–
G. humeralis	SA	T	D	I	–	–	+	+	+
Lepidoblepharis sanctaemartae	SA	G,L	D	I	–	+	–	–	–
L. xanthostigma	SA	G,L	D	I	+	–	–	–	–
Pseudogonatodes guianensis	SA	L	D	I	–	–	–	–	+
Sphaerodactylus homolepis	SA	G,L	D	I	+	–	–	–	–
S. lineolatus	SA	T	D	I	–	+	–	–	–
S. millipunctatus	SA	G,L	D	I	+	–	–	–	–
Thecadactylus rapicauda	SA	T	N	I	+	+	+	+	+
Iguanidae									
Anolis auratus	MA	B	D	I	–	+	–	–	–
A. biporcatus	MA	T	D	I	+	+	–	–	–
A. capito	MA	T	D	I	+	+	–	–	–
A. carpenteri	MA	B	D	I	+	–	–	–	–
A. chrysolepis	MA	B,G	D	I,L	–	–	+	+	+

A. frenatus	SA	T	D	I	–	+	–	–	–
A. fuscoauratus	MA	B,G	D	I	–	–	+	+	+
A. humilis	MA	G,L	D	I	+	–	–	–	–
A. lemurinus	MA	T	D	I	+	–	–	–	–
A. limifrons	MA	B	D	I	+	+	–	–	–
A. lionotus	MA	R	D	I	+	+	–	–	–
A. ortonii	MA	B,G,T	D	I	–	–	+	–	+
A. pentaprion	MA	T	D	I	+	+	–	–	–
A. philopunctatus	SA	T	D	I	–	–	+	–	–
A poecilopus	MA	R	D	I	–	+	–	–	–
A. punctatus	SA	T	D	I	–	–	–	+	+
A. trachyderma	MA	B,G	D	I	–	–	–	–	+
A. transversalis	SA	T	D	I	–	–	–	–	+
A. tropidogaster	MA	B	D	I	–	+	–	–	–
A. vittiger	MA	T	D	I	–	+	–	–	–
A. sp.	MA	B	D	I	–	+	–	–	–
Basiliscus basiliscus	MA	R	D	H,I,P	–	+	–	–	–
B. plumifrons	MA	B,T	D	H,I	+	–	–	–	–
B. vittatus	MA	B	D	H,I	+	–	–	–	–
Corytophanes cristatus	MA	B,T	D	I	+	+	–	–	–
Enyalioides cofanorum	SA	G	D	E,I	–	–	–	–	+
E. laticeps	SA	T	D	I	–	–	–	–	+
E. sp.	SA	T	D	?	–	–	–	+	–
Iguana iguana	MA	T	D	H	+	+	+	–	–
Ophryoessoides aculeatus	SA	G	D	?	–	–	–	+	–
Plica plica	SA	T	D	I(A)	–	–	+	+	–
P. umbra	SA	B,T	D	I(A)	–	–	+	+	+
Polychrus gutturosus	SA	T	D	H,I	+	+	–	–	–
P. marmoratus	SA	B,T	D	I	–	–	–	–	+
Uracentron azureum	SA	T	D	I(A)	–	–	+	–	–
U. flaviceps	SA	T	D	I	–	–	–	–	+
Uranoscodon superciliosum	SA	B,T	D	I	–	–	+	–	–

Taxon	Origin	Habitat	Diel	Food	Site				
					LS	BCI	MCSE	Manu	SC
Scincidae									
Mabuya bistriata	ON	G,L	D	I	–	–	+	+	+
M. unimarginata	ON	G,L	D	I	+	+	–	–	–
Sphenomorphus cherriei	ON	G,L	D	I	+	–	–	–	–
Teiidae									
Alopoglossus atriventris	SA	L	D	?	–	–	–	–	+
A. carinicaudatus	SA	L	D	I	–	–	+	–	–
A. copii	SA	L	D	I	–	–	–	–	+
Ameiva ameiva	SA	L	D	E,I,L	–	+	+	+	+
A. festiva	SA	G,L	D	I,L	+	+	–	–	–
A. leptophrys	SA	G,L	D	I	–	+	–	–	–
A. quadrilineata	SA	G,L	D	I	+	–	–	–	–
Arthrosaura reticulata	SA	L	D	I	–	–	+	–	+
Bachia cophias	SA	F	D	I	–	–	+	–	–
B. panoplia	SA	F	D	I	–	–	+	–	–
B. trisanale	SA	F	D	E,I	–	–	–	+	+
Dracaena guianensis	SA	R	D	G	–	–	–	–	+
Gymnophthalmus speciosus	SA	F,L	D	I	–	+	–	–	–
Iphisa elegans	SA	L	D	I	–	–	–	–	+
Kentropyx altamazonicus	SA	G	D	I	–	–	–	+	–
K. calcaratus	SA	G	D	I	–	–	+	–	–
K. pelviceps	SA	G	D	F,I,L	–	–	–	–	+
Leposoma guianense	SA	L	D	I	–	–	+	–	–
L. parietale	SA	L	D	I,L	–	–	–	–	+
L. percarinatum	SA	L	D	I	–	–	+	–	–

L. southi	SA	L	D	I	−	+	−	−	−
Neusticurus bicarinatus	SA	R	D	I(A),Al	−	−	+	−	−
N. ecpleopus	SA	R	D	I,L	−	−	−	−	+
Prionodactylus argulus	SA	L	D	I	−	−	−	−	+
P. manicatus	SA	L	D	I	−	−	−	−	+
Ptychoglossus brevifrontalis	SA	L	D	?	−	−	−	−	+
Tretioscincus agilis	SA	L	D	I	−	−	+	−	−
Tupinambis teguixin	SA	G	D	O	−	−	+	+	+
Xantusiidae									
Lepidophyma flavimaculatum	ON	L	N	I	+	+	−	−	−
AMPHISBAENIA									
Amphisbaenidae									
Amphisbaenia fuliginosa	SA	F	D	E,I	−	+	+	−	+
SERPENTES									
Aniliidae									
Anilius scytale	SA	F,L	B	E,Am	−	−	+	−	+
Anomalepidae									
Anomalepis mexicana	SA	F,L	?	I	−	+	−	−	−
Liotyphlops albirostris	SA	F,L	?	I	−	+	−	−	−
Boidae									
Boa constrictor	MA	G,T	B	B,L,M	+	+	+	−	+
Corallus annulatus	SA	T	N	B	+	+	−	−	−
C. caninus	SA	T	N	B,M	−	−	+	+	+
C. enydris	SA	B,T	N	B,M	−	−	+	+	+
Epicrates cenchria	SA	B,G,T	B	B,M	−	+	+	+	+
Eunectes murinus	SA	A	B	M	−	−	−	−	+

Taxon	Origin	Habitat	Diel	Food	Site				
					LS	BCI	MCSE	Manu	SC
Colubridae									
Amastridium veliferum	MA	G,L	D	F	+	+	−	−	−
Apostolepis pymi	SA	F	B	Am,C,S	−	−	+	−	−
Apostolepis sp.	SA	F	?	?	−	−	+	−	−
Atractus elaps	SA	G,L	D	E	−	−	−	−	+
A. latifrons	SA	F	B	?	−	−	+	−	−
A. major	SA	F,G,L	B	E	−	−	+	+	+
A. occipitoalbus	SA	G,L	D	E	−	−	−	−	+
A. snethlage	SA	F	?	?	−	−	+	−	−
A. torquatus	SA	F	?	?	−	−	+	−	−
A. sp.	SA	F	?	?	−	−	+	−	−
Chironius carinatus	ON	G,T	D	B,F,M	−	+	+	−	+
C. cinnamomeus	ON	G,T	D	F	−	−	+	−	−
C. exoletus	ON	?	D	?	−	−	−	+	−
C. fuscus	ON	G,T	D	B,F,M	−	−	+	−	+
C. grandisquamis	ON	G,T	D	F	+	+	−	−	−
C. multiventris	ON	G	D	?	−	−	−	−	+
C. scurrulus	ON	G	D	F	−	−	−	+	+
Clelia clelia	SA	G	B	L,M,S	+	+	+	+	+
Coniophanes fissidens	MA	G,L	D	F,L,S	+	+	−	−	−
Dendrophidion dendrophis	ON	G,T	D	F,I,M	−	−	+	−	+
D. percarinatum	ON	G,L,T	D	F,L	+	+	−	−	−
D. vinitor	ON	B	D	F	+	−	−	−	−
Dipsas catesbyi	MA	B,T	N	G	−	−	+	+	+
D. indica	MA	B,T	N	G	−	−	+	−	+
D. pavonina	MA	B,T	N	G	−	−	−	−	+

D. variegata	MA	B,T	N	G	–	+	–	–	–
Drepanoides anomalus	SA	G	N	?	–	–	–	+	+
Drymarchon corais	ON	G	D	L,S	+	+	+	+	–
Drymobius margaritiferus	ON	G	D	F	+	–	–	–	–
D. melanotropis	ON	G	D	F	+	–	–	–	–
D. rhombifer	ON	G	D	L	+	–	–	–	+
Drymoluber dichrous	ON	G,T	D	F,L	–	–	+	–	+
Enulius flavitorques	SA	G,L	D	?	–	+	–	–	–
E. sclateri	SA	G,L	D	?	+	+	–	–	–
Erythrolamprus aesculapii	SA	G	D	F,L,S	–	–	+	–	+
E. bizonus	SA	G,L	D	?	–	+	–	–	–
E. mimus	SA	G,L	D	S	+	–	–	–	–
Geophis hoffmanni	MA	G,L	D	I	+	+	–	–	–
Helicops angulatus	SA	A	B	Al,F,L,P	–	–	+	+	+
H. hagmanni	SA	A	B	Al,F,P	–	–	+	–	–
H. petersi	SA	A	B	?	–	–	–	–	+
H. polylepis	SA	A	?	?	–	–	–	+	–
Hydromorphus concolor	MA	R	D	?	+	–	–	–	–
Hydrops triangulatus	SA	A,R	?	E,P	–	–	+	–	–
Imantodes cenchoa	MA	B,T	N	F,L	+	+	+	+	+
I. gemmistratus	MA	B,T	N	L	–	+	–	–	–
I. inornatus	MA	B	N	F,L	+	+	–	–	–
I. lentiferus	MA	B,T	N	F	–	–	–	–	+
Lampropeltis triangulum	ON	G	D	B,L,M,S	+	+	–	–	–
Leptodeira annulata	MA	B,G,T	N	Ae,F,L	+	+	+	+	+
L. septentrionalis	MA	B,T	N	F	+	+	–	–	–
Leptophis ahuetulla	ON	B,T	D	B,F	+	+	+	+	+
L. depressirostris	ON	B,T	D	F	+	–	–	–	–
L. nebulosus	ON	B	D	F	+	–	–	–	–
Liophis breviceps	SA	G	D	F	–	–	+	–	–

Taxon	Origin	Habitat	Diel	Food	Site				
					LS	BCI	MCSE	Manu	SC
L. cobella	SA	G	B	F	−	−	+	−	+
L. epinephalus	SA	G,L	D	F	+	+	−	−	−
L. miliaris	SA	G	B	F	−	−	+	−	−
L. poecilogyrus	SA	G	D	F	−	−	+	−	−
L. reginae	SA	G	D	Al,F	−	−	−	+	+
L. typhlus	SA	G	D	Al,F	−	−	+	−	−
L. sp.	SA	G	D	?	−	−	−	+	+
Mastigodryas boddaerti	ON	G,T	D	F,I,L,M	−	−	+	−	−
M. melanolomus	ON	G,L,T	D	F,L,M	+	+	−	−	−
Ninia hudsoni	MA	G,L	D	?	−	−	−	−	+
N. maculata	MA	G,L	D	?	+	+	−	−	−
N. sebae	MA	G,L	D	?	+	−	−	−	−
Nothopsis rugosus	MA	G,L	D	F	+	−	−	−	−
Oxybelis aeneus	ON	B,T	D	F,L	+	+	+	−	−
O. argenteus	ON	B,T	D	F,L	−	−	+	+	+
O. brevirostris	ON	B	D	L	+	−	−	−	−
O. fulgidus	ON	B,G,T	D	B,L	+	+	+	−	−
Oxyrhopus formosus	SA	G	B	L	−	−	+	+	+
O. melanogenys	SA	G	B	L,M	−	−	−	−	+
O. petola	SA	G,L	B	L,M	+	+	−	+	+
O. trigeminus	SA	G	N	M	−	−	+	−	−
Philodryas viridissimus	SA	T	D	B,F,L,M	−	−	+	−	−
Pliocercus euryzonus	MA	G,L	B	F,P,U	+	+	−	−	−
Pseudoboa coronata	SA	G	B	L,M	−	−	−	+	+
P. neuwiedii	SA	G,L	D	L	−	+	−	−	−
Pseustes poecilonotus	ON	G,T	D	B,M	+	+	+	−	−
P. sulphureus	ON	G,T	D	B,M	−	−	+	−	+

Rhadinaea brevirostris	MA	G	D	F,L	−	−	+	+	+
R. decipiens	MA	G,L	D	F	+	−	−	−	−
R. decorata	MA	G,L	D	F,L,U	+	+	−	−	−
R. fulviceps	MA	G,L	D	U	−	+	−	−	−
R. guentheri	MA	G,L	D	F	+	−	−	−	−
Rhinobothryum lentiginosum	ON	B,T	N	F	−	−	+	+	−
Scaphiodontophis venustissimus	ON	G,L	D	F,L	+	−	−	−	−
Sibon annulata	MA	B	N	G	+	−	−	−	−
S. longifrenis	MA	B	N	G	+	−	−	−	−
S. nebulata	MA	B	N	G	+	−	−	−	−
Siphlophis cervinus	SA	G,L,T	N	B,L	−	+	+	+	+
Spilotes pullatus	ON	G,T	D	B,F,L,M,S	+	+	+	−	−
Stenorrhina degenhardtii	ON	G,L	D	I	−	+	−	−	−
Tantilla albiceps	ON	G,L	D	?	−	+	−	−	−
T. armillata	ON	G,L	D	I	−	+	−	−	−
T. melanocephala	ON	G,L	D	I	+	−	+	−	+
T. reticulata	ON	G,L	D	?	+	−	−	−	−
Tretanorhinus nigroluteus	MA	R	?	?	+	−	−	−	−
Trimetopon barbouri	MA	G,L	D	?	−	+	−	−	−
T. pliolepis	MA	G,L	D	F,L	+	−	−	−	−
Tripanurgos compressus	SA	B,T	N	L	−	−	+	+	+
Xenodon rhabdocephalus	SA	G	B	F	+	+	+	−	−
X. severus	SA	G	B	F	−	−	+	+	+
Xenopholis scalaris	SA	L	D	F	−	−	−	+	+
Elapidae									
Micrurus alleni	MA	G,L	B	S	+	−	−	−	−
M. averyi	SA	F,G	N	?	−	−	+	−	−
M. hemprichii	SA	F,G	N	Am	−	−	+	−	−
M. langsdorffi	SA	G	D	?	−	−	−	−	+
M. lemniscatus	SA	F,G	B	Am,C,S	−	−	+	+	+

Taxon	Origin	Habitat	Diël	Food	Site				
					LS	BCI	MCSE	Manu	SC
M. mipartitus	SA	G,L	B	L,S	+	+	−	−	−
M. narduccii	SA	F	?	?	−	−	−	−	+
M. nigrocinctus	MA	G,L	B	C,L,S	+	+	−	−	−
M. spixii	SA	F,G	D	Am,C	−	−	+	−	+
M. surinamensis	SA	A,R	D	P	−	−	+	+	+
Letotyphlopidae									
Leptotyphlops septemstriatus	SA	F	?	A,Te	−	−	+	−	−
L. tenella	SA	F	?	A,Te	−	−	+	−	−
Tropidophiidae									
Ungaliophis panamensis	MA	T	?	?	+	−	−	−	−
Viperidae									
Bothrops asper	SA	G	B	F,L,M	+	+	−	−	−
B. atrox	SA	B,G,T	B	F,L,M	−	−	+	+	+
B. bilineatus	SA	T	N	F,L,M	−	−	+	−	+
B. castelnaudi	SA	G	N	?	−	−	−	−	+
B. schlegeli	SA	B,T	B	B,F,L,M	+	+	−	−	−
Lachesis muta	SA	G	N	M	+	+	+	+	+
Porthidium nasutum	MA	G,L	D	L,M	+	−	−	−	−
TESTUDINES									
Cheliidae									
Chelus fimbriatus	SA	A	B	P	−	−	−	−	+
Mesoclemmys gibba	SA	A	D	?	−	−	−	−	+

Phrynops geoffroanus	SA	A	D	?	−	−	−	−	+
P. nasutus	SA	A,G	D	?	−	−	−	+	−
Platemys platycephala	SA	A,G	D	?	−	−	+	+	+
Chelydridae									
Chelydra acutirostris	ON	A	?	?	−	+	−	−	−
Emydidae									
Rhinoclemmys annulata	ON	G	D	?	+	+	−	−	−
R. funerea	ON	A	D	?	+	+	−	−	−
Trachemys scripta	ON	A	D	O	−	+	−	−	−
Kinosternidae									
Kinosternon angustipons	ON	R	D	?	+	−	−	−	−
K. leucostomum	ON	A,R	B	?	+	+	−	−	−
K. scorpioides	ON	A	B	?	−	−	+	+	+
Pelomedusidae									
Podocnemis unifilus	SA	A	D	H	−	−	−	+	−
Testudinidae									
Geochelone denticulata	SA	G	D	H	−	−	+	+	+
CROCODYLIA									
Crocodylidae									
Caiman crocodilus	SA	A	N	B,M,T	+	+	+	+	+
Crocodylus acutus	MA	A	N	M	−	+	−	−	−
Melanosuchus niger	SA	A	N	?	−	−	−	+	−
Paleosuchus trigonatus	SA	A	N	?	−	−	+	−	+

PART VI

Forest Dynamics

25

Forest Dynamics at La Selva Biological Station, 1969–1985

DIANA LIEBERMAN, GARY S. HARTSHORN, MILTON LIEBERMAN, AND RODOLFO PERALTA

The first forest plot inventory at La Selva, Costa Rica, was made in 1969–70, inaugurating a long-term record of tree growth, demography, and forest dynamics. The main body of data is based on all woody plants 10 cm diameter at breast height (DBH) or greater and includes the species identification, map position, and diameter of each stem. The most recent plot enumeration was completed in 1989. Comparable data have also been collected in a subsample of stems 2–10 cm DBH and seedlings; all size classes from newly germinated seedlings to the oldest trees are thus represented.

In this chapter, we summarize the major dynamic attributes of the forest at La Selva, including rates of mortality and recruitment, stand turnover and gap-phase dynamics, species growth behavior and longevity, and seed germination and seedling survivorship.

Methods

Our information is derived from repeated inventories of three permanent plots established at La Selva in 1969–70 by the University of Washington College of Forest Resources. The plots, totaling 12.4 ha of primary forest, vary in soils, topography, and drainage. Plot 1 (4.4 ha) is on an old alluvial terrace of the Rio Puerto Viejo and includes a small swamp area of less than 0.5 ha. Plot 2 (4.0 ha) is on old alluvium and

We are grateful to the Organization for Tropical Studies for logistical support. Fieldwork was carried out with the able assistance of Gerardo Vega and Danilo Brenes (1982–83 inventory) and Ing. Manuel Víquez and Ing. Maria de los Angeles Molina (1985 inventory). This study was supported by National Science Foundation grants BSR-8117507 and BSR-8414968. Preparation of the manuscript was supported by NASA grant NAGW-1033.

colluvium, and half the plot bears swamp forest. Plot 3 (4.0 ha) is on steep hills with residual soils overlying basalt parent material. The plots are described in detail elsewhere (Hartshorn 1983; Lieberman et al. 1985a,b,c; Lieberman and Lieberman 1987, 1990).

At the time of the original plot inventory in 1969–70, all tree and liana stems greater than 10 cm DBH were tagged, mapped, and measured.

The second complete inventory was carried out in 1982–83, thirteen years later. We retagged, remeasured, and identified all surviving stems; tagged, mapped, measured, and identified all new recruits to the 10-cm diameter class; and recorded the circumstances of death, if possible, for individuals that had died between 1969 and 1982. Diameter measurements were made to the nearest millimeter at breast height and above buttresses or stilt roots, if present. The new tags were fixed to the trees exactly 10 cm above the height of measurement.

The third complete inventory was conducted two and a half years later, in 1985, using the same protocols as in the second inventory.

As an adjunct to the tree censuses, smaller size classes were studied in subsamples located at random throughout the plots. The density and species composition of viable seeds in the soil were assessed by recording germination of seeds from soil samples collected from the field and spread out in flats in a shade house. Each sample was taken from an area of 0.25 m × 0.25 m × 3 cm.

Germination and survivorship of seedling cohorts were assessed by repeated censuses of tagged individuals in permanent transects within the three plots. In 1983, 48 seedling transects measuring 10 m × 0.5 m were established at random locations in the plots. Results are presented for the first twenty months of the seedling studies, during which time over 6,000 seedlings were tagged, identified, mapped to the nearest centimeter, and measured to the nearest centimeter of height.

Individuals 2–10 cm DBH were tagged, mapped, and measured in 1969 in a random subsample representing 10% of the area of each plot. Survivors were retagged, remeasured, and identified in 1983–84, and new recruits to the 2 cm DBH class were tagged, mapped, measured, and identified at that time.

Results

Mortality and recruitment. Patterns of mortality and recruitment during the thirteen-year period 1969–82 have been reported previously

TABLE 25.1 Dynamics of three permanent plots at La Selva, 1969–1985, based on density of stems ≥ 10 cm DBH

	Plot 1	Plot 2	Plot 3	Total
Density (stems/ha)				
1969	431.4	397.3	534.0	453.4
(−) Deaths	−94.3	−98.0	−144.8	−111.8
(+) Ingrowth	+80.7	+95.3	+139.3	+104.3
1982	417.7	394.4	528.5	446.0
Annual mortality λ	1.80%	2.01%	2.24%	2.03%
(−) Deaths	−21.4	−24.8	−30.3	−25.3
(+) Ingrowth	+23.9	+28.0	+35.8	+29.0
1985	420.2	397.8	534.0	449.7
Annual mortality λ	2.10%	2.59%	2.36%	2.34%
Net change, 1969–1985				
Stems/ha	−11.2	+0.5	0.0	−3.7
Percentage	−2.60%	+0.13%	0.00%	−0.82%

(Lieberman et al. 1985a; Lieberman and Lieberman 1987) and are summarized below.

Mortality rate was independent of tree size among individuals greater than 10 cm DBH, indicating a logarithmic loss of stems over time.

Of 5,623 trees and lianas tagged in 1969, 23.2% had died by 1982. Thirty-one percent of the dead stems were found to have fallen, 26% were standing, 7% were buried under treefalls, and 37% had decomposed completely, leaving no trace (Lieberman et al. 1985a).

Rates of mortality and recruitment, expressed in terms of stem density, are presented for 1969–82 and 1982–85 in table 25.1. Stem density changed less than 1% during 1969–85, as recruitment into the 10-cm DBH class (133.3 stems/ha) balanced losses from the stand due to death (137.1 stems/ha) over the 15.5-year period. The annual mortality rate λ, calculated as the slope of $\log_e$ survivorship versus time, was quite high, indicating a very dynamic stand; values were 2.03% during 1969–82 and 2.34% during 1982–85. The stand half-life, or number of years in which 50% of the individuals are expected to die, is 34 years, based on 1969–82 mortality data, and 30 years, based on 1982–85 mortality data. Annual stem mortality (fig. 25.1) was consistently lowest in Plot 1 (1.80% during the first census period and 2.10% during the second period), and was higher in Plot 2 (2.01% and 2.59%) and Plot 3 (2.24% and 2.36%). All three plots experienced higher mortality rates during the second census period than during the first.

Changes in basal area and biomass are shown in table 25.2 for 1969–

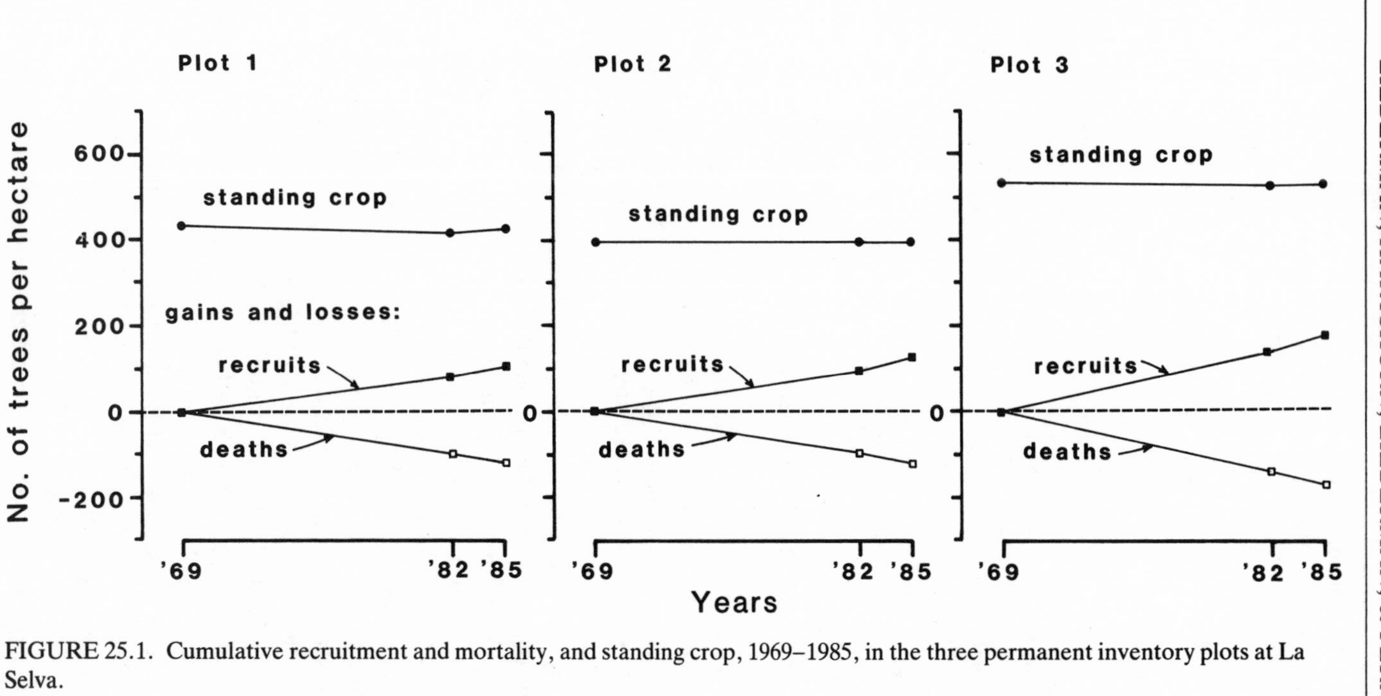

FIGURE 25.1. Cumulative recruitment and mortality, and standing crop, 1969–1985, in the three permanent inventory plots at La Selva.

TABLE 25.2 Dynamics of three permanent plots at La Selva, 1969–1982, based on basal area and biomass of stems ≥ 10 cm DBH

	Plot 1	Plot 2	Plot 3	Total
Basal area (m^2/ha)				
1969	28.49	32.97	28.95	30.08
(−) Deaths	−5.45	−6.48	−7.91	−6.40
(+) Stem growth	+3.57	+2.29	+2.54	+2.83
(+) Ingrowth	+1.10	+1.45	+1.95	+1.49
1982	27.71	30.23	25.54	27.82
Basal area increment, m^2/ha, 13 years	4.67	3.74	4.49	4.32
Biomass (metric tons/ha)				
1969	254.7	294.7	258.8	268.9
(−) Deaths	−48.7	−57.9	−70.6	−58.8
(+) Stem growth	+31.9	+20.5	+22.7	+25.3
(+) Ingrowth	+9.8	+12.9	+17.4	+13.3
1982	247.7	270.3	228.3	248.7
Biomass increment, metric tons/ha, 13 years	41.7	33.4	40.1	38.6
Annual loss of basal area or biomass, λ	1.63%	1.68%	2.54%	1.84%
Net change, 1969–1982	−2.74%	−8.31%	−11.78%	−7.51%

82. Basal area dropped by 7.5% during 1969–82. Losses due to death of stems were 6.5 m^2/ha, while gains were 1.5 m^2/ha from recruitment and 2.8 m^2/ha from growth of surviving individuals. The annual loss of basal area (logarithmic model) was 1.8%.

Estimates of biomass are derived from the allometric relation between stem basal area and total above-ground dry biomass. Biomass data were obtained from a sample of one hundred trees harvested a few kilometers southeast of La Selva by the Departamento de Ingenería Forestal, Instituto Tecnológico de Costa Rica, Cartago, under contract to the National Aeronautics and Space Administration's Earth Resources Laboratory; the relation is represented by the linear regression equation $y = 0.89x - 136$, where y is dry biomass (kg) and x is stem basal area (cm^2) (Lieberman and Lieberman 1990). The standing crop in the plots in 1969 was 269 metric tons/ha, which fell to 249 by 1982. Death removed 59 tons/ha, which was replaced by 13 from recruitment and 25 from growth of surviving stems. Annual biomass mortality percentages and net biomass change are necessarily equal to those reported for basal area because biomass was calculated as a linear function of basal area.

TABLE 25.3 Comparison of estimates of forest dynamics at La Selva based on mortality of stems ≥ 10 cm DBH (after Lieberman et al. 1985a) and accumulation of area in treefall gaps (after Hartshorn 1978)

	Plot 1	Plot 2	Plot 3
Stem mortality			
Total stems, 1969	1,898	1,589	2,136
Years observed	13	13	13
Death (%)	21.9	24.7	27.1
Annual loss (%) λ	1.80	2.01	2.24
Stand half-life $t_{.5}$ (years)	38.5	34.4	30.9
Gap formation			
Total area (m^2), 1970–71	40,000	40,000	40,000
Years observed	6	5	5
Gap area (%)	7.5	3.9	3.7
Annual loss (%) λ	1.31	0.80	0.76
Stand half-life $t_{.5}$ (years)	53.1	87.1	91.7
Ratio of stem half-life to gap area half-life	0.73	0.40	0.34

Stand turnover and gap dynamics. Stand turnover times for the permanent plots were estimated independently using both stem mortality rates and rates of gap formation. Hartshorn (1978) monitored the occurrence of tree falls in the plots from 1970 to 1976, mapping the accumulation of area covered by canopy gaps over that period. Rates of gap area increment (1970–76) reported by Hartshorn were recalculated using a logarithmic model, making them directly comparable with rates of stem mortality (1969–82). The comparison of the two turnover estimates is highly informative (table 25.3).

The annual loss of stems within the three plots ranged from 1.8% to 2.2%, representing stand half-life values of 31–39 years. The annual loss of canopy area was much lower, ranging from 0.8% to 1.3%, with the stand half-life ranging from 53 to 92 years. On average, the stem mortality turnover rate was around twice as fast as the gap area turnover rate. Thus, approximately half the trees greater than 10 cm DBH die without forming or contributing to a recognizable gap in the canopy. The death of small trees that have their crowns below canopy level has no recognizable effect on canopy closure, for example, and the gradual dieback of canopy tree crowns is sometimes matched by lateral growth of neighboring branches into the available space.

Growth, suppression, and longevity. To estimate lifetime growth curves for each species of tree, we developed a growth simulation tech-

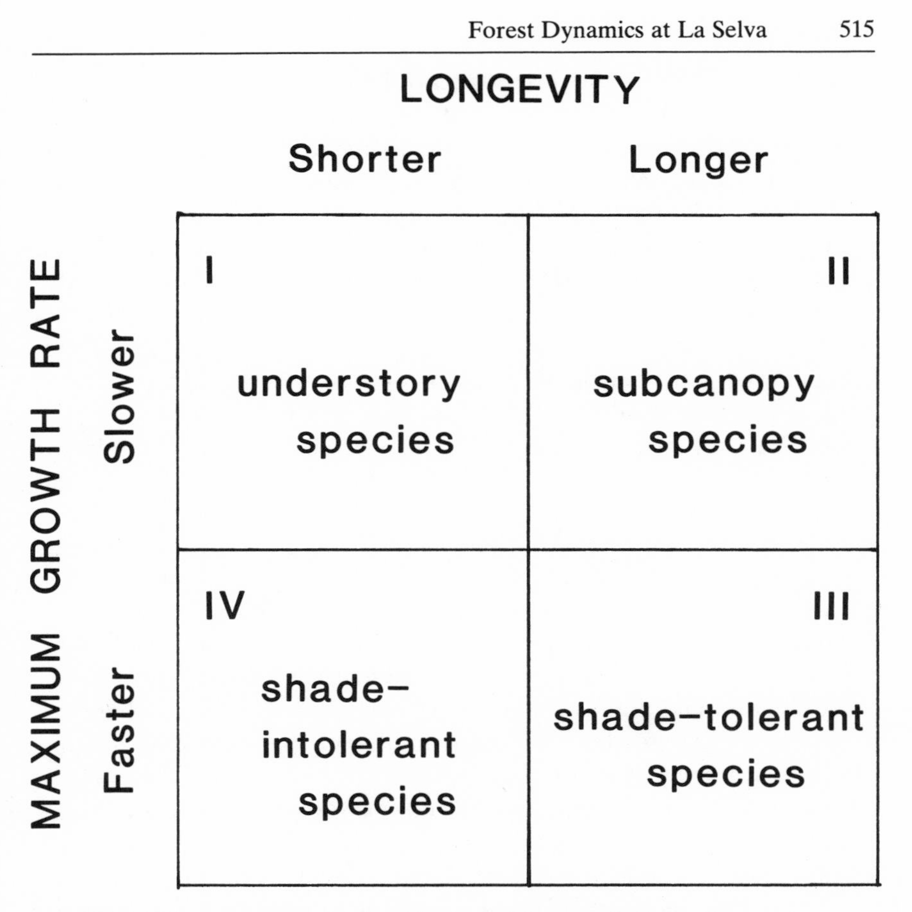

FIGURE 25.2. Diagrammatic classification of tree species according to maximum growth rate (slow v. fast) and projected lifespan (short v. long). *Group I*: understory species, *Group II*: slow-growing subcanopy species, *Group III*: fast-growing, shade-tolerant canopy and subcanopy species that respond opportunistically to increased light levels; *Group IV*: putative shade-intolerant canopy and subcanopy species.

nique, a form of bootstrapping (Lieberman and Lieberman 1985). This technique generates growth trajectories as follows: diameter increments of measured trees are arranged in a list ordered according to initial DBH. Increments are repeatedly sampled from the list in a stratified random manner (stratified by DBH), and these increments, each representing measured growth over the thirteen-year period 1969–82, are assembled into a model trajectory. The curve ends when the model tree is as large as the largest individual in the list. For each species, we projected 1,700 trajectories, and from the distribution of imputed trajectories we determined the minimum, maximum, and median curves, projected lifespan (median age at maximum DBH), and minimum age of largest tree.

Growth was analyzed in this way in 47 abundant species (Lieberman et al. 1985b, 1988; Lieberman and Lieberman 1987). Different patterns of growth tended to predominate in various groups of species (fig. 25.2). Understory species showed slow maximum growth and had short lifespans. Subcanopy species also generally had slow maximum growth rates but with long lifespans. Most canopy species (and a few subcanopy species) showed rapid maximum growth rates and long lifespans. Other canopy species showed very rapid maximum growth rates and short lifespans. We have hypothesized that these are relatively shade-intolerant species—that is, relatively unlikely to complete their life cycle and reproduce in deep shade.

For purposes of simplicity, the species groups and growth patterns are represented in fig. 25.2 as discrete entities. It is clear, however, that species are distributed continuously over the full range of growth behavior (fig. 25.3). Intergradations occur between understory and subcanopy species, subcanopy and shade-tolerant canopy species, and shade-tolerant and putative shade-intolerant species, although there are no true intermediates between shade-intolerant species and understory species (Lieberman et al. 1985b).

Intraspecific variability in growth rates was slight both in understory trees, which grew at consistently slow rates, and in shade-intolerant species, which grew at consistently rapid rates. Subcanopy and shade-tolerant canopy species, by contrast, were quite variable in their growth rates. There is a positive correlation between the variance of log-transformed diameter increments within species and the projected lifespan of the species; longer-lived species showed more variability in growth rates (Lieberman et al. 1985b).

Minimum projected growth curves were very slow or nearly level in

FIGURE 25.3. Relation between maximum growth rate, projected lifespan (starting at 10 cm DBH), and maximum size in 45 abundant tree species at La Selva. ●, understory trees; ◒, subcanopy trees; ○, canopy trees. *Group I*: understory species—Ac, *Anaxagorea crassipetala*; Cp, *Capparis pittieri*; Cc, *Casearia comersoniana;* Ce, *Cassipourea elliptica*; Cs, *Colubrina spinosa*; Ft, *Faramea talamancarum*; Gr, *Guarea rhopalocarpa*; Gi, *Guatteria inuncta*; Lo, *Lonchocarpus oliganthus*; Om, *Ocotea meziana*; Pc, *Protium costaricense*; Re, *Rheedia edulis*; Rip, *Rinorea pubipes*. *Group II*: slow-growing subcanopy

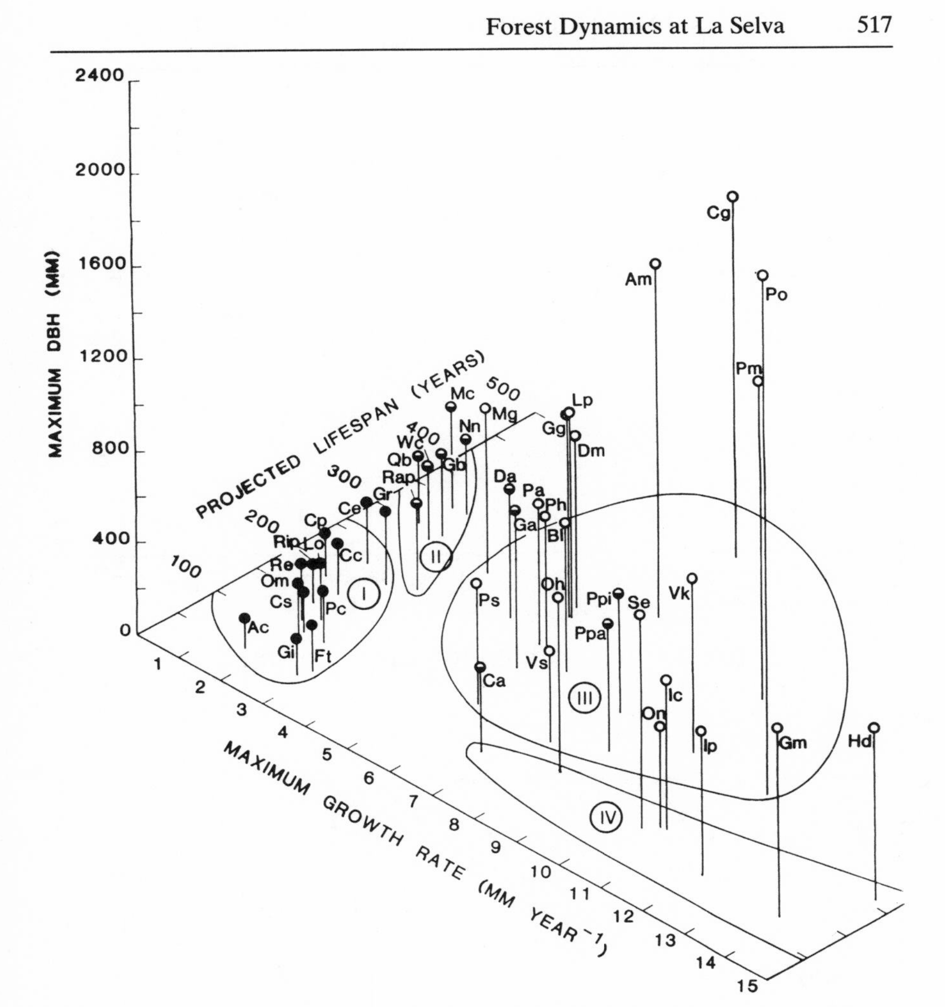

species—Gb, *Guarea bullata*; Mc, *Macrolobium costaricense*; Nn, *Naucleopsis naga*; Qb, *Quararibea bracteolosa*; Rap, *Rauvolfia purpurascens*; Wc, *Warscewiczia coccinea*. *Group III*: fast-growing, shade-tolerant canopy and subcanopy species—Am, *Apeiba membranacea*; Bl, *Brosimum lactescens*; Cg, *Carapa guianensis*; Da, *Dendropanax arboreus*; Dm, *Dussia macroprophyllata*; Gg, *Guarea glabra*; Ga, *Guatteria aeruginosa*; Lp, *Laetia procera*; Mg, *Minquartia guianensis*; Pm, *Pentaclethra macroloba*; Pa, *Pourouma aspera*; Ps, *Pouteria standleyana*; Ppa, *Protium panamense;* Ppi, *Protium pittieri*; Ph, *Pterocarpus hayesii*; Po, *Pterocarpus officinalis*; Vk, *Virola koschnyi*; Vs, *Virola sebifera*. Group IV: putative shade-intolerant canopy and subcanopy species—Ca, *Casearia arborea*; Gm, *Goethalsia meiantha*; Hd, *Hernandia didymanthera*; Ic, *Inga coruscans*; Ip, *Inga* cf. *pezizifera*; Oh, *Ocotea hartshorniana*; On, *Otoba novogranatensis*; Se, *Stryphnodendron excelsum*. (From Lieberman et al. 1985b.)

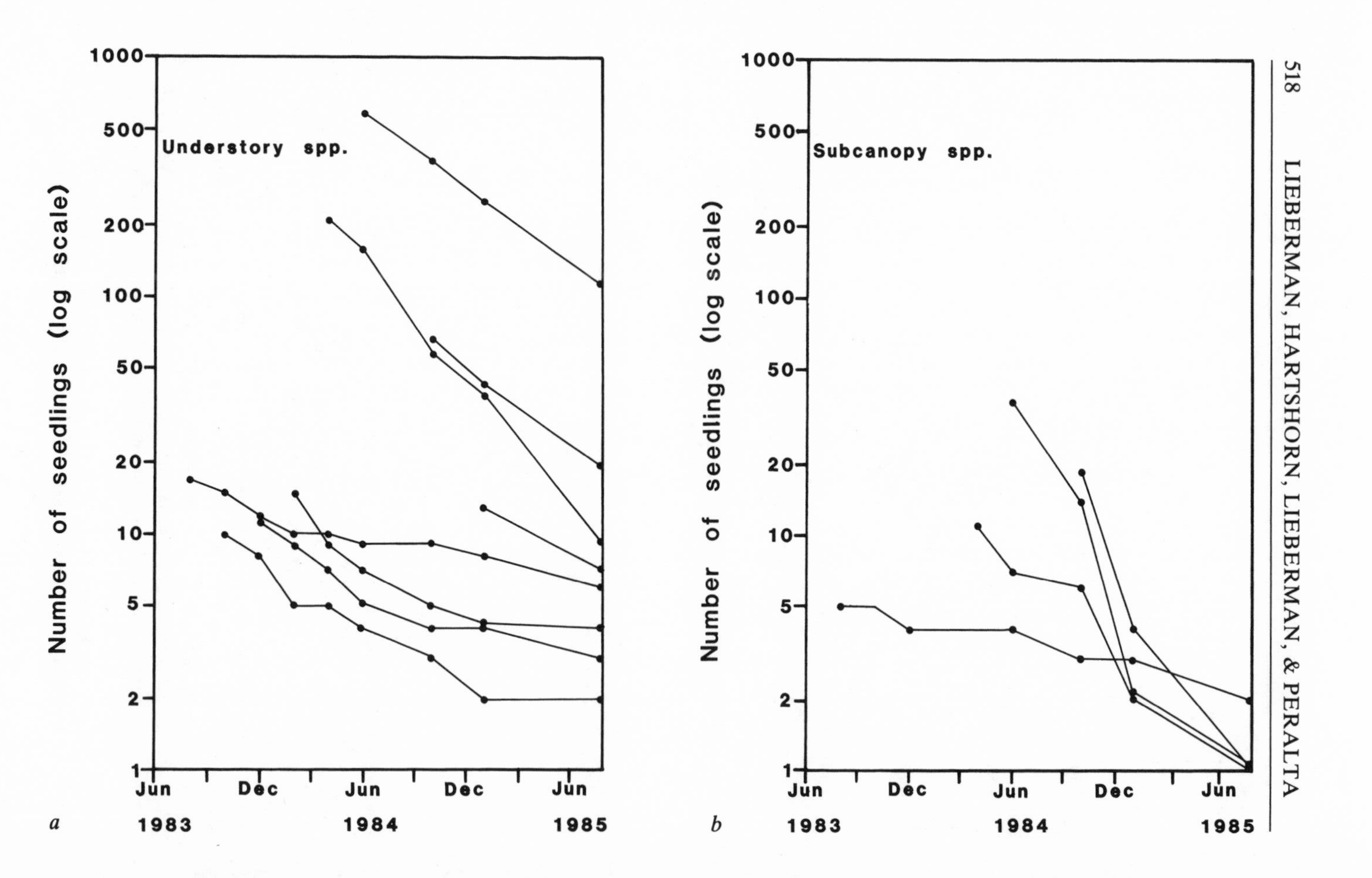
Understory spp.
Number of seedlings (log scale)
1000
500
200
100
50
20
10
5
2
1
Jun 1983
Dec
Jun 1984
Dec
Jun 1985
a
Subcanopy spp.
Number of seedlings (log scale)
1000
500
200
100
50
20
10
5
2
1
Jun 1983
Dec
Jun 1984
Dec
Jun 1985
b

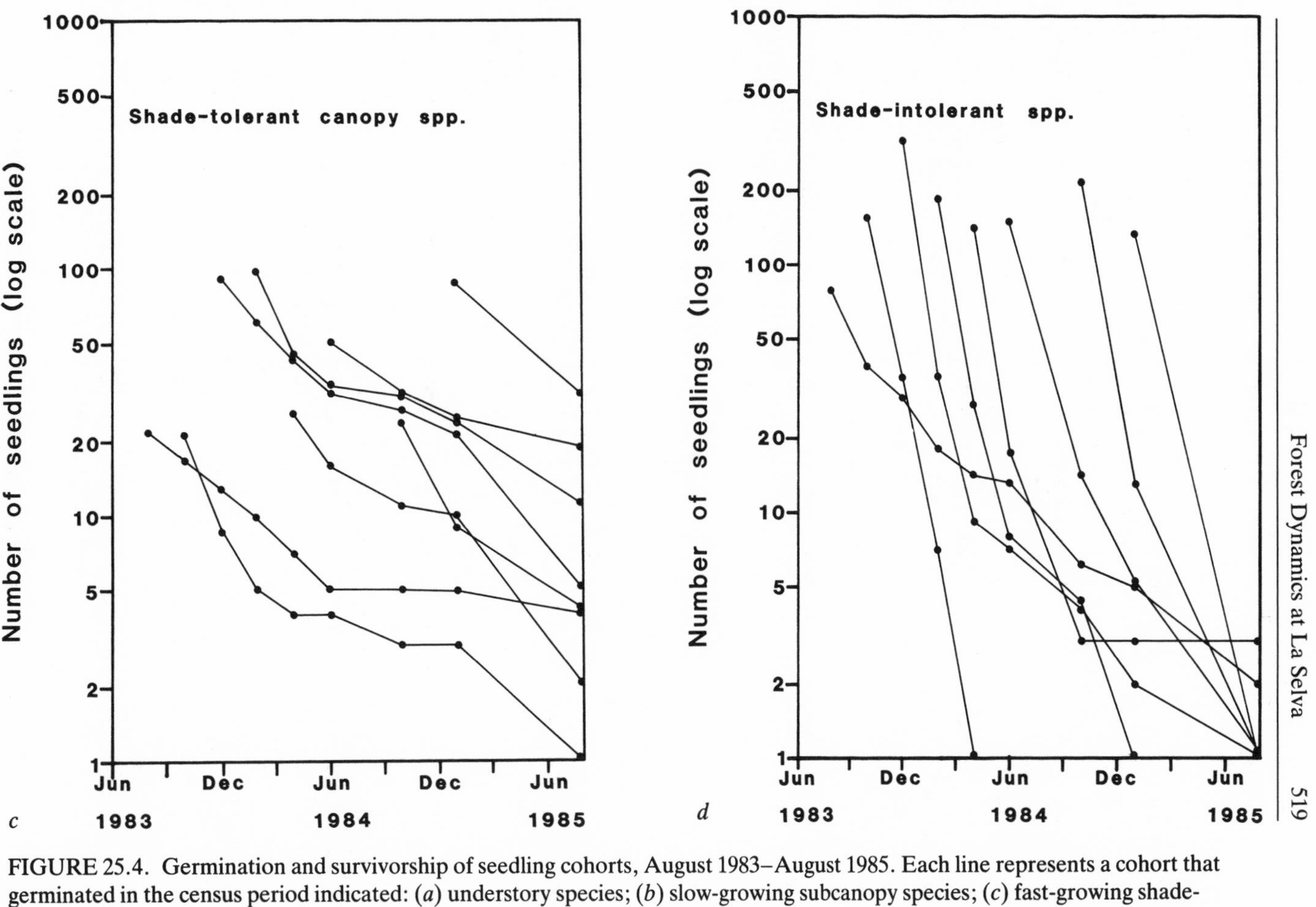

FIGURE 25.4. Germination and survivorship of seedling cohorts, August 1983–August 1985. Each line represents a cohort that germinated in the census period indicated: (*a*) understory species; (*b*) slow-growing subcanopy species; (*c*) fast-growing shade-tolerant canopy and subcanopy species; (*d*) putative shade-intolerant species. Species in each category are pooled for analysis.

most species, with many individual surviving trees showing no increase at all in diameter over a thirteen-year period. This pattern was common to understory species, subcanopy species, and shade-tolerant canopy species, and we infer that these trees frequently experience and survive prolonged periods of suppression. In contrast, shade-intolerant species had relatively rapid minimum growth curves, and all surviving trees showed measurable growth over the thirteen years (Lieberman et al. 1985b; Lieberman and Lieberman 1987).

Longevity was negatively correlated with recruitment rate; species with shorter projected lifespans produced more recruits to the 10-cm DBH class per unit time (Lieberman et al. 1985a). Thus, populations generally appear to be maintaining themselves.

Seedling germination and survivorship. Based on results of soil seed bank studies (Lieberman and Lieberman 1987), the mean density of viable buried seeds in the top 3 cm of soil was $307/m^2$, if all species are considered. If only those species capable of reaching 10 cm DBH are included, the mean density was $232/m^2$.

In the permanent transects, seedlings of understory species germinated abundantly and survived well (fig. 25.4a). Subcanopy tree species produced few seedlings, and their survivorship was poor (fig. 25.4b). Shade-tolerant canopy species had moderate amounts of germination, and the seedlings survived well (fig. 25.4c). Fast-growing, short-lived canopy species, those thought to be shade-intolerant, had very abundant germination in every census, and survivorship was extremely poor (fig. 25.4d).

It is interesting to note that shade-intolerant species were observed to germinate under a wide range of canopy conditions, from large openings to rather dense shade. Survivorship, however, was almost nil, and many of the seedlings died before reaching a height of 3 mm. This contradicts the notion that shade-intolerant species do not germinate until suitable conditions occur, an idea that has arisen both because soil seed banks contain high densities of these species and because established individuals are commonly noticed growing in open conditions.

Overall, the short-lived species (understory and shade-intolerant canopy trees) produced most of the seedlings, and did so consistently in every census period; the longer-lived species (subcanopy and shade-tolerant canopy trees) produced fewer.

The La Selva forest is characterized by overall stability in stand structure in spite of rapid turnover and high rates of mortality. Tree density remained unchanged over a 15.5-year period as recruitment matched mortality. Basal area and biomass also appear to be generally stable over time. There is good stocking in all size classes, from seedlings to the largest trees, and regeneration is abundant. Most species appear to survive lengthy periods of suppression; only fast-growing trees hypothesized to be shade-intolerant species do not.

The half-life of the stand is approximately 30–34 years, with some variation from plot to plot, which is equivalent to an annual mortality rate of 2.0%–2.3%. Comparable calculations based on rates of gap formation show a much longer turnover rate; the annual rate of stem mortality is about twice that of gap area increment.

Species groups defined on the basis of stature (understory, subcanopy, canopy) and apparent level of shade-tolerance differ from one another in several important ways. Tree longevity, maximum and minimum growth trajectories, variance in growth rate, and rate of recruitment differ among groups, as do seedling abundance and survivorship rates.

REFERENCES

Hartshorn, G. S. 1978. Tree falls and tropical forest dynamics. In P. B. Tomlinson and M. H. Zimmermann (eds.), *Tropical Trees as Living Systems*. Cambridge Univ. Press, Cambridge, pp. 617–638.

———. 1983. Plants: introduction. In D. H. Janzen (ed.), *Costa Rican Natural History*. Univ. of Chicago Press, Chicago, pp. 118–157.

Lieberman, D. and M. Lieberman. 1987. Forest tree growth and dynamics at La Selva, Costa Rica (1969–1982). *J. Trop. Ecol.* 3: 347–358.

Lieberman, D., M. Lieberman, R. Peralta, and G. S. Hartshorn. 1985a. Mortality patterns and stand turnover rates in a wet tropical forest in Costa Rica. J. Ecol. 73: 915–924.

Lieberman, D., M. Lieberman, G. S. Hartshorn, and R. Peralta. 1985b. Growth rates and age-size relationships of tropical wet forest trees in Costa Rica. J. Trop. Ecol. 1: 97–109.

Lieberman, M. and D. Lieberman. 1985. Simulation of growth curves from periodic increment data. Ecol. 66: 632–635.

———. 1990. Patterns of density and dispersion of forest trees. In L. A. McDade, K. S. Bawa, H. A. Hespenheide, and G. S. Hartshorn (eds.), *La Selva: Ecology and Natural History of a Neotropical Rainforest*. Univ. of Chicago Press, Chicago.

Lieberman, M., D. Lieberman, G. S. Hartshorn, and R. Peralta. 1985. Small-scale altitudinal variation in lowland wet tropical forest vegetation. J. Ecol. 73: 505–516.

Lieberman, M., D. Lieberman, and J. H. Vandermeer. 1988. Age-size relationships and growth behavior of the palm *Walfia georgii*. Biotropica 20: 270–273.

26

Structure, Dynamics, and Equilibrium Status of Old-Growth Forest on Barro Colorado Island

STEPHEN P. HUBBELL AND ROBIN B. FOSTER

Since the creation of Barro Colorado Island (BCI) about 75 years ago, the age and successional status of the tropical moist forest on the island have been the subject of considerable debate (Standley 1927; Bennett 1963; Knight 1975; Foster and Brokaw 1982). So-called old-growth forest covers approximately half of the 15 km^2 island. The remainder of BCI is a patchwork of secondary forest of varying age dating from precanal days. Based on comparisons of the structure and species composition of a series of 0.8–2 ha plots in both secondary and old-growth forest, Knight (1975) concluded that the old-growth forest was at least 130 years old. Recent extensive paleoecological studies by Piperno (1989) demonstrate that pre-Columbian hunter-gatherers used the site beginning about A.D. 500 and occasionally made small clearings (less than 0.1 ha) in the forest, possibly for seasonal hunting camps. Radiocarbon dating indicates that the last of these clearings was cut more than 550 years ago, and the phytolith record demonstrates that the forest was never cleared for maize agriculture.

Before this study, a series of smaller-scale studies of forest structure and dynamics were carried out on BCI. Knight's (1975) plots were reexamined after a decade (Lang and Knight 1983), and Putz and Milton

We acknowledge with gratitude the financial support of the National Science Foundation, the World Wildlife Fund, the Smithsonian Institution, the Geraldine R. Dodge Foundation, and the Center for Field Studies for supporting the BCI Forest Dynamics Project, most recently the recensus effort. We thank the many assistants and volunteers who have made the project a success. We single out Jeff Klahn, Sue Williams, Brit Minor, David Hamill and Steve Hewett for special thanks in overseeing the field crews and in assisting with the data base management and analysis.

(1982) recensused plots established five years earlier by Thorington et al. (1982). Brokaw (1982, 1985a,b) has followed the fate of plants in a cohort of treefall gaps of varying size since the gaps formed. Many other investigators have pursued studies on the biology of particular species or groups of species of trees and shrubs in the BCI forest (e.g., Augspurger 1979, 1983a,b; Howe and Van de Kerckhove 1979; Howe 1980, 1983; Howe et al. 1985; Hamill 1987; Schupp 1987).

Our own research has focused on the structure and dynamics of a large plot of old-growth forest on BCI. We established a permanent 50-ha plot in 1980 with the goal of monitoring tree recruitment, growth, and mortality over a period sufficiently long to construct horizontal life tables for many BCI tree species (Hubbell and Foster 1983). The primary census and map of woody plants in the plot was completed in 1982. This census included all free-standing woody plants over 1 cm diameter at breast height (DBH). We recensused the plot in 1985 and will do so again in 1990 and at five-year intervals thereafter. In addition, a detailed record of changes in canopy structure and treefall gaps has been made in an annual canopy census throughout the plot (Hubbell and Foster 1986a).

One of the main organizing questions for this research has been whether the BCI forest community is at or near species equilibrium. The general question of the equilibrium or nonequilibrium status of tropical forests has been receiving increasing attention over the past decade. At issue is the strength of equilibrating forces tending to maintain particular assemblages of tropical tree species under a given environmental regime. The issue takes on special significance because of its implications for tropical forest conservation and management (Hubbell 1984). Connell (1978, 1979) first drew notice to the question by arguing that rainforest tree communities were open, nonequilibrium systems. Hubbell (1979) and Hubbell and Foster (1986b) suggested that many tropical tree species might be nearly equivalent competitors that coexist for long periods in drifting relative abundance. Using circumstantial evidence from the first BCI census, Hubbell and Foster (1986b,c, 1987a,b) concluded that many of the most abundant BCI tree species were habitat and regeneration niche generalists and that the BCI forest was not in species equilibrium.

The recensus data reported here show that equilibrating forces are stronger than were apparent from the primary census. The evidence for nonequilibrium of the current species mix remains strong, however (Hubbell and Foster, in press-a,b), and there is still a question about the efficacy of the equilibrating forces to bring about species equilibrium. Here we discuss the changes in species abundances that have occurred over the recensus interval and their implications for the equilibrium status of the old-growth forest on BCI.

Study Site and Field Methods

The 50-ha plot on BCI was established on the central plateau of the island at a mean elevation of about 150 m above sea level. A grid of 1,250 20 m × 20 m quadrats was surveyed and permanently marked. Each of the 20 m × 20 m quadrats was further sub-divided into 5 m × 5 m sub-quadrats. The site is relatively level, aside from moderate slopes on the eastern and southern margins of the plot with grades between 10% and 20%. A small seasonal swamp with acidic, anaerobic soils in the wet season (May–December) occupies about 2 ha in the center of the plot. The soils on the plateau are poor, similar to Amazonian oxisols (Keller, p.c.), and vary little throughout the plot except at the swamp (Smith, p.c.). The slopes contain numerous seepage zones and springs and remain wetter than the flatter terrain of the plateau during the dry season (Becker, p.c.).

The island annually receives about 2,600 mm of precipitation, seasonally distributed, with a four-month dry season from mid-December through mid-April (Dietrich et al. 1982; Rand and Rand 1982). During the intercensus interval, 1982–85, however a long-term climatic extreme event occurred. The dry season of the year of El Niño, 1983, was the most intense on BCI on record in over sixty years (Windsor 1989). In most years occasional good rains fall during the dry season. In 1983, the dry season was six weeks longer than usual (mid-November–May), and dry season rainfall was only 15% of normal.

Census methods were as follows: All free-standing woody plants with a stem diameter of at least 1 cm DBH (130 cm height or above buttresses when present) were tagged, identified to species, and mapped to the nearest 0.5 m within each 5 m × 5 m subquadrat. The first census took two and a half years and was completed in 1982; the recensus went more quickly and was completed during 1985. We are confident of the accuracy of species identifications. Of the individual plants censused, only two could not be identified from vegetative characters even to family; both unknown plants were of the same species. The survival and growth of all plants present in the 1980–82 census was determined in 1985. Annual growth rates were computed based on dates of the first and second measurements of each plant. In addition, all new plants recruited into the plot were recorded if they had attained the minimum census diameter of 1 cm DBH.

Plants with stem diameters of 1 cm DBH or more in 1982 that suffered the loss of their principal stem during the intercensus interval and survived in 1985 with no stems larger than 1 cm DBH were recorded as "present." We are continuing to follow their fate, but they were not counted in the 1985 plot total of plants over 1 cm DBH. We have no comparable records for first-census plants whose prior history of stem loss is unknown.

Results

Species Composition of the Fifty-Hectare Plot

A total of 306 species, from shrubs to trees, were recorded in the 50-ha plot in one or both censuses. The plot contained about three-fourths of the approximately 400 woody species that we estimated from Croat's (1978) BCI flora would be able to grow to a size of 1 cm DBH. This total excludes lianas, which have not yet been censused and mapped in the plot.

Sixty-one families were represented in the plot; of these, 10 accounted for 48% of all species. If the subfamilies of legumes are lumped, then Leguminosae, with 36 species, was the most speciose family in the plot. If the legume subfamilies are considered separately, then the biggest family in number of genera, species, and individuals was Rubiaceae, with 16 genera, 30 species, and 42,748 individuals (1982 census), of which more than half (54.9%, 23,465 individuals) were of a single species, *Faramea occidentalis*. The second biggest family in numbers of individuals was Violaceae (42,478 individuals), but this ranking was due entirely to the extreme abundance of *Hybanthus prunifolius*; this family was represented by only 2 species. Nearly one-fourth (15) of the 61 families were represented by one species in the plot.

The number of species per family follows a truncated lognormal reasonably well (fig. 26.1). We used this lognormal to estimate the number of unsampled families of free standing woody plants (i.e., lianas excluded) that contain species with a stem diameter of 1 cm DBH or more on BCI. From the area under the lognormal to the left of the veil line (Preston 1948, 1962), we estimated that 11–13 additional families would be found if the sample size were increased to cover the island. The num-

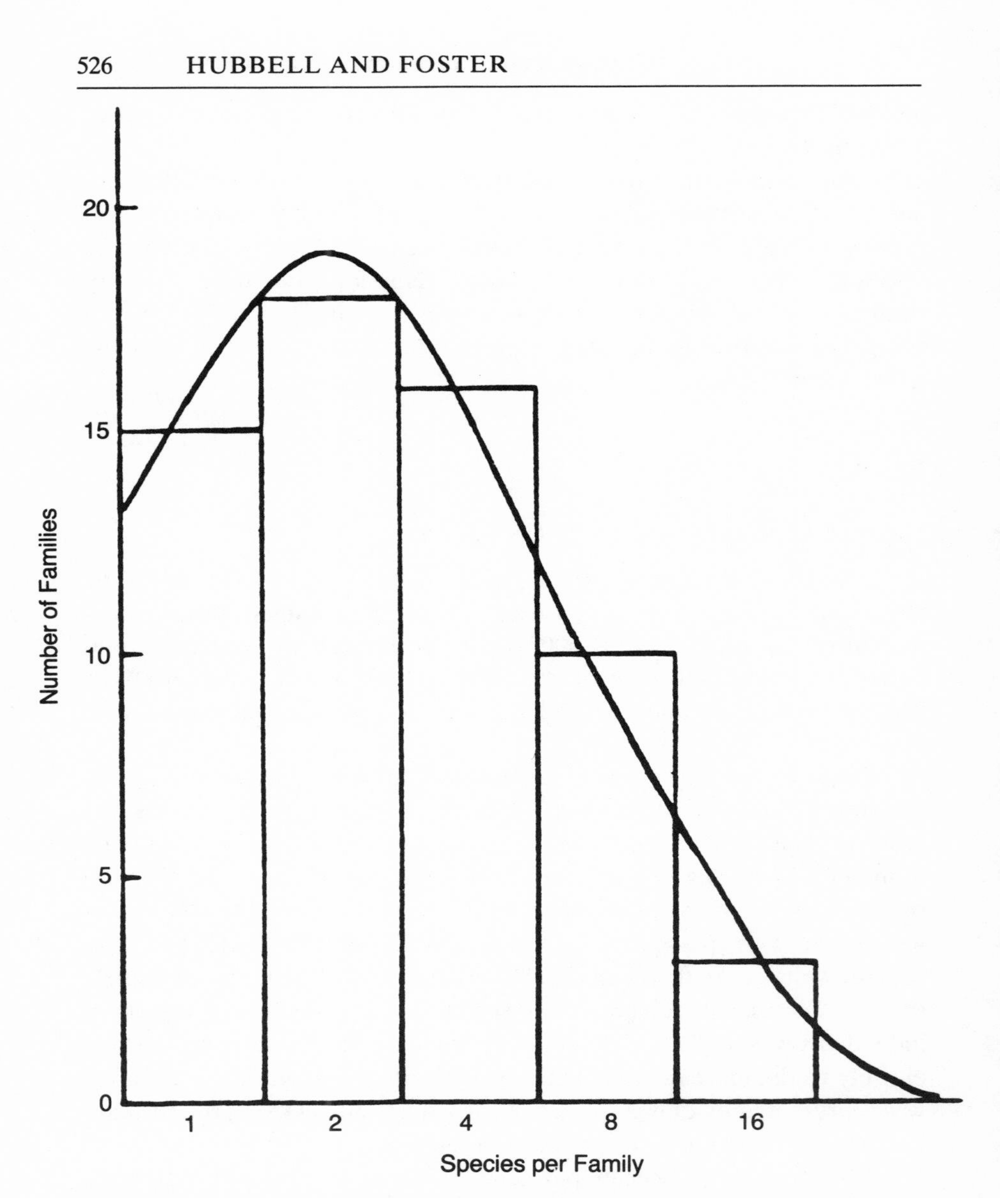

FIGURE 26.1. Truncated lognormal distribution of the number of species per family in the 50-ha plot on BCI. The position of the "veil line" (Preston 1948, 1962) indicates that we might expect to find an additional 11–13 families if the sample size were increased to the size of BCI. The actual number of families that are known from BCI but not found in the 50-ha plot and that have at least one woody species of censusable size is 11 (cf. Croat 1978).

ber of remaining families known to contain trees or shrubs of censusable size on BCI is, in fact, 11 (Croat 1978).

Stand Structure and Dynamics

At the completion of the first census in 1982, there were 235,895 recorded plants with a stem diameter of 1 cm or more alive in the 50-ha plot. Overall stand size structure is shown in fig. 26.2. The mean number of stems per hectare drops rapidly with increasing diameter class, following a remarkably tight log-log relation. The relation is significantly curvilinear and is relatively constant—regardless of differences in species composition—from one hectare to the next (cf. standard deviations among hectares, fig. 26.2). Interestingly, in spite of curvilinearity, the line has an average regression slope of about $-\frac{3}{2}$. The resemblance of this curve to the limiting curve for the "self-thinning" law of monospecific stands is probably not accidental, but exactly why and how it arises in a species-rich mosaic of gap-phase and mature-phase forest is not yet clear.

Based on the mean height of adult plants, we classified species into four arbitrary categories: shrubs and treelets (adults less than 4 m tall), understory trees (adults 4–10 m tall), midstory trees (adults 10–20 m tall), and canopy plus emergent species (adults 20–45 m tall). Of the species in the plot, 58 species were classified as shrubs and treelets, 61 as understory trees, 73 as midstory trees, and 114 as canopy and emergent species. Because of differences in adult plant size, shrubs and treelets could become more numerous than large trees. We therefore expected shrubs and treelets to be overrepresented among the very common species, but this was not the case. The proportions of shrubs and treelets, and of understory, midstory, and canopy trees, among the most common species were the same as the proportions of these growth forms among all species in the plot ($x^2 = 0.689$, 3 df., $p > 0.86$) (cf. Hubbell and Foster 1986c).

By 1985, the number of living plants with stems 1 cm DBH or more had risen to 242,218, a net increase of 2.7%. This small net increase masked a considerable turnover of plants in the plot. There were 20,705 (8.78%) deaths and 33,126 (14.04%) new recruits over the 1982–85 interval. The difference between number of recruits and number of deaths is greater than the net increase of plants 1 cm DBH or more in the plot. This difference was because 6,098 (2.6%) of the 1982 plants, though still alive, suffered a drastic reduction in stem diameter, falling below the 1-cm DBH minimum census diameter in 1985. In most these plants the main

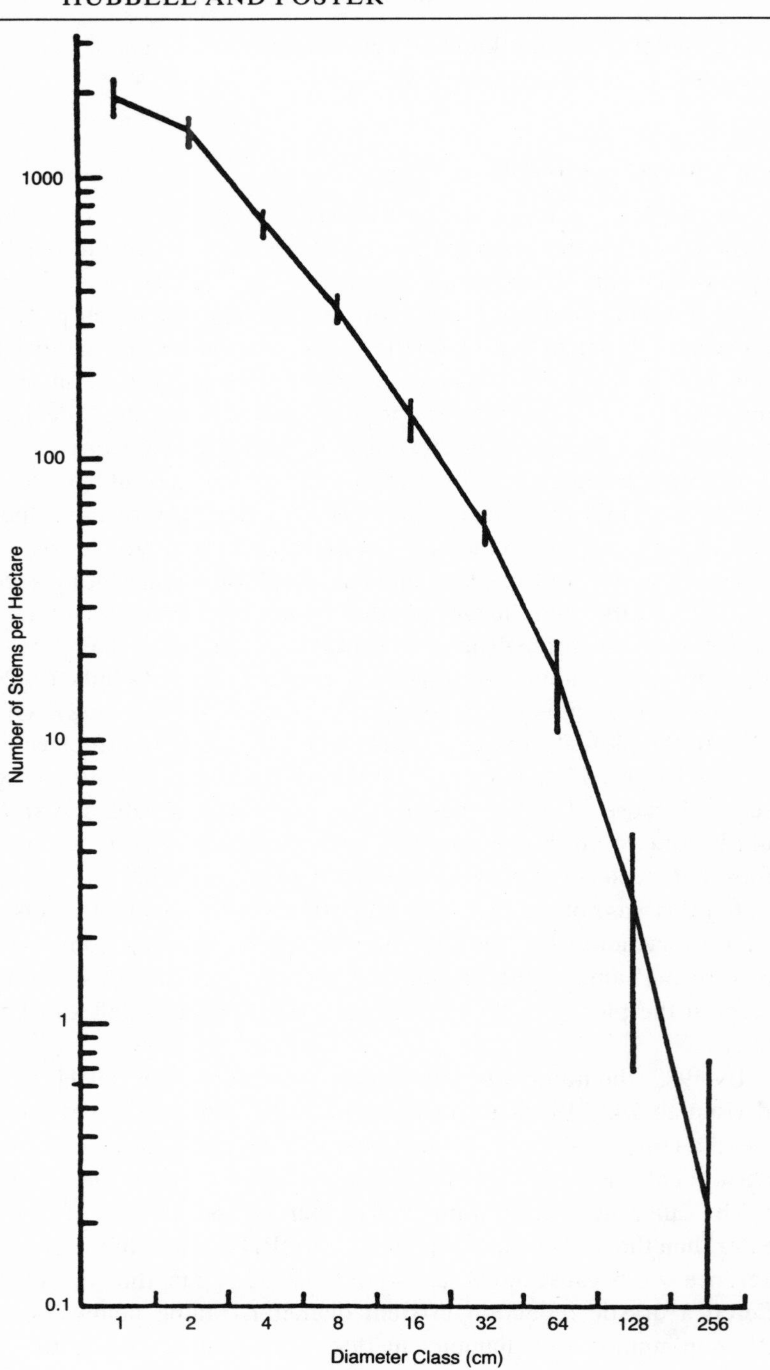
1000
100
10
1
0.1
Number of Stems per Hectare
1
2
4
8
16
32
64
128
256
Diameter Class (cm)

TABLE 26.1 Changes in total number of stems by doubling diameter classes over the recensus interval 1982–85

Diameter class DBH cm	Stems 1982	Deaths 1982–85	Outgrowth[1] 1982–85	Recruits[2] 1985	Ingrowth[3] 1985	Stems[4] 1985	Change (%)
Present[5]	—	—	—	—	6,098	6,098	—
1–1.9	98,909	8,373	17,661	31,577	1,608	106,060	+7.2
2–3.9	74,351	6,680	11,360	1,182	13,874	71,367	−4.0
4–7.9	34,889	3,231	3,442	289	8,325	36,830	+5.6
8–15.9	17,031	1,260	983	70	2,672	17,530	+2.9
16–31.9	7,034	720	457	8	883	6,748	−4.1
32–63.9	3,003	336	73	0	438	3,032	+1.0
64 and up	678	105	0	0	78	651	−4.0
Total	235,895	20,705	33,976	33,126	27,878[6]	242,218[6]	+2.7

Notes: The total for the Stems in 1985 column does not include plants in the Present category because these plants had 1985 diameters less than 1 cm DBH. The same is true for the total for the Ingrowth column.

[1] Includes both decreases and increases of stem diameter class

[2] Plants newly tagged in 1985

[3] Includes Outgrowth minus Present

[4] Computed as follows: Stems in 1982 − Deaths − Outgrowth + Recruits + Ingrowth

[5] Plants in 1982 with stem ≥ 1 cm DBH but which suffered loss of main stem(s) and whose resprouts were < 1 cm DBH in 1985

[6] Does not include Present

stem broke off below 1.3 m (DBH measurement height). Plants that survived the loss of their main stem either coppiced from their roots or produced stem suckers below the break. Most of these cases consisted of plants that were of small diameter in 1982. The 1–2-cm DBH class accounted for 3,989 (65.4%) of all cases of lost main stems. These plants were excluded from the 1985 count of plants over 1 cm DBH.

Changes in the numbers of plants per diameter class over the intercensus interval are shown in table 26.1. The 1–2-cm DBH size class registered the largest change, increasing from 98,909 to 106,060 stems

FIGURE 26.2. Mean per-hectare stand structure of the 50-ha BCI plot, for 9 diameter classes of free-standing woody plants (excluding lianas). Vertical bars are 1 SD about the mean for the given stem-diameter class across hectares. Many lianas are free-standing shrublike plants early in their ontogeny. If these woody liana "shrubs" are included in the count, the number of 1–2-cm DBH count is slightly elevated, and the curve with the lianas longer shows the bend at the 2–4-cm DBH diameter class.

TABLE 26.2 Most common taxa (over 1,000 individuals per species in 1982 or 1985) in the fifty-ha plot on Barro Colorado Island

Species	Family	Growth Form[1]	Abundance in 1982	Abundance in 1985	Change (%)
Acalypha diversifolia	Euphorbiaceae	S	1,582	1,218	−23.0
Alseis blackiana	Rubiaceae	T	7,587	8,029	+5.8
Beilschmiedia pendula	Lauraceae	T	2,387	2,679	+12.2
Capparis frondosa	Capparidaceae	S	3,548	3,681	+3.8
Cordia lasiocalyx	Boraginaceae	M	1,703	1,662	−2.4
Coussarea curvigemmia	Rubiaceae	U	1,494	1,650	+10.4
Cupania sylvatica	Sapindaceae	U	970	1,043	+7.5
Desmopsis panamensis	Annonaceae	U	11,752	12,139	+3.3
Drypetes standleyi	Euphorbiaceae	T	2,174	2,265	+4.2
Eugenia galalonensis	Myrtaceae	U	959	1,145	+19.4
E. oerstedeana	Myrtaceae	M	2,079	2,200	+5.8
Faramea occidentalis	Rubiaceae	U	23,465	25,118	+7.0
Guarea guidonia	Meliaceae	M	1,784	1,826	+2.4
Guarea sp. nov.	Meliaceae	M	1,558	1,491	−4.3
Guatteria dumetorum	Annonaceae	T	1,576	1,516	−3.8
Hasseltia floribunda	Flacourtiaceae	M	1,167	1,033	−11.5
Hirtella triandra	Chrysobalanaceae	M	4,142	4,649	+12.2
Hybanthus prunifolius	Violaceae	S	39,911	41,091	+3.0
Lacistema aggregatum	Lacistemaceae	U	1,551	1,656	+6.8
Maquira costaricana	Moraceae	M	1,409	1,435	+1.9
Mouriri myrtilloides	Melastomataceae	S	6,984	7,748	+10.9
Ocotea skutchii	Lauraceae	T	1,149	962	−16.3
Oenocarpus mapora[2]	Palmae	M	1,807	1,730	−4.3
Ouratea lucens	Ochnaceae	S	1,125	1,244	+10.6
Picramnia latifolia	Simaroubaceae	U	1,170	1,175	+0.4
Piper cordulatum	Piperaceae	S	3,160	3,723	+17.8
Poulsenia armata	Moraceae	T	3,437	2,680	−22.0
Pouteria unilocularis	Sapotaceae	T	1,645	1,706	+3.7
Prioria copaifera	Legum.: Caesalp.	T	1,357	1,402	+3.3
Protium panamense	Burseraceae	M	2,713	2,838	+4.6
P. tenuifolium	Burseraceae	M	2,671	2,919	+9.3
Psychotria horizontalis	Rubiaceae	S	6,175	6,450	+4.5
Pterocarpus rohrii	Legum.: Papil.	T	1,569	1,602	+2.1
Quararibea asterolepis	Bombacaceae	T	2,402	2,379	−1.0
Randia armata	Rubiaceae	U	1,126	1,151	+2.2
Rheedia edulis	Guttiferae	M	3,654	4,026	+10.2
Rinorea sylvatica	Violaceae	S	2,567	2,604	+1.4
Simarouba amara	Simaroubaceae	T	1,241	1,249	+0.6
Sorocea affinis	Moraceae	S	3,302	3,364	+1.9
Swartzia simplex v. grandiflora	Legum.: Caesalp.	U	2,255	2,419	+7.3
Swartzia simplex v. ochnacea	Legum.: Caesalp.	U	2,670	2,786	+4.3
Tabernaemontana arborea	Apocynaceae	T	1,280	1,319	+3.1
Tachigalia versicolor	Legum.: Caesalp.	T	2,922	2,969	+1.6

Tetragastris panamensis	Burseraceae	T	3,233	3,680	+13.8
Trichilia tuberculata	Meliaceae	T	12,942	13,158	+1.7
Virola sebifera	Myristicaceae	M	2.404	2,271	−5.5

Notes: Together these 46 species constitute 4/5 (80.4%) of all stems > 1 cm DBH, but only 15.4% of all species, in the BCI plot
[1] S (shrub, adults < 4 m tall); U (understory treelet, adults 4–10 m tall); M (midstory tree, adults 10–20 m tall); T (canopy plus emergents, adults > 20 m tall)
[2] Formerly Oenocarpus panamanus

(+7.2%). This was partly the result of increasing numbers of treefall gaps in the plot, reflected in the 15.4% decline in large trees (105 deaths from among 678 trees with stem dismeters of 64 cm DBH or more).

Species Abundance Patterns

Tree species abundances spanned 4.6 orders of magnitude. Twenty-one species were represented by single plants. At the other extreme was the shrub *Hybanthus prunifolius* (Violaceae), mentioned earlier, which had 42,146 individuals in the 1985 recensus and represented nearly 1 in every 6 plants (16.95%). The number of species classified by abundance to orders of magnitude were: 1–9 individuals (60 species in 1982; 62 species in 1985); 10–99 (86 in 1982; 91 in 1985); 100–999 (112 in 1982; 103 in 1985); 1,000–9,999 (40 in 1982; 43 in 1985) and 10,000+ (4 species in both 1982 and 1985). Two species became locally extinct (at least in the size classes of 1 cm DBH or more) over the intercensus period, and 3 new species, all rare, were added in 1985. Fifteen percent (46) of the species comprised 80% of all stems in the plot (table 26.2).

The changes in abundance 1982–85 were great in many species. Given that only three years elapsed between censuses, we considered a 10% change in abundance to be relatively great. By this criterion, 40% (122) of all species exhibited great changes in abundance, most of them negative. Some species were particularly hard hit by the El Niño drought and declined precipitously. Most adversely affected were moisture-loving species found in the wetter sites in the plot. Two dramatic examples among common species were *Acalypha diversifolia* (Euphorbiaceae), a shrub species found commonly along the streamsides in the plot, and *Poulsenia armata* (Moraceae), a tree species found mainly on the moist, steeper slopes (table 26.2). These species suffered 22% and 23% declines, respectively.

Five-eighths of all species (191, 62.4%) showed decreases or no change in abundance. Rare species suffered a disproportionately large number of the losses, and common species displayed a disproportionate

TABLE 26.3 Direction of change in species abundance 1982–1985 in relation to initial species abundance

1982 abundance class	Increase in 1985	Decrease or no change in 1985	Species with > 10% change (%)
1–9	13 (23.8)	50 (39.2)	46.0
10–99	25 (32.0)	60 (53.0)	45.9
100–999	43 (42.6)	70 (70.4)	38.1
1,000–9,999	30 (15.1)	10 (24.9)	27.5
10,000–99,999	4 (1.5)	0 (3.5)	0.0
TOTALS	115	190	

Note: Expected numbers of increasing and decreasing species are shown in parentheses under the null hypothesis of independence of the direction of population change and initial population abundance. The χ^2 value is 45.6, which is significant at $p < .0001$ (4 df). Rare species are much more likely to decrease, and common species to increase, than expected by chance.

number of the increases (table 26.3). Three-fourths (50, 79.4%) of the rare species with fewer than 10 individuals in the plot declined or remained unchanged. In contrast, three-fourths (34, 77.3%) of the very common species, with 1,000 or more individuals, increased in abundance. The nonindependence of population trajectory and initial species abundance is highly significant ($p < 0.0001$).

Mortality

The mortality of all plants in the BCI plot over the three years was high (8.8%), and spot field checks in 1983 indicated that much of this mortality occurred during the year of El Niño. Overall rates of mortality were not constant across diameter classes. In the larger diameter classes, mortality rates increased with tree size above the 8–16-cm diameter class (fig. 26.3): plants in the understory survived the year of El Niño better than did the overstory trees. Of trees over 64 cm DBH, 15.5% died during the intercensus interval.

Mortality patterns among individual tree species differed markedly. Two common patterns of mortality were apparent, although intermediates were present. A large group of abundant, shade-tolerant species exhibited high and virtually constant survival over all juvenile and young adult size-classes in the census, from 1-cm DBH saplings up to mid-adult sizes. Examples of such species are *Alseis blackiana* (Rubiaceae), *Drypetes standleyi* (Euphorbiaceae), *Prioria copaifera* (Leguminosae), and

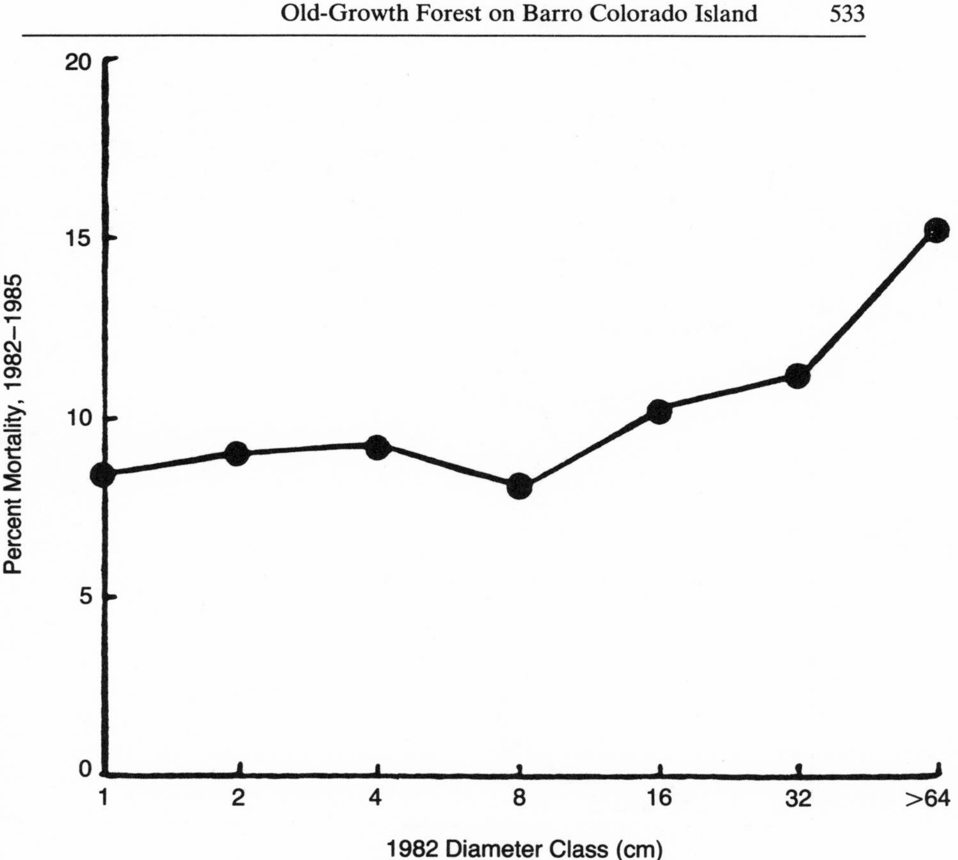

FIGURE 26.3. Percent mortality over the intercensus interval 1982–1985 as a function of 1982 diameter class. Percent mortality is nearly constant at about 8% up to the 8–16-cm DBH diameter class. However, percent mortality rises steadily in the higher size classes, reaching over 15% among trees above 64 cm DBH. Much of this mortality was concentrated in the El Niño dry season of 1983.

Quararibea asterolepis (Bombacaceae) (fig. 26.4). In the largest size classes (32 cm DBH or more), survival rate decreased in some species (e.g. the two most common canopy species, *Trichilia* and *Alseis*) but not in others (e.g., *Prioria*). In contrast, shade-intolerant gap species, such as *Cecropia insignis* (Moraceae), *Spondias mombin* (Anacardiaceae) and *Luehea seemannii* (Tiliaceae), exhibited decreasing mortality rates with increasing diameter class. Adult survival in these gap species approached the high survival rates seen in the shade-tolerant species, presumably as the larger trees entered the high-light environment of the canopy. Species

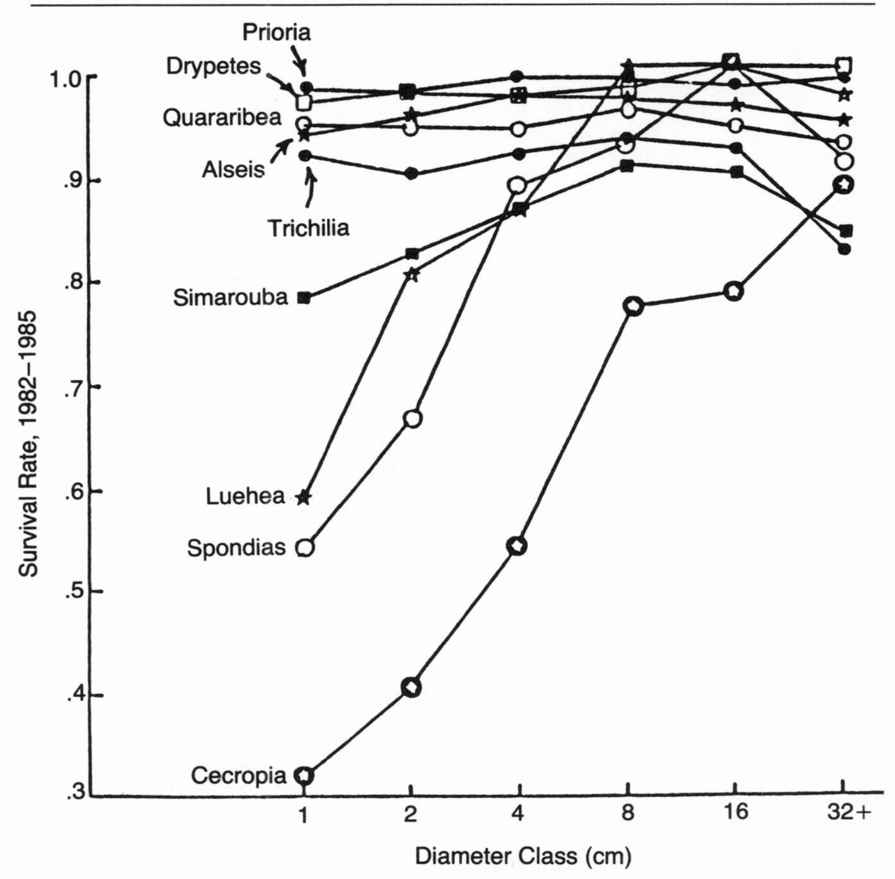

FIGURE 26.4. Fraction of plants surviving over the intercensus interval 1982–1985 as a function of diameter class for sample tree species in the BCl plot. Many shade-tolerant species showed high and relatively constant rates of survival at all size classes except the largest. Examples of such species are: *Prioria, Drypetes*, *Quararibea*, *Alseis* and *Trichilia*. In contrast, shade-intolerant species such as *Luehea*, *Spondias*, and *Cecropia* all show increasing survival with increasing diameter, approaching the high survival rates of the shade-tolerant species once the shade-intolerant species reach the high-light environment of the canopy.

TABLE 26.4 Per capita recruitment as a function of 1982 abundance class

1982 abundance class	Median per capita recruitment, 1982–1985 (%)	Species with no recruitment	
		N	(%)
1–9	0.0	41	65.1
10–99	8.3	18	21.2
100–999	11.4	1	0.9
1,000–9,999	11.7	0	0.0
10,000–99,999	12.6	0	0.0

with intermediate patterns, such as *Simarouba amara* (Simaroubaceae), exhibited a more gradual increase in survival with diameter. These results do not conflict with the overall increase in mortality in large trees because gap species constituted only a small fraction of total trees.

Recruitment

Almost all (94.4%, 31,577) new recruitment occurred into the 1–2-cm DBH diameter class. A few individuals also recruited into higher diameter classes (table 26.1). Some of these large-diameter recruits were individuals of very fast-growing species and were not present in 1982. However, an unknown but probably large fraction of the 367 "recruits" above 4 cm DBH were probably present but missed in 1982. Spot checks of some subquadrats after the 1982 census indicated that fewer than 0.2% of plants 1 cm DBH or more were missed.

The decline in abundance of many rare species was due not only to higher mortality rates in these species but also to poorer per capita recruitment (table 26.4). Per capita recruitment rates for each species were calculated as the ratio of new recruits to population abundance in 1982. Sixty-five percent (41) of the species in the 1–9 abundance category failed to recruit, whereas all species with 1982 abundances of 1,000 or more recruited. The median per capita recruitment increased monotonically with population size, from 0% in species with fewer than 10 individuals in 1982 to about 12% in species with more than 100 individuals in 1982 (table 26.4).

Conspecific Effects on Growth and Survival

The recensus results showed that, for many species in the BCI plot, growth and survival of saplings was worse if the nearest large tree was of

TABLE 26.5 Proportion of 1–3.9-cm DBH saplings of BCI overstory tree species that survived beneath a conspecific or a heterospecific large tree (≥ 20 cm DBH) over the 1982–1985 interval

	Survival rate beneath:			
Dataset	*Conspecific*		*Heterospecific*	Two-Sided P-value
All species	0.857 (1040)	<	0.905 (32,254)	.0001
Common species	0.861 (985)	<	0.919 (21,212)	.0001
Rare species	0.782 (55)	<	0.878 (11,042)	.037

Note: Common species included 11 species for which at least 10 saplings were present in the conspecific category for each species. Rare species were all other canopy species. Sample sizes are shown in brackets. A two-sided test was applied because the results could have gone either way. After Hubbell and Foster (1988a).

TABLE 26.6 Mean annual growth rate (cm DBH/yr) of 1–3.9 cm DBH saplings of BCI overstory tree species that were beneath a conspecific or a heterospecific large tree (> 20 cm DBH), over the 1982–1985 interval

	Growth rate beneath:			
Dataset	Conspecific		Heterospecific	2-Sided P-value
All species	0.065 (839)	<	0.085 (27834)	.0003
Common species	0.064 (798)	<	0.076 (18636)	.0327
Rare species	0.080 (41)	<	0.103 (9198)	.3608

Note: Common species included 11 species for which at least 10 saplings were present in the conspecific category for each species. Rare species were all other canopy species. Sample sizes are shown in brackets. A two-sided test was applied because the results could have gone either way (cf. Hubbell and Foster 1988a).

the same species than if it was of a different species (Hubbell and Foster, in press-a,b). The negative effects of being near a conspecific could be detected in most common species tested individually and in pooled common and pooled rare species, where the test was "next to conspecific" versus "next to heterospecific" large tree (tables 26.5, 26.6). In 1–4-cm DBH saplings of canopy tree species, three-year survival was depressed next to a conspecific by a mean of 5.8% in a set of eleven common species and by 9.6% in the remaining "rare" species (table 26.5). Growth rates of 1–4-cm DBH saplings were depressed by 17.1% in these same common species and by 22.2% in the remaining species (table 26.6), although the latter reduction was not significant, probably because of small sample size and the short intercensus interval. The negative conspecific effects were not just detectable in the small saplings but persisted into size classes of much larger subadult trees (8–16 cm DBH) in many species. Quantitatively similar effects occurred in midstory and understory trees as well (Hubbell and Foster, in press-a,b).

Tests ruled out four alternative hypotheses that might have explained these conspecific effects. If, for example, conspecific adults were closer to the tested saplings or larger than the adults in the heterospecific tests, then the effects might simply be due to competition with larger or closer trees. However, conspecific adults were not larger or closer to the test saplings than heterospecific trees. Conspecific test sites also had neither deeper mean canopy shade levels nor higher stand densities (cf. Hubbell and Foster, in press-a,b).

Discussion

With the 1985 census results in hand, it is clear that negative conspecific effects on sapling growth and survival are more pervasive throughout the BCI tree community than primary census results indicated. Whether these effects are quantitatively significant in maintaining the observed tree species richness of the old-growth forest on BCI remains a question, especially in light of the strong evidence that the BCI forest is not—or at least not yet—in species equilibrium. As a group rare species are declining and common species increasing in abundance.

The results suggest that the forest is successional, although in a very late stage, perhaps approaching an equilibrium for the currently most abundant species in the plot. If this interpretation is correct, what disturbance is forest recovering from? The adjacent half of the island was largely deforested in the last century, and so it remained until BCI became an island biological reserve. We hypothesize that many of the rarest species in the BCI plot are actually weedy or secondary forest shrubs and trees that invaded the older forest at that time. These species are not self-maintaining in the old forest, and their populations were once sustained at low to moderate levels of abundance by a continual subsidy of immigrants from neighboring deforested areas. Now that these source areas have regrown and are covered with seventy-year-old secondary forest, however, the immigration subsidy has been cut off. Populations of many of these rare species may therefore be heading toward local extinction. The discovery of a strong correlation between species rarity and degree of heliophily for species in the plot reinforces this hypothesis (see Hubbell and Foster 1986a, 1987a).

It is also possible that disturbances by pre-Columbians may still have some influence. The rare giant emergent trees of the BCI forest, for example—such species as *Ceiba pentandra* (Bombacaceae), *Cavanillesia plantanifolia* (Bombacaceae), and *Hura crepitans* (Euphorbiaceae)—could be old enough to have survived for five hundred years. These heliophilic but long-lived species require large gaps to regenerate, but few large gaps are being created in the BCI forest today.

Without doubt the most exciting discovery of the 1985 census was the pervasiveness of negative conspecific effects on sapling growth and survival. From the original census data, static evidence for density-dependence in canopy trees was found in the two most common species, *Alseis* and *Trichilia* (Hubbell and Foster 1986b, 1987b). The recensus results confirmed the negative density-dependence in these two species but also showed that similar effects occur in many, though not all, tree species in the plot. So far we have only demonstrated the existence of the effects phenomenologically; we have not yet identified causal factors. Negative density-dependent effects due to seed and seedling predation have been proposed before as limiting factors for tropical trees (Janzen 1970; Connell 1971), but we are obviously not dealing with seed or seedling predation here, since the smallest plants in the census were 1 cm DBH and averaged more than 3 m in height. Our current suspicion is that herbivores and pathogens are mainly responsible, but further investigations are needed to test these possibilities, and the factors may be diverse.

The importance of these density-dependent effects in maintaining tree species richness in the BCI forest is unclear at present. One reason to question their importance is that the effects on sapling growth and survival are highly local in nature and can be detected only a few meters (5–8 m) from the trunk of the conspecific tree, no further than the edge of the tree crown (Hubbell and Foster, in press-b). It is thus uncertain whether the effects are strong enough to prevent a grove of a single-species competitive dominant from developing. Nevertheless, we are able to detect conspecific growth rate suppression of 10–20% and lowered survival of 5–10% in large juvenile trees, in some species as big as the 8–16-cm DBH class. Thus, these effects can persist over a substantial fraction of the immature life history of some species. Given that many BCI trees spend several decades in their juvenile stages, the accumulated differences in growth and survival among sapling cohorts growing next to conspecific versus heterospecific trees could become large.

Although we do not discuss them in this chapter, the 1985 census results also supported the conclusion from the primary census that many of the more common tree species of BCI are generalists that broadly overlap in their habitat and regeneration niche requirements (Hubbell and Foster 1986c, in press-a). We have noted here, for example, that many shade-tolerant species have high and nearly identical survival rates over all juvenile diameter classes above 1 cm DBH. Although these common species could probably persist in drifting relative abundance as nearly equivalent competitors for long periods (Hubbell and Foster 1986b), their continued coexistence in close sympatry can only be enhanced by negative density dependence, even by weak or by the very local effects reported here. Recent theoretical investigations using a modified version of the Shmida and Ellner model (1984) have suggested that spatial density and frequency dependence need not be strong to have an important effect on the maintenance of tropical tree diversity (Thrall and Pacala, p.c.). We can look forward to a very exciting decade of research to test these and other ideas for the maintenance of tree species richness in tropical rainforests.

REFERENCES

Augspurger, C. K. 1979. Irregular rain cues and the germination and seedling survival of a Panamanian shrub (*Hybanthus prunifolius*). Oecologia 44: 53–59.

———. 1983a. Offspring recruitment around tropical trees: changes in cohort distance with time. Oikos 20: 189–196.

———. 1983b. Seed dispersal of the tropical tree *Platypodium elegans*, and the escape of its seedlings from fungal pathogens. J. Ecol. 71: 759–771.

Bennett, C. F. 1968. A phytophysiognomic reconnaissance of Barro Colorado Island, Canal Zone. Smithsonian Inst. Misc. Coll. 145(7): 1–8.

Brokaw, N. V. L. 1982. Treefalls: frequency, timing, and consequences. In E. G. Leigh, Jr., A. S. Rand, and D. M. Windsor (eds.), *The Ecology of a Tropical Forest: Seasonal Rhythms and Long-Term Changes*. Smithsonian Inst. Press, Washington, D.C., pp. 101–108.

———. 1985a. Gap-phase regeneration in a Neotropical forest. Ecol. 66: 682–687.

———. 1985b. Treefalls, regrowth, and community structure in tropical forests. In S. T. A. Pickett and P. S. White (eds.), *The Ecology of Natural Disturbance and Patch Dynamics*. Academic Press, New York, pp. 53–69.

Connell, J. H. 1971. On the role of natural enemies in preventing competitive exclusion in some marine animals and in rain forest trees. In P. J. den Boer and G. R. Gradwell (eds.), *Dynamics of Numbers in Populations*, pp. 298–312. Proc. Advanced Study Inst., Osterbeek Wageningen, Neth. Centre Agric. Publ. Doc.

———. 1978. Diversity in tropical rain forests and coral reefs. Science 199: 1302–1310.

———. 1979. Tropical rain forests and coral reefs as open non-equilibrium systems. In R. M. Anderson, B. D. Turner, and L. R. Taylor (eds.), *Population Dynamics*, pp. 141–163. Blackwell, Oxford.

Croat, T. B. 1978. *The Flora of Barro Colorado Island*. Stanford Univ. Press, Stanford.

Dietrich, W. E., D. M. Windsor, and T. Dunne. 1982. Geology, climate, and hydrology of Barro Colorado Island. In E. G. Leigh, Jr., A. S. Rand, and D. M. Windsor (eds.), *The Ecology of a Tropical Forest: Seasonal Rhythms and Long-Term Changes*. Smithsonian Inst. Press, Washington, D.C., pp. 21–46.

Foster, R. B. and N. V. L. Brokaw. 1982. Structure and history of the vegetation of Barro Colorado Island. In E. G. Leigh, Jr., A. S. Rand, and D. M. Windsor (eds.), *The Ecology of a Tropical Forest: Seasonal Rhythms and Long-Term Changes*. Smithsonian Inst. Press, Washington, D.C., pp. 67–81.

Hamill, D. N. 1987. Selected studies on dispersion, distribution, and species diversity. Ph.D diss., Univ. of Iowa.

Howe, H. F. 1980. Monkey dispersal and waste of a Neotropical fruit. Ecol. 61: 944–959.

———. 1983. Annual variation in a Neotropical seed dispersal system. In S. J. Sutton, T. C. Whitmore, and A. C. Chadwick (eds.), *Tropical Rain Forest: Ecology and Management*. Brit. Ecol. Soc. Spec. Publ. 2, Blackwell, Oxford, pp. 211–227.

Howe, H. F. and G. A. Van de Kerckhove. 1979. Fecundity and seed dispersal of a tropical tree. Ecol. 60: 180–189.

Howe, H. F., E. W. Schupp, and L. Westley. 1985. Early consequences of seed dispersal for a Neotropical tree (*Virola surinamensis*). Ecol. 66: 781–791.

Hubbell, S. P. 1979. Tree dispersion, abundance and diversity in a tropical dry forest. Science 203: 1299–1309.

———. 1984. Methodologies for the study of the origin and maintenance of tree diversity in tropical rain forest. Biol. Int. (IUBS) 6: 8–13.

Hubbell, S. P. and R. B. Foster 1983. Diversity of canopy trees in a Neotropical forest and implications for the conservation of tropical trees. In S. J. Sutton, T. C. Whitmore, and A. C. Chadwick (eds.), *Tropical Rain Forest: Ecology and Management*. 41. Brit. Ecol. Soc. Spec. Publ. 2, Blackwell, Oxford, pp. 25–41.

———. 1986a. Canopy gaps and the dynamics of tropical rain forest. In M. J. Crawley (ed.), *Plant Ecology*. Blackwell, Oxford, pp. 75–95.

———. 1986b. Biology, chance and history and the structure of tropical rain forest tree communities. In J. M. Diamond and T. J. Case (eds.), *Community Ecology*. Harper and Row, New York, pp. 314–319.

———. 1986c. Commonness and rarity in a Neotropical forest: implications for tropical tree conservation. In M. E. Soulé (ed.), *Conservation Biology: The Science of Scarcity and Diversity*. Sinauer, Sunderland, Mass., pp. 205–231.

———. 1987a. The spatial context of regeneration in a Neotropical forest. In M. J. Crawley, A. Gray, and P. J. Edwards (eds.), *Colonization, Succession and Stability*. Blackwell, Oxford, pp. 395–412.

———. 1987b. La estructura en gran escala de un bosque neotropical. Rev. Biol. Trop. 35(suppl. 1):7–22.

———. In press-a. The fate of juvenile trees in a Neotropical forest: implications for the natural maintenance of tropical tree diversity. In K. S. Bawa (ed.), *Reproductive Biology of Tropical Plants*. UNESCO/MAB Publ.

———. In press-b. *Structure and Dynamics of a Neotropical Forest*. Princeton Univ. Press, in prep.

Janzen, D. H. 1970. Herbivores and the number of tree species in tropical forests. Amer. Nat. 104: 501–528.

Knight, D. H. 1975. A phytosociological analysis of species-rich tropical forest on Barro Colorado Island, Panama. Ecol. Monogr. 45: 259–284.

Lang, G. E. and D. H. Knight. 1983. Tree growth, mortality, recruitment and canopy gap formation during a 10-year period in a tropical forest. Ecol. 64: 1075–1080.

Piperno, D. 1989. Phytoliths, archaeology, and prehistoric changes in the vegetation of a fifty-hectare plot on Barro Colorado Island. In E. G. Leigh, Jr. (ed.), *Ecologia de un bosque tropical*. Smithsonian Tropical Research Inst., Balboa, Panama.

Preston, F. W. 1948. The commonness, and rarity, of species. Ecol. 29: 254–283.

———. 1962. The canonical distribution of commonness and rarity. Ecol. 43: 185–215, 410–432.

Putz, F. E. and K. Milton. 1982. Tree mortality rates on Barro Colorado Island. In E. G. Leigh, Jr., A. S. Rand, and D. M. Windsor (eds.), *The Ecology of a Tropical Forest: Seasonal Rhythms and Long-Term Changes*. Smithsonian Inst. Press, Washington, D.C., pp. 95–100.

Rand, A. S. and W. M. Rand. 1982. Variation in rainfall on Barro Colorado Island. In E. G. Leigh, Jr., A. S. Rand, and D. M. Windsor (eds.), *The Ecology of a Tropical Forest: Seasonal Rhythms and Long-Term Changes*. Smithsonian Inst. Press, Washington, D.C., pp. 47–59.

Schupp, E. 1987. Studies on seed predation of *Faramea occidentalis*, an abundant tropical tree. Ph.D. diss., Univ. of Iowa.

Shmida, S. and S. P. Ellner. 1984. Coexistence of plant species with similar niches. Vegetatio 58: 29–55.

Standley, P. 1927. The flora of Barro Colorado Island. Smithsonian Inst. Misc. Coll. 78(8): 1–32.

Thorington, R. W., Jr., B. Tannenbaum, A. Tarak, and R. Rudran. 1982. Distribution of trees on Barro Colorado Island: a five-hectare sample. In E. G. Leigh, Jr., A. S. Rand, and D. M. Windsor (eds.), *The Ecology of a Tropical Forest: Seasonal Rhythms and Long-Term Changes*. Smithsonian Inst. Press, Washington, D.C., pp. 83–94.

27

Composition and Dynamics of the Cocha Cashu "Mature" Floodplain Forest

ALWYN H. GENTRY AND JOHN TERBORGH

Cocha Cashu field station in Manu National Park, Peru, has become world famous as one of the best examples of pristine forest in upper Amazonia. The density and diversity of large birds and mammals at Cocha Cashu is unusually high in comparison with other parts of Peruvian Amazonia and is thought to reflect the state of much of Amazonia before the advent of exploitative over-hunting that accompanied expanding European settlement. Game animals were over-hunted elsewhere, and only in remote sites like Manu has the fauna remained intact. Manu is thus often taken as a paradigm for what Amazonia "should" be like. The rich fauna has attracted much recent attention and important studies of the Manu fauna have been undertaken (e.g., Terborgh 1983; Emmons 1987).

Alternatively, the unusually rich alluvial soil at Cocha Cashu may make this forest intrinsically different from most other Amazonian forests (Gentry 1985, 1986; Foster, this volume). For example, rich soils may be correlated with an unusually high representation of high-quality mammal-dispersed fruits. Perhaps the uncommonly rich fauna owes its existence more to floristic peculiarities associated with the nutrient-rich soil and probable high productivity than to lack of hunting. If so, the intrinsic interest of Manu as a unique preserve of biotic diversity would be no less but the general applicability to Amazonia of results gained from studies at Cocha Cashu might have to be re-examined.

This study is a somewhat peripheral outgrowth of studies of the flora (AG) and fauna (JT) of Amazonian Peru and funded by a series of grants from the National Science Foundation (JT, AG), National Geographic Society (AG), New York Zoological Society (JT), and USAID (AG). We thank the Peruvian Ministry of Agriculture for authorization to conduct research in Manu National Park.

Until recently, data on Amazonian vegetation and its floristic composition have been generally lacking. Though studies of the floras and vegetation dynamics of a number of Central American and Southeast Asian forests (e.g., Nicholson 1965; Wyatt-Smith 1966; Hartshorn, 1975, 1978; Knight 1975; Putz and Milton 1982; Lang and Knight 1983; Lieberman et al. 1985a,b) have accumulated, relatively little has been published on Amazonian forests (but see Uhl 1982; Uhl et al. 1988). The only Amazonian areas where anything more than rudimentary checklists or forestry inventories have been carried out are the San Carlos area of Venezuelan Amazonia (e.g., Jordan 1982; Klinge and Herrera 1983) and the Manaus area of Brazil (e.g., Brunig and Klinge 1976; St. John, p.c.). Both have much poorer soils than and a very different flora from Cocha Cashu.

We report here the first long-term data on forest dynamics for any site with nutrient-rich soil in Amazonia. We also provide one of the most comprehensive analyses of the structure and floristic composition of such a forest that has been published to date (see also Pires and Prance 1977; Campbell et al. 1986; Balslev et al. 1987). In spite of some unusual features, we find the Cocha Cashu forest to be unexceptional as compared to other lowland tropical forests. Although the flora is quite different from that of nearby areas of Amazonia with poorer soils, the turnover rate (and presumably productivity) of the Cocha Cashu forest is very similar to that of other lowland tropical moist forests for which comparable data are available.

Study Site and Methods

The study area is in what appears to be mature phase forest of the alluvial floodplain of the Rio Manu at the Cocha Cashu Biological Station. It is on well-drained soil along Trail 3 just north of the laboratory clearing (fig. 27.1). This part of the forest is physiognomically spectacular, approaching the classic (but rarely encountered) "cathedral-like" rain forest. The understory is open, and the high, 25–30-m tall canopy is frequently broken by emergents that reach 50 m or more in height. This forest and the Cocha Cashu Biological Station are described in detail by Terborgh (1983, this volume), and its floristic composition is discussed by Gentry (1985) and Foster (this volume).

Our study area consists of two distinct, though largely overlapping, plots. The first plot system was laid out by JT in 1974–75. It consists of two subplots, one of which was laid out and measured in 1974 and the

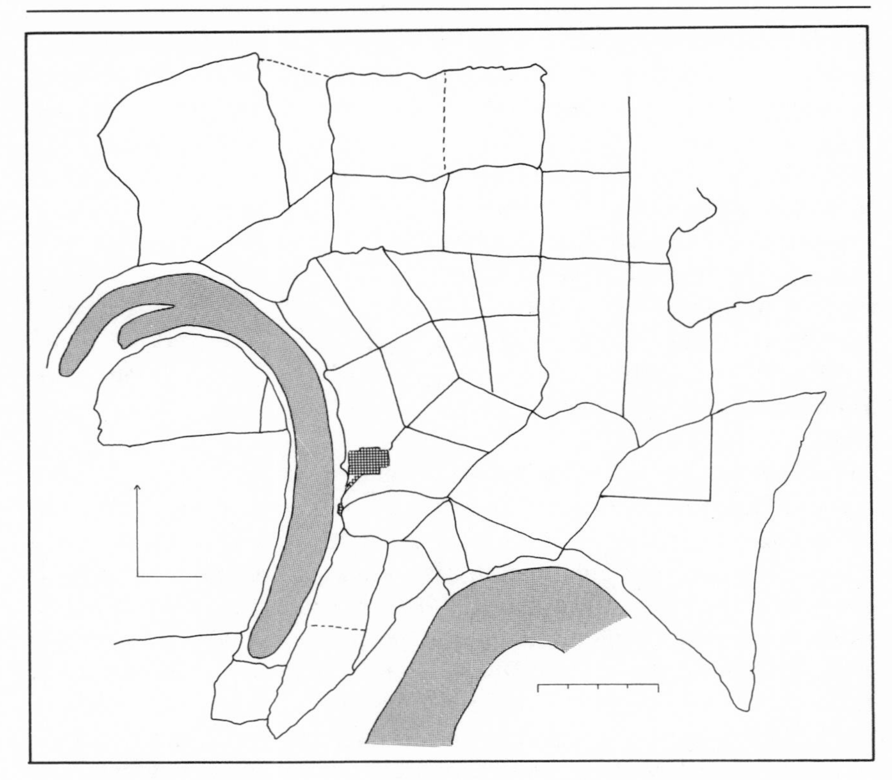

FIGURE 27.1. Cocha Cashu study area. Cross-hatched area = 1-ha tree plot; small dark area = laboratory.

second of which was similarly set up in 1975. The first subplot, 150 m × 30 m (5,230 m^2), is a narrow rectangle that parallels and is longitudinally bisected by Trail 3. The second subplot, 60 m × 75 m (4,500 m^2), is a broader rectangle adjacent to and slightly overlapping the first subplot. The overlapping corner consists of 300 m^2, so the total sample area is 9,430 m^2, or 0.943 ha.

In both subplots all trees 10 cm DBH (diameter at breast height) or more were permanently marked with numbered aluminum tags attached to the trunk by aluminum nails placed approximately at breast height (1.37 m above the ground). When the plot was set up, the position of each tree in it was mapped and the circumference of each dicot tree (but not palms and lianas) was measured at the level of the nail attaching its numbered plaque (i.e., at DBH). In addition to measuring its girth, the crown diameter and height of each tree was estimated, and the numbers of woody vines and termite nests it supported and of superimposed crowns shading it were recorded. Palms were similarly recorded, but their initial diameter was not measured; lianas were not measured or individually censused. The trees were recensused in 1977 and again in 1984 (first subplot) and 1985 (second subplot). At each recensus all dead trees that could be located were recorded as to whether they had fallen or had died standing. A careful search was also made for any new trees that had attained 10 cm DBH since the previous census in the plot. These recruits were tagged, recorded, measured, and mapped. Except for palms, the trees were not identified.

The second plot system was laid out by AG in August 1983, largely coincident with the first plot system. As originally measured, it consisted of a square hectare subdivided into twenty 10 m × 50 m subplots. Because the original plot had an irregular outline, the square hectare plot did not include its extremes. Therefore, the three northernmost subplots of the square hectare were relocated so as to overlap more fully with the original plot system (fig. 27.1). In the hectare plot all trees 10 cm DBH or more and all lianas 10 cm maximum diameter or more were measured and permanently marked with numbered aluminum tags. Most trees thus ended up with two tags, one from each survey. Tags were available only with numbers from 0 to 600; the last 71 trees (numbers 601–71) were tagged with an extra set of 300-series tags marked with an asterisk.

Each numbered plant in the 1-ha plot was positively identified or vouchered with a herbarium specimen, except tree number 370, a mimosoid legume possibly referrable to *Enterolobium*, from which it proved impossible to obtain a specimen. Voucher specimens were obtained either from the ground by using a telescoping aluminum clipper pole with six 2-m-long segments or by climbing either the tree to be vouchered or an adjacent one with a tree-climbing bicycle and then using the aluminum clipper pole to obtain a specimen. Duplicate specimens are deposited in the herbaria of the Missouri Botanical Garden, the Universidad Nacional de Amazonía Peruana in Iquitos, and the Universidad Nacional de San Marcos in Lima; another set of vouchers was distributed to taxonomic specialists and/or deposited in the Field Museum herbarium.

Complete identifications are thus available for the complete 1-ha plot (and for that part of the 0.943-ha one that it overlaps). A few additional identifications of trees in the nonoverlapping portion of the 0.943-ha plot were also made, resulting in a data set of 432 identified trees for which

1974–75 measurements, and thus the diameter increase over the ten-year period, are available. Few trees that died before 1983 were identified, and our data on tree death can thus not be categorized by species.

Except for the data on growth rate and turnover, our discussion of forest structure and composition is based on the 1-ha plot and on a supplemental survey of all plants 2.5 cm DBH or more in a 0.1-ha subsample area, using the technique of Gentry (1982, 1988b). The later data have been reported in more detail elsewhere (Gentry 1985).

Results

Structure and Physiognomy

More large trees grow in the Cocha Cashu hectare than in any of the nine other Amazonian tree plots for which comparable data are available (fig. 27.2; table 27.1). The Cocha Cashu plot has 25 trees larger than 60 cm DBH, compared with 5–22 in the other plots. Whereas other, relatively rich soil plots have only marginally fewer large trees, the plots on poor sandy soil have many fewer (5–15). If only trees 70 cm DBH or more are considered, the Manu hectare is even more outstanding, with 17 trees versus 4–11 (average 8.9/ha) in the other plots. Again, for the largest emergent trees, 100 cm DBH or more, which average 1.9/ha over the ten study sites, Manu has more individuals (5) than any other site.

Many of the large emergents are strangler figs. Perhaps the most striking physiognomic (and floristic) feature of the Cocha Cashu study area is the prevalence of these large stranglers. In the tree plot no fewer than ten fully developed stranglers have reached the ground and developed trunks exceeding 10 cm DBH. This contrasts with the almost complete lack of fully developed stranglers in plots studied elsewhere in the Amazon. Two other tree plots include single fully developed stranglers and seven have none at all. Smaller stranglers are common as hemiepiphytes in most forests of Amazonian Peru, but they only rarely reach the soil and develop into large trees. Perhaps there is a correlation between the rich soil and predominance of stranglers at Cocha Cashu, since the Rio Palenque Field Station in coastal Ecuador (Dodson and Gentry 1978), also with unusually rich soil, is the only other site where we have seen stranglers as abundant as at Cocha Cashu. Figs are an important food source for many mammals, and their unusual prevalence at Cocha Cashu may relate to the high mammal biomass at the field station. Strangler figs also impart a striking physiognomy to the forest, some be-

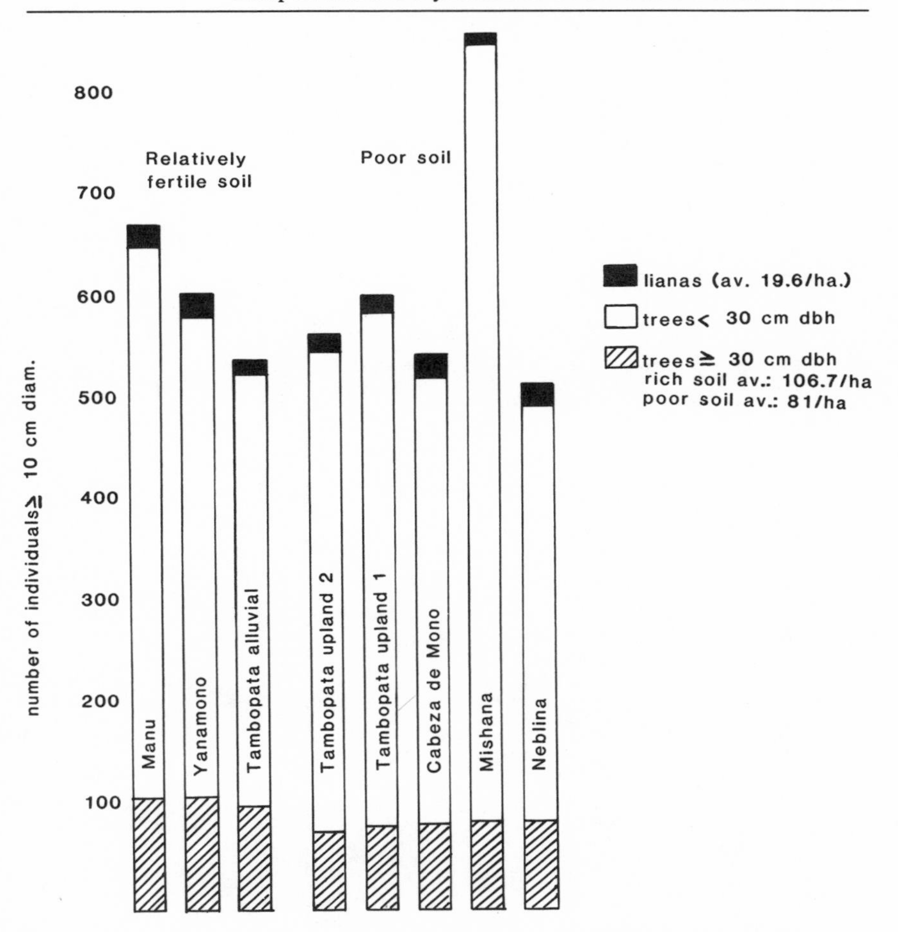

FIGURE 27.2. Forest structure at Cocha Cashu compared with some other upper Amazonian tree plots. Note greater prevalence of large trees ≥ 30 cm DBH in forests on better soils. (From Gentry 1988b.)

coming truly massive and strangling several host trees simultaneously. Some stranglers have such extensive gaps between their root columns that trails actually pass through their bases.

Palms are another unusually conspicuous element of the Cocha Cashu forest and, like stranglers, impart a distinctive physiognomy to the forest. Three of the five most common tree species in the plot are palms (*Astrocaryum, Iriartea, Scheelea*), and palms constitute altogether almost 1/6 (103 of 673) of all stems 10 cm DBH or more. This high density of palms seems a feature shared with many other Neotropical sites, especial-

TABLE 27.1 Size-class distribution of trees and lianas with DBH ≥ 10 cm in a 1-ha sample of physiognomically mature forest at Cocha Cashu, compared with similar samples from other sites in northwest South America

Class size DBH (cm)	Cocha Cashu	Tambopata alluvial	Yanamono	Cabeza de Mono	Mishana	Cerro Neblina	Bajo Calima	Tambopata sandy 1	Tambopata sandy 2	Tambopata clay 2
10–15	294	182	242	257	409	207	325	283	245	216
15–20	131	113	127	107	189	95	157	138	123	94
20–25	86	72	75	60	128	66	78	64	71	100
25–30	52	73	52	39	49	61	33	37	51	70
30–40	62	61	45	34	45	37	41	39	34	46
40–50	13	18	28	16	26	14	29	17	17	15
50–60	10	6	14	16	7	19	6	12	11	11
60–70	8	6	12	4	1	7	2	3	3	4
70–80	7	2	7	2	1	4	1	3	4	3
80–90	4	3	1	2	2	2	1	2	2	2
90–100	1	1	2	3	0	0	2	2	2	2
≥ 100	5	3	0	4	1	1	0	2	2	1
Total*	673	540	605	544	858	513	675	602	565	564
Lianas ≥ 10 cm. diam.	23	14	26	24	16	20	11	17	17	19

Sources: Amazonian data from Genty (1988a,b); Bajo Calima, original data
Note: * Includes lianas

ly on nutrient-rich to moderately infertile soils. Some Amazonian sites on clayey soils have even more palms—up to a quarter of the trees 10 cm DBH or more in one plot at Tambopata, Peru. At La Selva, Costa Rica, also on rich soil, palms similarly constitute one-fourth of all stems 10 cm DBH or more (Lieberman and Lieberman 1987).

The open quality of the understory is another subjective characteristic of the Cocha Cashu plot. This openness is due largely to the relative lack of trees and treelets 2.5–10 cm DBH. There are actually more small trees in the 10–15 cm DBH size-class in the Cocha Cashu tree plot than at most other sites. However, there are only 203 trees 2.5–10 cm DBH in the 0.1 ha subsampled for smaller stems, one of the lowest such values for any Amazonian forest.

Lianas, in contrast, are well represented at Cocha Cashu. The 0.1-ha sample contained 79 lianas 2.5 cm diameter or more, representing 43 species. This is just over the average value for a series of similarly sampled Amazonian forests. When only large lianas (greater than 10 cm diameter) are considered, the Cocha Cashu forest again has just slightly more individuals than the average for ten Amazonian hectares.

Floristic Composition

The floristic composition of the Cocha Cashu area is discussed in depth by Foster in chapter 7 (see also Foster 1985; Gentry 1985). Our plot data confirm the high diversity of the area's woody flora, with over 200 species greater than 10 cm diameter in our 1-ha plot, including representation of 47 plant families. We find the same families to be most diverse in our plot (fig. 27.3) as does Foster (this volume) for the tree flora as a whole. In both plot and tree lists Leguminosae are the most speciose family, and the next four most speciose families are Annonaceae, Moreaceae, Lauraceae, and Sapotaceae, followed by Meliaceae, Bombacaceae, Euphorbiaceae, and Myristicaceae. We thus feel that our 1-ha plot provides a representative sample of the local tree flora. It is also noteworthy that even a 0.1-ha subsample of plants 2.5 cm diameter or more gives the same result: when lianas are excluded from that data set, exactly the same six families are indicated as most speciose. As Foster (this volume) and AG (this volume) have emphasized the Cocha Cashu flora is unusual in the context of Amazonia in being a rich-soil flora, in many ways more similar to Central American than to most other Amazonian forests. Although many typically Central American species are present, the high species diversity of the Cocha Cashu forest is a characteristically upper Amazonian phenomenon (Gentry 1988a).

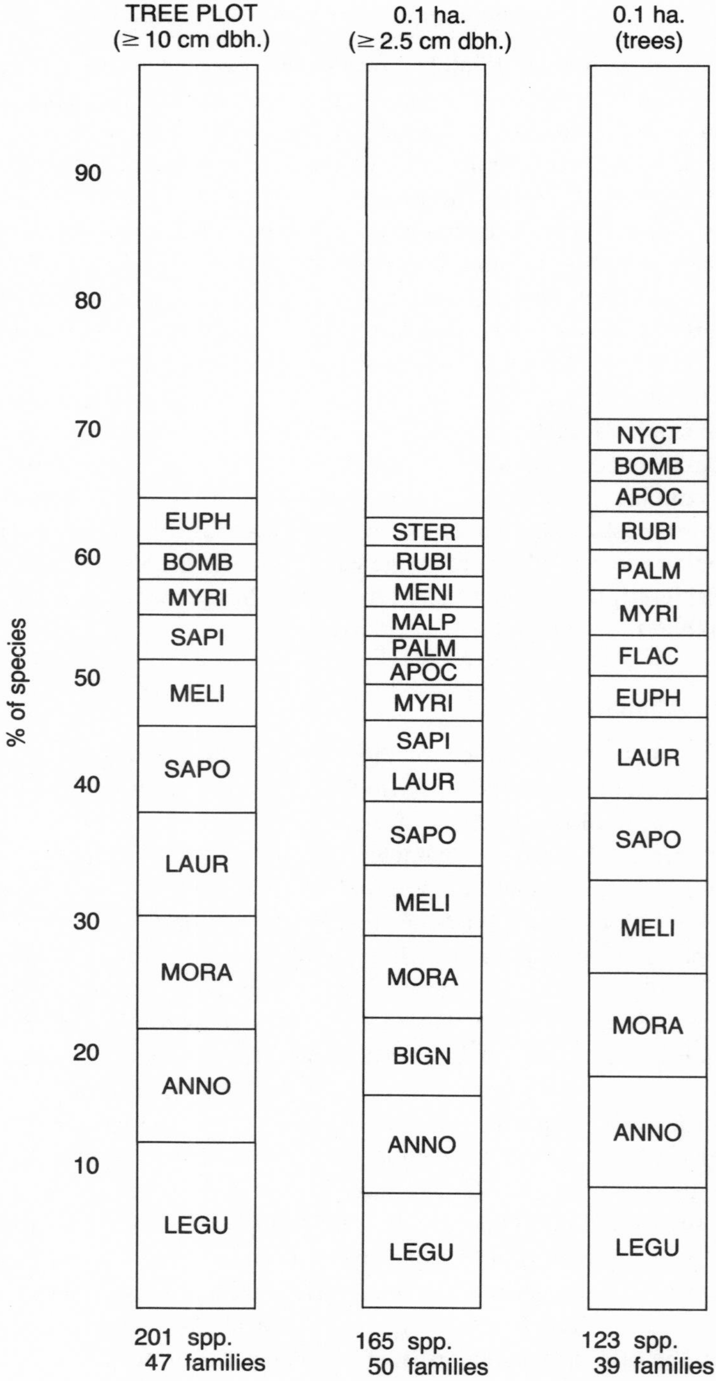

% of species

90
80
70
60
50
40
30
20
10

TREE PLOT
(≥ 10 cm dbh.)

EUPH
BOMB
MYRI
SAPI
MELI
SAPO
LAUR
MORA
ANNO
LEGU

201 spp.
47 families

0.1 ha.
(≥ 2.5 cm dbh.)

STER
RUBI
MENI
MALP
PALM
APOC
MYRI
SAPI
LAUR
SAPO
MELI
MORA
BIGN
ANNO
LEGU

165 spp.
50 families

0.1 ha.
(trees)

NYCT
BOMB
APOC
RUBI
PALM
MYRI
FLAC
EUPH
LAUR
SAPO
MELI
MORA
ANNO
LEGU

123 spp.
39 families

TABLE 27.2 Tree mortality by size class in 0.943-ha plot*

Diameter class (cm)	Initially in plot	Dying in ten years	Mortality/yr. (%)
10–20	372	62	1.7
20–30	144	26	1.8
≥30	51	5	1.0
TOTAL	567	93	1.6

$\chi^2 = 1.6$, n.s.

Note: * Data base has unresolved discrepancies. Data presented above are from computerized records. A tree-by-tree hand tabulation of ostensibly the same data gives an initial stock of 537 trees ≥ 10 cm DBH, of which 88 died in ten years. Both tabulations, however, produce an identical 1.6% annual mortality estimate.

Dynamics

During the ten years for which we collected data, 93 out of the 587 originally tagged trees died, for a mortality rate of 1.58% per year. This implies an average life of 63.3 years (Putz and Milton 1982) and a stand half-life of 40.2 years (Lieberman et al. 1985b; Hartshorn, this volume). Mortality rates in the two subplots were similar, with 46 of 305 trees dying in the west section for an annual mortality of 1.51% and 47 of 282 originally tagged trees dying in the east section for an annual mortality of 1.67%. Excluding the 36 trees slightly less than 10 cm DBH that were tagged and included in the analysis gives a total of 537 "legitimate" trees originally present in the 0.943-ha sample. Of these, 88 were lost, giving the same 1.6% a year average mortality. In general mortality rates did not differ among size classes. We compare the mortality for trees in 10–20, 20–30, and greater than 30 cm size classes in table 27.2. Although slightly fewer larger trees died, this difference is not significant ($\chi^2 = 1.6$, $p > .05$).

Recruitment

During the ten years of our census only 45 trees were recruited into our study population by attaining 10 cm minimum diameter while 88 trees

←

FIGURE 27.3. Floristic composition of Cocha Cashu 1-ha plot (plants ≥ 10 cm diameter) compared with that of 0.1-ha subplot (≥ 2.5 cm DBH) and with that of 0.1-ha subplot excluding climbers.

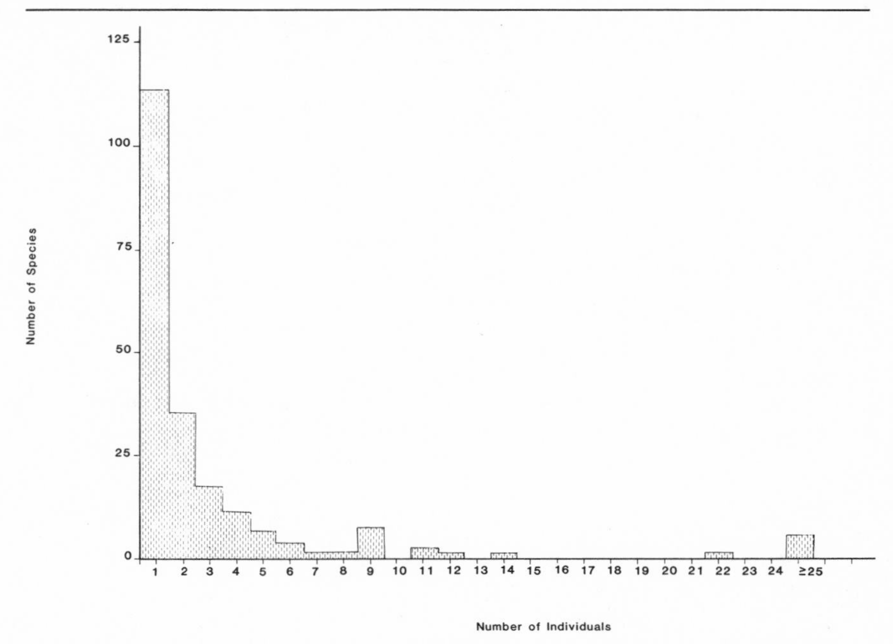

FIGURE 27.4. Number of species with given numbers of individuals in Cocha Cashu 1-ha tree plot.

died. Our data thus show a large excess of mortality over recruitment in our study area. As a result, in 1984–85 the 0.943-ha plot had only 494 trees 10 cm DBH or more, as compared to 537 in 1974–75.

Population Structure

Our data are generally insufficient to analyze interspecific differences in growth rates and mortality directly. Since only those trees still alive near the end of the ten-year period were identified, we can say nothing about interspecific differences in mortality. Moreover, since the forest is so rich in species, we have growth data for few individuals of most species. In the 1-ha plot only 19 species are represented by more than 6 individuals and only 10 by more than 9 individuals (figs. 27.4, 27.5). There are only 16 species with 5 or more individuals for which we have

FIGURE 27.5. Number of individuals per species for most common species in 1-ha tree plot.

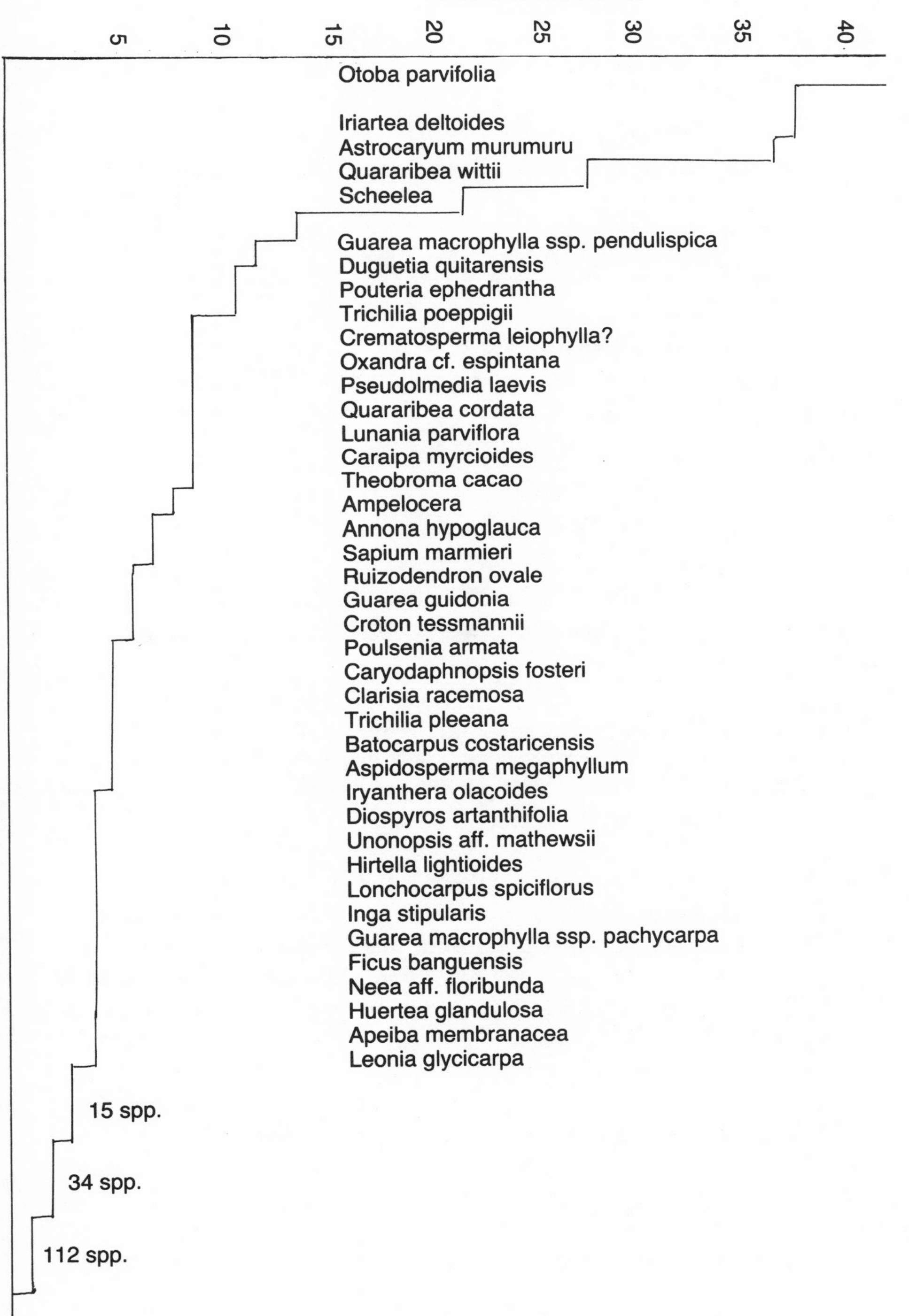
Number of Individuals
5
10
15
20
25
30
35
40
Otoba parvifolia
Iriartea deltoides
Astrocaryum murumuru
Quararibea wittii
Scheelea
Guarea macrophylla ssp. pendulispica
Duguetia quitarensis
Pouteria ephedrantha
Trichilia poeppigii
Crematosperma leiophylla?
Oxandra cf. espintana
Pseudolmedia laevis
Quararibea cordata
Lunania parviflora
Caraipa myrcioides
Theobroma cacao
Ampelocera
Annona hypoglauca
Sapium marmieri
Ruizodendron ovale
Guarea guidonia
Croton tessmannii
Poulsenia armata
Caryodaphnopsis fosteri
Clarisia racemosa
Trichilia pleeana
Batocarpus costaricensis
Aspidosperma megaphyllum
Iryanthera olacoides
Diospyros artanthifolia
Unonopsis aff. mathewsii
Hirtella lightioides
Lonchocarpus spiciflorus
Inga stipularis
Guarea macrophylla ssp. pachycarpa
Ficus banguensis
Neea aff. floribunda
Huertea glandulosa
Apeiba membranacea
Leonia glycicarpa
15 spp.
34 spp.
112 spp.

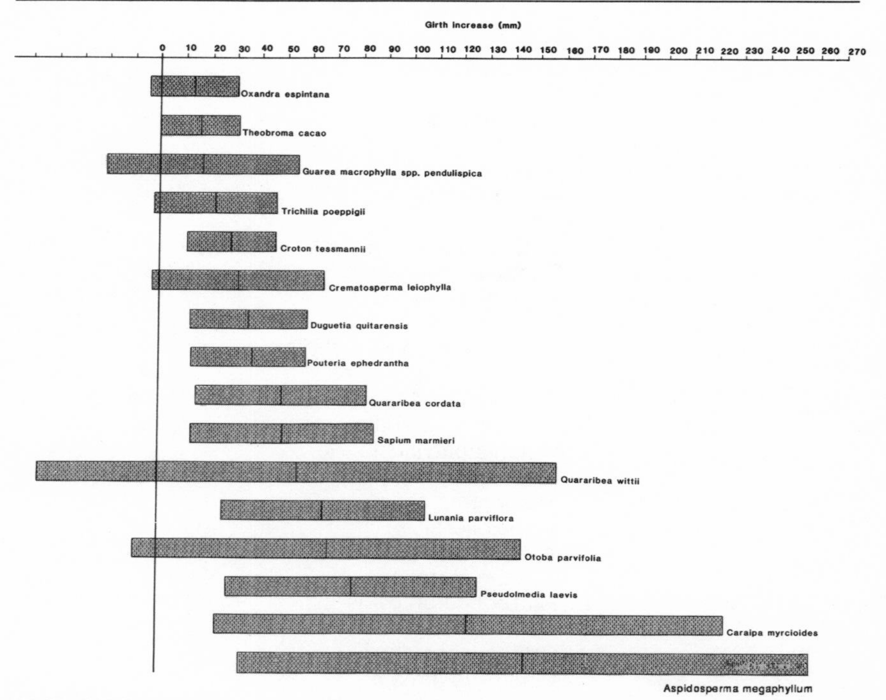

FIGURE 27.6. Average ten-year girth increases for 16 most common tree species in 0.9-ha plot. Bar = 1 SD. Species with 7 cm/yr average girth increase are shade intolerant as indicated by population structure. Species with less than 7 cm/yr average girth increase are shade tolerant as indicated by population structure except *Pouteria ephedrantha, Quararibea cordata*, and *Sapium marmieri* with intermediate population structure.

both identifications and growth data over the ten-year period (fig. 27.6). Three of the more common species are palms, which lack secondary thickening and are thus excluded from this analysis. The average diameter increases of the most common (more than five individuals) dicot tree species are shown in fig. 27.6. Although small sample size and high interspecific variability obscure the picture, some marked interspecific differences in growth rates are clear. Such undercanopy species as *Oxandra espintana*, *Theobroma cacao*, and *Guarea macrophylla* ssp. *pendulispica* average an increase in diameter of less than 6 cm over the ten years, or only about 0.5 mm a year. An extreme case is *Guarea macrophylla* ssp. *pendulispica*, where half of the ten individuals for which we have data decreased in diameter over the ten years! In contrast, many of the more common canopy species are relatively fast-growing, with species like

Caraipa myrcioides and *Aspidosperma megaphyllum* averaging well over 40 mm of increase in diameter over the ten years.

It has been suggested that larger trees tend to grow faster than smaller ones, reflecting suppression of smaller-diameter trees, with faster growth as the crown expands on attaining the canopy (Nicholson 1965; Longman and Jenik 1974; Lang and Knight 1983; Whitmore 1984). Our data for Cocha Cashu show a pronounced tendency for larger individuals to grow more rapidly. For the sixteen common species listed in fig. 27.6, the trees that started in the 10–20 cm DBH class averaged 13 mm increase in diameter, those in the 20–30 DBH class averaged 27 mm, and trees greater than 30 cm DBH averaged 51 mm.

Given the limited time span of our data set, we can also indirectly analyze differences in growth rates and reproductive strategies by analyzing size-class frequency distributions of different species of trees as advocated by several authors (Knight 1975; Hartshorn 1980). Although most species, especially emergent and canopy taxa, are too rare for meaningful analysis, many of the more common tree species in our plot clearly show a characteristic reverse-J population structure, with many individuals in smaller size classes and few in larger diameter classes (e.g., Annonaceae in fig. 27.7).

Several other common species are always small at maturity; a good example is *Trichilia poeppigii* (fig. 27.7), which never exceeds 15 cm DBH. That such species are reproducing is shown by the frequent representation of plants in the 2.5–10 cm diameter classes in the 0.1-ha subplot.

Many of the rarer emergent species are mostly or exclusively represented by large-diameter individuals (fig. 27.7, *Ceiba* and *Chorisia* in fig. 27.8). The large trees in our plot are so diverse, however, that few individuals of any of these species are represented in our data set. Of the 68 species with trees greater than 30 cm DBH, only 19 have more than one individual in this size class. The 25 trees 60 cm DBH or more belong to 20 species, and 10 of these are not represented by a single smaller individual. A few species, including *Pouteria ephedrantha* (fig. 27.8) and the "dominant" *Otoba parvifolia* (fig. 27.7), are intermediate, with an occasional large individual but concentration of individuals in intermediate size classes. Nor are the emergent species represented in the smaller size classes included in the 0.1-ha subsample. Of the 20 potentially emergent species with trunks attaining 60 cm DBH or larger in our plot, only 4 have individuals 2.5–20 cm DBH. Three species—*Pouteria ephedrantha*, *Otoba parvifolia*, and *Aspidosperma megaphyllum*—are

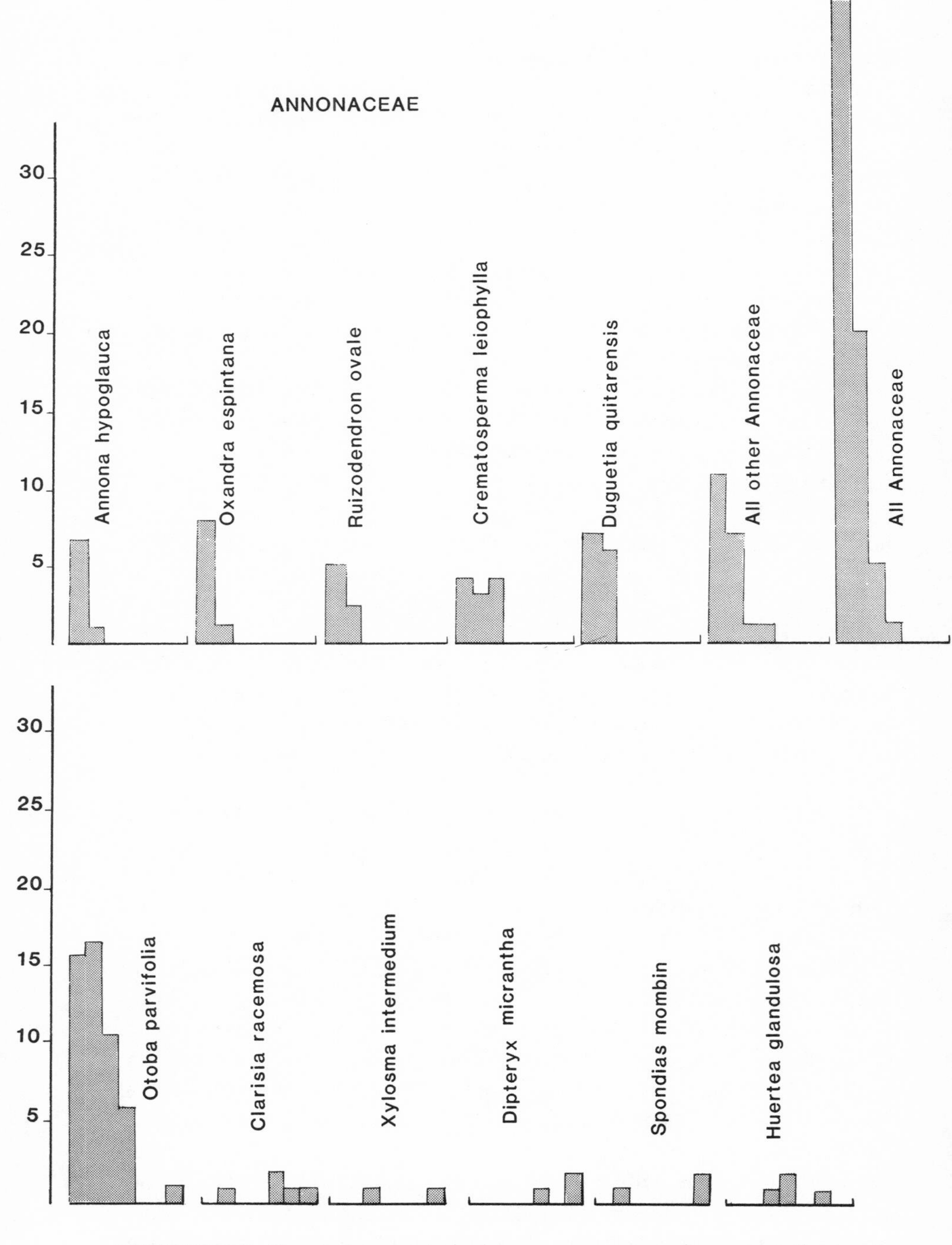

FIGURE 27.7. Population structures for some common and representative tree species. Each horizontal line represents 7 diameter classes: 10–15 cm DBH, 15–20 cm DBH, 20–30 cm DBH, 30–40 cm DBH, 40–50 cm DBH, 50–60 cm DBH, ≥ 60 cm DBH.

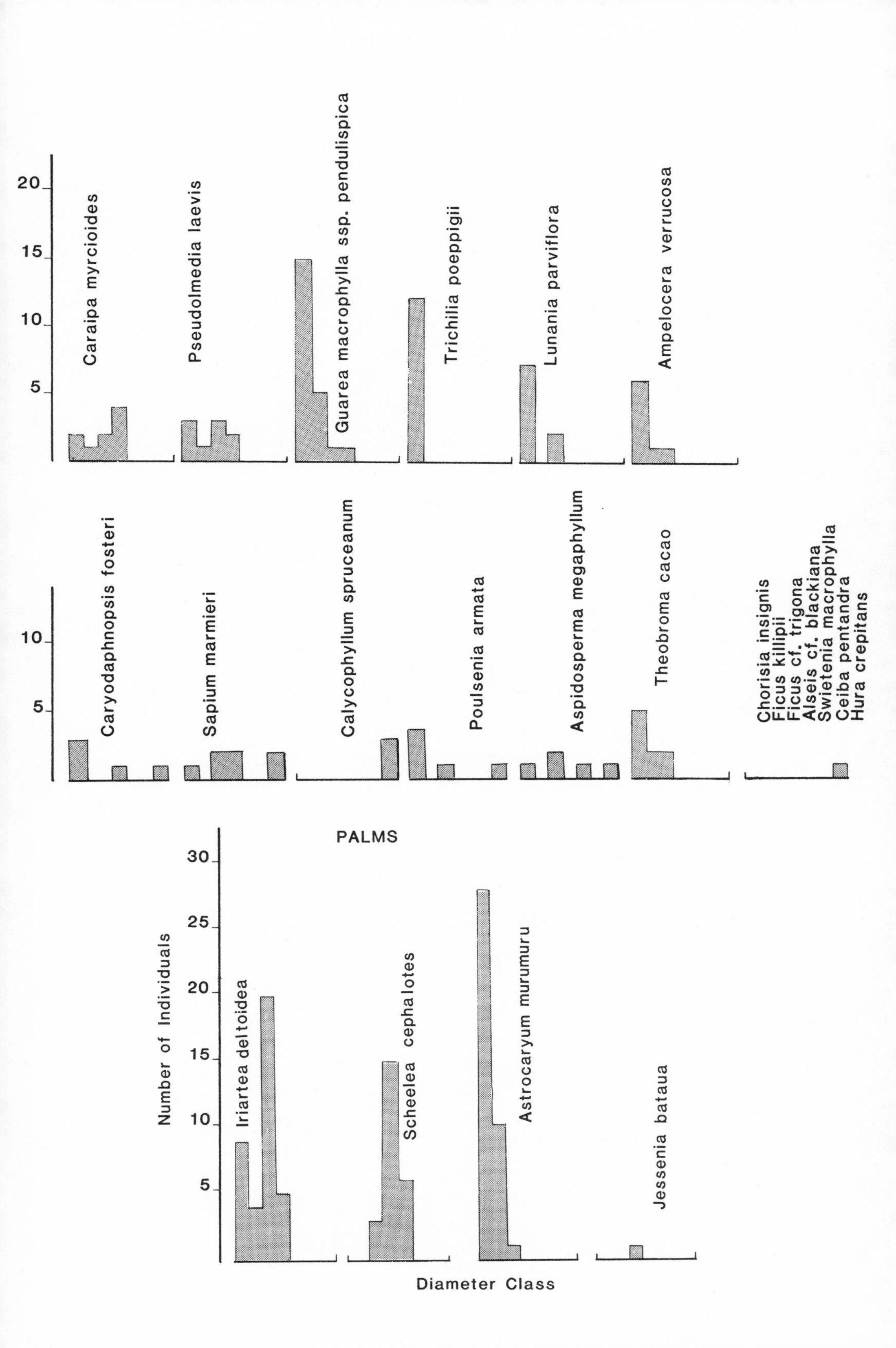

20
15
10
5
Caraipa myrcioides
Pseudolmedia laevis
Guarea macrophylla ssp. pendulispica
Trichilia poeppigii
Lunania parviflora
Ampelocera verrucosa
10
5
Caryodaphnopsis fosteri
Sapium marmieri
Calycophyllum spruceanum
Poulsenia armata
Aspidosperma megaphyllum
Theobroma cacao
Chorisia insignis
Ficus killipii
Ficus cf. trigona
Alseis cf. blackiana
Swietenia macrophylla
Ceiba pentandra
Hura crepitans
PALMS
30
25
20
15
10
5
Number of Individuals
Iriartea deltoidea
Scheelea cephalotes
Astrocaryum murumuru
Jessenia bataua
Diameter Class

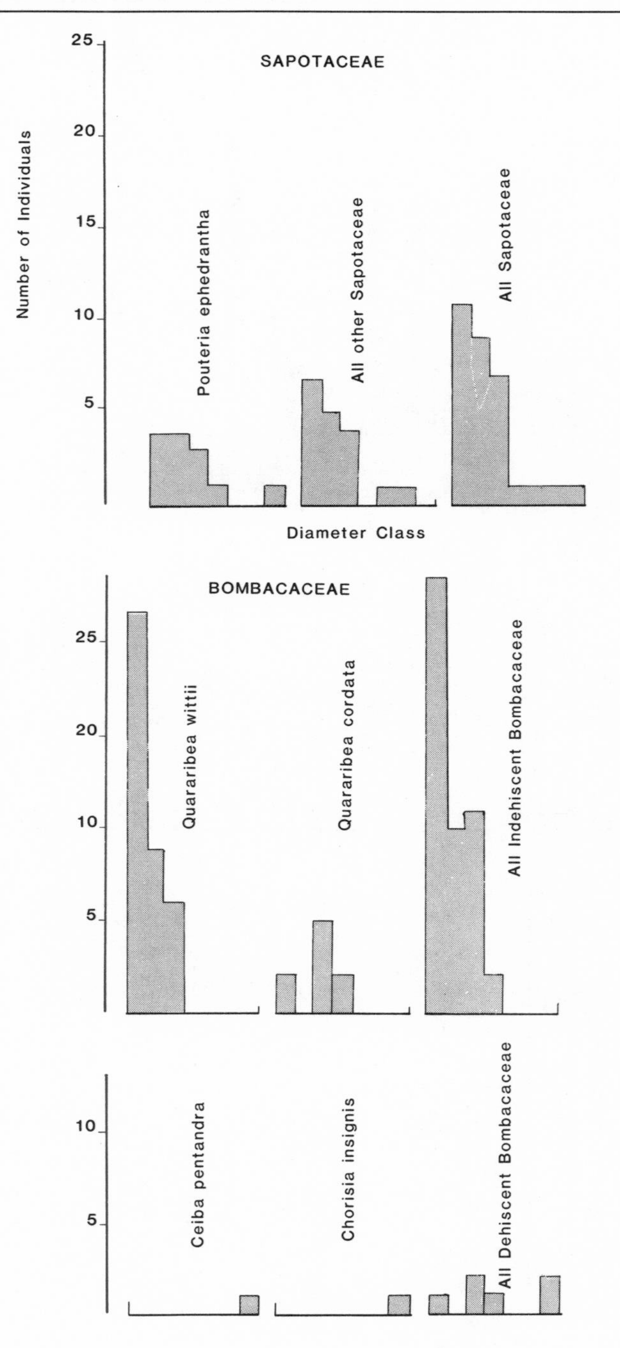
SAPOTACEAE
Number of Individuals
25
20
15
10
5
Pouteria ephedrantha
All other Sapotaceae
All Sapotaceae
Diameter Class
BOMBACACEAE
25
20
10
5
Quararibea wittii
Quararibea cordata
All Indehiscent Bombacaceae
10
5
Ceiba pentandra
Chorisia insignis
All Dehiscent Bombacaceae

represented by single saplings less than 4 cm DBH, and *Poulsenia armata* has a 7-cm DBH juvenile.

Discussion

Our data are entirely consistent with the emerging picture of tropical forests as dynamic systems with high rates of replacement and turnover. For a wide range of tropical forests tree mortality is 1–2% per year (Hartshorn 1980, this volume; Lieberman et al. 1985b; Swaine et al. 1987). The 1.6% a year mortality we record here for the mature forest at Cocha Cashu is thus decidedly ordinary. The possibility that the unusually rapid 2% a year mortality at La Selva is related to greater turnover on richer soil would not be supported by the slower rate on similarly rich soil at Cocha Cashu. At least as indicated by rate of tree mortality, the forest of Cocha Cashu is a typical tropical forest.

The most unusual aspect of forest dynamics at Cocha Cashu is the large excess of mortality over recruitment in our study plot. In a climax tropical forest, recruitment should almost exactly balance mortality, as it does at La Selva (Lieberman et al. 1985b). Though stand density may fluctuate significantly with time (e.g., Manokaran and Kochummen 1987), the imbalance in our data seems too large to result from random fluctuations, especially since there are no obvious vestiges of large-scale disturbances like a big blowdown that might temporarily increase mortality. If the Cocha Cashu forest is still undergoing successional change, then one might expect a thinning out of the smaller size-classes as the forest reaches maturity, but we find no evidence for preferential loss of individuals in any size class. Perhaps the most likely explanation is that the physiognomically "mature" forest that we studied has entered a kind of preclimax phase of the extremely long-term succession described by Foster (this volume). If so, we are witnessing a forest many of whose present constituents, especially among the large emergents, entered it perhaps several hundred years ago during an early successional stage. As shorter-lived individuals senesce and die they are replaced only incompletely by recruitment of largely shade-tolerant species under current closed-canopy conditions. Unfortunately, our lack of identifications for the trees that died precludes us from determining whether species composition has

←

FIGURE 27.8. Population structures for Sapotaceae and Bombacaceae. Each horizontal line represents 7 diameter classes: 10–15 cm DBH, 15–20 cm DBH, 20–30 cm DBH, 30–40 cm DBH, 40–50 cm DBH, 50–60 cm DBH, ≥ 60 cm DBH.

changed in a way that might support this interpretation. As envisioned by Foster, a true self-sustaining climax might only be reached as the long-lived ancient emergent individuals of species like *Ceiba pentandra*, *Calycophyllum spruceanum*, and *Swietenia macrophylla* finally die, opening the canopy sufficiently that many faster-growing light-gap species can regenerate. Our data thus seem entirely consistent with Foster's long-term successional hypothesis.

Although a decade of data is more than is available for most tropical forests, ten years is an inadequate time to study patterns that occur over centuries. We can, however, attempt to extrapolate from the current composition of the forest to longer time intervals (Knight 1975). Thus, it has been suggested that there are distinctive population structures that correlate with different roles in the forest dynamics. The two predominant population patterns are the reverse-J of shade-tolerant (or frequently reproducing sensu Knight 1975) species and the large-diameter concentration and lack of small-diameter replacements of shade-intolerant successional or light-gap species (infrequent reproducers sensu Knight 1975). Knight thought that the presence of a high proportion of the former should indicate climax conditions, whereas numerous infrequent reproducers indicated an early successional status, but Hartshorn (1980), Denslow (1980), and others have emphasized the importance of light-gap strategists even in mature forest.

At Cocha Cashu we can extrapolate only from species' behavior in other areas or from their morphological adaptations which of our species are "infrequent reproducers" or light-gap specialists. All of the large emergent trees showing little or no replacement are potential light-gap species. Such species have been suggested to be largely wind- or bat-dispersed (Hartshorn 1980; Foster et al. 1986), and it is noteworthy that all but one of the wind-dispersed large trees in our plot are represented only by large individuals. The exception is *Aspidosperma megaphyllum* (fig. 27.7), which has several individuals in smaller diameter classes. The lack of small-diameter individuals of tree species could be due either to the infrequency of their successful establishment or to rapid growth through small-diameter classes. Our population data would strongly seem to support the suggestion of Foster et al. (1986) that these species require extremely large disturbances for successful establishment, at Cocha Cashu generally establishing only in young seral vegetation on recently deposited alluvium. On the other hand, our growth data, though showing much between-individual variation, do show a definite correlation between rapid growth and regeneration niche, with the putatively shade-tolerant species with reverse-J population structures slow-growing and the few putative light-gap specialists with a preponderance of large trees in their populations that are represented by enough individuals to analyze relatively fast-growing (fig. 27.6).

Two other population patterns were adduced by Knight (1975) for tree species on Barro Colorado Island, and both are recognizable in our plot. The first (Knight's third) is that of small subcanopy species that barely reach 10 cm DBH and never achieve large diameters. The frequency in the small-diameter classes of the 0.1-ha data set of individuals of such species as *Rinorea viridifolia*, *Leonia glycicarpa*, *Trichilia poeppigii*, *Guarea macrophylla* (both subspecies), *Oxandra espintana*, *O. acuminata*, *Annona hypoglauca*, *Lunania parviflora*, and *Quararibea wittii* indicates that basically the small tree I-pattern species differ from the reverse-J shade-tolerant species only in skewing the J farther left into smaller diameter size-classes.

Knight's fourth pattern, which he regarded as something of a mystery, is a population structure with most individuals in the medium size-classes but with a few large and small individuals. This pattern, which is intermediate between (and which blurs the oft-cited distinction of) the shade-tolerant and shade-intolerant strategies, has generally been ignored. Several of the most common Cocha Cashu species show this pattern. Of these, most striking are *Otoba parvifolia*, the most common species at Cocha Cashu (fig. 27.7), and *Pouteria ephedrantha* (fig. 27.8), both of which had only a single 2.5-cm DBH sapling in the 0.1-ha small-diameter size classes. Knight suggested that this pattern might result from rapid growth through the small-diameter classes, but our growth-rate data do not support this suggestion. *Pouteria ephedrantha* is a consistently slow-growing species, and neither *Otoba parvifolia* nor *Quararibea cordata*, another common species that fits this pattern, grow unusually quickly. What this population structure indicates remains an open question.

A few supplementary notes relating floristic composition to stand structure and population structure are appropriate here. Several families are characterized by specific kinds of population structures shared by most of their species. For example, most Annonaceae (fig. 27.7) and many Sapotaceae (fig. 27.8) species have J-shaped (or I-shaped) distributions with few or no large individuals and many smaller ones. Bombacaceae (fig, 27.8) are especially interesting in that indehiscent-fruited animal-dispersed species are mostly shade-tolerant with reverse-J population curves, whereas dehiscent-fruited wind-dispersed species are shade-intolerant light-gap or successional specialists. Most of the dehiscent-fruited Bombacaceae are large emergent trees and exemplify the tendency for emergents to be wind-dispersed (see also Prance, this volume). The key point is that the strong tendency for species of the same family or subfamilial taxon to share the same general regeneration niche would appear to limit the potential for niche differentiation between close relatives of the kind suggested by Denslow (1980). The likely corollary is that, in the constantly changing dynamic tropical forest, many simultaneously occurring closely related species may be ecological equivalents whose co-occurrence is more a function of stochastic processes than of niche specificity (Shmida and Ellner 1984; Hubbell and Foster 1986).

In many ways Cocha Cashu fits what is emerging as typical of tropical forests in its dynamic nature. Cocha Cashu is, however, unusual in the apparent imbalance between mortality and recruitment, hinting that an equilibrium has not yet been reached and that very long-term successional changes are still taking place, as Foster suggests in chapter 28. It is possible that subjectively determined physiognomically mature forests may be no more than a transient stage whose continuity through time is made possible only by repetition of the riverine recycling processes.

REFERENCES

Balslev, H., J. Luteyn, B. Øllgaard, and L. B. Holm-Nielsen. 1987. Composition and structure of adjacent unflooded and floodplain forest in Amazonian Ecuador. Opera Bot. 92: 37–57.

Brunig, E. F. and H. Klinge. 1976. Comparison of the phytomass structure of equatorial "rain forest" in Central Amazonas, Brazil, and in Sarawak, Borneo. Garden's Bull., Singapore 29: 81–101.

Campbell, D. G., D. C. Daly, G. T. Prance, and U. N. Maciel. 1986. Quantitative ecological inventory of terra firme and várzea tropical forest on the Rio Xingu, Brazilian Amazon. Brittonia 38: 369–393.

Denslow, J. 1980. Gap partitioning among tropical rainforest trees. Biotropica 12(suppl.): 47–55.

Dodson, C. and A. Gentry. 1978. Flora of the Rio Palenque Science Center. Selbyana 4: 1–628.

Emmons, L. H. 1987. Comparative feeding ecology of felids in a Neotropical rainforest. Behav. Ecol. Sociobiol. 20: 271–283.

Foster, R. B. 1985. Plantas de Cocha Cashu. In M. A. Ríos (ed.), *Reporte Manu*. Centro de Datos para la Conservación, Univ. Agraria, La Molina, Peru, pp. 5/1–5/20.

———. (This volume). The floristic composition of the Manu floodplain forest.

———. (This volume). Long-term change in the successional forest community of the Rio Manu floodplain forest.

Foster, R. B., J. Arce B., and T. S. Wachter. 1986. Dispersal and the sequential plant communities in Amazonian Peru floodplain. In A. Estrada and T. H. Fleming (eds.), *Frugivores and Seed Dispersal*. W. Junk, Dordrecht, pp. 356–370.

Gentry, A. H. 1982. Patterns of Neotropical plant species diversity. Evol. Biol. 15: 1–84.

———. 1985. Algunos resultados preliminares de estudios botánicos en el Parque nacional del Manu. In M. A. Ríos (ed.), *Reporte Manu*. Centro de Datos para la Conservaci6n, Univ. Agraria, La Molina, Peru, pp. 2/1–2/24.

———. 1986. Sumario de patrones fitogeográficos neotropicales y sus implicaciones para el desarrollo de la Amazonía. Rev. Acad. Col. Cien. Ex. Fis. 16: 101–116.

———. 1988a. Tree species richness of upper Amazonian forests. Proc. Nat. Acad. Sci. 85: 156–159.

———. 1988b. Changes in plant community diversity and floristic composition on environmental and geographical gradients. Ann. Mo. Bot. Gard. 75: 1–34.

———. (This volume.). Floristic similarities and differences between southern Central America and Upper and Central Amazonia.

Hartshorn, G. S. 1975. A matrix model of tree population dynamics. In F. Golley and E. Medina (eds.), *Tropical Ecological Systems: Trends in Terrestrial and Aquatic Research*. Springer-Verlag, New York, pp. 41–51.

———. 1978. Tree falls and tropical forest dynamics. In P. B. Tomlinson and M. H. Zimmerman (eds.), *Tropical Trees as Living Systems*. Cambridge Univ. Press, London, pp. 617–638.

———. 1980. Neotropical forest dynamics. Biotropica 12 (suppl.): 23–30.

———. (This volume). An overview of Neotropical forest dynamics.

Hubbell, S. P. and R. B. Foster. 1986. Biology, chance and history and the structure of tropical rain forest tree communities. In J. Diamond and T. Case (eds.), *Community Ecology*. Harper and Row, New York, pp. 314–319.

Klinge, H. and R. Herrera. 1983. Phytomass structure of natural plant communities on spodosols in southern Venezuela: the tall Amazon caatinga forest. Vegetatio 53: 65–84.

Knight, D. 1975. A phytosociological analysis of species-rich tropical forest on Barro Colorado Island, Panama. Ecol. Monogr. 45: 259–284.

Lang, G. E. and D. H. Knight. 1983. Tree growth, mortality, recruitment, and canopy gap formation during a ten-year period in a tropical moist forest. Ecol. 64: 1075–1080.

Lieberman, D. and M. Lieberman. 1987. Forest tree growth and dynamics at La Selva, Costa Rica (1969–1982). J. Trop. Ecol. 3: 347–358.

Lieberman, D., M. Lieberman, G. Hartshorn, and R. Peralta. 1985a. Growth rates and age-size relationships of tropical wet forest trees in Costa Rica. J. Trop. Ecol. 1: 97–109.

———. 1985b. Mortality patterns and stand turnover rates in a wet tropical forest in Costa Rica. J. Ecol. 73: 915–924.

Longman, K. and J. Jenik. 1974. *Tropical Forest and Its Environment*. Longman, London.

Manokaran, N. and K. Kochummen. 1987. Recruitment, growth, and mortality of tree species in a lowland dipterocarp forest in Peninsular Malaysia. J. Trop. Ecol. 3: 315–330.

Nicholson, D. I. 1965. A study of virgin forest near Sandaken, North Borneo. Symp. Ecol. Res. in Humid Trop. Veg. Govt. of Sarawak and UNESCO Sci. Coop. Off. for SE Asia, Kuching, Sarawak.

Pires, J. M. and G. T. Prance. 1977. The Amazon forest: a natural heritage to be preserved. In G. Prance and T. Elias (eds.), *Extinction Is Forever*. New York Botanical Garden, New York, pp. 158–194.

Prance, G. T. (This volume). The floristic composition of the forests of central Amazonian Brazil.

Putz, F. E. and K. Milton. 1982. Tree mortality rates on Barro Colorado Island. In E. G. Leigh, Jr., A. S. Rand, and D. M. Windsor (eds.), *The Ecology of a Tropical Forest: Seasonal Rhythms and Long-Term Changes*. Smithsonian Inst. Press, Washington, D.C., pp. 95–100.

Shmida, S. and S. P. Ellner. 1984. Coexistence of plant species with similar niches. Vegetatio 58: 29–55.

Swaine, M. D., D. Lieberman, and F. E. Putz. 1987. The dynamics of tree populations in tropical forests: a review. J. Trop. Ecol. 3: 359–366.

Terborgh, J. 1983. *Five New World Primates: A Study in Comparative Ecology*. Princeton Univ. Press, Princeton, N.J.

———. (This volume). An overview of research at Cocha Cashu Biological Station.

Uhl, C. 1982. Tree dynamics in a species rich tierra firme forest in Amazonia, Venezuela. Act. Cien. Venez. 33: 72–77.

Uhl, C., K. Clark, N. Dezzeo, and P. Maquirino. 1988. Vegetation dynamics in Amazonian treefall gaps. Ecol. 69: 751–763.

Whitmore, T. C. 1984. *Tropical Rain Forests of the Far East*. 2d ed. Oxford Univ. Press, Oxford.

Wyatt-Smith, J. 1968. Ecological Studies on Malayan forest. I. Composition of and dynamic studies in lowland evergreen rain forest in two five-acre plots in Bukit Lagong and Sungei Menyala Forest Reserve, 1947–1959. For. Res. Inst. Res. Pamph. 52, Kepong, Malaysia.

28

Long-Term Change in the Successional Forest Community of the Rio Manu Floodplain

ROBIN B. FOSTER

Most studies of dynamics in a tropical forest assume it is at some sort of equilibrium or steady-state condition (see Hubbell and Foster 1986 for discussion). Short-term dynamic studies are then interpreted as revealing the processes that maintain the status quo. In this short chapter, I present some information to challenge that idea, at least for well-drained floodplain forests and perhaps for any tropical forest that is not more than five hundred years old.

The choice of tropical forests for studies of community dynamics is rarely if ever based on any knowledge of the history of that forest. The criterion is usually that it is the most typical or best-looking forest around, having a lot of large trees and what seems to be a lot of species—and preferably no large areas of blowdowns. Certainly no one would pick a young forest dominated by such obviously successional species as *Cecropia* or *Ochroma*, but beyond that just about any tropcial forest with large trees is usually considered "mature."

The Manu River in Manu National Park, southeastern Peru, provides an unusual opportunity to study upper Amazon floodplain with little recent human disturbance (see Terborgh 1983 for more description). The continuous meander of the river over its floodplain and the long, low rows of sediment left behind (levees, dikes, or restingas) create conditions whereby transects perpendicular to the levees traverse a time

For field assistance I thank Javier Arce B., Barbara d'Achille, and Jesus Huaman Valle. For assistance in processing data and specimens I thank Tyana Wachter. For logistical assistance I thank John Terborgh, Magda Chanco, and Emma Cerrate. I thank the Direcion General Forestal y de Fauna of the Peruvian Ministerio de Agricultura for permission to work in Manu National Park.

sequence of successional development. In this chapter I describe such a transect to reveal some of the processes at work in forest succession, reveal the possible instability of what otherwise might appear to be "mature" tropical forest, and set the basic stage for interpreting a plot in the Manu "old forest" (Gentry and Terborgh, this volume).

Study Site

The Cocha Cashu Biological Station at 350 m elevation in Manu Park, has been described in detail in Terborgh (1983). The location and layout of the transects near the station has been described in Foster et al. (1986). The first transect (fig. 28.1) is perpendicular to the current riverbed on a slowly expanding minor bend of the river with very closely spaced and pronounced levees 1–3 m in height. It ends where the levees run up against the outflow of an old ox-bow lake (Cocha Cashu). The second transect was chosen to be similar in terrain to the first but begin at a successional stage equivalent to the end of the first transect. It runs perpendicular from the edge of another ox-bow lake, Cocha Totora, 2–3 km from the first transect.

Both segments contain several levees with intervening depressions (swales). The youngest levees are inundated during the highest floods each year, the oldest levees remain dry while the swales between them become inundated. In a year of very high flood, even the older levees on the transect get covered with water. Consequently, all the transect is subject to some deposit of sediment by the river. But the frequency, level, and duration of the inundations varies considerably by year.

The two transects combined may not form a perfectly continuous age sequence, but the difference between the end of the first and the beginning of the second is estimated to be within 25 years (based on sizes of the dominant tree species). Many of the same successional tree species, notably *Ficus insipida*, are found in known-age forest in Panama (Foster and Brokaw 1982). The trees in the younger half of the second transect are in the same size range as those in 100- to 180-year-old forest in Panama. The age of the oldest part of the second Manu transect, in which *Ficus insipida* has almost died out, is estimated to be about 200 years (±50).

Methods

Each transect is 100 m wide. The first is 200 m long, following a line perpendicular to the river, beginning where the first trees (*Cecropia*) with

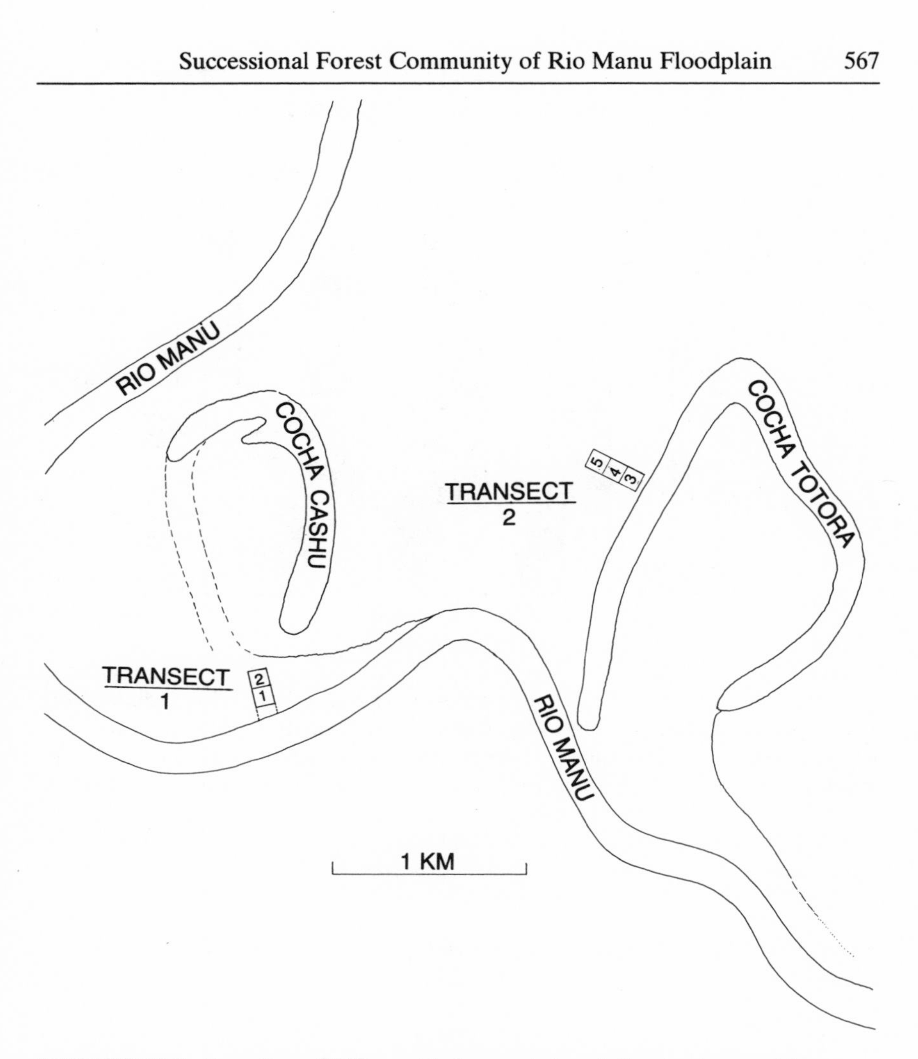

FIGURE 28.1. Location of 100-m wide transects in relation to river and two meander-formed lakes in Manu National Park. The 100-m square blocks are numbered in order of increasing age.

a diameter (at 1.3 m height) greater than 30 cm are encountered. This is considered the start of forest after the transition from the riverside thickets. The second transect is 300 m long; thus combined, the two form an effective transect of 5 sequential ha (fig. 28.1).

All trees over 30 cm diameter were mapped, measured, and identified. Juveniles and smaller trees and shrubs were considered in a separate study using part of the same transect (Foster et al. 1986). The largest

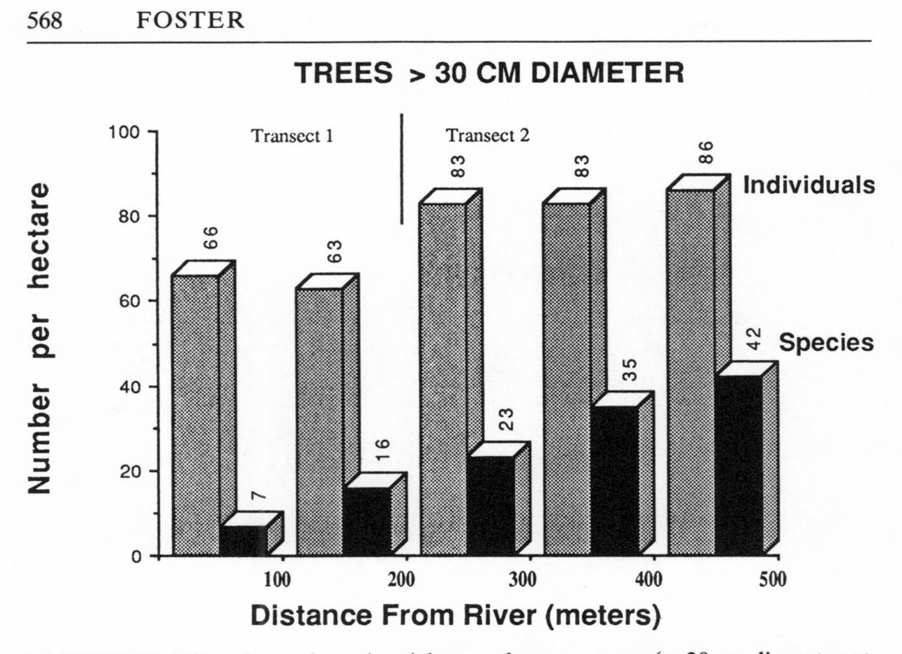

FIGURE 28.2 Density and species richness of canopy trees (> 30 cm diameter at 1.3 m) per hectare along a successional age sequence on the floodplain of Rio Manu. Gray columns are number of individuals; black columns are number of species for each 1-ha segment of the transect. Vertical line distinguishes the two noncontiguous transects. The Transect 2 distance from river is a conjecture, assuming a continuous sequence from Transect 1.

trees, with diameters greater than 100 cm, were also analyzed independently for this study.

Results

The number of tree stems larger than 30 cm diameter increases rapidly to three-fourths its maximum within 100 m of the river (fig. 28.2). The density of trees is fairly constant in the oldest 300 m. Over this same age progression, the density of the largest trees (those greater than 100 cm diameter) doubles between the second and fourth hectare, then drops precipitously in the fifth (oldest) hectare (fig. 28.3).

In contrast, the number of tree species greater than 30 cm diameter increases steadily over the same 500 m of increasing age (fig. 28.2). The species richness of the largest trees remains low with no sign of increasing until the oldest hectare (fig. 28.3).

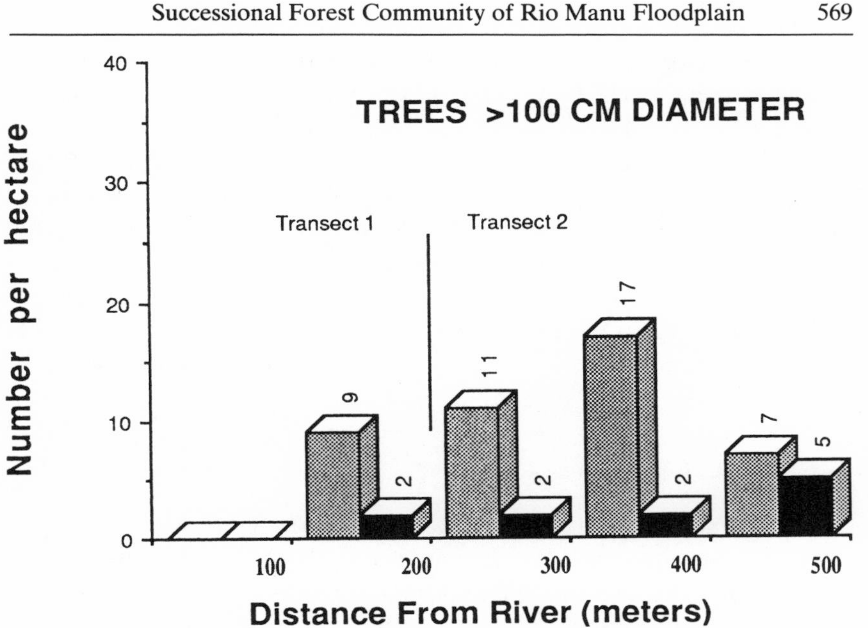

FIGURE 28.3. Density and species richness of largest canopy trees (> 100 cm diameter at 1.3 m) per hectare along a successional age sequence on the floodplain of Rio Manu. Gray columns are number of individuals; black columns are number of species (see fig. 28.1). There were no trees this size in the hectare of forest nearest the river.

In the first (youngest) hectare, the forest is dominated by 20–30 m tall *Cecropia membranacea* (Moraceae) and *Guarea guidonia* (Meliaceae). The second, third, and fourth hectares are dominated by emergent *Ficus insipida* (Moraceae) and *Cedrela odorata* (Meliaceae), both reaching 40 m in height and with large buttresses often several meters across. These are all species whose seedlings are found abundantly in direct sun on the upper beach in most years. It is only in the fifth hectare that *Ficus* and *Cedrela* are being supplanted by species that became established in the thin shade of the *Cecropia* forest next to the river or in the thickets preceding it.

The transition from the first phase of forest dominated by *Cecropia* to the second phase dominated by *Ficus* and *Cedrela* is very smooth and occurs without obvious disruption. The smaller first-phase trees have little impact on their surroundings when they die, and the second-phase trees are already filling in above them when they disappear.

The contrast between the fourth and fifth hectare is dramatic. Many of the giant *Ficus* and *Cedrela* emergents appear sick and start dying off in the fourth hectare. The fifth hectare has large clearings created by the fall of such trees; two had occurred recently enough to recognize the species that crashed. Some large *Ficus* and *Cedrela* are still present in the last hectare; but it is species like *Luehea cymulosa* (Tiliaceae), *Clarisia racemosa, Brosimum lactescens*, and *Poulsenia armata* (Moraceae), *Matisia cordata*, and *Ceiba pentandra* (Bombacaceae) that are becoming increasingly prominent. These will be giant emergents in the next phase (the "third wave"?), but they have clearly been growing there since the earliest stages of succession—showing up as seedlings in the youngest forest.

Discussion

Only one "age transect" is described here. Though considerably greater in area than most transects in tropical forests, and reinforced by numerous informal walking transects in other parts of the floodplain, it is still an extremely limited base from which to make sweeping conclusions. It represents a compact and well-defined set of rolling levees. However, much of the forest area of the floodplain is on shallow, widely spaced levees, broad depressions with poor drainage, or slowly filling ox-bow lakes and swamps. One can expect these areas to follow somewhat different patterns of long-term change. In support of the transect results presented here, it should be noted that this better-drained terrain is typical of what supports the most high-stature, high-diversity tropical floodplain forest. It can be seen as an extreme, not representative of most of the floodplain.

It should be noted further that the cutting and building process of the Manu River is very active (J. Salo et al. 1986). Judging from the forest seen on the cutting riverbank along the length of the river, forest that is older than the oldest hectare of this study is in a distinct minority. In other words, the river usually comes back and cuts down a forest before it gets older than approximately two hundred years.

The results from this 100 m × 500 m transect of increasing age forest indicate that, even after approximately two hundred years of succession, this floodplain forest is still undergoing dramatic changes in its composition and structure. The species richness of canopy trees grows unabated until some stage older than the end of this transect. The few species of emergent large trees that dominate the forest from its first stages finally start to give way to a more diverse set of very large emergents.

The forest structure progresses smoothly without obvious interruption from the *Cecropia* phase of small tree dominants to the ever increasing size of the emergent tree dominants of the *Ficus-Cedrela* phase. Only with the death of these few species of second-phase giants is there significant disturbance in the forest structure. If this handful of species were to be absent in the successional process, through some failure of reproductive effort, dispersal, or establishment, the conditions for the establishment of dominants in the following phases of forest development may be considerably altered. In other words, large, emergent, successional trees create large, open light-gaps, and the absence of such gaps may change the spectrum of species that eventually dominate the forest.

Dominating the forest during the first phases of succession are species of trees that rely on the conditions of the open beach to become established. Even the next more diverse array of large species that begins to replace the first sets seems to be of species that become established and survive to maturity if they germinate in the earliest stages of succession, with its thin canopy and presumably low level of root competition. These conditions can also be met (though less frequently) in older forest bordering river banks that have stabilized after an abrupt change in the course of the river.

The environmental conditions that promote growth and survival for juveniles of the canopy trees will not be the same for the next generation in the same place. Limited information on the distribution of juveniles suggests that this does not stop with the *Ficus-Cedrela* dominated forest for the first two centuries but continues with the longer-lived large trees for perhaps two to four more centuries. Beyond that, the forest may reach relative stability, in that all the species becoming canopy trees will have regenerated in the treefall, light-gap conditions inherent in a steady-state forest. The stability is only relative because of progressive change in alluvial soils from leaching, erosion, and progressive change in exposure to wind as the present river channel wears down to a lower elevation.

REFERENCES

Foster, R. B. and N. V. L. Brokaw. 1982. Structure and history of the vegetation on Barro Colorado Island. In E. G. Leigh, Jr., A. S. Rand, and D. M. Windsor (eds.), *The Ecology of Tropical Forest: Seasonal Rhythms and Long-Term Changes*. Smithsonian Inst. Press, Washington, D.C., pp. 67–81.

Foster, R. B., J. Arce B., and T. S. Wachter. 1986. Dispersal and sequential plant communities in Amazonian Peru floodplain. In A. Estrada and T. H. Fleming (eds.), *Frugivores and Seed Dispersal*. W. Junk, Dordrecht, pp. 357–370.

Gentry, A. H. and J. Terborgh. (This volume). Composition and dynamics of the Cocha Cashu "mature" floodplain forest.

Hubbell, S. P. and R. B. Foster. 1986. Biology, chance and history and the structure of tropical rainforest tree communities. In J. Diamond and T. J. Case (eds.), *Community Ecology*. Harper and Row, New York, pp. 314–329.

Salo, J. R. Kalliola, I. Häkkinen, Y. Mäkinen, P. Niemelä, M. Puhakka, and P. D. Coley. 1986. River dynamics and the diversity of Amazon lowland forest. Nature 322: 254–258.

Terborgh, J. 1983. *Five New World Primates: A Study in Comparative Ecology*. Princeton Univ. Press, Princeton, N.J.

29

Tree Mortality and Recruitment over a Five-Year Period in Undisturbed Upland Rainforest of the Central Amazon

JUDY M. RANKIN-DE-MERONA, ROGER W. HUTCHINGS H., AND THOMAS E. LOVEJOY

The upland rainforests of the Amazon Basin are being fragmented by deforestation brought on by planned and unplanned small-scale agricultural colonization, cattle ranching, tree plantations, and the expanding road networks that bring such activities to previously inaccessible regions. Evaluation of the impact of forest fragmentation and forest isolation by clear-cutting depends on detailed knowledge of the intact continuous forest community. The study described in this chapter was undertaken to determine tree mortality and recruitment for trees 10 cm DBH (diameter at breast height = 1.3 m) or more in undisturbed upland rain forest of the Central Amazon. The conditions of dead trees and the probable causes of death, the number of dead trees per treefall or other mortality-related event, the changes in mortality with tree size, stand turnover time, and the relation between family abundance and mortality and recruitment were also examined.

Study Area

This study was conducted in the Minimum Critical Size of Ecosystems (MCSE) project forest reserves maintained by the Instituto Nacional de Pesquisas da Amazonia and the World Wildlife Fund in the Distrito Agropecuario da SUFRAMA, approximately 90 km north of Manaus, in Amazonas, Brazil. It is part of a larger ongoing study of undis-

The authors wish to thank all field crew and project volunteers who aided in the collection of the data. This is contribution number 47 in the series of papers on the Minimum Critical Size of Ecosystems Project, a bilateral project of the World Wildlife Fund–US and the Instituto Nacional de Pesquisas da Amazonia in Manaus, Brazil.

turbed native rainforest and the impacts of isolation by clear-cutting of otherwise undisturbed forest fragments. The forest on the site is non-inundated or *terra firme* upland forest. It is characterized by a high density of small-diameter trees with few trees greater than 60 cm DBH and by remarkable species richness coupled with low population densities. The average number of trees 10 cm DBH/ha or more is 647 ± 50.5 for a sample of 15 ha, and basal area is 30.55 ± 3.5 m^2/ha. The forest is dominated numerically by trees of the families Burseraceae, Sapotaceae, Lecythidaceae, and Leguminosae. Preliminary taxonomic results for 31 ha indicate a minimum of 621 species of trees present in 55 families, or 791 taxa, if all morphospecies are included for this sample (Rankin-de-Merona et al., in press). Many families, as well as the tree community as a whole, are speciose and have low population densities: of the 60 species of Annonaceae encountered in 50 ha, 85% have abundance less than or equal to 1 tree/ha.

Methods

Five noncontiguous 1-ha plots, one each in reserves 1105, 1109, 1201.1, 1201.2, and 1201.3, were selected for this study. All trees greater than or equal to 10 cm DBH were tagged, measured for diameter, mapped, and a voucher sample of leaves or leaves and flower or fruits collected between February and August 1981. The plots were recensused between June and August 1986 for an average of five years between observations. All original inventory trees were remeasured and death or damage observed, and all trees having grown into the 10-cm DBH size class were measured, mapped, and collected. Uprooted trees, snapped trunks, standing dead trees, branch falls, and other damaged trees were remapped.

Results

Overall tree mortality for five years for the five 1-ha inventory plots was 176 trees, or 5.63% of the 3,125 trees in the original inventory (table 29.1). The number of dead trees in each hectare ranged from 27 to 42, for a mean of 35.2 ± 8.1 trees/ha. Overall there were 7.04 dead trees/ha/year, or 1.13% per year of the originally inventoried trees. This percentage of tree deaths per year is less than that found for studies with similar lower dbh limits in the Venezuelan Amazon (Uhl 1982) or Costa Rica (Lieberman et al. 1985) but slightly greater than that for similar forest approx-

TABLE 29.1. Summary of tree mortality and ingrowth during five years with probable cause of death and original inventory population for 5 ha of upland rainforest in the Central Amazon

Dead tree status	Dead trees in 1 ha. inventory plots *1105* N	*1109* N	*1201.1* N	*1201.2* N	*1201.3* N	Total by category N
Standing dead:						
no damage evident	6	6	3	8	12	35
damaged by another tree	4	0	0	0	2	6
unknown	0	0	0	0	3	3
Trunk snapped off:						
by wind	14	12	5	12	11	54
by another tree	8	4	1	3	1	17
unknown	1	1	3	1	3	9
Uprooted:						
by wind	7	7	7	3	3	27
by another tree	5	3	3	1	5	17
unknown	0	0	0	0	1	1
Lost (probably) dead:	0	1	5	0	1	7
Total dead trees by plot	45	34	27	28	42	176
Total trees in original inventory	638	591	637	626	633	3,125
% Dead in 5 years	7.05	5.75	4.24	4.47	6.64	5.63
Ingrowth 10-cm DBH class	25	23	30	37	23	138
Mortality 10-cm DBH class	17	13	11	11	20	72

imately 50 km from this study site (St. John, p.c.) when a 15-cm lower DBH limit is used for comparison (table 29.2).

The conditions of dead trees were divided into four categories: standing dead, trunk snapped off, uprooted, and lost, with each of these further subdivided by probable cause of death (table 29.1). Snapped trunks accounted for the greatest number of dead trees (46%), and of these, 68% were attributable to wind damage. Putz and Milton (1982) observed higher percentages of snapped trunks on Barro Colorado Island (BCI), Panama, presumably due to differences in forest composition, structure, and site. This difference may not be a function of their 20-cm lower DBH limit for study trees, since Putz et al. (1983) reported a negative correlation between tree size and the probability of snapping. Stand-

TABLE 29.2 Comparisons of tree mortality, recruitment, and turnover rates for six forest sites in three Neotropical regions

Region	Source	Lower DBH limit	Trees orig. invent. *N*	Dead trees *N*	Years observed	Ha.s	Dead trees ha/year	Dead trees/ year (%)	Turnover/ year	Recruited into DBH limit *N*	Recruited ha/year *N*	Recruited/ dead
Central Amazon	This study	10.0	3,125	176	5	5.00	7.04	1.13	89	138	5.52	0.80
Central Amazon	This study	15.0	1,704	104	5	5.00	4.16	1.22	82	—	—	—
Central Amazon	This study	20.0	1,072	61	5	5.00	2.44	1.14	88	—	—	—
Central Amazon	This study	25.0	688	33	5	5.00	1.32	.96	104	—	—	—
Central Amazon	St. John (p.c.)	15.0	226	10	4	.64	3.90	1.11	90	—	—	—
Central Amazon	Higuchi (1987)	25.0	1,893	159	5	12.00	2.65	1.68	60	—	—	—
Venezuelan Amazon	Uhl (1982)	10.0	786	39	5	1.00	7.80	2.93	101	56	11.20	1.43
Venezuelan Amazon	Uhl (1982)	20.0	238	10	5	1.00	2.00	.84	119	17	3.40	1.70
Central America (Panama)	Lang and Knight (1983)	2.5	4,668	934	10	1.50	62.26	2.00	50	438	29.20	0.47
Central America (Costa Rica)	Lieberman et al. (1985)	10.0	5,623	1,302	13	12.40	8.08	1.78	56	1,293	8.02	0.99
Central America (Panama)	Putz and Milton (1982)	20.0*	328	17	5	2.00	1.70	1.00	96	—	—	—

— = Not available

* Original data used girth = 60 cm

ing dead and uprooted trees were almost equally represented and accounted for about 25% each of all dead trees. Most standing dead trees showed no outward signs of damage or disease and were not near treefalls or other dead trees. The percentage of standing dead trees in this study ranged among plots from 11% to 40%. Other studies using a 10-cm DBH limit cite figures for standing dead trees of 10% (Uhl 1982) and 26% (Lieberman et al. 1985).

Ingrowth, or number of trees recruited into the 10-cm or larger DBH size class (trees not included in the original inventory, which during the five-year interval grew to be 10 cm DBH or greater), for the total inventory area of 5 ha was 138 trees (table 29.1). For individual hectares the number of trees recruited ranged from 23 to 37, for a mean of 27.6 ± 5.98 trees/ha in five years, or 5.52 trees/ha/year over all plots. There were originally 1,421 trees in the 10–14.9-cm DBH class; 72 died or were lost but were replaced by 138 recruits, for a net gain of 66 trees, or a 4.64% increase in this size class (table 29.1). Ingrowth has been greater in the two other Neotropical rainforest studies using a 10-cm minimum DBH limit: 11.20 trees/ha/year over five years for terra firme forest in Amazonian Venezuela (Uhl 1982), and 8.02 trees/ha/year over thirteen years for La Selva, Costa Rica (Lieberman et al. 1985).

Tree mortality for the 5 ha was examined by 5-cm diameter intervals to determine the distribution of dead trees relative to the original population size-class distribution. In general, the size-class abundances of the original inventory were followed by the size-class distribution of dead trees, but with a slightly greater proportion of trees deaths found in the 15-, 20-, and 25-cm size classes and a lesser proportion in the progressively larger classes (fig. 29.1). Considering only trees 15–25 cm DBH, 40% of the original inventory trees as compared to 48%of the dead trees occurred in this size range. Only 1.13%, or approximately half as many dead trees as expected based on original inventory data, were encountered in the 55-cm and larger classes, where some size classes had no mortality whatsoever (table 29.3). A Kolmogorov-Smirnov test of the maximum difference between the two frequency distributions shows a significant difference between the DBH distribution of trees in the original inventory and trees dying over the 5-year observation period (Alpha, 0.001; D Observation, 0.1511). This percentage of deaths is approximately the same as that found for trees greater than or equal to 57 cm DBH in old forest on BCI where only 1 of 80, or 1.25%, trees died over the same elapsed time (Putz and Milton 1982). These low percentages are probably due to the small numbers of trees in each size class coupled with a very low mortality rate related to the species, the site, or both. In contrast,

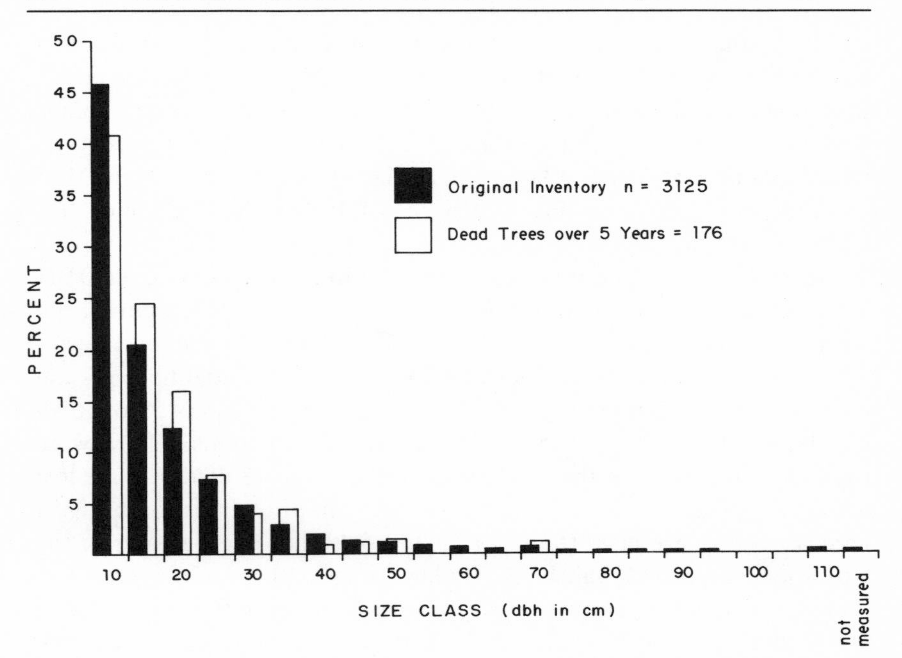

FIGURE 29.1. Size-class distribution for trees in the original inventory (black) and trees dead during the five subsequent years (white) in 5 ha of undisturbed upland rainforest in Central Amazon.

Higuchi (1987), working in a forest stand containing a much greater number of trees above 55 cm DBH than our study (an average of 16.4 versus 0.4 trees/ha) found greater mortality rates, due principally to deaths of larger trees. Finally, for relatively long-lived organisms such as trees, a five-year observation interval is probably inadequate to estimate mortality rates for the larger trees.

Turnover time, the number of years necessary for all of the originally inventoried trees to die, has been calculated as (# original trees)/(# dead trees/years of observation) (see Uhl 1982). This calculation assumes that regrowth balances mortality in a steady-state population. Of the seven studies listed in table 29.2, only four provide data on recruitment that permit a check on this assumption, excluding one study where the formula has been used (St. John, p.c.). Only two of these come close to meeting the assumption, our study, where recruitment/mortality = 0.80, and Lieberman et al. (1985), where recruitment/mortality = 0.99. This formula was used to calculate turnover time for various lower DBH limits

TABLE 29.3 Size-class distributions for original inventory and trees dead during five subsequent years in 5 ha of upland rainforest in the Central Amazon

	Original inventory		Five-year mortality	
Size class (cm DBH)	*Trees* (*N*)	*Trees in class* (%)	*Trees* (*N*)	*Trees in class* (%)
10	1,421	45.47	72	40.91
15	632	20.22	43	24.43
20	384	12.29	28	15.91
25	229	7.33	13	7.39
30	155	4.96	6	3.41
35	90	2.88	7	3.98
40	57	1.82	1	0.57
45	43	1.38	2	1.14
50	32	1.02	2	1.14
55	24	0.77	0	0
60	20	0.64	0	0
65	8	0.26	0	0
70	14	0.45	2	1.14
75	3	0.10	0	0
80	2	0.06	0	0
85	7	0.22	0	0
90	1	0.03	0	0
95	1	0.03	0	0
100	0	0	0	0
105	0	0	0	0
110	1	0.03	0	0
not measured	1	0.03	0	0
Totals	3,125		176	

in our study. Turnover times were also calculated for the other six studies listed in table 29.2.

The calculated turnover time for all trees greater than or equal to 10, 15, and 20 cm DBH ranged from 82 to 89 years (table 29.2). Increasing the minimum DBH limit for the calculation increases the turnover time to 104 years at 25 cm DBH, 115 years at 30 cm DBH, and 204 years at 55 cm DBH, as mortality rate decreases with size. For a superficially similar nearby forest 50 km away, St. John (p.c.) arrived at a similar figure of 85 years for the 15-cm DBH limit.

Calculations based on Higuchi (1987) for a site 50 km away give a much lower figure of 60 years for trees greater than 25 cm DBH. This later study had many more large trees in the original inventory and therefore more large and possibly senecent trees that died off during the 5-year

TABLE 29.4 Summary of the number of dead trees involved in tree falls and other events related to tree death during five years for 5 ha of upland rainforest in the Central Amazon

Dead trees per event (*N*)	Events related to tree death in one hectare inventory plots					Total by category
	1105 (*N*)	*1109* (*N*)	*1201.1* (*N*)	*1201.2* (*N*)	*1201.3* (*N*)	
1	21	18	13	21	23	96
2	5	2	3	3	2	15
3	1	0	1	0	0	2
4	1	4	0	0	0	5
5	1	0	0	0	1	2
6	0	0	0	0	0	0
7	0	1	0	0	0	1
8	0	0	0	0	1	1
Total by plot	29	25	17	24	27	122

observation period. Other figures calculated from the literature range from 56 years at La Selva to 119 years for the Venezuelan Amazon (table 29.2). There are significant problems in comparing such figures related to differences in forest structure, species composition, local causes of mortality, study DBH limits, and overall observation period. We consider the problem of turnover time estimation further in the discussion.

Counts of the events involving tree deaths—wind throws, branch falls, standing dead trees, crown decapitations, snapped trunks, and so on, only some of which produced significant openings in the canopy—were made for each of the five hectares (table 29.4). Over five years 122 such events occurred, for a mean of 24.4 ± 4.56/ha, or 4.88 events/ha/year. Almost 80% involved the death of a single tree, even though 75% of dead trees were snapped off or uprooted, thereby threatening neighboring trees. Many uprooted trees fell without killing other trees, and many heavily damaged trees (e.g., some snapped off just below the crown) were still alive and resprouting leaves at the time of the recensus. The fate of such damaged trees is still being studied. Few events involved more than two dead trees, though up to eight dead trees per event were observed. Events involving multiple tree deaths tended to open larger gaps in the canopy than those with fewer dead trees, but quantitative data, that would permit canopy gap number or gap surface/ha to be calculated are not yet analyzed.

The relation between family abundance, recruitment, and mortality was also examined. The number of trees per family, percentage family abundance, and rank abundance order (1 = most abundant, 2 = next most abundant, and so forth) was determined for all trees in the original inventory plus all recruited and dead trees (table 29.5). The original inventory shows Lecythidaceae, Sapotaceae, Burseraceae, and Leguminosae as the four most abundant families (table 29.5; for more detail on taxonomic composition of the forest see Rankin-de-Merona et al., in press), followed in rapidly diminishing numbers by 47 other families, 25% of which have densities of less than one family member per hectare. Of these 51 families, the greatest mortality and recruitment generally occurred in the most abundant families: the 10 most abundant families in the original inventory also ranked among the 10 most abundant in both mortality and recruitment, though not necessarily in the same rank order. These represent 73% of the original inventory trees and 69% and 71% of the dead and recruited trees, respectively. Seventeen families representing 4.8% of the original inventory remained static, with no dead or recruited trees during the five-year observation period; of these, 15 had original inventory rankings of greater than thirtieth (that is, not abundant). The most abundant family to remain static was the Apocynaceae, with 36, or 1.15%, of the original inventory trees. Tree deaths but no recruitments were recorded for 10 families, all of which ranked greater than fifteenth in abundance. These accounted for 6% of the original inventory and 10.8% of all dead trees. Five families originally ranked greater than fourteenth in abundance showed only recruitment and no tree deaths. These 2.8% of the original trees accounted for 5.1% of all recruited trees. Finally, a new family (Proteaceae) was added to the family list due to the recruitment of one individual. No families were lost due to tree death. In terms of absolute numbers of trees, recruitment was greater than mortality in 11 families, including 5 with only recruitment. Among the more abundant families, the Burseraceae, Chrysobalanaceae, Euphorbiaceae, and Palmae were remarkable in having high family-level mortality relative to recruitment, whereas recruitment was noticeably greater than mortality for the Lecythidaceae, Myristicaceae, and Melastomataceae. Recruitment and mortality were about equal for the Leguminosae, Moraceae, and Annonaceae. Species-level identifications will be included in future analyses.

TABLE 29.5 Family abundances for trees in the original inventory and trees dead or recruited during five years for 5 ha of upland rainforest in the Central Amazon

Family	Inventory		Dead		Recruited		Rank		
	Trees		*Trees*		*Trees*				
	(*N*)	(%)	(*N*)	(%)	(*N*)	(%)	Inventory	Dead	Recruited
LECYT	456	14.59	6	3.41	17	12.41	1	7	2
SAPOT	362	11.58	15	8.52	8	5.84	2	3	5
BURSE	347	11.10	37	21.02	16	11.68	3	1	3
LEGUM	310	9.92	21	11.93	23	16.79	4	2	1
MORAC	208	6.66	8	4.55	9	6.57	5	6	4
CHRYS	160	5.12	10	5.68	4	2.92	6	4	9
LAURA	128	4.10	4	2.27	6	4.38	7	9	7
EUPHO	113	3.62	9	5.11	3	2.19	8	5	10
ANNON	105	3.36	8	4.55	7	5.11	9	6	6
VIOLA	88	2.82	3	1.70	3	2.19	10	10	10
MYRIS	85	2.72	1	.57	4	5.11	11	12	9
PALMA	71	2.27	15	8.52	5	3.65	12	3	8
OLACA	58	1.86	2	1.14	1	.73	13	11	12
MYRTA	56	1.79	4	2.27	5	3.65	14	9	8
STERC	51	1.63			1	.73	15		12
BOMBA	46	1.47	1	.57			16	12	
MELIA	38	1.22	2	1.14	1	.73	17	11	12
UNKNO*	38	1.22	5	2.84	5	3.65	17	8	8
APOCY	36	1.15					18		
HUMIR	35	1.06	1	.57	2	1.46	19	12	11
MELAS	33	5.35	3	1.70	7	5.11	20	10	6
SAPIN	29	.93	5	2.84			21	8	
ANACA	28	.90	2	1.14	1	.73	22	11	12
VOCHY	25	.80					23		
ELAEO	24	.77	2	1.14			24	11	
COMBR	23	.74			1	.73	25		12
RUBIA	2	.70	1	.57			26	12	
FLACO	16	.51	1	.57			27	12	
QUIIN	14	.45	3	1.70			28	10	
GUTTI	13	.42	3	1.70			29	10	
CELAS	12	.38	1	.57			30	12	
MALPI	11	.35					31		
NYCTA	10	.32					32		
BIGNO	8	.26	1	.57			33	12	
CARYO	8	.26			1	.73	33		12
SIMAR	8	.26					33		
LACIS	7	.22	1	.57	1	.73	34	12	12
OCHNA	7	.22					34		
MONIM	6	.19			2	.73	35		9
DICHA	4	.13					36		
BORAG	3	.10	1	.57			37	12	
EBENA	3	.10					37		

ICACI	3	.10				37	
LINAC	3	.10				37	
RHIZO	3	.10				37	
TILIA	3	.10				37	
ERYTH	2	.06				38	
RUTAC	2	.06		3	2.19	38	10
DUCKE	1	.03				39	
HIPPO	1	.03				39	
POLYG*	1	.03				39	
RHABD	1	.03				39	
PROTE	0	0		1	.73		12
TOTALS	3,125		176	137			

* : UNKNO = family not identified; POLYG = Polygonaceae

Discussion

The turnover time for a forest stand calculated using the formula cited above will vary depending on the lower DBH limit used and may not be reliable when the assumption of a steady-state system is not met. It will also vary between forests even when the same DBH limit is used, depending on the original size-class structure of the stand and any differences in site, species composition, and the type of natural perturbations to which the community is subject, such as landslides, seasonal strong winds, and seasonal or sporadic inundations. Lang and Knight (1983) produced an estimate of 137 years for stand turnover on BCI based on the time necessary to cover the study area with gaps in the canopy caused by treefalls. When their data on tree deaths for trees 2.5 cm DBH or greater are used to calculate a turnover time, the figure is 50 years, markedly different from the gap area estimate and probably related to high mortality among small plants, as well as the marked difference in mortality versus recruitment (0.47).

Generalizations about trends in stand turnover remain elusive. Before conclusions are drawn about stand turnover time, estimates for both gap formation and tree mortality and recruitment should be produced and the effects of DBH limits examined. Emphasis should also be put on collecting data for both types of estimates for plots where long-term studies are underway so that better estimates of mortality for larger and long-lived trees can be obtained. This information will soon be available for the present study area and will, it is hoped, facilitate the estimate of true stand turnover time for this site, as well as comparisons with other studies.

REFERENCES

Higuchi, N. 1987. Short-term growth of an undisturbed tropical moist forest in the Brazilian Amazon. Ph.D. diss., Michigan State Univ.

Lang, G. E. and D. H. Knight. 1983. Tree growth, mortality, recruitment and canopy gap formation during a ten-year period in a tropical moist forest. Ecol. 64: 1075–1080.

Lieberman, D., M. Lieberman, R. Peralta, and G. S. Hartshorn. 1985. Mortality patterns and stand turnover rates in a wet tropical forest in Costa Rica. J. Ecol. 73: 915–924.

Putz, F. E. and K. Milton. 1982. Tree mortality rates on Barro Colorado Island. In E. G. Leigh, Jr., A. S. Rand, and D. M. Windsor (eds.), *The Ecology of a Tropical Forest: Seasonal Rhythms and Long-Term Changes*. Smithsonian Inst. Press, Washington, D.C., pp. 95–100.

Putz, F. E., P. D. Coley, K. Lu, A. Montalvo, and A. Aiello. 1983. Uprootings and shapings of trees: structural determinants and ecological consequences. Can. J. For. Res. 13: 1011–1020.

Rankin-de-Merona, J. M., G. T. Prance, M. F. de Silva, W. A. Rodrigues, M. E. Uehling, and R. W. Hutchings. In press. Preliminary tree inventory results from upland rain forest of the Central Amazon. Acta Amazônica.

Uhl, C. 1982. Tree dynamics in a species rich tierra firme forest in Amazonia, Venezuela. Act. Cien. Venez. 33: 72–77.

30

An Overview of Neotropical Forest Dynamics

GARY S. HARTSHORN

The study of tropical forest dynamics has attracted much attention over the past decade. Recent reviews (Brokaw 1985a; Martínez-Ramos 1985; Denslow 1987) emphasize the importance of canopy openings (gaps) in the structural and compositional dynamics of many tropical forests. In this chapter I compare forest dynamics in the four Neotropical rainforests addressed in this volume: Manu National Park, Peru; Barro Colorado Island (BCI), Panama; La Selva Biological Station, Costa Rica; and the Minimum Critical Size of Ecosystems (MCSE) project near Manaus, Brazil. Sparse or inadequate data make substantive comparisons among these four forests difficult. Because of these limitations I attempt here to place these four sites in a regional overview of Neotropical forest dynamics.

Research on tropical forest dynamics has focused on characteristics of gaps, including definition of boundary, frequency of creation, size of canopy opening, species dependence, internal heterogeneity, physiologic requirements, species packing, and equilibrium or nonequilibrium status, among others (e.g., Hartshorn 1978, 1980; Denslow 1980; Brokaw 1982a, 1985b; Orians 1982; Augspurger 1984; Bazzaz 1984; Hubbell and Foster 1983, 1986, 1987; Brandani et al. 1988). Few authors, however, have addressed practical applications (e.g., forest management, conservation of biodiversity) of the rapidly expanding body of scientific information on

This manuscript benefited from the comments of Lynne Hartshorn, Diana Lieberman, and an anonymous reviewer. The author acknowledges travel support from the Organization for Tropical Studies to attend the AIBS meetings and to participate in the symposium on Four Neotropical Forests.

tropical forest dynamics. I highlight here some potential development and conservation applications based on current understanding of Neotropical forest dynamics. Finally, I look at some research needs concerning tropical forest dynamics.

Stand Dynamics

The dynamics of canopy openings can be analyzed in several ways, such as rates of tree mortality, modes of gap formation, and stand turnover times; however, these components may not be directly comparable (Lieberman and Lieberman 1987). Tree mortality rates may vary among size classes, canopy position, or even with reproductive syndrome. Although a new gap is usually conspicuous, where to define the gap edge or boundary is still debated (Brokaw 1982b). The frequent disconformity of a canopy opening with disturbance and light regimes at ground level, expansion of a gap by subsequent treefalls, and the superposition of later treefalls onto older gaps further complicate the calculation of gap frequency and stand turnover. In spite of the definitional and interpretational problems with gaps, some authors use average annual area in new gaps to estimate stand turnover times. For these four Neotropical rainforests, each of these dynamic components is analyzed.

Tree Mortality

Information on tree mortality from long-term inventory plots in tropical America is considerable (Swaine et al. 1987). Annual tree mortality is mostly in the range of 1–2% (table 30.1), with the highest rates reported for El Verde, Puerto Rico, and BCI. Hubbell and Foster (this volume) attribute the high tree mortality (3%) on BCI to the prolonged and severe 1983 dry season associated with El Niño. The short interval (1982–83 to 1985) between inventories should make this a valuable data set for comparison with future inventories. Of 306 species inventoried, the densities of 40% had changed more than 10% and most of these large changes were negative (Hubbell and Foster, this volume). Much lower tree mortality rates on BCI were reported by Putz and Milton (1982) and Lang and Knight (1983) before the drought of 1983.

Except at BCI, trees show no apparent pattern of size-specific mortality. On the BCI plot there was a substantial increase in mortality of trees larger than 16 cm DBH (diameter at breast height). Species typical of moist sites were particularly susceptible to the severe El Niño drought

TABLE 30.1 Tree mortality and stand half-life for tropical American forests

Site/Country	Plot size (ha)	Minimum DBH(mm)	Time period	Total stems	Annual mortality (λ)	Stand half-life (yr)	Source
BCI, Panama	50.0	80	1982–85	27,746	3.04%	22.8	Hubbell and Foster, this volume
BCI, Panama	2.0	190	1975–80	328	1.06%	—	Putz and Milton 1982
BCI, Panama	1.5	100	1968–78	768	1.04%	—	Lang and Knight 1983
Blue Mountains, Jamaica	0.1	32	1974–84	507	1.16%	—	E. V. J. Tanner, p.c.
La Selva, Costa Rica	12.4	100	1969–82	5,623	2.03%	34.2	Lieberman et al. 1985
La Selva, Costa Rica	12.4	100	1982–85	5,530	2.00%	34.6	Lieberman et al., this volume
El Verde, Puerto Rico	0.72	100	1943–76	427	2.78%	—	Crow 1980
Luquillo, Puerto RicoTS2&3	0.8	40	1946–76	—	1.28%	—	Brown et al. 1983
Luquillo, Puerto RicoCS2&4	0.8	40	1946–76	—	0.80%	—	Brown et al. 1983
Manaus, Brazil	5.0	100	1981–86	3,125	1.16%	59.8	Rankin et al., this volume
Manu, Peru	1.0	100	1974–85	587	1.73%	40.2	Gentry and Terborgh, this volume
San Carlos, Venezuela	1.0	100	1975–80	786	1.02%	68.1	Uhl et al. 1988
San Carlos, Venezuela	1.0	100	1980–85	803	1.34%	51.8	Uhl et al. 1988

(Hubbell and Foster, this volume). The high mortality rates of trees in the large size-classes on BCI have not been documented at other tropical American sites. The more typical consequence of drought is high mortality among the small size-classes. In tropical wet forests (La Selva, Costa Rica; Palcazú Valley, Peru) I have observed high mortality of seedlings and saplings during long (4–8 weeks) rainless periods. At these wet sites, larger trees appear to be unaffected by short droughts, possibly because of their more extensive root systems.

Gap Formation

Surprisingly few authors of studies on tropical forest dynamics have documented gap formation patterns over a period of many years. I know of no long-term project in tropical America to specifically monitor gap formation and stand dynamics. Of our four Neotropical forests, information on gaps is available for La Selva and BCI, but not for Manu or the MCSE project. Canopy openings may be caused by any number of natural occurrences, such as bole, branch, or crown breakage, fall of liana tangles. "windthrow" (root tip-up), the domino-effect of a large tree taking down others, localized winds (*venturones* or *vendavales*), tropical cyclones (hurricanes), lightning, landslides, and simply the death of a standing tree.

In spite of differences or uncertainty in defining the boundary of a gap (Brokaw 1982b) and considerable range in minimum limit of gap area, gap formation appears generally to average about 1% per year and about 100 m^2 per gap (table 30.2). New gap formation, particularly of large multiple-tree gaps, probably varies considerably from year to year, but no data are available from long-term studies to support this impression. Gaps range from a few square meters to, rarely, a few hectares. The frequency distribution of gap sizes is log-normal, with a majority of small gaps, but a few large gaps contribute most of the total area in gaps (Sanford et al. 1986; Hubbell and Foster 1987). Using large-scale aerial photographs, Sanford et al. (1986) determined a median distance of only 12 m between gaps ($>$40 m^2) in the primary forest at La Selva.

Turnover Times

Some authors have used average annual area in gaps to project a forest turnover time (Hartshorn 1978; Brokaw 1982a). Turnover time is defined as the number of years to cover a unit area of forest (e.g., 1 ha)

TABLE 30.2 Gap formation and stand turnover times in tropical American forests

Site/country	Plot area (ha)	Time period (yrs)	Gap/yr (%)	Stand turnover (yrs)	Source
BCI, Panama					
young forest	14.6	3	0.63	159	Brokaw 1982a
young forest	1.5	10	0.73	137	Lang and Knight 1983
old forest	13.4	3	0.88	114	Brokaw 1982a
old forest	12.0	?	1.61	62	Foster and Brokaw 1982
La Selva, Costa Rica					
I (old terrace)	4.0	6	1.26	79	Hartshorn 1978
II (swamp)	2.0	5	0.84	119	Hartshorn 1978
II (upland)	2.0	5	0.73	137	Hartshorn 1978
III (hilly)	4.0	5	0.74	135	Hartshorn 1978
Palcazú, Peru	9.7	3	1.09	92	Hartshorn (unpubl. data)
San Carlos, Venezuela	1.0	5	0.96	104	Uhl and Murphy 1981

using the average annual area in gaps (table 30.2). Turnover time is also defined as the average number of years between successive treefalls at a given spot. This second interpretation seems less informative than the first because new gaps are commonly superposed over older gaps and later treefalls expand gap area. Averages, however, often have little ecological significance. For example, there are permanent canopy openings in the La Selva primary forest, known locally as *tacotales*, where trees of small stature fall at high frequency (Hartshorn 1983; Sanford et al. 1986).

Turnover times for heterogeneous Neotropical primary forests range from 60 to 140 years (table 30.2). Late secondary forests, such as the BCI "young" forest, have the longest projected turnover times simply because few large trees or treefalls occur in these successional forests. Foster's (this volume) projection of a 500-year successional sequence for the floodplain forest at Manu is particularly interesting in view of the much shorter turnover times estimated for other Neotropical forests (table 30.2). Upper Amazonian rivers have active floodplains (Salo et al. 1986; Gentry and Terborgh, this volume), usually preventing forest succession from attaining climax status. The abundance of emergent trees on recent alluvium, as at Cocha Cashu (e.g., *Ficus* spp., *Hura crepitans*, *Luehea seemannii*, *Terminalia oblonga*) may be related to unrestricted growth (L. R. Holdridge, p.c.) until the tree is undercut by a river meander, hit by lightning, or blown down. At La Selva, some large trees on fertile alluvial floodplains have diameter increments of 2–3 cm-year (G. Hartshorn, unpubl. data); thus we need diameter growth data or age determinations to corroborate Foster's estimate of 500-year-old trees.

Regional Perspectives

Tree mortality, gap formation, and stand turnover times appear to vary by a factor of two in the Neotropical forests studied to date (tables 30.1, 30.2). Too few data exist to discern patterns in Neotropical forest dynamics associated with vegetation types, site conditions, or geographic factors. Nevertheless, it seems that forest dynamics should be influenced by these and other extrinsic factors. Hurricanes can devastate forests in the outer tropics (or subtropics, sensu Holdridge 1967) of Meso-America and the Caribbean (e.g., Wadsworth and Englerth 1959; Hartshorn et al. 1981; Hartshorn 1988). Not only do hurricanes take down many trees or even large patches of forest, but they can cause tremendous damage to human communities and dams, roads, and irrigation systems when upper watersheds have been deforested.

In the inner tropics, local winds take down many trees and patches of forest, but there is little documentation of canopy openings attributable to local winds. Several blowdowns of a few hectares each have been documented on BCI (Foster and Brokaw 1982). I know of no such large blowdowns at La Selva; however, there is anecdotal evidence of a 50-ha blowdown not far away, near San Miguel, Sarapiquí (R. Chavarría, p.c.). Nor is there any evidence that Manu, the most subtropical of the four sites focused on here, has more large blowdowns than the three inner tropical sites.

Soil fertility, type, structure, and drainage should also be important factors in patterns of gap formation in tropical forests. Forests on infertile, white sand soils are smaller in stature, denser, and with fewer large trees (Uhl and Murphy 1981) than forests on fertile soils. Treefalls appear to be less frequent and less important as regeneration sites in the San Carlos, Venezuela, forests on white sand than on clay soils. Few windthrown trees and many standing dead trees at San Carlos suggest that the shade-tolerant regeneration mode may be more important than gap colonization by shade-intolerant species of trees in these types of forest (e.g., bana, caatinga, heath). It is also possible that turnover time is longer and tree mortality rates are lower in forests on infertile soils.

At the other extreme of soil conditions, forests on fertile, well-drained soils of alluvial floodplains are frequently disturbed by river meanders and floods. These site conditions may favor shade-intolerant tree species that are good colonizers. The abundance of large, shade-intolerant trees on the Manu plot (Gentry and Terborgh, this volume) and the probable late-secondary status of the Manu floodplain forest (Foster, this volume) suggest that there should be few shade-tolerant canopy tree species on excellent soils.

Intermediate between these extremes in soil conditions are the vast forest areas on relatively infertile, adequately drained ultisols and oxisols (*terra firme* forests). These clayey soils on rolling or hilly terrain usually support well-developed, structurally complex, floristically diverse forests where appreciable numbers of canopy tree species appear to be dependent on gaps for establishment and/or reaching the canopy. Several slower-growing, more shade-tolerant tree species also comprise the canopy, yet it is not known if these trees can reach the canopy without benefit of a gap or gaps. These typical forests of tropical America are dynamic, with tree mortality rates of 1–2%, median tree longevity (>10 cm DBH) of 30–50 years, and stand turnover times of 80–150 years (tables 30.1, 30.2).

The abundance of *Tachigalia* spp. in western Amazonian forests adds a striking component to the forest dynamics pattern, for many of these species are semelparous (Foster 1977). Most *Tachigalia* species are large canopy or emergent trees with broad, monolayered (sensu Horn 1971) crowns. The trees that massively flower and fruit only once before dying leave a sizeable opening in the canopy that differs from the more typical windthrow (e.g., Brandani et al. 1988). Shade-tolerant trees under the crown of a *Tachigalia* that survive the rain of falling branches may form a conspicuous component of the forest. Several of the gaps surveyed in the Palcazú forest (table 30.2) were caused by the death of large *Tachigalia* trees.

Some authors have attempted to categorize native tree species by regeneration modes or guilds (Knight 1975; Hartshorn1978; Hubbell and Foster, this volume; Swaine and Whitmore 1988). Given the richness of tree species in most tropical forests, one should expect a spectrum of species responses along the gradient between the extremes of full sun and dense shade (Hartshorn 1980). Many gap species and even some pioneer tree species grow better in light shade (10–30%) than in full sun (Oberbauer and Strain 1984; Oberbauer and Donnelly 1986; Lee 1987). Under natural gap conditions, many tree species grow faster near the gap edge than in the center (Denslow 1987).

Limited attempts to classify tree species by regeneration modes make it difficult to generalize on a regional basis. The few tree species occurring at two or more research sites also exacerbate attempts to generalize about regeneration patterns. Some widespread species (e.g., *Simarouba amara*) are enigmatic; Hubbell and Foster (this volume) classify this species as indifferent to gaps on their BCI plot, whereas it is considered strongly dependent on gaps at La Selva (Hartshorn 1978) and in the western Amazon.

Potential Applications

Studies of tropical forest dynamics have contributed much to our growing understanding of the role and importance of gaps in tree mortality and regeneration, stand turnover, and species packing and diversity. Comparatively little attention, however, has been given to potential applications of this information to such serious problems as the conservation and management of tropical forests. I offer here some possible applications of forest dynamics to sustained-yield forestry, wildlife management, and conservation of biological diversity.

Until a few years ago, it was generally accepted that management of heterogeneous tropical forests for sustained yield of timber was ecologically impossible and economically unjustifiable (Leslie 1977). Recently, Leslie (1987a,b) retracted his gloomy view of the economics of managing tropical forests. From the ecological perspective, information on forest dynamics can be used to design management systems for tropical forests (Hartshorn 1989a). The creation of artificial gaps to promote natural regeneration of fast-growing, shade-intolerant trees has been proposed as a management technique for sustained yield of timber from Neotropical forests (Hartshorn 1979, 1981). Called a strip-shelterbelt system, these long, narrow clear-cuts function as elongated gaps. By adequately spacing and rotating these narrow strip clear-cuts through a production forest, the forest matrix is the source of seeds for natural regeneration of recently harvested strips. This natural forest management system is being implemented by an Indian cooperative in the Palcazú Valley of the Peruvian Amazon (Tosi 1982; Hartshorn et al. 1987; Stocks and Hartshorn 1989; Hartshorn 1989b).

Although indigenous peoples have long used gaps for planting fruit trees in swidden fields for future harvest or hunting forays (Alcorn 1984; Denevan et al. 1984), scientists have only begun to investigate gaps as wildlife habitat. Seedlings and young saplings are attractive browse for terrestrial herbivores (deer, peccary, tapir, and so on); frequent natural gaps or scheduled strip clear-cuts could thus maintain important habitat or food resources for some animals. Older gaps (building phase, sensu Whitmore 1975) or regenerating strips probably offer preferred foliage and fruit for frugivores and arboreal herbivores. The planting of a few particularly desirable native fruit trees in natural gaps or harvested strips might enhance the value of natural forest as habitat for some species of wildlife.

Recent initiatives to conserve the biological diversity of tropical forests have focused almost exclusively on preservation of wildlands. Even if countries with tropical forests implement the world conservation strategy of protecting 10% of their wildlands (IUCN 1980), vast areas of tropical forests will fall to the advancing agricultural frontier (e.g., Myers 1986). Viable management systems for natural forests could greatly aid the critical efforts to conserve tropical biodiversity.

In the operational plan for the first production stand using the Palcazú forest management system, only 48% of the 500-ha stand is assigned for strip clear-cutting on a forty-year rotation (Sanchoma et al. 1986). This means that most of this natural forest will never be harvested and that the natural forest structural matrix should be maintainable. The excluded primary forest will be permanent seed sources for regenerating the harvested strips, as well as for providing medicinal plants, useful vines, edible fruits, herbs, and game.

Two demonstration strips clear-cut in 1985 for the Palcazú project show excellent natural regeneration of tree species. The first demonstration strip (20 m × 75 m) had about 1,500 juveniles representing 133 tree species at 15 months and 209 species at 30 months (Hartshorn 1989a); the former is more than double the number of tree species harvested from that strip. Many of the smaller stumps, furthermore, have vigorous sprouts, including several of the purportedly slow-growing, beautiful hardwoods (e.g., *Tabebuia obscura*, *Myrocarpus* sp., *Vatairea* sp.). The vegetative regeneration from stump sprouts complements and diversifies the natural regeneration of trees on the strip shelterbelts.

The strip-shelterbelt forest management system includes silvicultural treatments of the regenerating strips to favor particular trees or species and to control lianas or thin undesirable trees. Although the forest management system is based on natural regeneration, it does not prohibit planting of tree species to enhance the rebuilding forest. Thus it is not unrealistic to envision the planting of native fruit trees that are keystone species (Terborgh 1986) for threatened frugivores. Implementation of a natural forest management system that maintains or even enhances the floristic composition of production forests should also maintain wildlife habitat and help to conserve the biological diversity of tropical forests.

Some Research Needs

In spite of impressive advances in our understanding of tropical forest dynamics, more research is clearly needed. This overview of tree mortality, gap formation, and stand turnover indicates the importance and necessity of long-term data sets. It is somewhat shocking to realize that the meager twenty-year data sets for La Selva and BCI constitute the longest-term records available from the continental Neotropics. Permanent plot establishment and monitoring at BCI, La Selva, Manu, and MCSE must be continued. Such key research sites as Los Tuxtlas, Mexico, and more Amazonian forests should be included in a network of sites for long-term studies on Neotropical forest dynamics.

More detailed investigations of gap heterogeneity, microclimate, nutrient fluxes, and edge effects are needed to provide precise information on gap environments and the spectrum of physical conditions under which gaps are colonized. Most research efforts have focused on gaps and the first few years of development, whereas we know practically nothing about floristic or microenvironmental changes during forest development in older gaps.

As I have mentioned, gap researchers have serious differences in the recognized number and definition of regeneration modes or guilds (Swaine and Whitmore 1988). Much more experimental research will be needed to generate broader acceptance of species groupings and regeneration classes, as well as to work out the silvics of key species. Similarly, the selection of tree species for experimental, physiological, or autecological research should be based not only on local importance of a species but also on its occurrence at other research sites. Even if relatively few tree species have broad geographic distributions (e.g., *Calophyllum brasiliense*, *Ceiba pentandra*, *Simarouba amara*, *Terminalia amazonia*), many ecologically similar congeners merit detailed comparative study at several research sites (e.g., *Brosimum*, *Dipteryx*, *Guazuma*, *Parkia*, *Tachigalia*, *Virola*). Of course, locally or regionally important species should also be the focus of research (e.g., *Bertholletia excelsa*, *Cedrelinga catenaeformis*, *Goupia glabra*, *Pentaclethra macroloba*). With the great richness of tree species at any research site, it is often difficult to select a few for study. Nevertheless, closer collaboration among researchers in tropical forest dynamics and more comparisons and interchange of data from different regions as attempted in this volume could have an appreciable synergistic effect on our efforts to understand gap environments, regeneration guilds, species packing, and the possibilities of applying that knowledge to conservation and management of Neotropical forests.

REFERENCES

Alcorn, J. B. 1984. Development policy, forests, and peasant farms: reflections on Huastec-managed forests' contributions to commercial production and resource conservation. Econ. Bot. 38: 389–406.

Augspurger, C. K. 1984. Light requirements of Neotropical tree seedlings: a comparative study of growth and survival. J. Ecol. 72: 777–796.

Bazzaz, F. A. 1984. Dynamics of wet tropical forests and their species strategies. In E. Medina, H. Mooney, and C. Vásquez-Yanes (eds.), *Physiological Ecology of Plants of the Wet Tropics*. W. Junk, Boston, pp. 233–243.

Brandani, A., G. S. Hartshorn, and G. H. Orians. 1988. Internal heterogeneity of gaps and tropical tree species richness. J. Trop. Ecol. 4: 99–119.

Brokaw, N. V. L. 1982a. Treefalls: frequency, timing, and consequences. In E. G. Leigh, Jr., A. S. Rand, and D. M. Windsor (eds.), *The Ecology of a Tropical Forest: Seasonal Rhythms and Long-Term Changes*. Smithsonian Inst. Press, Washington, D.C., pp. 101–108.

———. 1982b. The definition of treefall gap and its effect on measures of forest dynamics. Biotropica 14: 158–160.

———. 1985a. Treefalls, regrowth, and community structure in tropical forests. In S. Pickett and P. White (eds.), *The Ecology of Natural Disturbance and Patch Dynamics*. Academic Press, New York, pp. 53–69.

———. 1985b. Gap-phase regeneration in a tropical forest. Ecol. 66: 682–687.

Brown, S., A. Lugo, S. Silander, and L. Liegel. 1983. Research history and opportunities in the Luquillo experimental forest. Inst. Trop. For. Tech. Rep. SO-44.

Crow, T. R. 1980. A rainforest chronicle: a thirty-year record of change in structure and composition at El Verde, Puerto Rico. Biotropica 12: 42–55.

Denevan, W. M., J. Treacy, J. Alcorn, C. Padoch, J. Denslow, and S. Flores Paitán. 1984. Indigenous agroforestry in the Peruvian Amazon: Bora Indian management of swidden fallows. Interciencia 9: 346–357.

Denslow, J. S. 1980. Gap partitioning among tropical rainforest trees. Biotropica 12(suppl.): 47–55.

———. 1987. Tropical rain forest gaps and tree species diversity. Ann. Rev. Ecol. Syst. 18: 431–451.

Foster, R. B. 1977. *Tachigalia versicolor* is a suicidal Neotropical tree. Nature 268: 624–626.

———. (This volume). Long-term change in the successional forest community of the Río Manu floodplain.

Foster, R. B. and N. V. L. Brokaw. 1982. Structure and history of the vegetation on Barro Colorado Island. In E. G. Leigh, Jr., A. S. Rand, and D. M. Windsor (eds.), *The Ecology of a Tropical Forest: Seasonal Rhythms and Long-Term Changes*. Smithsonian Inst. Press, Washington, D.C., pp. 67–81.

Foster, R. B. and S. P. Hubbell. (This volume). The floristic composition of the Barro Colorado Island forest.

Gentry, A. H. and J. Terborg. (This volume). Composition and dynamics of the Cocha Cashu mature floodplain forest.

Hartshorn, G. S. 1978. Tree falls and tropical forest dynamics. In P. B. Tomlinson and M. H. Zimmermann (eds.), *Tropical Trees as Living Systems*. Cambridge Univ. Press, London, pp. 617–638.

———. 1979. Preliminary management plan for Sarapiquí. Report to USAID, San José, Costa Rica.

———. 1980. Neotropical forest dynamics. Biotropica 12(suppl.): 23–30.

———. 1981. Forestry potentials in the Palcazú valley, Peru. In *Central Selva Natural Resources Management Project*. USAID, Lima, app. G.

———. 1983. Plants. In D. H. Janzen (ed.), *Costa Rican Natural History*. Univ. of Chicago Press, Chicago, pp. 118–157.

———. 1988. Tropical and subtropical vegetation of Meso-America. In M. J. Barbour and W. D. Billings (eds.), *North American Terrestrial Vegetation*. Cambridge Univ. Press, New York, pp. 365–390.

———. 1989a. Application of gap theory to tropical forest management: natural regeneration on strip clear-cuts in the Peruvian Amazon. *Ecol.* 70: 567–569.
———. 1989b. Gap-phase dynamics and tropical tree species richness. In L. B. Holm-Nielsen, I. C. Nielsen, and H. Balslev (eds.), *Tropical Forest: Botanical Dynamics, Speciation and Diversity*. Academic Press, London, pp. 65–73.
Hartshorn, G. S., G. Antonini, R. DuBois, D. Harcharik, S. Heckadon, H. Newton, C. Quesada, J. Shores, and G. Staples. 1981. *The Dominican Republic Country Environmental Profile: A Field Study*. JRB Assoc., McLean, Va.
Hartshorn, G. S., R. Simeone, and J. A. Tosi, Jr. 1987. Manejo para rendimiento sostenido de bosques naturales: un sinopsis del proyecto de desarrollo del Palcazú en la Selva Central de la Amazonía Peruana. In J. C. Figueroa Colón, F. H. Wadsworth, and S. Branham (eds.), *Management of the Forests of Tropical America: Prospects and Technologies*. Inst. Trop. For., Río Piedras, P.R., pp. 235–243.
Holdridge, L. R. 1967. *Life Zone Ecology*, rev. ed. Trop. Sci. Ctr., San José, Costa Rica.
Horn, H. S. 1971. *The Adaptive Geometry of Trees*. Princeton Univ. Press, Princeton, N.J.
Hubbell, S. P. and R. B. Foster. 1983. Diversity of canopy trees in a Neotropical forest and implications for conservation. In S. Sutton, T. Whitmore, and A. Chadwick (eds.), *Tropical Rain Forest: Ecology and Management*. Blackwell, Oxford, pp. 25–41.
———. 1986. Biology, chance, and history and the structure of tropical rain forest tree communities. In J. Diamond and T. Case (eds.), *Community Ecology*. Harper and Row, New York, pp. 314–329.
———. 1987. La estructura espacial en gran escala de un bosque neotropical. Rev. Biol. Trop. 35(suppl.): 7–22.
———. (This volume). Structure, dynamics, and equilibrium status of old-growth forest on Barro Colorado Island.
IUCN. 1980. *World Conservation Strategy*. Int. Union. Cons. Nat. and Nat. Res., Gland, Switzerland.
Knight, D. H. 1975. A phytosociological analysis of species-rich tropical forest on Barro Colorado Island, Panama. Ecol. Monogr. 45: 259–284.
Lang, G. E. and D. H. Knight. 1983. Tree growth, mortality, recruitment, and canopy gap formation during a ten-year period in a tropical moist forest. Ecol. 64: 1075–1080.
Lee, D. W. 1987. The spectral distribution of radiation in two Neotropical rainforests. Biotropica 19: 161–166.
Leslie, A. J. 1977. Where contradictory theory and practice coexist. Unasylva 29: 2–17.
———. 1987a. The economic feasibility of natural management of tropical forests. In F. Mergen and J. R. Vincent (eds.), *Natural Management of Tropical Moist Forests: Silvicultural and Management Prospects of Sustained Utilization*. Yale Univ. School For. and Env. Stud., New Haven, pp. 177–198.
———. 1987b. A second look at the economics of natural management systems in tropical mixed forests. Unasylva 39: 46–58.

Lieberman, D. and M. Lieberman. 1987. Forest tree growth and dynamics at La Selva, Costa Rica (1969–1982). J. Trop. Ecol. 3: 347–358.

Lieberman, D., M. Lieberman, R. Peralta, and G. S. Hartshorn. 1985. Mortality patterns and stand turnover rates in a wet tropical forest in Costa Rica. J. Ecol. 73: 915–924.

Lieberman, D., M. Lieberman, G. S. Hartshorn, and R. Peralta. (This volume). Forest dynamics at La Selva Biological Station, Costa Rica, 1969–1985.

Martínez-Ramos, M. 1985. Claros, ciclos vitales de los árboles tropicales y regeneracíon natural de las selva altas perennifolias. In A. Gómez-Pompa and S. del Amo R. (eds.), *Investigaciones sobre la regeneración de selvas altas en Veracruz, Mexico* II. INIREB, Xalapa, Mexico, pp. 191–239.

Myers, N. 1986. Tropical deforestation and a mega-extinction spasm. In M. E. Soulé (ed.), *Conservation Biology: The Science of Scarcity and Diversity*. Sinauer, Sunderland, Mass., pp. 394–409.

Oberbauer, S. F. and B. R. Strain. 1984. Photosynthesis and successional status of Costa Rican rain forest trees. Photosyn. Res. 5: 227–232.

Oberbauer, S. F. and M. A. Donnelly. 1986. Growth analysis and successional status of Costa Rican rain forest trees. New Phytol. 104: 517–521.

Orians, G. H. 1982. The influence of tree-falls in tropical forests on tree species richness. Trop. Ecol. 23: 255–279.

Putz, F. E. and K. Milton. 1982. Tree mortality rates on Barro Colorado Island. In E. G. Leigh, Jr., A. S. Rand, and D. M. Windsor (eds.), *The Ecology of a Tropical Forest: Seasonal Rhythms and Long-Term Changes*. Smithsonian Inst. Press, Washington, D.C., pp. 95–100.

Rankin-de-Morona, J. M., R. W. Hutchings, and T. E. Lovejoy. (This volume). Tree mortality and recruitment over a five-year period in undisturbed upland rainforest of the central Amazon.

Salo, J., R. Kalliola, I. Häkkinen, Y. Mäkinen, P. Niemelä, M. Puhakka, and P. D. Coley. 1986. River dynamics and the diversity of Amazon lowland forest. Nature 322: 254–258.

Sanchoma R., E. R. Simeone G., M. Velis, and H. Vílchez B. 1986. Plan de manejo forestal: bosque de producción de la comunidad nativa Shiringamazú, 1987–1989. Trop. Sci. Ctr. Rep. 105–C, San José, Costa Rica.

Sanford, R. L., Jr., H. E. Braker, and G. S. Hartshorn. 1986. Canopy openings in a primary Neotropical lowland forest. J. Trop. Ecol. 2: 277–282.

Stocks, A. and G. Hartshorn. 1989. The Palcazú project: forest management and native Amuesha communities. In S. Hecht and J. Nations (eds.), *The Social Dynamics of Deforestation in Latin America: Processes and Alternatives*. Univ. of California Press, Berkeley and Los Angeles.

Swaine, M. D. and T. C. Whitmore. 1988. On the definition of ecological species groups in tropical rain forests. Vegetatio 75: 81–86.

Swaine, M. D., D. Lieberman, and F. E. Putz. 1987. The dynamics of tree populations in tropical forests: a review. J. Trop. Ecol. 3: 359–366.

Terborgh, J. 1986. Keystone plant resources in the tropical forest. In M. E. Soulé (ed.), *Conservation Biology: The Science of Scarcity and Diversity*. Sinauer, Sunderland, Mass., pp. 330–344.

Tosi, J. A., Jr. 1982. Sustained yield management of natural forests: forestry subproject, Central Selva resources management project, Palcazú valley, Peru. Trop. Sci. Ctr. Rep., San José, Costa Rica.

Uhl, C. and P. G. Murphy. 1981. Composition, structure, and regeneration of a tierra firme forest in the Amazon Basin of Venezuela. Trop. Ecol. 22: 219–237.

Uhl, C., K. Clark, N. Dezzeo, and P. Maquirino. 1988. Vegetation dynamics in Amazonian treefall gaps. Ecol. 69: 751–763.

Wadsworth, F. H. and G. H. Englerth. 1959. Effects of the 1956 hurricane on forests in Puerto Rico. Carib. For. 20: 38–51.

Whitmore, T. C. 1975. *Tropical Rain Forests of the Far East*. Clarendon Press, Oxford.

Scientific Index

Subject Index